ELECTRON DENSITY AND BONDING IN CRYSTALS

ELECTRON DENSITY AND BONDING IN CRYSTALS

PRINCIPLES, THEORY AND X-RAY DIFFRACTION EXPERIMENTS IN SOLID STATE PHYSICS AND CHEMISTRY

V G TSIRELSON
R P OZEROV
Mendeleev University of Chemical Technology of Russia
Moscow, Russia

Institute of Physics Publishing
Bristol and Philadelphia

British Library Cataloguing-in-Publication Data

A catalogue record for this book is available from the British Library.

ISBN 0 7503 0284 4

Library of Congress Cataloging-in-Publication Data are available

Published by Institute of Physics Publishing, wholly owned by The Institute of Physics, London

Institute of Physics Publishing, Techno House, Redcliffe Way, Bristol BS1 6NX, UK

US Editorial Office: Institute of Physics Publishing, The Public Ledger Building, Suite 1035, 150 South Independence Mall West, Philadelphia, PA 19106, USA

Typeset by P&R Typesetters Ltd, Salisbury
Printed in the UK by J W Arrowsmith Ltd, Bristol

CONTENTS

PREFACE

This book is intended as an introduction to the subject of the physics and chemistry of crystals, which treats their electron structure and properties from the viewpoint of an analysis of a continuous ground state electron density distribution. The underlying concept is quite general. First, the electron density describes the behaviour of electrons in the field of nuclei in accordance with the laws of quantum mechanics. Second, it determines all properties of the ground electronic state and some properties of excited states of such a system and acts as a basic variable in the electron density functional theory. Finally, the electron density can be reconstructed from diffraction experiments using x-rays, and more recently γ-ray and synchrotron radiation.

The theory, in which the basic variable is an experimentally measured quantity, looks rather attractive. In particular, in studying the electronic states of substances it is possible, without losing generality, to avoid calculating the wave-functions—a procedure which remains rather tedious, expensive and of restricted accuracy to date. It is also of importance to establish experimentally the typical features of the electron distribution for various types of atomic interaction in a crystal. Besides, the electron density determines the peculiarities of a crystal field, and its characteristics appear as parameters in various physical models. Being able to determine these parameters by means of diffraction experiments, one can combine different experimental methods into a unique approach and sometimes new information, inaccessible to many individual methods, can be obtained in this way.

The use of the experimental electron density in the study of chemical bonds and the modelling of crystal properties seems to us especially promising. The purpose of this book is to convince a wide community of physicists, chemists, biologists, geochemists, and material science experts that the above approach is useful for them and may become a tool for obtaining information on the structure and properties of substances they are interested in at a new level of knowledge.

In the beginning of the book we will try to travel along a path 'from the Hamiltonian to the diffractometer' together with the reader, so that he or she can judge the possibilities of the approach discussed and appreciate all its 'weak points'. Particular attention is paid to describing the electron density properties, to methods of non-empirical calculation, to accurate procedures of indirect x-ray diffraction measurement of electron density, and to determination of the electron density characteristics and related quantities such as the electric field in a crystal etc. Whenever it is possible, a comparison of theory with experiment will be given.

The remaining part of the book is devoted to various aspects of using the electron density derived from the x-ray diffraction data. Of great importance here is the modern version of the chemical bond theory based on the electron density distribution analysis. Then, we consider how it is possible to use the experimental electron density in the evaluation of crystal physical properties such as the electric field gradient, the diamagnetic susceptibility, electronic static polarizabilities, and nonlinear optical characteristics etc.

Reading this book requires knowledge of quantum mechanics, solid state physics, and physical chemistry at the level of introductory university courses, as well as an acquaintance with the basics of x-ray structure analysis. To facilitate reading for adjacent-area specialists and students, we have included a few appendices to this book. This has allowed us, without overloading the main text with mathematical and physical details, to avoid excessive simplification, which would affect the treatment of the subject. The reader will judge to what extent we have managed to achieve clarity of presentation.

Some remarks on systems of units. We have tried to use the SI system as the basic one, but we could not keep to this throughout the book. This is because of the fact that the Atomic System of units dominates quantum chemistry, in which only wave-function parameters are determined, whereas the so-called electron units are used in crystallography where, for instance, the electron density is measured in units of e Å^{-3} (the number of electrons in Å^{-3}) rather than in C m^{-3} units (Cowley 1975). That is why we have deviated from the SI system in some cases while commenting on this in each specific instance. For the reader's convenience the relations between various units are given in appendix A.

The book owes much to the influence of the Sagamore community. The authors are very grateful to the organizers of at least the last four Sagamore Conferences, to Professors Ivar Olovsson, Luis Alta da Veiga, Wolf Weyrich, and Genevieve Loupias for supporting and promoting our participation. Some of our colleagues have given us pieces of valuable advice. We are especially grateful to Professors D Feil, P J Brown, B N Figgis, J B Forsyth, and A Fox and J Gerratt for this and for the amendments to the text. We appreciate very much receipt of preprints and reprints from Professors J Dunitz, K Hermansson, R G Parr, C Pisani, W Schülke, K Schwarz, and W H E Schwarz. Discussions with our colleagues Professors I V Abarenkov, R F W Bader, I D Brown, P Coppens, N K Hansen, K Ivon, I G Kaplan, A A Levin, L Massa, M A Porai-Koshits, B M Shchedrin, D S Schwarzenbach, R F Stewart, and K B Tolpygo influenced our imagination very much.

The authors are very pleased to express their deep gratitude to the people who have cooperated with them for many years, making it possible to write this book. Among these people are M Yu Antipin, E L Belokoneva, V K Belsky, L A Butman, R G Gerr, M M Mestechkin, U Pietsch, I M Reznik, Yu T Struchkov, V E Zavodnik, and P M Zorky. Many results which have been used in the book were obtained together with our colleagues and fellows:

Yu A Abramov, Yu V Alexandrov, T N Borovskaya, M Flugge, M Kapphahn, O V Korolkova, N N Lobanov, V A Streltsov, A G Tsarkov, A A Varnek, I V Voloshina, and E A Zhurova.

Chapters 1, 2, 3, 6, and 7, and appendices A–E and G–J were written by V G Tsirelson. Chapters 4, 5, and appendices F and K were written by R P Ozerov. Mrs Z V Shcherbakova performed a tremendous amount of work in preparing the manuscript for publication. We are very obliged to her. We also greatly appreciate the assistance of Dr J R Hester in the translation of some chapters of the text.

V G Tsirelson
R P Ozerov

GLOSSARY

AO	atomic orbital
APW	augmented plane wave
DZ	double zeta
DM	density matrix
DFT	density functional theory
DC	doubly substituted configuration
ED	electron density
EP	electrostatic potential
EFG	electric field gradient
e.s.d.	estimated standard deviation
GTO	Gaussian-type orbital
GVB	generalized valence bond
HF	Hartree–Fock
HFR	Hartree–Fock–Roothan
HFS	Hartree–Fock–Slater
LS	least squares
LCAO	linear combination of atomic orbitals
LDA	local density approximation
LAPW	linearized augmented plane waves
LMTO	linearized muffin tin orbitals
MO	molecular orbital
MS	Mössbauer spectroscopy
MT	muffin tin
MEM	maximum-entropy method
MD	magnetization density
NQR	nuclear quadrupole resonance
NMR	nuclear magnetic resonance
PDF	probability density function
PND	polarized neutron magnetic diffraction
QC	quadruply substituted configuration
RHF	restricted Hartree–Fock
SNP	sodium nitroprusside
SD	spin density

SCF	self-consistent field
STO	Slater-type orbital
SZ	single zeta
SC	singly substituted configuration
TZ	triple zeta
TC	triply substituted configuration
TDS	thermal diffuse scattering
UHF	unrestricted Hartree–Fock
VB	valence bond
VBM	valence bond method

Мир электрона

Быть может, эти электроны -
Миры, где
пять материков
Искусства, знанья,
войны, троны
И память
сорока веков!

Еще, быть может,
каждый атом -
Вселенная,
где сто планет:
Там все, что здесь,
в объеме сжатом
Но также то,
чего здесь нет.

В Я Брюсов, 1922

The world of an electron

Maybe these electrons
Are the worlds with
five continents,
Arts, knowledge,
wars, thrones
And the memory
of forty centuries!

Furthermore, maybe
each atom
Is a universe of
hundred planets,
Where there is everything which
is here, in compressed form,
But also what is
absent here.

V Ya Brjusov, 1922

[The English version was written by Sheila Mackay.]

1

INTRODUCTION. THE ELECTRON DENSITY CONCEPT IN PHYSICS AND CHEMISTRY

The electron structure, the chemical bond, and crystal properties will be considered in this book from the viewpoint of electron density (ED) distribution in crystals. As a rule, the consideration will deal with the ED reconstructed from single-crystal x-ray diffraction data, and the theory will widely be used for treatment of the results. Since this approach is not traditional, it is natural that, first of all, the underlying approximations be outlined and the expected results be sketched.

The modern theory considers a crystal at a particular temperature to be a long-ordered, equilibrium dynamical system of electrons and nuclei. Their interactions as well as the crystal structure and properties can be described only by using model representations. The ED concept is one of them. It is based on the Born–Oppenheimer approximation (Born and Huang 1954), which allows one to separate the motion of electrons and nuclei. In non-relativistic quantum mechanics the Hamiltonian (the energy operator) of a system of electrons and nuclei, which depends only on their interactions and kinetic energies and does not take into account their spins, has the form

$$\hat{\mathscr{H}} = \hat{T}_N + \hat{T}_e + \hat{V}_{NN} + \hat{V}_{ee} + \hat{V}_{Ne} = -\frac{\hbar^2}{2}\sum_a \frac{1}{M_a}\nabla_a^2 - \frac{\hbar^2}{2m}\sum_i \nabla_i^2 + \frac{1}{4\pi\varepsilon_0}\sum_{a<b}\frac{Z_a Z_b}{|\boldsymbol{R}_a - \boldsymbol{R}_b|}$$

$$+\frac{e^2}{4\pi\varepsilon_0}\sum_{i<j}\frac{1}{|\boldsymbol{r}_i - \boldsymbol{r}_j|} - \frac{e}{4\pi\varepsilon_0}\sum_a\sum_i \frac{Z_a}{|\boldsymbol{r}_i - \boldsymbol{R}_a|}. \qquad (1.1)$$

Here $\hat{T}_N$ and $\hat{T}_e$ are operators of kinetic energy of nuclei and electrons, and $\hat{V}_{NN}$, $\hat{V}_{ee}$, and $\hat{V}_{Ne}$ are the operators of potential energy of nuclei, electron, and electron–nuclear interactions, respectively. The coordinates of electrons $\boldsymbol{r}_i$ and of nuclei $\boldsymbol{R}_a$ originate from the centre of inertia of a system; e and Z_a are the charges of electrons and nuclei, respectively, and m and M_a are their masses; $\hbar = h/2\pi$, h is the Planck constant, and ε_0 is the permitivity of free space. If the operator of nuclei kinetic energy, $\hat{T}_N$, is assumed to be a small disturbance with respect to the system Hamiltonian with invariable ('frozen') nuclei positions, then in the zeroth approximation the problem of searching for the system's stationary state is reduced to solution of the Schrödinger equation of type

$$\hat{\mathscr{H}}_0 \Psi_n(\boldsymbol{r}, \boldsymbol{R}) = \Phi_n(\boldsymbol{R})\Psi_n(\boldsymbol{r}, \boldsymbol{R}). \qquad (1.2)$$

Here $\hat{\mathscr{H}}_0 = \hat{T}_e + \hat{V}_{NN} + \hat{V}_{ee} + \hat{V}_{Ne}$, $\Psi_n(\boldsymbol{r}, \boldsymbol{R})$, and $\Phi_n(\boldsymbol{R})$ are, respectively, electron wave-functions and energies of stationary electron states, which depend on some configuration of nuclei coordinates $\boldsymbol{R}$ as a parameter: $\boldsymbol{R}=\{\boldsymbol{R}_1, \boldsymbol{R}_2, \ldots, \boldsymbol{R}_a, \ldots\}$ is a multidimensional vector of nuclear coordinates. The solution to the Schrödinger equation for the system with total Hamiltonian $\hat{\mathscr{H}}$ (1.1), $\Psi^T(\boldsymbol{r}, \boldsymbol{R})$, is searched for in the form of a series over eigenfunctions of the Hamiltonian $\hat{\mathscr{H}}_0$:

$$\Psi^T(\boldsymbol{r}, \boldsymbol{R})=\sum_n \chi_n(\boldsymbol{R})\Psi_n(\boldsymbol{r}, \boldsymbol{R}); \tag{1.3}$$

the coefficients of expansion $\chi_n(\boldsymbol{R})$ depend on nuclei positions. Inserting (1.3) into the Schrödinger equation with the total Hamiltonian and taking into account orthogonal properties of a total set of functions $\Psi_n(\boldsymbol{r}, \boldsymbol{R})$, one obtains

$$(\hat{T}_N+\Phi_m(\boldsymbol{R})-E)\chi_m(\boldsymbol{R}) =\sum_n \left\{\hbar^2 \sum_a \frac{1}{M_a}\int \Psi^*_m(\boldsymbol{r}, \boldsymbol{R})[\nabla_a \Psi_n(\boldsymbol{r}, \boldsymbol{R})\nabla_a + \tfrac{1}{2}\nabla_a^2\Psi_n(\boldsymbol{r}, \boldsymbol{R})]\,\mathrm{d}\boldsymbol{r}\right\}\chi_n(\boldsymbol{R}) \qquad m=1, 2, 3, \ldots. \tag{1.4}$$

Operator ∇_a^2 acts on nuclear coordinates. This expression contains the exact solution of the Schrödinger equation for a system of moving electrons and nuclei. It represents an infinite series with functions $\chi_n(\boldsymbol{R})$ coupled with each other by means of the term on the right-hand side of expression (1.4) and corresponds to various electron states. The approximate solution of the system of equations (1.4) can be obtained with different degrees of accuracy. The simplest approximation consists of neglecting the right-hand side of expression (1.4), owing to the fact that $\nabla_a\Psi \ll \nabla_a\chi$ since the electron wave-functions are delocalized throughout a system. Then the system of equations (1.4) is separated into the set of independent equations with respect to functions $\chi_m(\boldsymbol{R})$ which describe nuclei motion

$$[\hat{T}_N+\Phi_m(\boldsymbol{R})]\chi_{mk}(\boldsymbol{R})=E_{mk}\chi_{mk}(\boldsymbol{R}) \tag{1.5}$$

and the wave-function of the total system (1.3) takes the form

$$\Psi^{\mathrm{T}}_{mk}(\boldsymbol{r}, \boldsymbol{R})=\chi_{mk}(\boldsymbol{R})\Psi_m(\boldsymbol{r}, \boldsymbol{R}). \tag{1.6}$$

The latter equations express the essence of the Born–Oppenheimer approximation: when the electrons are in the quantum state Ψ_m, the nuclear subsystem can exist in k states χ_{mk}.

The motion of nuclei is characterized by potential energy $\Phi_m(\boldsymbol{R})$, which is a sum of the electron's energy for some nuclear configuration $\boldsymbol{R}$ and the nuclear interaction energy. In a crystal the nuclear states with the lowest energy have maximum statistical weighting, and their wave-functions are localized in the vicinity of points of minima of potential Φ_m. Thus, the Born–Oppenheimer approximation results in the notion of the equilibrium nuclear configuration that is characterized by multidimensional vector $\boldsymbol{R}_0$ and corresponds to the potential energy minimum. The potential can be further expanded into a series over small displacements of nuclei, $\boldsymbol{u}$, with respect to their equilibrium positions

$$\Phi(\boldsymbol{R}) = \Phi_m(\boldsymbol{R}_0) + \sum_{\alpha,\beta} \left(\frac{\partial^2 \Phi(\boldsymbol{R})}{\partial R_\alpha \, \partial R_\beta} \right) u_\alpha u_\beta + \ldots \qquad \alpha, \beta = x, y, z \tag{1.7}$$

and the electron wave-functions, the solutions of equation (1.2), can be represented, in first-order perturbation theory, in the following form:

$$\Psi_m(\boldsymbol{r}, \boldsymbol{R}) = \Psi_m(\boldsymbol{r}, \boldsymbol{R}_0) + \sum_{n \neq m} \frac{W_{nm}}{E_m - E_n} \Psi_n(\boldsymbol{r}, \boldsymbol{R}_0). \tag{1.8}$$

Here W_{nm} is the matrix element of the quadratic (with respect to nuclei displacements) term in expression (1.7). Inserting expression (1.8) into (1.4), one can easily see that the right-hand side of (1.4) can be neglected only if the energies of both ground and excited electron states, E_m and E_n, differ noticeably. The vibrational character of nuclei displacements allows one to obtain the sufficient condition for the Born–Oppenheimer approximation validity: $\hbar\omega_N \ll E_m - E_n$. The nuclear vibration frequency has an order of $\omega_N \sim 10^{13}\,\mathrm{s}^{-1}$; electron transition frequencies are of the order of $10^{16}\,\mathrm{s}^{-1}$, so that the latter condition is valid, as a rule. This is not the case for metals, however: the Born–Oppenheimer approximation remains applicable for them also. If the motion of electrons and nuclei cannot be separated, then the vibronic interactions have important practical consequences (Bersuker 1984). This case will not be further dealt with.

Bringing the Born–Oppenheimer approximation to its logical completion one assumes that the electrons 'rigidly' follow the vibration nuclei, preserving their quantum states. This allows one to finally write the wave-function of system (1.6) in the form

$$\Psi^T_{mk}(\boldsymbol{r}, \boldsymbol{R}) = \chi_{mk}(\boldsymbol{u}) \Psi_m(\boldsymbol{r}, \boldsymbol{R}_0). \tag{1.9}$$

Since the mean value of the zero-point vibration amplitude of a nucleus, $[\hbar/(2M\omega_N)]^{1/2}$ (see appendix B, formula (B.11)), is much smaller than the characteristic distances between the minima of the potential $\Phi_m(\boldsymbol{R}_0)$, $(1.5\text{–}2.0) \times 10^{-10}$ m, the uncertainty resulting from using approximation (1.9) should not be taken into account.

Thus, within the Born–Oppenheimer approximation framework there exists the possibility of studying the electronic states of electron–nuclear systems by using a set of electron wave-functions, $\Psi_m(\boldsymbol{r}, \boldsymbol{R}_0)$, which correspond to the equilibrium nuclear configuration $\boldsymbol{R}_0$. The motion of nuclei is then considered as their displacement with respect to this equilibrium configuration and is described by means of wave-functions $\chi_{mk}(\boldsymbol{u})$. It is this approximation which allows one to introduce the notion of molecular and atomic crystal structure. However, the description of interactions between electrons still remains rather complicated. Even if the ground electron state is assumed to be sufficient for the majority of structural problems, the wave-function $\Psi_0(\{\boldsymbol{r}_i, s_i\}, \boldsymbol{R}_0)$ that describes this state depends on the spatial coordinates $\boldsymbol{r}_i$ and the spins s_i of all N electrons of the system. Fortunately, the description can essentially be simplified by introducing the one-particle function of electron ground state density, ρ (McWeeny and Sutcliffe 1976). Using the coordinate-dependent operator of electron local density $e\sum_i^N \delta(\boldsymbol{r}-\boldsymbol{r}_i)$ and taking into account the indistinguishability of electrons, ρ can be written as follows:

$$\rho(\boldsymbol{r}, \boldsymbol{R}_0)=e\int_{r_1}\cdots\int_{s_1}\cdots\Psi_0^*(\boldsymbol{r}_1, s_1,\ldots,\boldsymbol{r}_N, s_N; \boldsymbol{R}_0)\sum_i^N\delta(\boldsymbol{r}-\boldsymbol{r}_i)$$
$$\times\Psi_0(\boldsymbol{r}_1, s_1,\ldots,\boldsymbol{r}_N, s_N; \boldsymbol{R}_0)\,\mathrm{d}\boldsymbol{r}_1\,\mathrm{d}s_1\cdots\mathrm{d}\boldsymbol{r}_N\,\mathrm{d}s_N$$
$$=Ne\int_{r_1}\cdots\int_{s_1}\cdots\Psi_0(\boldsymbol{r}_1, s_1,\ldots,\boldsymbol{r}_N, s_N; \boldsymbol{R}_0)$$
$$\times\Psi_0(\boldsymbol{r}_1, s_1,\ldots,\boldsymbol{r}_N, s_N; \boldsymbol{R}_0)\,\mathrm{d}\boldsymbol{r}_1\,\mathrm{d}s_1\cdots\mathrm{d}\boldsymbol{r}_N\,\mathrm{d}s_N \tag{1.10}$$

(the integration over spin variables is equivalent here to summation over all possible directions of spins: the minus sign is not taken into account explicitly, but it is included in the electron charge e). The function ρ is the average over interelectron interactions, a density function of N electrons, the electron density. It depends on the coordinates $\boldsymbol{r}$ of position space, i.e. it is a much more simple construction than that of the wave-function. Note that ρ, as the electron distribution probability density function, has the same sign for all $\boldsymbol{r}$ values (or it is equal to zero). In crystallography one assumes conventionally that $\rho \geqslant 0$.

We can conclude that the presentation of a molecule and a crystal as a 'sea' of continuous ED with immersed nuclei vibrating about equilibrium positions has a clear physical basis. The lowering of the dimension of the electron structure description problem down to three space variables, x, y, z, is an important argument in favour of the ED-based construction of such a description. Another argument consists in the fact that the Fourier transformations of ED are the coherent structure scattering amplitudes. These quantities contribute in experimentally measured scattering intensities of the x-ray reflections, which also contain the contributions of incoherent scattering and other effects. Applying special experimental data processing techniques (which will be discussed in detail below), one can separate out the coherently scattered radiation and present the crystal ED as a Fourier series. The lost reflection phases can be measured by special multiwave diffraction methods (Chang 1987) or by calculation based on some particular crystal structural model. Thus, with some assumptions, the ED can be thought to be a measured quantity.

The reconstruction of all ED details from the x-ray, γ-ray, and synchrotron radiation diffraction data is not a simple procedure. It requires the proper statistical precision of reflection intensity measurements, the correction of measured values with respect to various factors influencing the coherent scattering, the determination and elimination (as far as possible) of the experiment-inherent limitation effects, the evaluation of the accuracy of results obtained, and their representation in a form suitable for interpretation and comparison with the theory. Each of these problems is not trivial and appropriate procedures are not routine.

In passing from the wave-function to ED, some of the information about the system under study is lost due to integration (i.e. averaging over electron coordinates). Let us ask ourselves the question of whether the remaining information is sufficient for describing the electron states of crystals? Fortunately, the answer is 'yes'. Hohenberg and Kohn (1964) have shown that the ED is a fundamental characteristic of atoms, molecules, and crystals. They considered the system of interacting electrons moving in a potential field, $v(\mathbf{r})$, of motionless nuclei $v(\mathbf{r})=\sum_i v(\mathbf{r}_i)=(4\pi\varepsilon_0)^{-1}\sum_a (z_a/|\mathbf{r}_i-\mathbf{R}_a|)$. The corresponding Hamiltonian $\hat{H}_e=\hat{T}_e+\hat{V}_{ee}+\hat{V}_{Ne}$ has among eigenfunctions only the function Ψ_0 corresponding to the electronic ground state, which is considered to be non-degenerate. They supposed then that there exists some other local potential $v'\neq v$ $(v-v'\neq\text{constant})$, which is included in the Hamiltonian $\hat{H}'_e\neq\hat{H}_e$ and results, for a system with the same number of electrons N, in the ground state $\Psi'_0\neq\Psi_0$ with energy $E'_0\neq E_0$, but with the same electron density $\rho'=\rho$. The ground state energy must be a minimum. Using the Ψ'_0 value, which is not an eigenfunction for $\hat{H}_e$, as a sample function and applying the variational principle to $\hat{H}_e$, they obtained

$$E_0<\int \Psi'^*_0\hat{H}_e\Psi'_0\,\mathrm{d}\mathbf{r}=\int \Psi'^*_0(\hat{H}'_e+\hat{V}_{Ne}-\hat{V}'_{Ne})\Psi'_0\,\mathrm{d}\mathbf{r}=E'_0+\int (v-v')\rho'(\mathbf{r},\mathbf{R}_0)\,\mathrm{d}\mathbf{r}. \tag{1.11}$$

The same procedure applied to the Hamiltonian $\hat{H}'_e$ and the sample function Ψ_0, leads to the expression

$$E'_0<E_0+\int (v'-v)\rho(\mathbf{r},\mathbf{R}_0)\,\mathrm{d}\mathbf{r}. \tag{1.12}$$

Summing up both inequalities, Hohenberg and Kohn arrived at a contradictory result (since $\rho=\rho'$): $E_0+E'_0<E_0+E'_0$. They concluded, therefore, that only one local potential v corresponds to some specific ED. The number of electrons in a system is strictly related to the ED by the relationship

$$\int \rho(\mathbf{r})\,\mathrm{d}\mathbf{r}=N \tag{1.13}$$

(the integration is carried out over the total space of a system). Hence, for the given number of electrons the wave-function of a system, its energy, and other characteristics are all functionals of ρ. That is, ρ uniquely defines the ground state of a many-electron system. In particular, the energy of the N-electron system in the given outer potential field $v(\boldsymbol{r})$ can be presented as

$$E(\rho)=\int v(\boldsymbol{r})\rho(\boldsymbol{r})\,\mathrm{d}\boldsymbol{r}+\frac{1}{2}\int\frac{\rho(\boldsymbol{r})\rho(\boldsymbol{r}')}{|\boldsymbol{r}-\boldsymbol{r}'|}\,\mathrm{d}\boldsymbol{r}\,\mathrm{d}\boldsymbol{r}'+G[\rho]. \tag{1.14}$$

Here $G[\rho]$ is the universal (i.e. common for all many-electron systems) functional of ρ that includes the kinetic energy and the non-classical part of an electron–electron interaction: the exchange and correlation of electrons. It can be proved that the exact ED of the ground state for the given v and N minimizes functional (1.14).

This result, known as the Hohenberg–Kohn theorem, has stimulated a rapid development of the theory of many-electron systems in which the measured value, electron density $\rho(\boldsymbol{r})$, rather than the wave-function, is the basic variable. The exact functional, which associates the ED with characteristics of many-electron systems, has not been found yet (this would be equivalent to the exact solution of a many-electron problem); however, the principles of the electron density functional theory have been formulated rather quickly (see e.g. Lundqvist and March 1983, Parr and Yang 1989, Dreizler and Gross 1990, Kryachko and Ludeña 1990, March 1992, Ellis 1995, Seminario and Politzer 1995). In initial steps the basic efforts were concentrated on searching for efficient approximations for $G[\rho]$ from (1.14) and on developing and theoretically substantiating various approximate techniques. The possibility of verifying computational schemes by comparing just the basic variable, the calculated ED, with that measured by diffraction methods, has greatly stimulated their development. The relativistic effects (Rajagopal 1980, March 1989), the finite temperatures (Mermin 1965) and the non-locality of potentials (Gilbert 1975) have also been taken into consideration. Teophilou (1979) generalized the Hohenberg–Kohn theorem for excited states, and Gunnarsson and Lundqvist (1976) for degenerate states. Later quantum hydrodynamics arose as the generalization of a density functional theory (March and Deb 1987). As a result, it became possible to describe the non-stationary electronic effects. The number of applications has grown rapidly following the development of the theory as well. During this process an important advantage of the method was implemented, namely, the possibility of its application for studying complicated many-nuclear systems. One of the outcomes of this (Dahl and Avery 1984, Dreizler and da Providencia 1985) is the modelling of crystal properties. This approach proved to be especially useful in absorption physics, in studying the effects related to crystal structure defects and impurity atoms and in studying band structure and properties of unstable multicomponent intermetallic compounds and still unsynthesized substances. The chemical applications of the density functional theory have also been developed. A rather solid basis was laid under such concepts, widely used in chemistry as the atom in a molecule or in a crystal and the molecular structure (Bader 1990), the electronegativity, the hardness, and the softness (Sen and Jorgensen 1987, Sen 1993). Progress in the explanation of some substance properties in terms of electrostatic potential, related to the total charge density by the Poisson equation (Politzer and Truhlar 1981), in the analysis of energy characteristics of molecules on the ED basis (Fliszar 1983), and in the development of efficient methods for obtaining interionic interaction potentials in crystals (Allan *et al* 1990) was achieved as well.

Of great importance is the progress in the chemical bond theory resulting from the establishment of the ED role in describing the ground state of many-electron systems (Bader 1975a, b, 1981, Bader and Essen 1984, Bamsai and Deb 1981, Becker 1980, Coppens 1977, Coppens and Hall 1982, Cremer and Kraka 1984a, Hall 1986, Hirshfeld and Rzotkiewicz 1974, Schwarz *et al* 1985, 1989, Spackman and Maslen 1986, Tsirelson and Antipin 1989, Low and Hall 1990, Hirshfeld 1991, Tsirelson and Ozerov 1992, Tsirelson 1993). The chemical bond concept allows one to replace the problem of description of interelectronic interactions in a many-electron and many-nuclear system by a problem on interatomic interactions. The different types of interatomic interaction are characterized by the wide energy range from 40 to 1000 kJ mol^{-1}. Correspondingly, they are said to concern covalent, ionic, donor–acceptor, metallic, and hydrogen bonds. Such a subdivision in bond types is a consequence of the existence of stably reproduced ED features (or other related characteristics of interatomic interactions) in various compounds. The association of observed peculiarities of ED with specific types of bonding is one of the main problems to be considered in this book.

The most popular method of specific ED feature investigation is the use of difference functions of various types. All of them will be considered in detail below. We mention here only one of these functions, the deformation ED function $\delta\rho$, introduced by Roux *et al* (1956) and widely used now:

$$\delta\rho(\boldsymbol{r})=\rho(\boldsymbol{r})-\tilde{\rho}(\boldsymbol{r}). \tag{1.15}$$

This function represents the difference between the ED of the system, ρ, and the ED of a set of spherically averaged non-interacting atoms, $\tilde{\rho}$, placed at the same positions as the atoms of the system. The value $\tilde{\rho}$ is called a promolecule or a procystal. Function $\delta\rho$ describes the redistribution of electrons, which takes place in a system when the latter is formed from free atoms. It is obvious that ρ and $\tilde{\rho}$ values must be calculated in a unique manner, and the integration over the whole system should give zero: $\int \delta\rho(\boldsymbol{r})\,\mathrm{d}\boldsymbol{r}=0$. The $\delta\rho$ profiles and sections allow one to establish what the ED value is at some specific points of a crystal (for example, in the internuclear space, in the hydrogen bond region, etc), whether the chemical bonds are 'bent', where the maxima, related to the chemical bond and electron lone pairs, are located, etc. On the $\delta\rho$ maps, the points with the same excessive and deficient ED are connected by isolines, and several maps allow one to judge many of the features of the electron distribution of the compound under consideration.

There are a few lines of applications in which the crystal EDs are used. One of them is in searching for relations between electron structure and properties by means of various model schemes whose parameters are determined from ED. These schemes are really quite versatile. Some of them are rather simple, for example, the procedure of calculating the electric field gradient, which is the integral characteristic of a total charge distribution (electronic plus nuclear). In other schemes, such as the approximate method for calculating nonlinear optical characteristics based on a pseudopotential theory, the ED characteristics are included in the model in a rather complicated manner. It is important that these methods allow one to evaluate with a small single-crystal specimen a few of the potential properties of a substance under study, based on the x-ray diffraction experiment only. A wide field of opportunity is opened here for applying these approaches to the search for new perspective technological materials.

We have outlined a range of problems in solid state physics and chemistry in which the ED plays a key role. Now we shall consider them in detail.

2

THEORY

The ED distribution in the position space of a crystal can be obtained in two different ways. The first one consists of the solution, under some approximation, of the Schrödinger equation or corresponding equations of the electron density functional theory. The second approach is based on the ED reconstruction from the scattering of x-rays, γ-rays, or synchrotron radiation. First, we shall consider the theoretical approach to ED investigation.

2.1 Methods of ground state quantum chemistry

2.1.1 *The self-consistent field method for molecules*

Let us consider the methods for solution of the electronic Schrödinger equation with a Hamiltonian $\hat{H}_e = \hat{T}_e + \hat{V}_{ee} + \hat{V}_{Ne}$ which do not use the ED-based approximations. These methods are attributed conventionally to the classical quantum chemistry of molecules and solids. We restrict our analysis to those methods which do not use additional empirical data, the so-called *ab initio* methods. The wave-functions are calculated in these methods using the variational principle (Epstein 1974) with minimal introduction of approximations.

The Hamiltonian $\hat{H}_e(\boldsymbol{R})$ characterizes the motion of N electrons in the field of motionless nuclei (the vector $\boldsymbol{R}$, which characterizes nuclear configuration, will be omitted hereafter). The interactions in such a system have a two-particle character; therefore, $\hat{H}_e$ can be represented as a sum of one-electron and two-electron parts:

$$\hat{H}_e = \sum_i^N \hat{h}_i + \sum_{i<j} \hat{g}_{ij} \tag{2.1}$$

$$\hat{h}_i = -\frac{\hbar^2}{2m}\nabla_i^2 - e\sum_a \frac{z_a}{4\pi r\varepsilon_0 |\boldsymbol{r}_{ia}|} \tag{2.2}$$

$$\hat{g}_{ij} = e^2/4\pi\varepsilon_0 |\boldsymbol{r}_{ij}|. \tag{2.3}$$

Subscripts i and j refer to the ith and jth electrons spaced at distance $|\boldsymbol{r}_{ij}|$; $|\boldsymbol{r}_{ia}|$ is the distance between nucleus a and electron i. The presence of the two-electron operator $\hat{g}$ in the Hamiltonian makes the exact solution of the differential Schrödinger equation impossible, since one cannot separate the spatial coordinates of electrons. This difficulty can be overcome by applying the quasi-independent particle approximation. In this approximation the system's wave-function is presented as a combination of products of orthonormal one-electron functions φ_i, called spin orbitals. The latter depend on spatial, $\boldsymbol{r}_i$, and spin s_i, coordinates of each electron moving in the effective average field of rest electrons and nuclei. The electron velocities are large and the state of each electron depends more upon the average field than on the exact spatial arrangement of the rest electrons. Therefore, the approximation discussed has a physical basis. To take into account the indistinguishability of electrons and the requirement of the N-electron wave-function antisymmetry (the Pauli principle), this function is approximated by a set of determinants composed of $\varphi_i(\boldsymbol{x}_i)$ functions, the Slater determinants. Each of these has the form $\Psi=(N!)^{-1/2}\det\{\varphi_1(\boldsymbol{x}_1)\varphi_2(\boldsymbol{x}_2)\ldots\varphi_N(\boldsymbol{x}_N)\}$, where $\boldsymbol{x}_i=\{\boldsymbol{r}_i, s_i\}$: the permutation of two electrons is equivalent to a change of sign of the Slater determinant. Using the single determinant, applying the variational principle to the mean value of the energy, $E=\langle\Psi|\hat{H}_e|\Psi\rangle$, under the additional condition of orthonormalization of φ_i functions

$$\langle\varphi_i|\varphi_j\rangle=\delta_{ij} \tag{2.4}$$

and assuming that the first-order variation of the energy is zero, one can obtain the set of integrodifferential equations which are satisfied by spin orbitals φ_i corresponding to stationary values of electronic energy. Such an approach, which provides optimal, in the sense of an average energy minimum, approximate wave-functions is called the Hartree–Fock (HF) method. The HF equations can be written as follows (McWeeny and Sutcliffe 1976):

$$\hat{F}\varphi_i=\varepsilon_i\varphi_i \qquad i=1, 2, 3, \ldots, N. \tag{2.5}$$

Here $\hat{F}$ is the one-electron effective Hamiltonian, the Fock operator. Matrix elements, F_{ii}, of this operator have the form

$$F_{ii}=\langle i|\hat{h}|i\rangle+\sum_{k}^{N}\{\langle ik|\hat{g}|ik\rangle-\langle ik|\hat{g}|ki\rangle\} \tag{2.6}$$

i.e. the presence of one-electron and two-electron parts is conserved, as in the exact Hamiltonian $\hat{H}_e$ (2.1). The essential distinction from $\hat{H}_e$ consists in the fact that the operator $\hat{F}$ is a functional of unknown functions φ_i, that is the HF equations are nonlinear. For this reason the solutions of HF equations are constructed in an iterational manner. Having specified some set of spin orbitals, one constructs the Fock operator and then the HF equations are solved. The solution corresponding to the energy minimum (the ground state) is used again for constructing the new operator $\hat{F}$ and so on. This procedure is repeated until a 'self-consistency' is achieved, i.e. until φ_i functions differ only slightly from those obtained at a preceding iteration. Such an approach to solving equations is called a self-consistent field (SCF) method. The averaging over electronic interactions is achieved during such a procedure of obtaining solutions because the self-consistent potential in the operator $\hat{F}$ for each function φ_i is obtained from the rest of the spin orbitals of the same set.

The two-electron terms of type

$$\langle ik|\hat{g}|ik\rangle = \frac{e^2}{4\pi\varepsilon_0}\iint \varphi_i^*(\boldsymbol{x}_1)\varphi_k^*(\boldsymbol{x}_2)\frac{1}{|\boldsymbol{r}_{12}|}\varphi_i(\boldsymbol{x}_1)\varphi_k(\boldsymbol{x}_2)\,\mathrm{d}\boldsymbol{x}_1\,\mathrm{d}\boldsymbol{x}_2 \tag{2.7}$$

which appear in expression (2.6) describe the mean Coulomb repulsion of electrons, associated with spin orbitals φ_i and φ_k, regardless of spin direction. The two-electron integrals

$$\langle ik|\hat{g}|ki\rangle = \frac{e^2}{4\pi\varepsilon_0}\iint \varphi_i^*(\boldsymbol{x}_1)\varphi_k^*(\boldsymbol{x}_2)\frac{1}{|\boldsymbol{r}_{12}|}\varphi_k(\boldsymbol{x}_1)\varphi_i(\boldsymbol{x}_2)\,\mathrm{d}\boldsymbol{x}_1\,\mathrm{d}\boldsymbol{x}_2 \tag{2.8}$$

due to orthogonality of spin components of wave-functions φ_i and φ_k differ from zero only when the spins of electrons are parallel. These terms take into account the correlation in motion of electrons with the same spins caused by the Pauli principle. Integrals (2.7) are usually called the Coulomb integrals and integrals (2.8) the exchange integrals. The Coulomb potential has local character, whereas the exchange potential is non-local.

Wave-functions φ_i, the solutions of HF equations, are delocalized throughout a position space. Since the symmetry of the Fock operator $\hat{F}$ must correspond to the symmetry of the system studied, the φ_i, as eigenfunctions of the $\hat{F}$ operator, must be transformed over irreducible representations of some particular group of symmetry. However, an undesirable wave-function symmetry violation takes place sometimes through approximating the electronic interaction (Mestechkin 1990).

The eigenvalues of Fock's operator, ε_i, are interpreted as energies of spin orbitals φ_i. These eigenvalues include contributions of the kinetic energy of electrons, described by the φ_i function, of the potential energy of electron attraction to nuclei, and of the mean potential energy of repulsion from the rest electrons taking account of exchange. According to Koopmans theorem (McWeeny and Sutcliffe 1976), energy ε_i gives an approximate value of the ionization potential of an electron, described by the ith spin orbital, if one neglects the ionization-related relaxation of a system. The latter approximation for stationary states is satisfied for wave-functions to an accuracy of first order in a perburbation series.

The use in the HF method of a many-electron wave-function in the form of a single determinant gives rise to a number of specific features and properties. There are differences in the description of a system with filled (the total spin is zero) and open electron shells. In the latter case one must ensure that the electronic wave-function is an eigenfunction of total spin operators $\hat{S}_Z$ and $\hat{S}^2$. Therefore, the modifications of the HF equations for systems with even and odd numbers of electrons are different. Besides, for systems with closed shells, which will be considered further, the Brillouin theorem

$$\langle \Psi | \hat{H}_e | \Psi_S \rangle = 0 \tag{2.9}$$

is valid. Here Ψ and Ψ_S are single-determinant wave-functions; one of them, Ψ, describes the ground electronic state, in each of the others, Ψ_S, one spin orbital, occupied by electrons, is replaced by a single-excited-state spin orbital. The latter is orthogonal to all other orbitals. The consequence of this theorem is the fact that the HF many-electron wave-function Ψ provides a description of ED that is accurate to second order in energy (the question of how essential second-order corrections are will be considered later). Besides, the HF wave function Ψ is invariant with respect to unitary transformations of a set of occupied spin orbitals. As a result, the latter are not determined uniquely. This circumstance can be used for the spatial localization of different spin orbitals based on different criteria (Chalvet *et al* 1975). As a result the formal description can be used to provide the 'pictorial' concepts of electron 'lone pairs', 'bond' electrons, 'core' and 'valence' electrons, etc. In this case the wave-function of a system and the total ED remain unchanged, though the mean value of operators, which depend on individual orbitals (ionization potentials, for example), are changed. The correspondence between (indistinguishable!) electrons and spin orbitals, existing in the HF method, essentially simplifies the conceptual treatment of ED in the system mentioned above. That is why it is the single-determinant approximation which is used (or implied), as a rule, in the analysis of chemical bonds and in the interpretation of various experimental results.

The presence in the Hamiltonian of term (2.3), describing the Coulomb interaction of electrons, implies that the system energy is minimized when electrons are far in position space from each other. In other words, the motion of electrons must be correlated. The latter circumstance is ignored in the HF method: the repulsion of electrons is considered here on average only. As a result, an error in energy minimization arises: the underestimation of the kinetic energy with respect to its actual value and the overestimation of the potential energy due to interaction between electrons with opposite spin situated in the same spatial orbital. This error is evaluated by a fictitious correlation energy $E_c < 0$. In order to take the electron correlation into account, one must escape from the HF method framework. For this purpose the many-electron wave-function Φ is presented as the set of Slater determinants. One of them, Ψ_0, is the HF determinant for the electronic ground state; in the rest, Ψ_k, the occupied spin orbitals have been successively substituted by virtual spin orbitals. The latter are orthogonal to all remaining orbitals:

$$\Phi = c_0 \Psi_0 + \sum_{k \neq 0} c_k \Psi_k. \tag{2.10}$$

The expansion can contain all symmetry-allowed single- double- triple-, . . ., n-fold-substituted determinants, describing different electronic configurations.

Such a description is called complete and the approach is described as the configuration interaction method. In fact, the expansion includes a limited, but large number of determinants (10^6–10^9). The strategy of their selection has been discussed by Smith (1980).

One should note that the many-electron wave-function approximated by Slater determinants, in contrast to the exact wave-function, does not have singilarities at $|\boldsymbol{r}_i - \boldsymbol{r}_j| \to 0$. This is a general drawback of the HF method and the method of configuration interaction.

The solutions of the one-electron equations taking into account many configurations can be obtained either by solving a secular problem (Shavitt 1977, Roos and Siegbahn 1977), or by means of the Rayleigh–Schrödinger perturbation theory, in which the single-determinant approximation serves as the zeroth-order approximation (Møller and Plesset 1934, Pople *et al* 1976). Having studied the capabilities of both methods, Handy (1984) came to the conclusion that both of them allow one to take into account up to 99% of the correlation energy value (for small molecules) and indeed lead to the same results. Wang and Boyd (1989) have arrived at a similar conclusion independently. At the same time, they have found that the use of the perturbation theory saves as much as two to four times the computer time and is therefore preferable. Morrison and Froese-Fischer (1987) suggested the combination of the configuration interaction method with many-particle perturbation theory using a multiconfigurational wave-function as the initial approximation. It is difficult to say now whether this approach will provide any computational advantage compared to the standard techniques available.

The estimations show (Iwata 1982), that the error in the energy determination is at least 1% in the HF method. This accuracy is often insufficient for calculating the molecule dissociation energies. The error in the energy spectrum of excited states of crystals can be about 100%. As far as the dissociation energies are concerned, Cook and Karplus (1987) have shown that in calculating these energies for molecules within the framework of the HF method there exists some error in a molecular wave-function with respect to atomic wave-functions. This error, which depends on atomic states varying with changing molecule geometry, makes a major contribution to the potential surface error at high internuclear distance especially. The value of error noticeably decreases when a large number of configurations is taken into consideration. It is important that the error discussed does not have a marked influence on the ED of stable molecules or crystals, which are determined to an accuracy of a few per cent in the HF method. As a result, this method allows one to embrace all main chemical bond features, as will be demonstrated below.

In the HF and configuration interaction methods there exist some computational problems related to symmetry violations, solution instability, the presence of false minima and saddle points in a minimized energy functional, etc. The problems listed and techniques for their solution are discussed by McWeeny and Pickup (1980), Koelling (1981), Mougenot *et al* (1988), Mestechkin (1990) and Pisani (1995).

2.1.2 *The Hartree–Fock method for solids*

Spatially localized atomic orbitals are suitable for calculating the electronic structure of atoms and molecules, as well as solids consisting of weakly coupled atomic groups, for example, molecular crystals. For metals, covalent semiconductors, and ionic and framework inorganic crystals the basis set functions must obey the translational periodicity of a crystal. The periodicity is taken into account by using the Bloch theorem (Ashcroft and Mermin 1976), according to which the solutions of the one-electron equations for a system with a periodic crystal potential depend on the electron wave vector $\boldsymbol{k}$ and have the form

$$\varphi_{\boldsymbol{k}}(\boldsymbol{r})=\sum_j c_j(\boldsymbol{k})u_{\boldsymbol{k}j}(\boldsymbol{r}). \tag{2.11}$$

Here $u_{\boldsymbol{k}}(\boldsymbol{r})$ is a periodic function in a lattice, the Bloch function. As a result, the electrons in a crystal are distributed over energy bands, which represent the sets of permitted energy values characterized by vector $\boldsymbol{k}$. In order to obtain the ED in position space, one must integrate the crystal orbitals over $\boldsymbol{k}$ within the first Brillouin zone and then sum the results over all occupied bands. This rather tedious procedure can be essentially facilitated by summing over special symmetry points of the Brillouin zone (Baldereschi 1972). The special points formalism allows one to replace, to a very good accuracy, the integration over $\boldsymbol{k}$ by summation over a specially chosen, limited number of points determined by symmetry. This greatly simplifies the transition from crystal orbitals to the ED distribution.

Functions (2.11) can be constructed in different ways. One of them (the tight-binding approximation) consists in approximating basis set functions, $u_{\boldsymbol{k}j}$, by atomic orbitals (AOs) in the form

$$u_{\boldsymbol{k}j}(\boldsymbol{r})=n^{-1/2}\sum_{\boldsymbol{R}}\exp(\mathrm{i}\boldsymbol{k}\cdot\boldsymbol{R})\Phi_j(\boldsymbol{r}-\boldsymbol{R}) \tag{2.12}$$

where n is the number of translationally equivalent AOs. For the given value of vector $\boldsymbol{k}$ and fixed value of j, each function $u_{\boldsymbol{k}j}$ is transformed according to the same irreducible representation of a group of translations. Thus, for a crystal with continuous electron distribution the HF equations should be solved in a basis set of orbitals $u_{\boldsymbol{k}j}$ for every $\boldsymbol{k}$ value. Harris (1976), Pisani *et al* (1988), and Dovesi *et al* (1992) have thoroughly developed the HF method for calculating the electron structure of periodic solids and characterizing the results obtained. Pisani *et al* (1988) carried out systematic calculations of a large number of crystals by using the HF method. These authors managed to calculate the band structure and ED distributions for not only the simplest monoatomic and diatomic crystals (Orlando *et al* 1990, Mackrodt *et al* 1992, Causa *et al* 1991, Prencipe *et al* 1995), but also for solid polyacetylene (Dovesi 1984), quartz (Dovesi *et al* 1987, Nada *et al* 1990), corundum-type crystals (Causa *et al* 1987, Gatti *et al* 1995), spinels (Gatti *et al* 1994), perovskites (D'Arco *et al* 1993, Ricart *et al* 1995, Towlet *et al* 1995), etc. In simple compounds the calculated ED agrees well with the experimental ED.

Analysis of the results of the calculations of Pisani *et al* (1988) and Mackrodt *et al* (1992) has shown that HF binding energies represent approximately 70% of the experimental ones; all energy gaps are 1.5–2 times wider than the experimental ones, lying in the range of 1–25 eV. As will be shown later, the local density approximation of density functional theory yields the opposite results for energy gaps. Monkhorst and Pack (1979) and Tschinke and Ziegler (1990) have explained this as the result of excessive non-locality of the exchange potential (2.8) in the HF method due to taking account of Pauli correlations only.

2.1.3 *The wave-functions*

As is known, the quantum mechanical description of many-electron systems does not lead to the analytical solution of the Schrödinger equation. The exceptions are the simplest cases of one-electron systems: a hydrogen atom and a molecular hydrogen ion for which the solutions in analytical form can be obtained (Bethe and Salpeter 1957). The hydrogen atom nuclear potential has spherical symmetry and the wave-functions in the total set have certain values of angular momentum. Therefore, these functions are usually classified according to the symmetry properties of atomic electronic states. One distinguishes s, p, d, f, etc states, which correspond to orbital quantum number values $l=0, 1, 2, 3$, etc. For states with $l \geqslant 1$ the complex wave-functions form degenerate sets, whose terms differ by values of magnetic quantum number m which determine the momentum projection on the axis. Using these functions one can construct real linear combinations, which are solutions of the Schrödinger equation as well. Real wave-functions are separated in position space by nodal surfaces on the regions where the wave-functions have a constant sign (the number of surfaces increases with the increasing number of energy levels in the discrete spectrum) and have a rather simple interpretation in the Cartesian coordinate system. These functions are especially valuable for clarifying the role of particular wave-functions in forming a chemical bond.

In describing many-electron atoms by the HF method the one-electron problem scheme is conserved: one assumes the approximation of a central field whose potential is constructed by the variational principle and the self-consistent field method. In this approximation the variables in the wave-function of each electron in spherical r, θ, φ coordinates, in atomic spin orbitals, can be separated. The spin orbitals can then be presented as the product

$$\Phi_{nlms} = N(n, l)R_{nl}(r)Y_{lm}(\theta, \varphi)\eta(s). \tag{2.13}$$

The radial part $R_{nl}(r)$ depends on the principal quantum number n; the angular part $Y_{lm}(\theta, \varphi)$, a spherical harmonic, is characterized by quantum numbers l and m; and spin function $\eta(s)$ is determined by spin quantum number $m_s = \pm\frac{1}{2}$; $N(n, l)$ is a normalizing factor. Hereafter we will consider the systems with closed electron shells only, unless otherwise stated. In these cases it is sufficient to use spinless atomic wave-functions $\Phi_{nlm}(\eta(s)=1)$, AOs.

The presentation of orbitals in the form of $\Phi \sim R_{nl}Y_{lm}$ is valid for atoms in fully symmetrical states only. However, the Fock operator symmetry is often lower than the true one. As a result, the eigenfunctions of this operator can have a form which is distinct from the product $R_{nl}Y_{lm}$. Spherical harmonics $Y_{lm}(\theta, \varphi)$ describe the 'shape' of the AO and are the same for all atoms. Real combinations of these functions are uniquely oriented with respect to coordinate axes of the Cartesian system (figure 2.1). Complex and real spherical harmonics with $1 \leqslant 2$ are given in table 2.1.

The approximate wave-functions for multinuclear systems are sought by modifying the solution of the one-electron problem. The molecular orbitals (MOs), φ_i, are often written as linear combinations of atomic orbitals (LCAOs), Φ_j, centred on the nuclei of the system and possessing proper

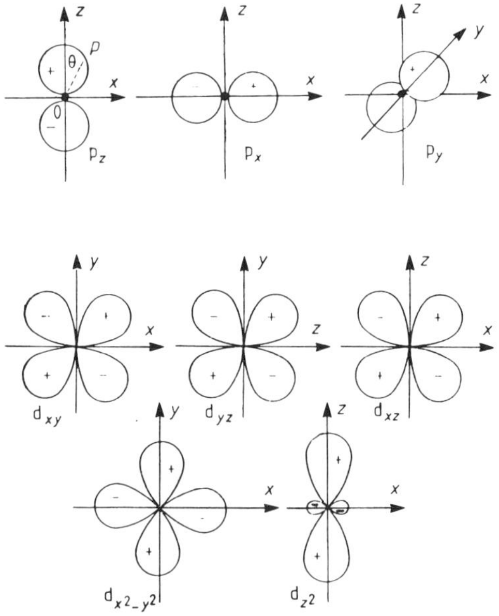

Figure 2.1 Space orientations of angular parts of AOs.

Table 2.1 Normalized spherical harmonics $Y_{lm}(\theta, \varphi)$ and real spherical harmonic functions $y_{lm\pm}$ for $l \leqslant 2$. $Y_{lm}(\theta, \varphi) = (-1)^{(m+|m|)/2}[[(2l+1)/4\pi](1-|m|)!/(1+|m|)!]^{1/2} P_l^{|m|}(\cos\theta)\, e^{im\varphi}$, $P_l^{|m|}$ is the associated Legendre polynomial (Arfken 1985), $y_{lm+} = (1/\sqrt{2})[(-1)^m Y_{lm} + Y_{l-m}]$, $y_{lm} = -(i/\sqrt{2})[(-1)^m Y_{lm} + Y_{l-m}]$, $i = \sqrt{-1}$.

Wave-function type	Orbital quantum number l	Magnetic quantum number m	Y_{lm}	$y_{lm\pm}$
s	0	0	$\frac{1}{2}\pi^{-1/2}$	$\frac{1}{2}\pi^{-1/2}$
p	1	0	$\frac{1}{2}(3/\pi)^{1/2}\cos\theta$	$\frac{1}{2}(3/\pi)^{1/2}\cos\theta$
		± 1	$\mp\frac{1}{2}(3/2\pi)^{1/2}\sin\theta\, e^{\pm i\varphi}$	$\frac{1}{2}(3/\pi)\sin\theta\begin{cases}\cos\varphi\\ \sin\varphi\end{cases}$
d	2	0	$\frac{1}{4}(5/\pi)^{1/2}(3\cos^2\theta - 1)$	$\frac{1}{4}(5/\pi)^{1/2}(3\cos^2\theta - 1)$
		± 1	$\mp\frac{1}{2}(15/2\pi)^{1/2}\sin\theta\cos\theta\, e^{\pm i\varphi}$	$\frac{1}{4}(15/2\pi)^{1/2}\sin 2\theta\begin{cases}\cos\varphi\\ \sin\varphi\end{cases}$
		± 2	$\mp\frac{1}{2}(15/2\pi)^{1/2}\sin^2\theta\, e^{\pm 2i\varphi}$	$\frac{1}{4}(15/\pi)\sin^2\theta\begin{cases}\cos\varphi\\ \sin\varphi\end{cases}$

symmetry (the MO LCAO approximation):

$$\varphi_i = \sum_j c_{ij}\Phi_j. \tag{2.14}$$

Equations (2.5), modified in the LCAO approximation, are called the Hartree–Fock–Roothaan (HFR) equations. The set of AOs used in the expansion is called the basis set.

The choice of analytical functions and construction of a basis set from them is of great importance. This choice must provide the basis set with flexibility sufficient for describing the charge transfer that accompanies the molecule or crystal formation from free atoms. On the other hand the size of a basis set should be limited to provide the efficient calculation of the integrals required. A proper local AO behaviour near the nucleus and far from it should also be provided. The electron–nuclear potential in (1.1) has singularities on nuclei. As Kato (1957) has shown, the accurate many-electron wave-function, Ψ, in order to satisfy the Schrödinger equation on nuclei, should obey some specific cusp condition

$$(\partial\bar{\Psi}/\partial r)_{r=\boldsymbol{R}_a} = -Z_a\bar{\Psi}_{r=\boldsymbol{R}_a}. \tag{2.15}$$

Here $\bar{\Psi}$ is the wave-function Ψ averaged over a small sphere in the vicinity of the singularity point. If Ψ is written as a linear combination of MOs and the latter, in turn, are expanded over some basis set functions, then condition (2.15) must be met for basis functions of s type, which have a non-zero value at a nucleus. In addition, Handy *et al* (1969) have shown that at large distances from nuclei the relation

$$\Phi_j \sim \exp[-(2I_1)^{1/2} r] \tag{2.16}$$

is met for HF orbitals (I_1 is the first ionization potential). The relations (2.15) and (2.16) can be used to test the quality of orbitals obtained.

In the central field approximation, it is sufficient to find, using the variational principle, only the radial parts of the AO, which for the given quantum numbers n and l depend on the distance from the nucleus r and on the nuclear charge. This greatly simplifies the problem, of course.

The radial parts of the AO are usually approximated by functions of the type $r^{n-1}\exp(-\xi r)$ or $r^{n-1}\exp(-\alpha r^2)$, where n is the principal quantum number. The former functions are called the Slater-type orbitals (STOs), the latter the Gaussian-type orbitals (GTOs). These nodeless functions, which are often called basis set functions, have simple form and their linear combinations well approximate the correct wave-functions, in general (Wilson 1987). The GTOs are most convenient from the viewpoint of the calculation of multicentred integrals (Huzinaga 1984) though the quadratic dependence on the distance to the nucleus in the exponent requires one to use larger expansions compared to STOs. The STO expansions over certain fixed sets (the 'contracted' Gaussian basis sets) are often used in calculations as well as mixed STO and GTO combinations (Raffenetti 1973). The matter of AO approximations is considered in more detail in appendix C.

Various authors have proposed their own versions of STO and GTO wave-function expansions of different types. The most complete summaries of expansion parameters are given by Clementi and Roetti (1974), McLean and McLean (1981), Huzinaga (1984), and Bunge *et al* (1993). Davidson and Feller (1986) and Wilson (1987) have considered the advantages and drawbacks of various basis sets.

Thus, the computational work is fed by reliable initial data. As an example of an 'accurate' approximation to wave-functions, figure 2.2 presents the radial HF AO for the Cr atom calculated from the data of Clementi and Roetti (1974). The shell structure of the atomic electronic cloud is clearly seen.

A complete set of linearly independent basis functions must lead, after some variational procedure, to the lowest energy value that is possible within the HF method framework; in this case the HF limit is reached. When a limited basis set is used, one can only approach the above-mentioned boundary by varying the expansion coefficients c_{ij} in the formula (2.14) and exponential parameters in the radial parts of basis AOs in the course of solution of HFR equations. In such a manner the basis set limit can be achieved. The basis set functions are described as energy optimized in this case.

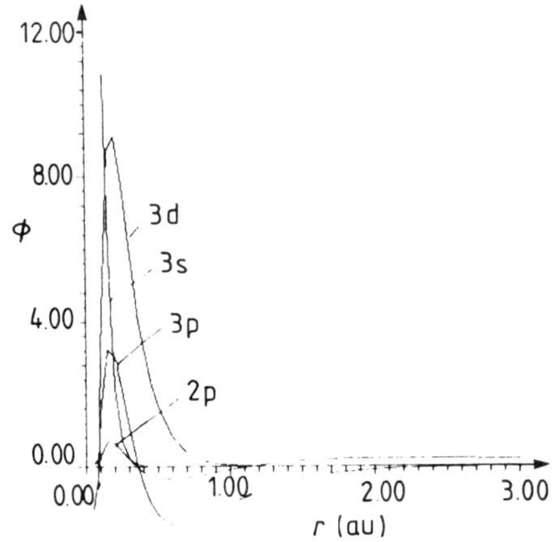

Figure 2.2 Radial parts of AOs for the Cr atom calculation from Clementi and Roetti (1974) wave-functions.

The degree of basis set limitation may be different. In the simplest case the basis set is composed of AOs which are occupied by electrons in free atoms. Such a basis set is called minimal, if the radial part of the AO is approximated by a single function (single zeta, SZ). For example, the SZ basis set in the diamond crystal consists of 1s, 2s, $2p_x$, $2p_y$, and $2p_z$ AOs of the C atom. The basis set is called double zeta (DZ) or triple zeta (TZ) if the AO radial part is approximated by two or three sizes of functions, respectively. The basis set flexibility can be extended with addition of polarization AOs corresponding to the non-occupied levels in the free atoms in their ground states. The H atom is well described in this case by two functions of s type and by one set of p functions; the atoms from lithium to neon are better described with six additional d functions, etc. When using polarization functions, the abbreviated notation is often used: for instance DZ+DP means that the two-exponential basis set is augmented by d-type polarization functions. The diffuse (large) high-angular-momentum functions are added to the basis set as well (Foresman and Frisch 1993). The polarization and diffuse functions not only better reproduce the ED deformations of atoms in a molecule, but also they are useful for satisfying the atom position symmetry requirements, if the AOs corresponding to occupied levels are insufficient for this purpose. A discussion of the chemical essence of these functions was given by Smith (1977).

Sometimes, the polarization functions are replaced by so-called bond functions centred in the interatomic region. Breitenstein *et al* (1983) have tabulated the bond function characteristics for a great number of bonded pairs of atoms of the first and second rows of the periodic table. They have also found that the 4−31G+BF basis set yields results close to those achieved with a basis set at the HF limit.

The application of extended basis sets has raised a number of questions related to calculation of properties observed and separation of ED in a system into interpretable parts (atomic charges etc). In particular, it is important to know how many functions of a particular type of symmetry should be included in the calculation. Considering this problem, Mulliken (1962) introduced the notion of a balanced basis set. He distinguished a physically balanced basis set, which results in 'reasonable' observed quantities, and a formally balanced basis set, that gives rise to 'reasonable' atomic charges. Such a separation is obviously empirical. Smith (1977) has formulated a more substantiated requirement for a balanced basis set: the energy of each orbital should be the same percentage of its energy as at the HF limit. This criterion provides a rule for selection of a number of functions of each type (s, p, d, etc) on atoms and is especially useful for calculating the electronic structure of systems containing atoms of different sorts.

It should be pointed out that the many-electron wave-function at the HF limit cannot approach the exact wave-function if the self-consistent potential differs from the exact one, for example, due to the lowering of symmetry. Therefore, the use of a carefully selected basis set still does not guarantee that one obtains good EDs.

2.1.4 *Relativistic effects*

The necessity of taking into account the relativistic effects in the Hamiltonian arises when systems with heavy atoms are studied. The main effects are the dependence of electron mass on velocity, which is most important for electrons of inner (core) atomic shells, and the spin–orbit interaction and the Darwin interaction arising from small fluctuations of the moving electron with respect to its average position (Snijders and Baerends 1982, Malli 1983, Pyykkö 1988, Schwarz 1990).

Relativistic effects have influenced the calculated values of bond lengths, force constants, dissociation energies, ionization potentials, ED, and band structures of crystals. Relativity manifests itself in ED in the form of radial compression and energetic stabilization of s and p electron subshells, in the spin–orbit splitting of energy levels, and some expansion and energetic stabilization of valence d and f electron shells. All these effects have the same order of magnitude, proportional to Z^2. Specific changes in ED are illustrated in figure 2.3, which presents hydrogen-like radial ED 3s orbitals of the Hg atom ($Z=80$), calculated in the relativistic and non-relativistic approximations by Burke and Grant (1967). The compression of 1s and 2s electron subshells is comparable in magnitude with the compression of the 3s subshell, whereas it is essentially smaller for p subshells.

In the Dirac–Fock method, the relativistic analogue of the HF method, the total many-electron function is presented as a linear combination of antisymmetrized products of one-electron (four-component) Dirac bispinors.

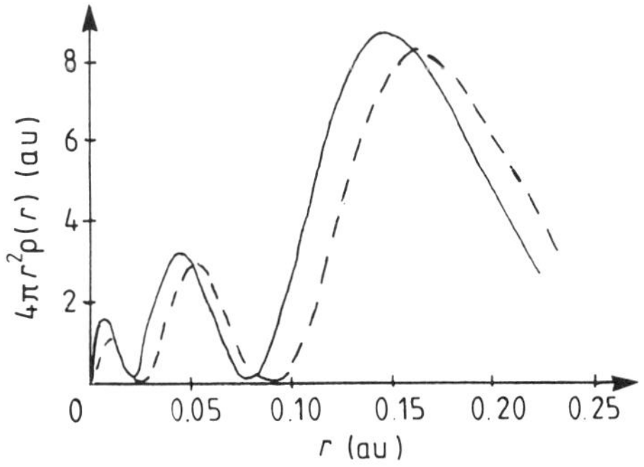

Figure 2.3 Radial densities for 3s state of a hydrogen-like atom with $Z=80$ (Hg). The solid curve is the relativistic calculation and the dashed cuve is the non-relativistic calculation (Burke and Grant 1967).

The latter can be presented as an expansion over some basis set, in the LCAO approximation in particular (Lee and McLean 1982), and the solution of corresponding equations can be obtained in an iterative self-consistent manner. The implementation of all-electron calculations for complicated systems is a very difficult problem, however. Sometimes it is simplified (Bersuker *et al* 1977, Pyykkö 1988) by applying the Pauli quasirelativistic approximation that is valid to an accuracy of terms of the order of $(v/c)^2$ and is applicable to valence electrons. In this case the wave-functions can be approximated by two-component spinors, which reduces the computation work. Bersuker and Ogurtsov (1979) have emphasized that this approximation, which is justified in semiempirical and pseudopotential methods, is insufficient in a non-empirical approach since the electrons of outer atomic s subshells have velocities comparable with the velocity of light.

Having analysed the vast number of empirical data concerning the influencc of rclativity on the structure and properties of substances, Pyykkö (1988) pointed out that in compounds containing Cu, Ag, and Au atoms the relativistic effects manifest themselves to the highest degree. These effects are also noticeable in compounds of lanthanides.

2.1.5 *Pseudopotentials*

Calculations of the crystal electronic structure are often performed using the model in which the ED is initially separated into the core and valence parts. Further, the all-electron problem is replaced by the problem of describing the valence electrons in a self-consistent field of atomic cores, having taken into account the Coulomb and exchange interactions between electrons of both subsystems (Szacz 1985). Thus, the nuclear potentials are replaced by pseudopotentials and the equations of the method are written for valence and conductivity electrons only. This essentially lowers the dimension of the problem. The solutions of these equations, pseudo-wave-functions, are found using the variational method. These functions are close in form to the all-electron wave-function outside the core, but inside the core region they are smooth, unlike the all-electron wave-functions. The eigenvalues of these equations, the energies of valence electronic levels, are close to those obtained by solving the all-electron Schrödinger equation. Relativistic effects can easily be taken into account in pseudopotentials (see e.g. Balasubramanian 1989).

The pseudopotentials are usually approximated by some analytical expressions, which include parameters either fitted to spectral experimental data (Heine and Weaire 1970) or found by optimization during the solution of the atomic problem (Zunger 1979, Pickett 1989a). The latter (first-principles) approach is preferable, apparently. Firstly, there exists some ambiguity in the pseudopotential selection: if a pseudo-wave-function is modified by addition of any linear combination of wave-functions of a core, then the orthogonalization of both pseudofunctions to yield core functions results in the same 'accurate' wave-function. The variation of pseudopotential parameters allows one to find the 'optimal' pseudopotential. Secondly, in this way the so-called norm conserving pseudopotentials can be generated (Bachelet *et al* 1982). For these potentials the total charge inside the sphere comprising the atomic core, divided by the value of the wave-function squared inside this sphere, should be equal to the same quantity for the all-electron wave-function. It results in the same electrostatic potentials and valence wave-functions beyond the sphere. Thirdly, the parameter optimization can be made in such a way that the first and second radial-logarithmic energy derivatives of pseudo- and all-electron wave-functions are the same on the sphere mentioned (Shirley *et al* 1989). In the latter case improvement in band structure calculation is achieved.

The pseudopotential field is non-local and, therefore, it differs from the Coulomb field in the all-electron Hamiltonian. Therefore, the basis set functions in the pseudopotential method must take into account properly a repulsive term in a potential for those electronic states whose wave-functions penetrate into the core. There exist appropriate basis sets (see e.g. Gaspar and Gaspar 1986) and calculations with a large number of molecules involved have shown that the PS-31G basis, composed of pseudo-wave-functions, gives results equivalent to all-electron, non-empirical calculation in the 6-31G basis (Bouteiler *et al* 1988).

As has been shown by Bachelet *et al* (1989) the non-local pseudopotentials can be transformed into local angular-momentum-dependent pseudo-Hamiltonians by Monte Carlo modification of the kinetic energy operator.

Topiol *et al* (1977) and Zunger and Cohen (1978) obtained first-principles pseudopotentials for a large number of atoms by variational calculation in the local and non-local approximations. They have found that the moments of atomic orbitals $\int r^n \rho_i(r)\,\mathrm{d}r$ ($n = -1, 0, 2$; ρ_i is the ED of the ith orbital) agree to an accuracy of $\sim 0.01\%$ with the results of all-electron calculations.

Table 2.2 Computed and experimental binding energies (eV), lattice parameters (Å) and bulk moduli (GPa) from all-electron (AE) and pseudopotential (PS) HF calculations. (Reprinted with permission from Silvi *et al* 1989.)

Method	Diamond	Silicon	MgO	Lithium
Binding energy				
AE	4.93	2.43	7.40	0.46
PS	4.64	2.39	6.99	0.48
Experimental	7.62	4.60	10.45	1.66
Lattice parameters				
AE	3.58	5.57	4.19	3.55
PS	3.59	5.54	4.31	3.54
Experimental	3.57	5.43	4.21	3.51
Bulk moduli				
AE	467	94	184	26.3
PS	444	93	176	22.4
Experimental	442	99	156	11.5

Shirely *et al* (1990) have found that pseudopotentials which are generated with the use of HF (non-local) exchange in combination with accounting for electron correlation in the local density approximation (see below) are very transferable and lead to better values of physical characteristics in comparison with other types of first-principles potentials.

Silvi *et al* (1989) have described the pseudopotential version of the HF method for crystals and compared its results with the data of all-electron HF calculations for different crystals, such as covalent diamond and silicon, ionic MgO, and metallic lithium. The results obtained are listed in table 2.2. One can see that the error in binding energies is rather high for all compounds in both versions. This is a typical feature of the HF method, as was mentioned above. The lattice parameters and bulk moduli are closer to each other and to the experimental values (the exceptions are bulk moduli for MgO and, to a greater extent, for metallic Li). The band structure was reproduced much more adequately in both versions of the HF method, both for occupied and for virtual states. This is illustrated in figure 2.4, where the results for a band structure of silicon along two highly symmetrical lines in the Brillouin zone are presented. Thus, the pseudopotential version is not inferior to the all-electron HF method with respect to the accuracy of various calculated characteristics, whereas the calculation time according to Silvi *et al* (1989) decreases significantly by factors of 1.5–2. This allows one to apply the pseudopotential HF method to complicated crystals such as silicates and germanates (Silvi *et al* 1992, 1993).

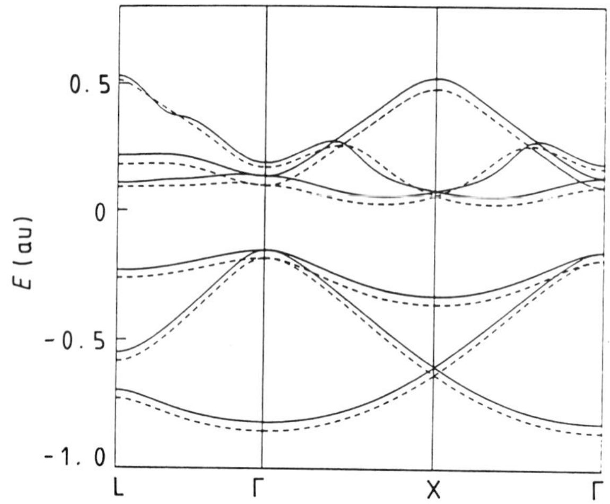

Figure 2.4 The HF band structure for silicon: full lines are the all-electron calculation; dashed lines are the pseudopotential calculation (Silvi *et al* 1989).

2.1.6 *The density matrix*

A very general description of the structure and properties of many-particle systems can be obtained with the N-particle density operator (Davidson 1976):

$$\hat{\rho}=\exp(-\beta\hat{H})/\mathrm{Tr}_N[\exp(-\beta\hat{H})]. \tag{2.17}$$

This includes the N-particle Hamiltonian and depends on temperature, T ($\beta=(kT)^{-1}$, k is the Boltzmann constant). Tr_N denotes the trace over a complete N-particle basis. For non-degenerate N-electron wave-functions at zero temperature this operator can be replaced after integrating over all populated electronic levels by another operator, the so-called electronic density matrix (DM)

$$\Gamma^N(\boldsymbol{X}_1, \boldsymbol{X}_2, \ldots, \boldsymbol{X}_N; \boldsymbol{X}_1', \boldsymbol{X}_2', \ldots, \boldsymbol{X}_N')=\Psi(\boldsymbol{X}_1, \boldsymbol{X}_2, \ldots, \boldsymbol{X}_N)\Psi^*(\boldsymbol{X}_1', \boldsymbol{X}_2', \ldots, \boldsymbol{X}_N'). \tag{2.18}$$

The DM (2.18) completely describes physical states of the N-electron system and allows one to calculate the expectation values of any N-particle operator. However, all interactions in a system of electrons have one- and two-particle character. Hence, for quantum mechanical description of many-electron systems it is necessary to use both the one-electron Γ^1 and two-electron Γ^2 DMs. They are obtained by integration of (2.18) over coordinates and spins $(N-1)$ and $(N-2)$ of electrons, respectively. For example, the expectation value of any one-electron operator $\hat{A}(\boldsymbol{X}_1)$ can be described by the relation

$$\langle\hat{A}\rangle=\int\hat{A}(\boldsymbol{X}_1)\Gamma^1(\boldsymbol{X}_1;\boldsymbol{X}_1')_{\boldsymbol{X}_1'=\boldsymbol{X}_1}\,\mathrm{d}\boldsymbol{X}_1=\mathrm{Tr}\,\hat{A}\hat{\Gamma}^1 \tag{2.19}$$

and that of the two-electron operator $\hat{B}(X, X)$ by the relation

$$\langle \hat{B} \rangle = \iint \hat{B}(X_1, X_2)\Gamma^2(X_1, X_2; X_1', X_2')_{\substack{X_1' = X_1 \\ X_2' = X_2}} \, dX_1 \, dX_2 = \mathrm{Tr}\, \hat{B}\hat{\Gamma}^1. \quad (2.20)$$

Here the symbol Tr means 'integral trace': the integration over 'diagonal elements' of the DM after the action of appropriate operators on the DM. Using these DMs one can write the expectation value of the electronic Hamiltonian (2.1) for a closed shell system in the form

$$\langle \hat{H}_e \rangle = 2 \int \hat{h}\Gamma^1(X_1; X_1')_{X_1' = X_1} \, dX_1 + \int \hat{g}\Gamma^2(X_1, X_2; X_1', X_2')_{\substack{X_1' = X_1 \\ X_2' = X_2}} \, dX_1 \, dX_2. \quad (2.21)$$

Thus, the application of the DM allows one to replace the $4N$-dimensional integration over spatial and spin coordinates of all electrons with the 16-dimensional integration for four electrons. This circumstance, along with the others, has stimulated the development of the theory of molecules in terms of DMs (Lowdin 1955, McWeeny 1960, Coleman 1963, 1972, 1978, Davidson 1976, Mestechkin 1977, Smith and Absar 1977, Erdahl and Smith 1987).

In the HF method (the single-determinant approximation) of special importance is the one-electron Fock–Dirac density matrix (DM1) $\Gamma_D^1(X_1; X_1')$, which can be expressed in terms of N occupied natural spin orbitals in diagonal form:

$$\Gamma_D^1(X_1; X_1') = \sum_{i=1}^{N} \varphi_i(X_1)\varphi_i^*(X_1'). \quad (2.22)$$

The population numbers of these spin orbitals are obviously equal to unity. The DM1 is idempotent

$$\int \Gamma_D^1(X_1; X_2)\Gamma_D^1(X_2; X_1') \, dX_2 = \Gamma_D^1(X_1; X_1') \quad (2.23)$$

and invariant with respect to unitary spin orbital transformation. Besides,

$$\Gamma_D^2(X_1, X_2; X_1', X_2') = \tfrac{1}{2}\{\Gamma_D^1(X_1; X_1')\Gamma_D^1(X_2, X_2') - \Gamma_D^1(X_1; X_2')\Gamma_D^1(X_2; X_1')\}. \quad (2.24)$$

This means that all one- and two-electron properties of a system can be expressed in terms of only DM1 in the HF method. This allows one to consider the Fock–Dirac matrix to be fundamental invariant of this method and to write the HF equations directly in terms of DM1, rather than wave-functions (Lowdin 1955). In the MO LCAO approximation the matrix HFR equation has the form

$$\hat{F}\hat{\Gamma}_D^1 = \hat{\Gamma}_D^1\hat{F}. \quad (2.25)$$

Here $\hat{F}$ is Fock's operator (2.6). Corresponding to $\hat{\Gamma}_D^1$, the spinless matrix for systems with closed shells has the form

$$\hat{\Gamma}_D(\boldsymbol{r}, \boldsymbol{r}) \equiv 2\hat{\gamma}(\boldsymbol{r}, \boldsymbol{r}) = 2 \sum_{\mu\nu} P_{\mu\nu}\Phi_\mu(\boldsymbol{r})\Phi_\nu^*(\boldsymbol{r}) = 2 \operatorname{Tr} \hat{\Phi}\hat{P}\hat{\Phi}^+. \tag{2.26}$$

Here $\hat{\Phi}$ is the matrix row composed of AOs Φ_μ, the superscript + means the Hermitian conjugation. The $P_{\mu\nu}$ matrix is called the charge–bond order matrix. It completely represents DM1 in the AO basis set and its elements can be used for interpretation of ED from the chemical bond point of view, when $\boldsymbol{r} = \boldsymbol{r}'$. It is essential that, in transition from one basis set to another, the $P_{\mu\nu}$ elements are changed, but not the $\hat{\gamma}$-matrix itself.

Redendo (1989) has found the many-particle Hamiltonian of the ground state to be a universal functional of DM1. Thus, the significance of DM1 is not restricted by the HF framework.

A number of important questions arise when the theory is developed in DM terms. It is possible (Coleman 1963) that the direct application of a variational principle to DM1 may lead to a lower value of energy than the ground state energy obtained with the wave-function. To eliminate this nonsense the DM1 must be assumed to obey Fermi–Dirac statistics, i.e. it must obey some limitations. Therefore, one has to formulate the conditions which would ensure for the N-electron DM the existence of an antisymmetric N-electron wave-function, from which the given DM could be constructed according to (2.18). Such an DM is referred to as N representable. The fulfilment of the simple requirements of normalization, Hermitivity, antisymmetry, and positive semidefiniteness by the DM are insufficient for this purpose. The search for such conditions gave rise to a fundamental problem that is usually called the N-representability problem (Coleman 1963, 1978, Mestechkin 1977, Smith and Absar 1977). This problem has not yet been solved generally: the solutions have been found for some special cases only. It was established, in particular, that the validity of relation (2.23), the idempotency condition, is a necessary and sufficient condition of N representability for $\hat{\Gamma}_D^1$. Thus, in the matrix HFR method, where one- and two-electron properties are functionals of $\hat{\Gamma}_D^1$, obtaining correct physical results is strictly ensured.

The direct interpretation of the DM1 matrix is not a simple matter. Fortunately, Schmider *et al* (1992a) have found that the case of linear molecules is a rare exception, if one is interested in the DM1 along the bond axis only. The two-dimensional plane through a spinless DM1 can be displayed in this case. Such planes for Li_2, F_2, and LiF molecules are presented in figure 2.5 (only the σ component of the bond is visualized in these pictures). In the Li_2 molecule the off-diagonal peaks reflect the presence of a $\sigma_{s\text{–}s}$ bond, while minima correspond to overall negative interference between two centres. In the F_2 molecule the nodal structures of the (p–p) bonding orbital manifest themselves in positive regions which are directed from off-diagonal

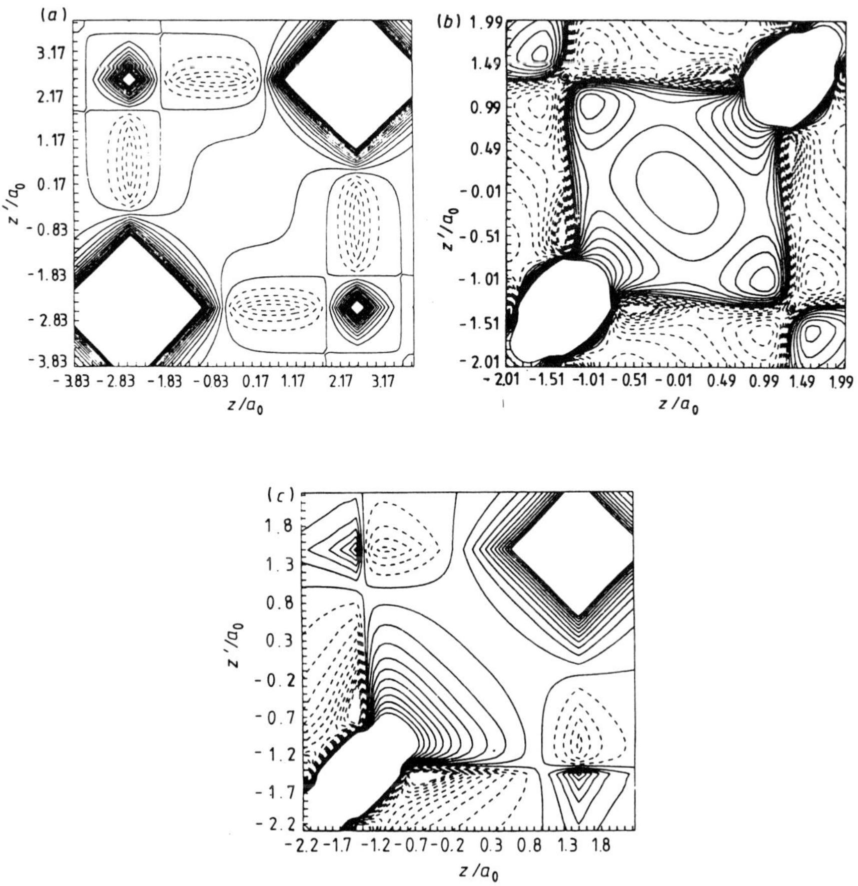

Figure 2.5 The sections of spinless DM1 along the bond axis z in some linear molecules (non-empirical calculations in 6-31G basis set): (*a*) Li_2; the nuclei are at $z=z'=\pm 2.55\,a_0$; (*b*) F_2; the nuclei are at $z=z'=1.3389\,a_0$; (*c*) LiF; the Li nucleus is at $z=z'=1.4778\,a_0$, the F nucleus is at $z=z'=-1.4778\,a_0$. The full lines correspond to positive interference and dashed lines correspond to negative interference between centres (Schmider *et al* 1992a).

peaks to the diagonal. As results from comparison with the DM1 component for a pure $\sigma_{p\text{-}p}$ bond, the nodal pattern symmetry is due to the contribution of the bonding $\sigma_{s\text{-}s}$ MO. In LiF the picture near the Li nucleus is very close to that in Li_2, and the picture near the F nucleus to that in F_2. It is also interesting that core contributions do not have counterparts in the off diagonals. Thus, the covalent chemical bond can be, in principle, interpreted directly in terms of DM1.

2.1.7 *The valence bond method*

The idea of the valence bond (VB) method was formulated by Heitler and London (1927) and further developed by Pauling (see e.g. Pauling 1960). This method supposes that AOs or their symmetry-adapted linear combinations, hybrid orbitals, generally conserve their individual features, and that the chemical bond can be considered to be a perturbation. The many-electron wave-function Ψ is represented here as an antisymmetrized product of sets of AOs of atoms and a combination of spin functions chosen according to a special group theory technique (McWeeny and Sutcliffe 1976). As a result, Ψ is described by expressions similar to (2.10), i.e. it is represented as a linear combination of some linearly independent function products, termed VB structures. Each VB structure corresponds to a set of atoms with some particular electronic configurations. The weights of VB structures and parameters of orbitals are determined by the variational principle. The feature of the method discussed is the fact that the electron structure of a system can often be described by a set of several VB structures only. The ED of a system is an average value over all VB structures providing their weight factors are taken into account. The mixture of electronic configurations related to the given atom in each of the VB structures is called a valence state for this atom. The notion of an atomic valence state is a valuable subsidiary (not observed spectroscopically) notion that often simplifies the chemical bond analysis.

The representation of a many-electron function in the VB method by means of a set of functions corresponding to various electronic configurations allows one to take electron correlation effects into account fairly completely. This approach is, in fact, equivalent to the configuration interaction method in the MO LCAO SCF theory described above. In order to take into consideration all chemical bond effects, the number of VB structures must be large. One cannot often choose the 'basic' structure *a priori* and must consider the remaining ones in perturbation theory procedures; this is a disadvantage of the VB method, of course. At the same time, the number of possible structures can be decreased by using the hybridization concept. Another drawback lies in the fact that AOs used for construction of VB structures are non-orthogonal, causing some computational problems (see e.g. McWeeny and Pickup 1980).

Goddard (1967) generalized the VB method by using linear combinations of STOs or hybrid orbitals as basis set functions. In this case the equations obtained are very similar to the ones resulting from the independent particle approximation since they contain some self-consistent potential. In this respect the generalized VB method is a generalization of the MO method as well.

Gerratt (1971) and Cooper *et al* (1984, 1987) have constructed a spin-coupled valence bond method. In this method one electron is assigned to each orbital and non-orthogonal orbitals are represented as an expansion over linear combinations of AOs. The parameters of functions are determined by means of a variational procedure without restrictions on the degree of orbital locality. The optimization is performed in such a way that all possible spin coupling is taken into consideration and the true total spin of a many-electron system maintained. The value of energy achieved in this case is usually lower than that in the HF method; in particular, the molecular dissociation process is correctly described (Klein and Trinojstic 1990). This means that correlation effects are well taken into account. It is also important to emphasize that, as a result of non-restricted optimization, the ground state VB structure always possesses the highest weight. The spin-coupled VB theory is useful for analysing fine chemical bond effects for example in studying hypervalent valence states of atoms (Cooper *et al* 1989) and decomposition of ED into interpretable parts (Low and Hall 1990). The modern applications of this theory were recently described by Cooper *et al* (1991).

2.2 Methods of the density functional theory

2.2.1 *General statements*

Let us consider the methods of the study of many-electron systems, in which ED plays a principally important role. The general approach is based on the statement that all properties of the ground state (that is the only state which we are interested in) of an arbitrary many-electron system are functionals of ED, $\rho(\boldsymbol{r})$. For all systems there exists a universal functional that relates ED to the kinetic energy of electrons and to the energy of their exchange–correlation interaction. In addition, for the given outer local potential $v(\boldsymbol{r})$, the true ED provides the minimum of the energy functional (1.14). Thus, by minimizing functional $E[\rho]$ (1.14) with respect to ρ one can, in principle, obtain equations for ρ determination, provided that corresponding functional dependences of various energy components on ρ are known. This is not the case and one should establish first what are the necessary conditions for the ED which ensure that the ED obtained as a result of minimization describes the distribution of electrons in the ground state. In other words, what are the limitations which ρ must satisfy in order to be N representable?

This question was answered independently by Harriman (1981) and by Kolmanovich and Reznik (1981) for a one-dimensional system of fermions. These authors showed that for the given function ρ there always exists an antisymmetric wave-function Ψ that can be chosen as a Slater-type determinant for a system of non-interacting particles. The practical procedure for doing this has also been proposed. We shall present here the proof of this statement in order to clarify the essence of the matter (Reznik 1982).

Consider a system with an even number of electrons, N, and with zero spin. Let $\{g_i(\xi)\}$ be a set of $N/2$ smooth and, generally speaking, complex functions g_i, $\xi \in [0, 1]$, for which the conditions

$$\begin{cases} 2\sum_{i=1}^{N/2} |g_i(\xi)|^2 \equiv N \\ \int_0^1 g_i^*(\xi) g_k(\xi)\, \mathrm{d}\xi = \delta_{ik} \end{cases} \tag{2.27}$$

are valid. An example of such functions is the plane waves. In order to provide a finite kinetic energy value, one must require the existence of an integral, related to the $\partial g/\partial\xi$ derivative, whose form can be constructed from the final result (2.30). If one constructs continuous functions

$$\varphi_i(y) = (\rho(y)/N)^{1/2} g_i(F(y)) \qquad F(y) \in [0, 1] \tag{2.28}$$

which have also continuous first derivatives, then for any choice of functions $F(y)$ the relation

$$2\sum_{i=1}^{N/2} |\varphi_i(y)|^2 \equiv \rho(y) \tag{2.29}$$

is valid. In order to consider φ_i as non-interacting electron gas orbitals filled with electrons of opposite spins, the orthonormalization condition needs to be satisfied. For this purpose it is sufficient to choose functions $F(y)$ in the form

$$F(y) = \frac{1}{N}\int_{-\infty}^{y} \rho(y)\, \mathrm{d}y. \tag{2.30}$$

In the three-dimensional case the orthonormalization condition is satisfied if

$$F(\mathbf{r}) = \frac{1}{N}\int_{D(\mathbf{r})} \rho(\mathbf{r}')\, \mathrm{d}\mathbf{r}' \tag{2.31}$$

where $D(\mathbf{r})$ is the area within the surface with $\rho = \rho(\mathbf{r})$. The expressions (2.28) and (2.29) are also valid in this case; however, in contrast to the one-dimensional case, these expressions together with (2.31) result in only one possible variant of ED decomposition.

Thus, the Slater determinant can always be constructed from functions φ_i expressed in terms of ρ: if a set of functions g_i is chosen, then some unique ED expansion of (2.29) type is specified for any ρ. These sets, however, can be chosen in an infinite number of ways. For example, one can specify some arbitrary system of $N/2$ orthonormal functions, construct their ρ and $F(y)$ and use equality (2.27) to find g_i. Thus, the ED expansion (2.29) is always possible, but it is not unambiguous: each set of $\{g_i\}$ functions resuts in its own expansion of the ED ρ.

The following important conclusion can be drawn: if there exists a single outer local potential $v(r)$ for the given ρ, then a single condition of N

representability of ρ by an antisymmetric wave-function is the conservation of the number of electrons in the system (1.13), e.g. $\int \rho(r)\,\mathrm{d}r = N$.

The physical meaning of each possible set of functions $\{g_i\}$ generally remains open. However, if the chosen set of functions $\{g_i\}$ contains some parameters, then they can be determined by the variational principle, which allows one to find the best kinetic energy functional given a set of functions (Levy 1979, 1982, 1990).

The simplest ED expansion over one-particle states of a homogeneous electron gas, the plane waves, was given by Kolmanovich and Reznik (1981) and by Harriman (1981). Obtaining satisfactory sets of three-dimensional $\{g_i\}$ functions is a much more difficult problem, however (see e.g. Reznik 1982, Zumbach and Maschke 1983, Ghosh and Parr 1985, Cioslowski 1988).

Thus, the Fermi–Dirac statistics requirements do not impose any extra limitation on ED. The question remains open, nevertheless, of whether any ρ can exist in a many-electron system as the ground state ED for some local potential v. The clarification of these corresponding conditions which should be imposed on ρ is called the v-representability problem. There is no complete solution of this problem. Nevertheless, Levy (1982) and Leib (1983) have shown that there exist some ρ functions which are free from any 'pathology' and, at the same time, do not represent the ground state ED for any v. For example, these ρ include spherically symmetrical ED for atoms with incomplete filling of electron shells. The absence of a solution to the v-representability problem does not confine, fortunately, practical density functional theory (DFT) applications, since in the usually applied local density approximation the extremal ED are v representable (the corresponding potential arises during the self-consistent solution of equations). In the statistical method the functional possesses more severe defects. Besides, Kohn (1983) has shown that if ρ is v representable then small perturbations do not carry it out of this region.

2.2.2 *Approximations of the density functional theory*

Attempts to construct a general rigorous DFT with the ED as a basic variable have failed. The main difficulty of the DFT lies in the construction of the approximation to functional $G[\rho)$ in the total energy functional $E[\rho]$ (1.14). $G[\rho]$ contains both kinetic $T[\rho]$ and exchange–correlation $F_{xc}[\rho]$ energies. Various DFT methods utilize various approximations for these energies, the fields of their applicability considerably differing from each other. All these methods can be subdivided into explicit ones, in which the values of functionals are determined directly from ρ, and implicit ones, in which this relation is indirect, via one-particle wave-functions, for instance. Approaches are also possible where some functionals are explicit and the others are not. It is clear that the possibility of excluding wave-functions from the ground state theory, which is postulated by the Hohenberg–Kohn theorem, is implemented completely using explicit functionals only. However, the construction of such functionals with an accuracy sufficient for describing equilibrium crystal properties is an extremely difficult problem.

In the case of sufficiently small deviations, $\Delta\rho$, from the homogeneous ED distribution ρ_0 ($\Delta\rho/\rho_0 \ll 1$) one can make use of a rapidly converging expansion

$$G[\rho] = G[\rho_0] + \iint g_2(\mathbf{r}_1 - \mathbf{r}_2)\,\Delta\rho(\mathbf{r}_1)\,\Delta\rho(\mathbf{r}_2)\,\mathrm{d}\mathbf{r}_1\,\mathrm{d}\mathbf{r}_2 + \cdots. \tag{2.32}$$

The stationarity of energy at $\rho = \rho_0$ and the absence of a perturbing potential result in the disappearance of the first-order term. $G[\rho_0]$ is a sum of kinetic, exchange, and correlation energies of a homogeneous electron gas of density ρ_0. There exists a series of simple expressions approximating $G[\rho_0]$ and ρ_0 in the explicit form (see e.g. Kohn and Vashishta 1983). The function g_2 can be obtained by the perturbation theory for an electron gas (March *et al* 1967) which is usually applied to a system of non-interacting electrons and describes the ED response to a weak variation of a total effective potential.

The translational invariance of ρ_0 in a crystal implies that $g_2 = g_2(|\mathbf{r}_1 - \mathbf{r}_2|)$. The consideration of the ED as a result of perturbation of a homogeneous electron distribution makes sense only for an electron gas in the field of a weak pseudopotential. Therefore, expansion (2.32) is most applicable to the simple metals, where the linear approximation can be used. In this approximation the response of ED to a superposition of spherically symmetric atomic or ionic pseudopotentials is reduced to a superposition of spherically symmetric charge distribution at lattice sites. This results in a rigorous extinction rule for x-ray reflections. Meanwhile, in covalent compounds such as Si crystals, for example, the forbidden reflections have noticeable intensity and the concentration of ED in the internuclear space that is described by these reflections is responsible for certain crystal properties. Hence, the linear approximation is not applicable for such crystals at all.

Taking into account the quadratic response terms removes the x-ray extinctions. According to Pickenheim and Milchev (1970), in this case the difference of ED obtained with the empirical pseudopotential from that calculated with one-particle equations does not exceed 6% at the midpoints of the internuclear space. Unfortunately, the non-empirical pseudopotentials are always far deeper. Therefore, one can hardly hope that the series in the perturbation theory will converge satisfactorily when semiconductors are calculated from first principles. This fact would appear to explain the absence of such calculations.

If the characteristic distance at which ED changes noticeably is rather large, for example, much larger than the reverse Fermi momentum $k_F^{-1} \sim \rho^{-1/3}$, then the local homogeneity principle (Kirzhnits *et al* 1975) can be applied. According to this principle, the energy density at each point $\mathbf{r}$ is supposed to be the same as that in a homogeneous electron gas, whose ED is equal to $\rho(\mathbf{r})$ everywhere. The limitations are not imposed on ED variations in this case. Such an approach to the functional approximation is widely used in the theory of electron structure of crystals because it allows one to apply directly to the inhomogeneous system theory the results obtained for a homogeneous electron gas. In describing the exchange–correlation functional $E_{xc}[\rho]$ it yields rather accurate results, even if the criteria of smooth ED variation are noticeably violated. The ED of such a gas is constant and is determined by the expression that follows from quantum statistics (March 1983):

$$\rho=(1/3\pi^2)k_F^3. \tag{2.33}$$

In the atomic units system ($e=m=\hbar=1$), applied hereafter in this section (see appendix A), functional $G[\rho]$ in the local homogeneity approximation can be written as follows:

$$G[\rho]=\int\left\{\frac{3^{5/3}\pi^{4/3}}{10}\rho^{5/3}(\boldsymbol{r})-\frac{1}{4}\nabla^2\rho(\boldsymbol{r})+\varepsilon_{xc}[\rho(\boldsymbol{r})]\right\}\mathrm{d}\boldsymbol{r}. \tag{2.34}$$

Here the first and second terms in the integrand describe the kinetic energy density; the third one is the exchange–correlation energy density.

If one neglects the term ε_{xc} and minimizes the total energy functional (1.14) with use of (2.34) under additional condition (1.13), one can obtain the Thomas–Fermi equation which associates ED with the potential $\varphi(\boldsymbol{r})$ of nuclei and electrons (March 1983):

$$\rho(\boldsymbol{r})=(1/3\pi^2)[2(\mu-\varphi(\boldsymbol{r}))]^{3/2}. \tag{2.35}$$

Here μ is the Lagrange multiplier that arises under energy variations taking into account the electron's number constancy in system (1.13); μ can be identified as the Fermi energy, i.e. it represents a chemical potential.

The description of the exchange–correlation energy density, ε_{xc}, in various approximations allows one to obtain this or that modification of the Thomas–Fermi method. For example, if only the exchange component is taken into account, then one arrives at the Thomas–Fermi–Dirac scheme. The correlation energy can be described in the extreme cases of high density (Gell-Mann and Bruckner 1957) and low density of electron gas (Wigner 1934). The real intermediate densities can be described by means of various interpolation procedures, which lead to similar results. A complete review of properties of such a statistical model was given by Kirzhnits *et al* (1975) and March (1983).

Despite the apparent simplification of the Thomas–Fermi method, the calculations of valence electron distribution in such crystals as diamond, Si, Ge, α-Sn, GaAs, ZnSe, and NaCl carried out by Baldereschi *et al* (1981) have yielded results which were in surprisingly good agreement with the evidence of more sophisticated methods. This circumstance stimulated Reznik (1988, 1993) to develop a method that combined the perturbation theory for the Fock–Dirac spinless density matrix $\hat{\gamma}(\boldsymbol{r}, \boldsymbol{r}')$ with the statistical method. Reznik started from the Dayson-type equation for the one-particle Green function describing the electron gas in the outer potential $v(\boldsymbol{r})$ (March *et al* 1967). This was presented in the form

$$v(\boldsymbol{r})=U(\boldsymbol{r})+[v(\boldsymbol{r})-U(\boldsymbol{r})] \tag{2.36}$$

where U is the result of averaging of the true potential $v(\boldsymbol{r})$ over the neighbourhood of some point $\boldsymbol{r}$, and the term in brackets is considered as a perturbation. The solution of the initial integral equation for the potential $U(\boldsymbol{r})$ coincides with the result of the Thomas–Fermi method, if the local Fermi momentum

$$\tilde{k}^2(\boldsymbol{r})=k_F^2-2U(\boldsymbol{r}) \tag{2.37}$$

is used instead of the constant momentum k_F. The general solution has the form of a series for Green functions analogous to the series of perturbation theory for the $\hat{\gamma}$-matrix (March *et al* 1967).

The average potential U is smoother than the initial potential v, which is presented as a superposition of some model pseudopotentials, the value of perturbation correction is small, and convergence of the series is quick. When terms of only first order are retained in this series, then one obtains for the ED

$$\rho(\boldsymbol{r})=\theta(\tilde{k}^2)\left\{\frac{\tilde{k}^3(\boldsymbol{r})}{(6\pi^2)}-\frac{\tilde{k}^2(\boldsymbol{r})}{(4\pi^3)}\int\frac{j_1[2\tilde{k}(\boldsymbol{r}')|\boldsymbol{r}-\boldsymbol{r}'|]}{|\boldsymbol{r}-\boldsymbol{r}'|^2}[v(\boldsymbol{r}')-U(\boldsymbol{r})]\,\mathrm{d}\boldsymbol{r}'\right\} \tag{2.38}$$

(this result should be doubled for systems with closed shells). Here θ is the Heaviside function that provides the approximation of the finite radius of an ion, the j_1 is a spherical Bessel function. The potential in (2.38) is constructed in a self-consistent manner for each value of parameter k_F (for exchange–correlation potential some approximations should be used). The Fourier components of U are described by $U(\boldsymbol{g})=F[\boldsymbol{g}/2\langle\tilde{k}\rangle]v(\boldsymbol{g})$, where F is the Lindhard function, if the average value of $\tilde{k}$ is used. Hence, components with small values of the reciprocal vector $\boldsymbol{g}$ contribute mainly to potential U. The calculations performed for diamond-type covalent semiconductors (Babenko *et al* 1988) have shown that one can achieve with this method rather good agreement of ED with more rigorous methods. Babenko *et al* (1988) have also applied this method to the study of high-temperature superconducing copper oxides. The mean ED in these compounds is close to that in semiconductors mentioned above, which substantiates the applicability of the method. Indeed, the valence ED, calculated for the $YBa_2Cu_3O_7$ crystal, is very close to the ED obtained successfully by a more rigorous quantum mechanical calculation. Some other interesting results have been obtained, such as the treatment of photoelectron spectra in $YBa_2Cu_3O_{7-x}$ crystals, the establishment of the dependence of the quadrupole splitting of lines in Mössbauer spectra upon the particular site the ^{57}Fe atoms occupy in the high-temperature superconductors by substitution, and the explanation of dependence of the orientation of axes of

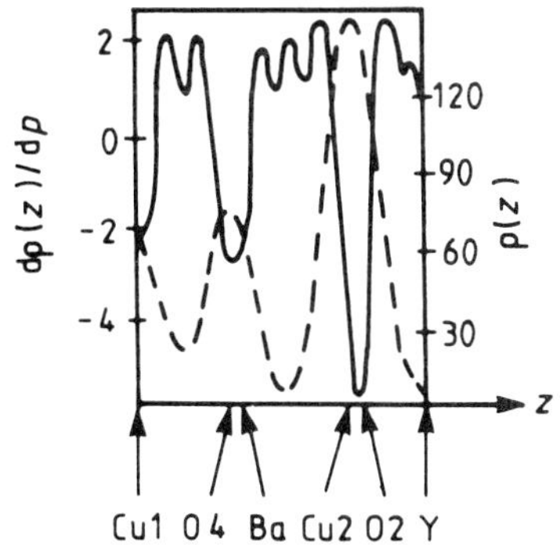

Figure 2.6 Linear valence electron density $\rho(z)=\int\rho(x, y, z)\,dx\,dy$ (– – –) and its derivative $d\rho(z)/dp$ (——) in $YBa_2Cu_3O_{6.6}$ (Babenko *et al* 1988). Units are e/(axis *c*) for $\rho(z)$ and c/(axis *c*) $kbar^{-1} \times 10^{-2}$ for $d\rho(z)/dp$. Arrows show positions of atomic layers in the crystal structure.

the electric field gradient tensor at Cu sites on a crystal stoichiometry. An important conclusion was achieved concerning the structure of superconductors studied. Many models of chemical bonding in the copper oxides are based on the approximation representing these compounds as CuO layer structures. The anisotropy of their physical properties is attributed to such a structure. Babenko *et al* (1988) have considered ED peculiarities by modelling hydrostatic compression of the $YBa_2Cu_3O_{6.6}$ crystal by varying the unit cell parameters. Even such a rough approximation allowed them to evaluate the compression behaviour of ED and its derivative with respect to pressure, $d\rho/dp$. As can be seen from figure 2.6, the layer structure of a crystal is actually noticeable. The application of pressure decreases ED in all layers, shifting it into the interlayer space. This effect is small, however; for the mean ED of 50 e/cell the shifting is about 0.02 e $kbar^{-1}$. This probably explains the fact that the pressure only weakly influences the superconducting properties of Y–Cu–Ba compounds.

The local homogeneity approximation has essential disadvantages as applied to the description of molecular and crystal properties under normal conditions regardless of specific choice of description of energy density ε_{xc}. These disadvantages arise from the impossibility of successful description of the electron shell structure and associated qualitative variation of atomic properties with changing electron numbers. The model (2.34) is satisfactory for the 'mean' density region, which is not too close to nuclei and not so far as the valence region, provided that the number of electrons is large enough. The properties of molecules or crystals, however, mostly depend on the valence electron interactions, where ED is essentially inhomogeneous. As Teller (1962) has shown, the local homogeneity approximation does not allow one to describe the bonding of atoms and the ED gradient must be taken into consideration.

The incorporation of ED gradient and higher-order derivatives is known as the gradient expansion (Hohenberg and Kohn 1964):

$$G[\rho]=G[\rho_0)+\iint g_2[\rho]|\Delta\rho|^2\,\mathrm{d}\boldsymbol{r}+\cdots. \tag{2.39}$$

Coefficients g_i are related to the polarizability of a homogeneous gas of interacting electrons. The gradient expansion of the kinetic energy $T[\rho]$ has been studied most. It is the functional which plays a key role in DFT. Indeed, by describing the exchange–correlation energy density ε_{xc} very roughly and even ignoring it, but calculating kinetic energy $T[\rho]$ with a quantum mechanical method (i.e. solving one-particle equations), we shall correctly describe electron shells associated with filling one-electron states. We may say, well in advance, that this fact is especially clearly revealed in the successes of the Kohn–Sham method as compared to the results of a statistical approach with the same description of energy $E_{xc}[\rho]$.

The ED homogeneity of the initial state results in the fact that all terms of the gradient expansion are of even order in the gradient operator. Three gradient (quantum) corrections to the statistical expression for kinetic energy are known, i.e. corrections of the second, fourth, and sixth orders. The first of them is

$$T_2[\rho]=\lambda\int\frac{|\nabla\rho|^2}{\rho}\,\mathrm{d}\boldsymbol{r}. \tag{2.40}$$

Two values of λ are obtained: $\lambda=\frac{1}{72}$ (Kirzhnits 1957) and $\lambda=\frac{1}{8}$ (Weizsaker 1935). Jones and Young (1971) have shown that the Weizsaker correction, which is valid from the viewpoint of the variational principle, works well in a short-wave limit in terms of ED expansion in the momentum space, i.e. for $k>k_F$. The Kirzhnits correction is valid for $k<k_F$. The presence of two different λ for these cases is a consequence of a singularity of the homogeneous electron gas polarizability at $k=2k_F$. Due to this fact, the expansions over degrees of wave number k do exist in the $k>2k_F$ and $k<2k_F$ regions, respectively.

The fourth-order correction was found by Hodges (1973). As Shin (1976) had found, the first two corrections brought the results closer, obtained with the aid of gradient expansion, to the data of the HF method. However, Murphy (1981) found later that the sixth-order correction diverges both in the nuclear region and in the system periphery. Oliver and Perdew (1979), based on dimensional considerations, have proposed that divergency exists in higher orders as well. However, the method of divergence summing, that would be equivalent to circular diagrams adding in a diverging perturbation theory by Gell-Mann and Bruckner (1957), has not been found yet.

Despite the difficulties in the theory, the account of the first correction in the kinetic energy gradient expansion, corresponding to the correct HF ED of the atoms, results in the error in energy being lower than $\sim 1\%$. Reznik and Shatalov (1981) have shown that this is due to the high contribution of exponential states: the error becomes larger as the distance from the nucleus increases. Nevertheless, the simplicity of a gradient ED expansion with the first correction has stimulated the development of a new method for describing crystal ground state properties (Gordon and Kim 1972). This method proceeds from functional $T[\rho]=T_0+T_2$ for kinetic energy and uses the local approximation for a density of exchange–correlation energy ε_{xc} and the presentation of crystal ED in the form of superposition of spherically averaged ED of atoms or ions. As applied to the description of ionic crystal properties such a model is fairly successful. However, when the ED deviation from a simple superposition of atoms is significant (as it takes place at a covalent bond formation), ignoring the covalency gives rise to errors (Mulhausen and Gordon 1981a, b). The crystal ED determined in a self-consistent manner from the condition of crystal total energy minimum, along with measurable quantities, is far from the actual ED (Chelikowski 1980). Reznik and Shatalov (1981) believe that this is due to the same factor as in the statistical method, i.e. the insufficiency of the gradient expansion for a correct description of electron shells. This means that the model under consideration can be applied in a self-consistent manner only in the region where ρ varies smoothly. This condition is fulfilled, for example, when the pseudopotentials are used. In this case the gradient expansion of $T[\rho]$ converges, but as the internuclear distance increases, the divergence becomes prominent. In other cases the use of *a priori* information on the crystal ED is necessary. For ionic crystals it is sufficient to use the superposition of ED of ions. For covalent crystals the results of x-ray diffraction may be used.

Harrison (1980) stated that the refinement of ED relative to the initial superposition model is inexpedient because small errors in ED result only in smaller errors in the ground state energy. This is correct for the near-true ED only when the accurate energy functional is used. However, as has already been mentioned, the extremes of approximate functionals are far from the correct ED. Therefore, the errors in energy are of the same order as those in ED.

In general, the problem of description of electron shells arises with any explicit form of functional $E[\rho]$. If this problem is not solved the description of a chemical bond in crystals fails. As we have seen above, it is quite difficult (though possible in principle, see Kolmanovich and Reznik (1984)) to determine shell features in a self-consistent manner using ED only. Therefore, models which take into account any evidence on shell structure are used or the initial problem is modified in such a way that only part of the ED, for which the methods considered are justified, is determined. Such a reformulation of the problem is possible, in particular, when a rapidly oscillating core component, calculated from the atomic wave-functions, is excluded from the total ED. The remaining part, the valence ED, is relatively smooth. In this case the replacement of potential by the pseudopotential may be done. However, while working with ED only, we do not know any wave-functions (for valence electrons at least) and, as a result, the extraction of a smooth part of ρ is not so obvious or substantiated. This is the essential distinction from the pseudopotential theory.

Reznik (1986) proposed the following general approach to construction of model expressions for explicit ED functionals. If there is a good initial approximation $G_0[\rho]$ for the functional $G[\rho]$, for example, in the form of a gradient expansion with the first correction (2.40), then the $G_0[\rho]$ values are nearly correct for ED close to the accurate one. The accurate ρ, however, are not extremes of the approximate energy functional $E_0[\rho]$ in which $G[\rho]$ is substituted by $G_0[\rho]$. Suppose that ρ_0, corresponding to the initial potential v_0, is known (from the independent quantum chemical consideration, for example) and expand the difference $G-G_0$ into the functional Taylor series in the vicinity of ρ_0:

$$G[\rho]-G_0[\rho]=G[\rho_0]-G_0[\rho_0]+\sum_{i=1}^{\infty}\int\cdots\int v_i(\boldsymbol{r}_1\cdots\boldsymbol{r}_i)$$

$$\times\,\Delta\rho(\boldsymbol{r}_1)\cdots\Delta\rho(\boldsymbol{r}_i)\,\mathrm{d}\boldsymbol{r}_1\cdots\mathrm{d}\boldsymbol{r}_i+\cdots \tag{2.41}$$

($\Delta\rho=\rho-\rho_0$). In order that the energy $E[\rho]$ will be extremal at $\rho=\rho_0$ and $v=v_0$ the condition

$$\delta\left[E[\rho]-\mu\int\rho(\boldsymbol{r})\,\mathrm{d}\boldsymbol{r}\right]\Bigg|_{\rho=\rho_0}=0 \tag{2.42}$$

should be fulfilled, where μ is a chemical potential. Using expressions (1.14) and (2.41), one has

$$\frac{\delta G_0[\rho]}{\delta\rho}\Bigg|_{\rho=\rho_0}+v_1(\boldsymbol{r})+v_0(\boldsymbol{r})+\varphi_0(\boldsymbol{r})=\mu. \tag{2.43}$$

Here φ_0 is the electrostatic potential produced by the charge distribution of ρ_0. Since the explicit expression for $G_0[\rho]$ is known, as both ρ_0 and v_0 are also known, equality (2.43) to a constant term μ determines v_1, the kernel of a term of the first order in $\Delta\rho$ in expansion (2.41). This term acts as the effective potential. The kernels of higher orders can be associated with system polarizabilities of corresponding orders, but their calculation is rather complicated. However, if only first-order terms in expression (2.41) are retained, then the variational procedure for the problem with the potential which is close to v_0 (or for a problem with the same potential, but with a close number of electrons) yields the electron shell structure close to the model problem's shell structure. Reznik (1986) proposed to consider first the problem for atoms, to calculate the corresponding atomic potential v_1^{at}, and then the crystal potential as a superposition of atomic potentials can be constructed. Such an approximation is justified due to a rather rapid decrease of atomic effective potentials. Calculation using such a scheme (Reznik 1989) has shown that in diamond and silicon the ED distribution between atoms is described qualitatively correctly, though the deformation ED peaks on the bonds are about five times lower than the experimental ones.

2.2.3 *One-particle equations in many-electron theory*

The explicit ED functionals exploit the DFT possibilities in the most complete manner. However, as we have seen, the approximations for the available functionals have some serious drawbacks. For this reason in modern crystal electron structure calculations one prefers a more sophisticated approach to describing this structure, as was first discussed by Kohn and Sham (1965). Thcy proposed, without any approximations, to use the kinetic energy of the ground state of a non-interacting electron gas with the given density as the kinetic energy functional $T[\rho]$. The existence of corresponding one-particle orbitals is assured by the N-representability theorem. It is assumed, of course, that the ground state density of interacting electrons is the same as that of non-interacting electrons. Thus,

$$T[\rho]=\tfrac{1}{2}\min_{\{\varphi_i\}}\sum_i\int|\nabla\varphi_i(\boldsymbol{r})|^2\,\mathrm{d}\boldsymbol{r} \tag{2.44}$$

under conditions (1.13) and the orthonormalization of orbitals $\varphi_i(r)$. Expressing the functional derivative $\delta T/\delta\rho$ in terms of derivatives with respect to orbitals φ_i, one obtains the Euler–Lagrange equations

$$\left[-\tfrac{1}{2}\nabla^2+\hat{v}(\boldsymbol{r})+\int\frac{\rho(\boldsymbol{r}')}{|\boldsymbol{r}-\boldsymbol{r}'|}\,\mathrm{d}\boldsymbol{r}'+\hat{v}_{xc}(\boldsymbol{r})\right]\varphi_i(\boldsymbol{r})=\varepsilon_i\varphi_i(\boldsymbol{r}) \tag{2.45}$$

in which $v_{xc}=\delta E_{xc}[\rho]/\delta\rho$. This system of equations is known as Kohn–Sham equations. The potentials in these equations depend upon the ED, thus, the many-particle problem is reduced to a one-particle problem with self-consistency. The properties of the ground state of an ensemble of interacting electrons are, in a formal way, accurately described by Kohn–Sham equations, though one need not construct the many-electron wave-function or the Slater determinant in this case. The approximations are introduced with a specific choice of potential $\hat{v}_{xc}$ only; the calculated results are far less sensitive to errors in the exchange–correlation energy than in the kinetic energy. All the above considerations regarding the choice of basis functions in the HF method are valid for the Kohn–Sham method as well.

The accurate exchange–correlation energy is a functional of the diagonal elements of the two-electron density matrix, not ED. However, the local homogeneity approximation is most frequently used for the E_{xc} and v_{xc} potential in (2.45). This method of solving the Kohn–Sham equations is known as the local density approximation (LDA). This approach originated from the work of Kohn and Sham (1965). There exist some approximate versions of LDA considered in detail by Kohn and Vashishta (1983). Von Barth (1986), having considered a large set of data, has come to the conclusion that for molecules the solution of equations (2.45) in the LDA leads to errors in the binding energy of about 1 eV, in equilibrium distance errors of about 0.1 Å, in vibrational frequencies errors of about 20%, and in ED errors of about 2%. The accuracy achieved is almost always higher than that in the HF method. At the same time, in contrast to HF theory, the LDA exchange potentials contain the self-interaction contribution. Due to this, they behave asymptotically in atoms as $-(z-N)/r$ whereas the HF exchange potential asymptocity is $-(z-N+1)/r$ (Perdew and Zunger 1981). This results in a lowering of the LDA eigenvalues respective to their HF analogues.

In addition to the LDA, in various versions (including the local spin density approximation) of the E_{xc} approximation one uses gradient-corrected functionals, as well as the approximations of mean and weighted density, and corrections for self-interaction, etc, are used. There are a number of comprehensive reviews where all aspects of this problem are analysed: see Perdew (1995), von Barth (1986), Becke (1989), Feil (1990), and Tsirelson *et al* (1992). We shall present here only the data on errors in dissociation energies for molecules of oxides of the first-row elements, obtained by Carrol *et al* (1987) using various exchange–correlation functions (figure 2.7). The dissociation energy is very sensitive to selection of the E_{xc} form; nevertheless, the results obtained using different E_{xc} approximations are close to each other and they are more accurate than the HF results. At the same time the

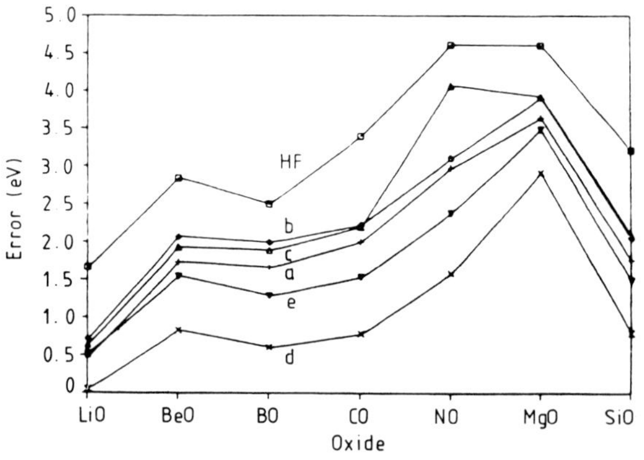

Figure 2.7 Errors in the dissociation energy calculations of first-row oxides: (*a*) Kohn–Sham method (LDA); (*b*), (*c*) LDA corrected for self-interaction; (*d*), (*e*) gradient-corrected functionals. For comparison the error of the HF method is given (Carroll *et al* 1987).

values of dissociation energy for the oxides mentioned lie within the interval of 4.16 eV for MgO to 11.9 eV for CO molecules. Keeping this in mind, we should admit that the relative error in the dissociation energy is still inadmissibly large; it varies from 30 to 108%.

Significant success in applying the Kohn–Sham method to crystals was reached by Zunger and colleagues. In early works Zunger and Freeman (1977a, b) solved the one-particle equations in which the local approximation for the exchange–correlation potential was applied, the non-spherical parts of the crystal potential were retained, and the numerical extended basis set of exact AOs was used. At the first stage of the equations' solution, 'self-consistency upon charge and configurations' is carried out: AOs and their electron populations were varied until the superposition of atoms reproduced the ED of a system in the best possible manner. This stage allowed one to take into account the electron intra-atomic redistribution when an atom is placed into a crystal. Then the multicentre Poisson equation was solved in reciprocal space iteratively with respect to the 'ignored' interatomic ED until the latter became smaller than some *a priori* specified quantity. In this way the ED and other characteristics of some binary semiconductors, including compounds of transition 3d and 4d metals, have been calculated. Later the method was further improved by Bendt and Zunger (1982) and Zunger (1983): the parameters of an effective potential in the one-electron equations were varied self-consistently. The numerical AOs, describing localized electronic states, and a large set of plane waves were used as a basis set covering all electrons. Using this approach, Jaffe and Zunger (1983) and Zunger (1986) have calculated semiconducting halcopyrites of $A^{I}B^{III}X^{VI}$-type ($CuGaS_2$, $CuInS_2$, $CuAlSe_2$, $CuGaSe_2$, $CuInSe_2$) and pseudobinary alloys. These compounds are large crystalline objects: they contain up to 292 electrons and eight atoms in the primitive unit cell. Figure 2.8 shows valence ED profiles for some of the compounds studied by Jaffe and Zunger (1983). Their analysis indicates that in $CuGaS_2$ and $CuGaSe_2$ the EDs in the A–X bonds are more polarized in the direction towards the anion than those in the B–X bonds. In general, the A–X and B–X bonds are more polar in the sulphides than in the selenides. However, considering the distribution of valence electrons, one cannot find a unique answer to the question of which of the above-listed compounds have more ionic bonds. Figure 2.6 has some features that are typical for all the compounds studied: the ED maxima in the internuclear space are localized on the B–X bonds between two distinct minima, whereas in the D–X bonds the ED minimum near the cation position is less pronounced.

Zunger (1979) combined the Kohn–Sham method with the pseudopotential theory. He calculated pseudopotentials from the first principles proceeding from the smoothness criterion for pseudo-wave orbitals, which are the solutions of the Kohn–Sham equations. Great accuracy was achieved with such pseudopotentials in reproducing the unit cell parameters, the crystal

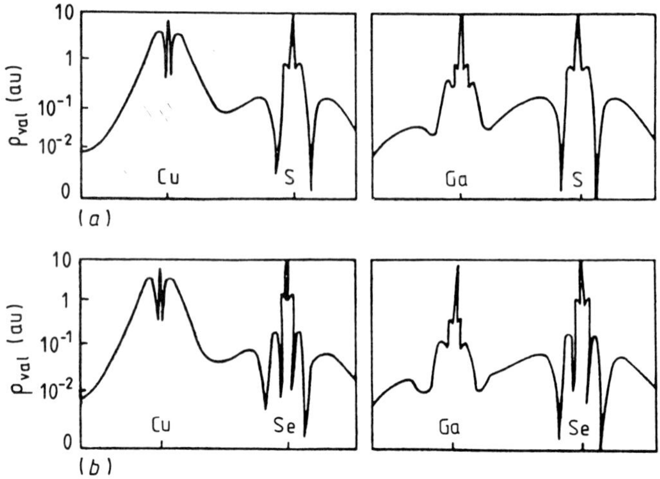

Figure 2.8 Valence ED distributions in (*a*) $CuGaS_2$ and (*b*) $CuGaSe_2$ crystals as a result of self-consistent calculations by the Kohn–Sham method in the LDA approximation with variable effective potential (Jaffe and Zunger 1983).

binding energies, and band structure as well as the chemical bond effects (see below).

Reznik and Tolpygo (1983) have emphasized that pseudopotentials in a many-electron theory are, in general, non-local and non-one-particle in character. For non-local pseudopotentials the Hohenberg–Kohn theorem is not valid: all ground state properties are functionals of the one-particle DM, rather than ED functional (Gilbert 1975). Nevertheless, the pseudopotentials are quite effective in DFT.

Very close in concept to the LDA, the X_α method (Slater 1951) appeared earlier than DFT and has been developed independently for many years. It uses the approximation of a non-local HF exchange potential with a local ED-dependent one:

$$v_x(\mathbf{r}) = -3\alpha((3/8\pi)\rho(\mathbf{r}))^{1/3}. \qquad (2.46)$$

Such a form of the potential results from the fact that the wave-functions for a homogeneous electron gas are plane waves and then the averaged exchange potential in corresponding HF equations is proportional to $\rho^{1/3}$. The α value is determined from the atomic or ionic data: $0.67 \leqslant \alpha \leqslant 1$. For systems with open shells, potential (2.46) is taken separately for electrons with spin 'up' and spin 'down'; the total ED is the sum $\rho = \rho_\uparrow + \rho_\downarrow$.

An important advantage of the X_α method is the weak dependence of computer time on the number of basis functions. This allows one to apply them to large, many-atomic systems. There exist pseudopotential (Snijders and Baerends 1977) and relativistic (Rosen and Ellis 1975) modifications for the X_α method. The original relativistic version of the method was further developed by Snijders and Baerends (1978) and by Snijders *et al* (1979). The core orbitals, where the relativistic effects are most significant, were taken from preliminary relativistic atomic calculations. The influence of relativity on valence orbitals was then taken into account by perturbation theory. It was this approach which was applied by Ros *et al* (1980) and Snijders and Baerends (1982) to address the question of how relativistic effects influence ED (see below).

The majority of non-empirical calculations of crystal electron structure have been performed either in the LDA or in the X_α approximation (often with the addition of some approximation for a correlation potential). The most direct way of using the X_α exchange potential is to introduce it into the HF equations instead of non-local potential. The corresponding modification of the method is called the Hartree–Fock–Slater (HFS) method (Slater 1951). There exist also other modifications of the X_α method which are usually applied to molecules. In the X_α method of scattered waves (Johnson 1973) the system studied is embraced by a large sphere (Watson sphere) with small atomic spheres and interatomic space inside. The one-electron equation with potential (2.46) is solved for each of these regions separately. Then the solutions are matched at their boundaries. Inside atomic spheres the potential is spherically symmetric and each one-electron function is presented here as an expansion over AOs with the radial part determined numerically. In the interatomic space such a function is expanded in series over spherical Bessel and Neumann functions. The solutions in the space between atoms can be interpreted as spherical waves scattered from atomic spheres and from an outer sphere; this provides the name of the method. The orbitals are determined by the SCF procedure. The other method, the discrete variational X_α method (Baerends *et al* 1973), uses the LCAO approximation (which allows us to use the extended basis sets) and the ED is often represented as a sum of spherical atomic contributions. In the solution of the method equations, the matrix elements are calculated numerically by applying a special point procedure in the three-dimensional integration. The LCAO approximation is also used in the third modification of the X_α method (Sambe and Felton 1975), in which the ED and potential v_{X_α} are represented, in addition, as an expansion over subsidiary analytical functions centred at nuclei.

The X_α method is also frequently applied to the investigation of crystal electron structure in the cluster approximation. Only the finite fragment of a crystal is considered; the remaining crystal is taken into account either by means of cyclic boundary conditions (Evarestov 1982), by compensating the bond breaking effects (Larkins 1971), or by introducing point charges into the Hamiltonian simulating the environment (Nagel 1985a). For the description of delocalized states with this approximation, Anisimov *et al* (1981) proposed to express one-electron functions in the form of an expansion over localized cluster orbitals and the Bloch functions. Such an 'embedding' cluster scheme has been successfully applied to the study of the electron structure of stoichiometric and defect compounds of transition metals with rock salt and corundum structures and of high-temperature superconducting copper oxides (Ellis 1990, 1995).

The ED quality given by the X_α method is no worse than that achieved by the HF method (see below). The method reproduces quite well not only binding energies, lattice parameters, and bulk moduli, but also the electron distribution in momentum space, measured in Compton scattering experiments. Therefore Heijser *et al* (1980a) came to the conclusion that the X_α method can be successfully applied for chemical bond feature investigations.

There are some methods in which traditional one-electron schemes are combined with DFT ideas. So, the solutions of one-electron equations for a homogeneous electron gas, the plane waves, form a full orthonormal set of basis functions with a true asymptotic behaviour. The truncated sets of plane waves, whose linear combinations form the crystal orbitals, are conveniently used in calculations. To describe the regions with sharp ED variations, a great number of plane waves is needed. For description of 1s electrons in silicon 2.5×10^4 plane waves are required per single point in $\boldsymbol{k}$-space (Pisani *et al* 1988). However, for description of comparatively smooth valence ED a relatively small number of plane waves is sufficient.

This situation is used in the augmented plane wave (APW) method (Locus 1967). The volume of a crystal is separated into parts, by means of non-overlapping spheres comprising the atoms. The potential is spherically symmetrical inside spheres and constant in the interspherical space. Such a muffin tin (MT) potential corresponds to wave-functions, APWs, which have the form of plane waves between the spheres and an atomic-like form inside the spheres. The functions for the whole crystal are found using the variational principle with APWs used as basis functions. The relativistic effects can easily be taken into consideration in the APW method (Takeda 1979).

Koelling and Arbman (1975) have suggested the modification of the APW method, the linearized method of augmented plane waves (LAPW), that is now widely applied. The energy dependence of the radial parts of functions within the limits of MT spheres is linearized in the LAPW method with respect to the energy for each angular momentum value corresponding to the symmetry-adapted spherical harmonics in the wave-function expression. This decreases the number of terms in the linear combination of such functions, which is matched with plane waves, using the requirement of continuity of basis functions and their derivatives on the surface of an MT sphere. There is also the full-potential variant of the LAPW method in which the MT approximation is not used: the basis function set consists of a mixture between plane waves and atomic-like wave-functions, obtained usually from the numerical solution of one-electron atomic equations. The number of basis functions per atom in APW and LAPW methods is approximately 10^2. The methods described make the calculations much faster (some 10^2 times). In addition, the approach gives all-electron wave-functions with the correct behaviour in the valence region and near the nucleus. In spite of some of the limitations discussed by Koelling (1981) and Calais (1991), this approach is widely used for electron structure calculation of nitrides, carbides, oxides, transition metal alloys, and molecular crystals (Schwarz and Blaha 1988, 1991, Margl *et al* 1993, Schwarz 1994). Its application to high-temperature superconductors has been described by Krakaner *et al* (1988) and Ambrosch-Draxl *et al* (1994).

The method of orthogonalized plane waves (Ashcroft and Mermin 1976) is somewhat like the APW method. The electron core states here are described by Bloch functions in the LCAO approximation, whereas the valence wave-functions are represented as plane waves orthogonalized to the core functions. The variational principle is used to solve the one-electron equations by an SCF method for core and valence potentials. The advantage here is in the absence of need to choose an MT potential, and the drawback consists of the assumption that the core potential is spherically symmetric; this is not the case in covalent compounds.

In the linearized muffin tin orbital (LMTO) method (Skriver 1983, Weyrich 1988) each atom in a crystal is placed into a sphere in which the potential is considered to be centralized. The radii of spheres are chosen to yield their total volume equal to that of the unit cell volume, therefore the space between spheres is not considered. By solving the Schrödinger equations for each sphere, one finds the set of symmetry-adapted energy-linearized local functions (MT orbitals) which are continuous and differentiable on the sphere. The linear combinations of these functions, satisfying the Bloch theorem, serve as a basis in LMTO calculations. The number of MT orbitals per atom does not exceed 16 usually and, therefore, the diagonalization of a secular equation matrix is 10^2–10^3 times faster than in the LAPW method (Pickett 1989b). The disadvantages of this method are the necessity to compensate for effects related to the overlapping of atomic spheres and the potential sphericity within each sphere.

The LMTO method is most effective for close-packed solids: it was applied for studying metals, compounds of d elements (Gubanov *et al* 1987), and high-temperature superconductors (Pickett 1989b). The relativistic version of the LMTO method is described by Nemoshkalenko *et al* (1983).

Returning to the Kohn and Sham equations (2.45), one should note that they provide more than the description of ground state properties only because these are one-particle equations with a self-consistent potential. It is worthwhile to compare the spectrum of Kohn–Sham Hamiltonian eigenvalues with the one-particle spectrum of the system studied. However, the general theory does not allow one to establish any relation between the eigenvalues ε_i with the one-particle excitation energies (Kohn and Vashishta 1983). Whereas in the HF theory the ε_i are the electronic ionization energies, provided the relaxation processes are neglected (the Koopmans theorem), in Kohn–Sham equations they have another meaning. If the theory is supposed to be valid for fractional occupation numbers $0 \leqslant n_i \leqslant 1$ of states φ_i, then one can show (Slater 1974) that

$$\varepsilon_i = \partial E / \partial n_i. \tag{2.47}$$

This relation is common for DFT, though it has been initially established for the X_α method only.

In order to establish ε_i values as close as possible to the spectrum of crystal energy bands and so that the $\varepsilon_i - \varepsilon_j$ difference can be interpreted as the energies of transitions between the levels, Slater (1974) suggested the 'transition state' concept in which $n_i = n_j = \frac{1}{2}$. In this case the error in energy of the $i \to j$ transition is of third order in the variation of occupation numbers and is rather small. Each transition requires individual consideration however.

McHenry *et al* (1987) have shown that the transition state concept provides good agreement between the ε_i values and ionization potentials for atoms ranging from He to Ar. The eigenvalues of energies of the highest occupied orbitals were found to be in good agreement with electronegativities, according to Mulliken (see e.g. Sen and Jorgensen 1987) defined as $\chi = (I_1 + A)/2$ (I_1 is the first ionization potential, A is the electron affinity). Thus, a quite different treatment of the Kohn–Sham problem eigenvalues is possible.

Comparing a rigorous approach to the description of the one-particle spectrum of the many-electron system by Green functions and the Dayson equation with the Kohn–Sham method, one can also conclude that the set of energies can be interpreted as the spectrum of one-particle excitations only if some particular approximations are valid (Kohn and Sham 1966). Among the approximations is the assumption of a short-range character of the exchange–correlation interaction and of the energy dependence of the potential, which is taken into consideration in the exact equation of Dayson type (Dreizler and Gross 1990). These conditions can sometimes be not valid: this is a defect of the one-particle interpretation of the LDA energy spectrum; however, the results can still be rather accurate. For example, according to Hedin and Lundqvist (1971), in metals at typical band energies, the proper energy part varies within a few per cent only as the energy changes.

The energy dispersion law in the LDA for the region of occupied states and near the Fermi level of metals shows good agreement with experiment. The description of the conduction band of dielectrics and semiconductors is less successful. The energy gap, E_g, is 30–50% lower than the experimental one in these cases and there are no correlations between the relative error and the character of the crystal chemical bond. This can be seen from table 2.3,

Table 2.3 The energy gaps E_g for some crystals calculated in the LDA and from experiment.

Crystal	Si[a]	GaAs[b]	AlAs[b]	CuCl[c]	LiF[d]	LiCl[e]
E_g, theory (eV)	0.60	1.09	1.34	2.0	9.8	5.81
E_g, experiment (eV)	1.17	1.53	2.30	3.4	14.2	9.4
Error (%)	48	29	42	41	31	38

[a]Zunger 1980.
[b]Ihm and Joannopoulos 1981.
[c]Zunger and Cohen 1979b.
[d]Zunger and Freeman 1977b.
[e]Heaton *et al* 1983.

which presents E_g values calculated in the LDA and measured in experiments for some simple semiconductors.

The results for the energy gap can be improved to some extent by a more careful choice of the exchange–correlation functional. However, this is, in DFT, only part of the true one-particle potential which can be a non-analytical functional of ED. Even if the exact $E_{xc}[\rho]$ functional is available, the calculated E_g values may be expected to be lower than experimental ones. Sham and Schüter (1983) and Perdew and Levy (1983) have found that the variational derivative of E_{xc} functionals undergoes discontinuity in the presence of a gap in a spectrum and the differences between total energies of the $(N+1)$-, N-, and $(N-1)$-electron systems do not reduce to a difference between one-electron eigenvalues. This discontinuity results in a difference in potential v_{xc} for a valence band and conduction band at least by a constant. It follows from this that the errors in the description of an energy gap arise not so much from the neglect of the v_{xc} non-locality in the LDA as from ignorance of the energy dependence of the potential. The problems arising due to this situation in non-empirical methods are very serious: if the discontinuity is large, attempts to take into account the non-locality of potentials cannot improve the results (Godby *et al* 1988).

Note that, according to Moullet and Martins (1990), the description of the correlation functional in the LDA weakly influences the static electron polarizabilities of atoms and molecules and, correspondingly, the virtual orbital. At the same time, the dissociation energy value essentially depends on the form of this functional.

An objective conclusion on advantages and defects of various theories can be drawn by comparing them not only with experiment, but also with each other. The results of non-empirical calculations of the electronic ground state of a crystal by DFT methods are usefully compared with the HF method which principally differs from DFT. The comparison can be carried out, in particular, for coherent structure amplitudes which are the Fourier

Table 2.4 Moduli of x-ray structure amplitudes for Si calculated by different methods and measured by x-ray (all values are given in units of electrons per primitive cell at 0 K).

	Structure amplitude values							
hkl	a	b	c	d	e	f	g	h
111	15.11	15.12	15.13	15.57	15.07	15.17	15.193(7)	14.89
220	17.26	17.28	17.14	17.58	17.32	17.33	17.318(12)	17.42
311	11.37	11.33	11.02	11.48	11.37	11.36	11.354(9)	11.55
222	0.25	0.34	0.38	0.22	—	0.34	0.386(2)	0
400	14.92	14.88	14.70	14.94	14.93	14.90	14.898(10)	15.02
331	10.17	10.20	9.94	10.22	10.23	10.22	10.260(9)	10.16
422	13.37	13.36	13.30	13.44	13.43	13.39	13.438(12)	13.40
333	9.07	9.02	8.92	9.14	9.08	9.06	9.091(9)	9.11
511	9.08	9.08	8.98	9.14	9.13	9.08	9.120(9)	9.11

[a] SCF LCAO LDA non-empirical calculation with extended GTO basis (Wang and Klein 1981).
[b] SCF APW calculation (Raccah *et al* 1970).
[c] SCF LDA calculation with first-principles pseudopotential (Zunger and Cohen 1979a, b).
[d] HF STO-3G calculation (Dovesi *et al* 1981).
[e] SCF HFS calculation of Si_8H_{18} cluster with very extended basis set (Velders and Feil 1989).
[f] SCF full-potential relativistic LAPW calculation with numerical wave-functions (Lu and Zunger 1992).
[g] The x-ray data of Treworte and Bonse (1984) corrected for anharmonic and harmonic atomic thermal vibrations and anomalous scattering and averaged by Spackman (1986c) over two experimental sets measured with Mo Kα and Ag Kα radiation: the 222 reflection is from Alkire *et al* (1982). The mean square deviations are in brackets.
[h] HF calculation of atomic superpositional model.

transformations of ED. These quantities calculated for the Si crystal by various methods are given in table 2.4. One can conclude that various techniques of ED calculation lead to concordant results which are close to the experimental data. In any case, whatever crystal (metal or dielectric) was considered, the ground state properties are described by the DFT methods in the LDA not worse, but sometimes better, than by the HF method (Krijn and Feil 1988c, Carrol *et al* 1987). Note that the HF method is inapplicable for describing the Fermi surface of metals (Monkhorst 1979, Assing and Monkhorst 1993): regardless of the specific properties of the Fermi surface, the static density on it disappears due to a singularity in the HF exchange term, contrary to experiment. Thus, the LDA has advantages in describing the crystal electron structure despite some of the drawbacks noted above. Moreover, if the LDA is slightly modified, say in the spirit of the X_α method, the method becomes accurate enough for calculating the elementary excitation spectrum as well.

It is interesting to note that in earlier works the local approach to the approximation of exchange and correlation was meant as an inevitable simplification of the HF exchange interaction, it seemed that a more accurate theory must be as close to the HF theory as possible. Therefore, the approximation (2.46) was assumed to be too crude *a priori*. Now it is clear that this is not the case. Cook and Karplus (1987) and Tschinke and Ziegler (1990) have shown that the HF method in the dissociation limit results in error due to neglect of electron correlation, which is different for molecules and for their atomic counterparts. This error, depending upon molecular geometry and the strongly distorting potential surface of a dissociating molecule, is cancelled when the wave-function is squared in calculating ED. Since the v_{xc} potential is approximated in the LDA by expressions including just ED, this approach only suffers slightly from this error.

Thus, we have seen that the efforts of a great number of people have made it possible to improve the DFT methods up to a level that allows one to apply them to analysis of the electron structure of rather complicated crystals, and this is in spite of the fact that many theoretical DFT problems remain yet unsolved and the efficiency of some computational procedures cannot be explained for certain. This gives rise to the danger that any method which is found to be effective for one class of compounds may be unsuitable for others. As a result, the choice of computational method within the DFT cannot be performed yet without taking into consideration the physical and chemical specificity of the problem under consideration.

2.3 The quality of the theoretical electron densities

As we have seen above, the theoretical calculations of the ED of molecules and crystals are based on the approximate solution of the Schrödinger equation or DFT equations via use of the variational principle. The method used and the level of approximation depend on the nature of the object, on the property studied, and on the available computer facilities. Unfortunately, in the general theory the stage-by-stage evaluation of calculation errors is not possible, unlike in diffraction measurements. In theoretical approaches only the errors in the calculation of the operator matrix elements can be evaluated (Bagus and Wahlgreen 1976, Sihai *et al* 1982). For this reason, when considering the reliability of calculation results, it is preferable to use the term ‘quality’ rather than ‘accuracy’.

Typically the quality of a calculation is judged by the coincidence of calculated expected values of one-electron operators with their experimentally measured values (Green 1974, Iwata 1982, Bartlett 1985). The theoretical and experimental ED can be compared also. However, to eliminate the disadvantages of both approaches one must also have some independent evaluations of reliability for each of them. For this reason we shall consider the factors influencing the results of ED calculations.

The majority of non-empirical ED calculations have been carried out in the non-relativistic approximation using either HF or DFT methods. The traditional computational tactic is such that the obtained wave-function would best reproduce the properties under study. In this case the accuracy of other property calculations can be ignored. However, the problem of ED calculation requires that the wave-function describe all system position space with equal accuracy. This can be approached by using an orbital-energy-balanced basis set (Smith 1977), which provides a proper behaviour of wave-functions, and by improving the methods for approximate solution of equations.

One should note here that the variational principle can give a good estimate of average energy of the system studied and, at the same time, yield bad wave-functions. Therefore, the achievement of the energy minimum cannot be a universal criterion for the wave-function quality. To an ever greater extent this concerns the quality of ED which does not depend directly on Hamiltonian operator $\hat{H}_e$. The indirect relation between $\hat{H}_e$ and ρ hinders the improvement of methods for accurate ED determination using system energy considerations only. In addition, the contribution of the electron–nuclear interaction energy to the average energy of a system is rather large and the application of a variational principle provides a better optimization of those wave-function parameters which describe the inner electronic shells. Hence, the reproduction of the valence ED requires special attention. For the same reason it is difficult to obtain a good ED in the close vicinity of a nucleus: small energy variations correspond to considerable ED deviations in these regions.

According to Cade (1972), Bader (1975a), Smith (1982), Kikkawa *et al* (1987), and Towler (1995) the errors introduced by applied basis sets dominate. So, the calculations of small molecules in a minimal basis set underestimate the charge accumulation in the internuclear space and, simultaneously, overestimate the density of electron lone pairs. The errors in these regions may reach ~ 0.35 e Å^{-3} (20% of molecular ED), and may exceed the deformation ED $\delta\rho$ (1.15) value. The inclusion of two exponential functions in radial parts of basis functions only slightly improves the situation (Bader 1975a); the non-compensated error of $\sim$10–15% ($\sim$0.2 e Å^{-3}) still remains. Extended basis sets, containing polarization functions, allow one, in general, to correctly reproduce the ED distribution in a system by lowering the error to 2–5%. As an example, we present the results of calculating the deformation ED of the CO molecule carried out by Heijser *et al* (1980b) for eight basis sets of various degrees of completeness (figure 2.9). One can see that the minimal SZ basis set does not accurately reproduce the ED features in the chemical bond region. As the basis set is extended, the $\delta\rho$ value in the interatomic space increases, its relative growth being approximately 100 times larger than the relative decrease of molecular energy. At the same time the $\delta\rho$ value decreases in other regions of the molecule.

Eisenstein (1979) has found that for the six-atom formamide molecule,

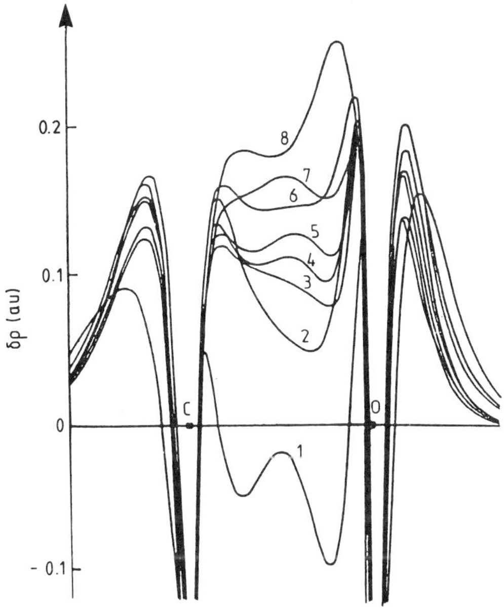

Figure 2.9 Deformation ED profiles calculated with different bases for the CO molecule. (Reprinted with permission from Heijser *et al* 1980b.) 1, single-exponential basis set; 2, double-exponential basis set; 3, triple-exponential basis set; 4, DZ+DP basis set (the polarization d-type function is added); 5, TZ+DP basis set; 6, TZ+2DP basis set; 7, TZ+DP+FP basis set; 8, TZ+2DP+2FP basis set (the polarization f-type function is added).

$CHONH_2$, the addition of polarization functions to the two-exponential basis set, which decreased the energy by only 0.0955 14 au (~0.057%), changes the values and directions of local atomic dipole moments. This implies considerable changes in the ED. These facts agree with the above remark concerning the limited role of the system energy as the single criterion for improving ED.

For the reliable reproduction of ED distribution one can recommend the application or at least a two-exponential basis set, extended by polarization functions in a balanced manner. Breitenstein *et al* (1983) have shown that the results obtained with such a basis set are well reproduced if s- and p-type functions centred at the chemical bond areas are used instead of polarization functions centred at the nuclei. In these cases the accuracy of $\delta\rho$ determination can be estimated as 0.05–0.07 e Å^{-3}. This estimation deals with molecules of two to six atoms with a small number of electrons (~20–25) only. For systems containing heavier atoms the above estimation should be doubled at least. This is especially true for calculations carried out at a lower level of approximation.

It should also be emphasized that when calculating the deformation ED

(1.15) the ρ and $\tilde{\rho}$ functions must be evaluated with the same basis set in order to avoid errors resulting from the use of different basis sets.

Silvi *et al* (1989) have compared the ED for diamond, silicon, lithium, and MgO crystals obtained in all-electron and pseudopotential versions of the HF method. The largest discrepancies in valence ED were observed for MgO; corresponding ρ_{val} values are presented in figure 2.10. The general agreement between the results obtained with both modifications of the method can be accepted as satisfactory. The discrepancy in ED near the Mg atom was explained by the authors as a result of overestimating the repulsion in the corresponding pseudopotential which is due to uncertainty in evaluating the effective charge of the Mg^+ ion core.

Consider now the errors arising due to the neglect of electron correlation effects in the HF method. The Brillouin theorem (2.9) states that only doubly substituted Slater determinants make a contribution to the first-order correction to the HF wave-function. This is why in earlier works the contribution to ED of singly substituted configurations (which are insignificant from the energy point of view) was neglected when correlation effects were taken into account. Further studies (Bader 1975a, Smith 1977, Meyer *et al* 1982) have shown, however, that the inclusion of these configurations in the calculation, along with the doubly substituted ones, provides a uniform mixing of occupied and free orbitals and improves the values of calculated dipole moments associated with the ED distribution. To clarify this point, one can modify expression (2.10), following Smith (1977), and write the wave-function in the form

$$\Phi = c_0\Psi_0 + c_S\Psi_S + c_D\Psi_D + \ldots \qquad (2.48)$$

Keeping in mind that ED is a one-electron quantity, it will be determined

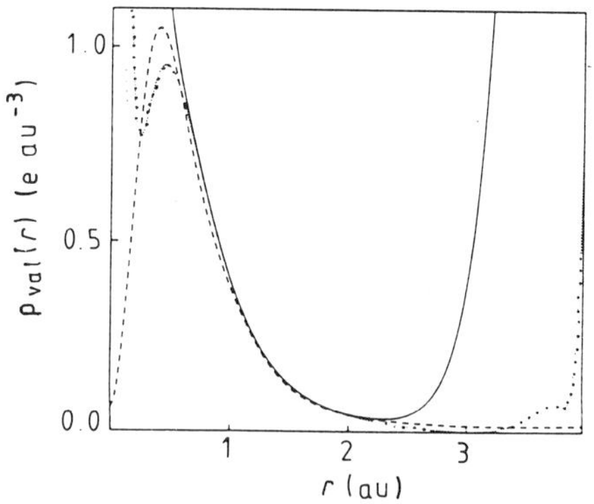

Figure 2.10 HF ED of an MgO crystal along the [100] direction (Silvi *et al* 1989). The solid line is the all-electron total density; the dotted line is the all-electron valence density; the dashed line is the pseudopotential valence electron density.

by the expression

$$\rho = N(\rho_{HF} + 2c_s\rho_{0s} + c_d^2\rho_{dd} + 2c_sc_d\rho_{sd} + c_s^2\rho_{ss} + \ldots). \tag{2.49}$$

In these expressions subscripts s and d relate to singly and doubly substituted Slater determinants, respectively, and N is a normalizing factor. Coefficients c_s and c_d in terms of perturbation theory are related to first- and second-order corrections to the wave-function. Then, to second-order terms inclusively, the ED is given by the expression

$$\rho \cong (1 + c_d^2)^{-1}(\rho + 2c_s\rho_{0s} + c_d^2\rho_{dd}). \tag{2.50}$$

Hence, in ED calculations one should take into account both singly and doubly substituted configurations simultaneously. This approach is easily extrapolated to the case of triply, quadruply, etc substituted configurations.

Estimations of the influence of various excited configurations on the molecular energy and on the deformation ED were obtained by Meyer *et al* (1982). One of the examples considered was the N_2 molecule. The HF calculation in the basis set including bond functions was performed. In addition, the calculations accounting for singly, doubly, triply, and quadruply substituted configurations (SC, DC, TC, QC, respectively) were carried out. The number of configurations taken into consideration and their relative contributions to the electron correlation energy and to the mean correction to deformation ED $\delta\rho$ due to this correlation account are given in table 2.5. The relative contributions to ED were evaluated by averaging over chemical bond and electron lone-pair regions as well as over the areas adjacent to nuclei. The DC makes a dominant contribution to the correlation energy; the next most dominant is the QC. However, their influence on corrections to $\delta\rho$ is quite different. In the case of the N_2 molecule, taking account of the SC causes the main changes in $\delta\rho$. In the general case, however, the SC effect has the same contribution as that of the DC. The higher-order configurations

Table 2.5 Relative contributions of different excited configurations to the correlation energies and to the correlation correction to deformation ED $|\rho_i/\delta\rho_{corr}|$ for the N_2 molecule. The $|\rho_i/\delta\rho_{corr}|$ value is averaged over different molecule regions and therefore has no dependence on the $\rho_i/\delta\rho_{corr}$ sign. (Reprinted with permission from Meyer *et al* 1982.)

Type and number of excited configurations (i)	E_i/E_{corr} (%)	$\|\rho_i/\delta\rho_{corr}\|$ (%)
Singly excited (21)	1.1	56
Doubly excited (896)	91.7	25
Triply excited (1989)	2.3	13
Quadruply excited (8201)	4.9	10

are responsible for 20–25% of the $\delta\rho$ changes. The corrections to the calculated values of other one-electron quantities due to contributions of higher excited configurations have approximately the same values.

Wang and Boyd (1989) found, while studying correction effects in molecules, that the role of the DC is most essential for ED near the nuclei, whereas the SC is more important in the chemical bond region. They have also found that taking account of electron correlation within the framework of the configuration interaction method and Moller–Plesset second-order perturbation theory leads to the same quatitative changes in ED in H_2, N_2, and CF_2 molecules. In general, however, taking account of electron correlation leads to increasing ED near the nuclei and to decreasing ED in the chemical bond region. The errors in the total molecular density due to the neglect of electron correlation are equal to 0.02–0.06 e Å^{-3} in the internuclear space, to 0.05–0.15 e Å^{-3} near the nuclei, and to 0.06–0.08 e Å^{-3} in the lone-pair region. These values do not exceed 2–3% of the total ED; however, they can reach 30% of the deformation ED (Stephens and Becker 1983). Therefore, this effect may be noticeable for light-atomic systems. One should emphasize that taking account of electron correlation does not compensate for errors due to limitations of the chosen basis (Green 1974).

Zunger and Freeman (1977c) have investigated the influence of different modifications of the same DFT method on the ED. They have performed three full self-consistent calculations for diamond using the extended basis set of a numerical AO. At first, they took into account only electron–electron and electron–nucleus potentials in the Kohn–Sham equations; then they added the local exchange potential, and, finally, the local correlation potential. The ED changes at the internuclear space due to exchange reached 40%. Subsequently taking account of correlation changed the ED in this region by only 0.1–0.3%. However, the combined account of exchange and correlation increased ED in the chemical bond region and near the nuclei due to electron displacement from 'non-bonding' regions of the crystal.

The comparison of ED in silicon obtained by various DFT calculations with others and with those reconstructed from very accurate x-ray diffraction data, is very instructive (figure 2.11). Strange as it may seem, the most approximate Thomas–Fermi method gives the best agreement with experiment at the centre of the internuclear space. More sophisticated DFT methods slightly underestimate the ED value here with the application of any version of computational techniques (Spackman 1986c, Lu and Zunger 1992). The reason for such an underestimation still remains obscure.

Ros *et al* (1980) have examined the influence of relativistic effects on ED by calculations for the $HgCl_2$ molecule with the HFS method in both the non-relativistic and relativistic approximations. Figure 2.12 shows how the relativity influences the deformation ED. The $HgCl_2$ molecule contains 10 σ, 12 π, and 4 δ valence electrons. The corresponding promolecule $\tilde{\rho}$, in which the Hg atom has the $d^{10}s^2$ configuration and the Cl atom has the s^2p^5

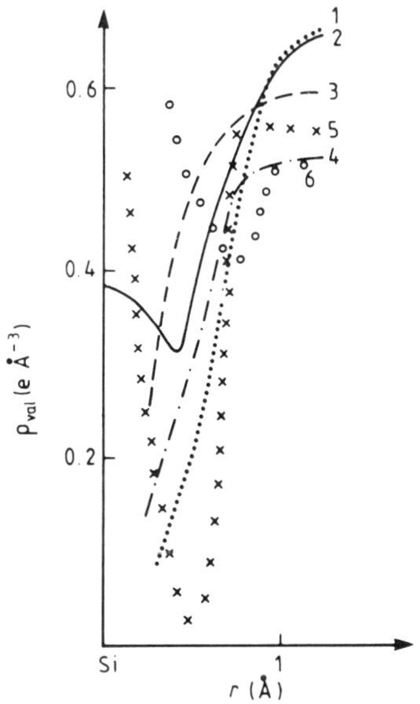

Figure 2.11 Valence electron densities for silicon calculated by different methods and measured by x-rays: 1, Thomas–Fermi model with local pseudopotential (Baldereschi *et al* 1981); 2, x-ray (Pietsch *et al* 1986a); 3, SCF LDA method with first-principles pseudopotential (Zunger and Cohen 1979a); 4, SCF LDA non-empirical calculation (Wang and Klein 1981); 5, all-electron LAPW LDA calculation (Lu and Zunger 1992); 6, full-potential LMTO pseudopotential calculation (Weyrich 1988).

configuration, contains $11\frac{1}{3}\,\sigma$, $10\frac{2}{3}\,\pi$, and $4\,\delta$ electrons in the valence shell. In the non-relativistic $\delta\rho$ maps one can observe the 'displacement' of $\frac{4}{3}$ electrons from the σ to the π subsystem. This implies a lack of a charge in the σ region and an excess in the π region, predominantly on p_π orbitals of the Cl atom. The extra stabilization of the 6s orbital of Hg due to relativistic corrections results in an electron 'returning' from the Cl to the Hg atom and increases the charge in the 6s subshell of the Hg atom. The maximum error found in the case described above is equal to 0.14 e Å^{-3}. Similar effects were observed by Ros *et al* (1980) in $ZnCl_2$ and $CdCl_2$ and by Ziegler *et al* (1981) in AuH, AuCl, and Au_2 molecules. The main influence of relativistic effects in these compounds is associated with stabilization of the 6s orbital of the Au atom in accordance with the above-mentioned conclusion of Pyykkö (1988).

Thus, ignoring the relativistic effects can lead to erroneous interpretations of the theoretical deformation ED, since the error of 0.15 e Å^{-3} can already exceed some typical features of $\delta\rho$.

In molecular crystals the ED is most frequently calculated using the HF

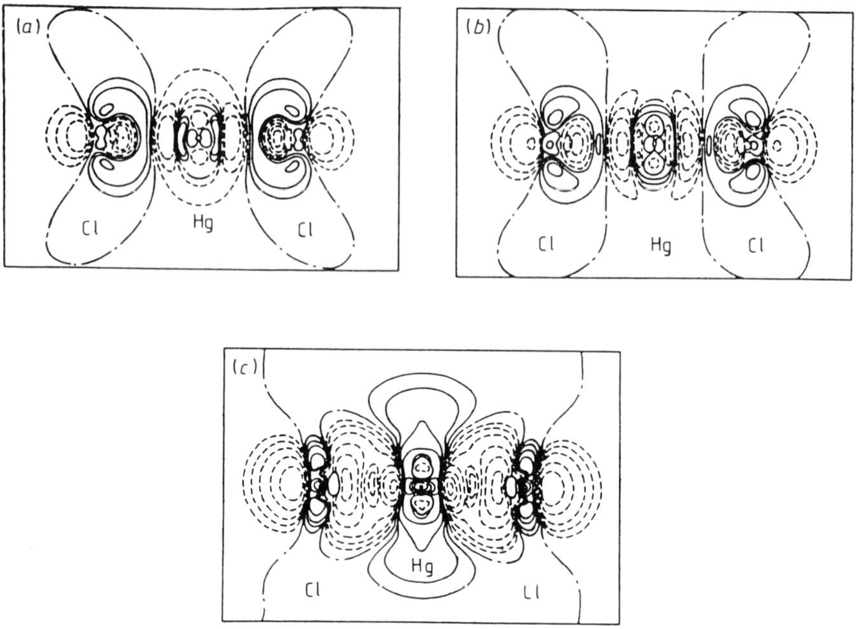

Figure 2.12 Non-relativistic (*a*) and relativisitic (*b*) deformation EDs calculated by the HFS method and the difference function $\delta\rho_{relat} - \delta\rho_{non\text{-}relat}$ (*c*) for the $HgCl_2$ linear molecule. Contour values (au): (*a*, *b*) 0; ± 0.0025; ± 0.005; ± 0.01; ± 0.025; ± 0.05; ± 0.1; (*c*) 0; 0.000 25; ± 0.0005; ± 0.001; ± 0.0025; ± 0.005; ± 0.01; ± 0.025. Here and below solid lines connect points with excess (positive) deformation ED; dashed (or dotted) lines apply to negative ED; and dash–dot lines concern zero-density points. (Reprinted with permission from Ros *et al* 1980.)

or X_α methods. To evaluate their efficiency we shall compare deformation ED for the H_2O molecule calculated by Breitenstein *et al* (1983) using the HF method with the $4-31G+BF$ basis set and by Krijn and Feil (1986) using the discrete variational X_α LCAO method with the following basis set: 1s, 2s, 2p, 3d, and 4f STOs for the O atom and 1s, 2p, and 3d STOs for H atoms. Both deformation EDs are shown in figure 2.13. They coincide, in fact, in the whole position space; the quantitative difference takes place only in the negative-deformation-ED region around the O atom. The ED of the N_2 molecule, calculated by the same methods, and the ED of the Cl_2 molecule, calculated by the X_α method of scattered waves and by the X_α discrete variational method (McMaster *et al* 1982) are also in close agreement.

Summarizing, one can mention that the most sophisticated theoretical methods allow one to achieve a qualitative agreement of the ED distribution. The quantative estimation of error in results obtained by various methods is about 0.1 e $Å^{-3}$ for deformation ED in the chemical bond region of a

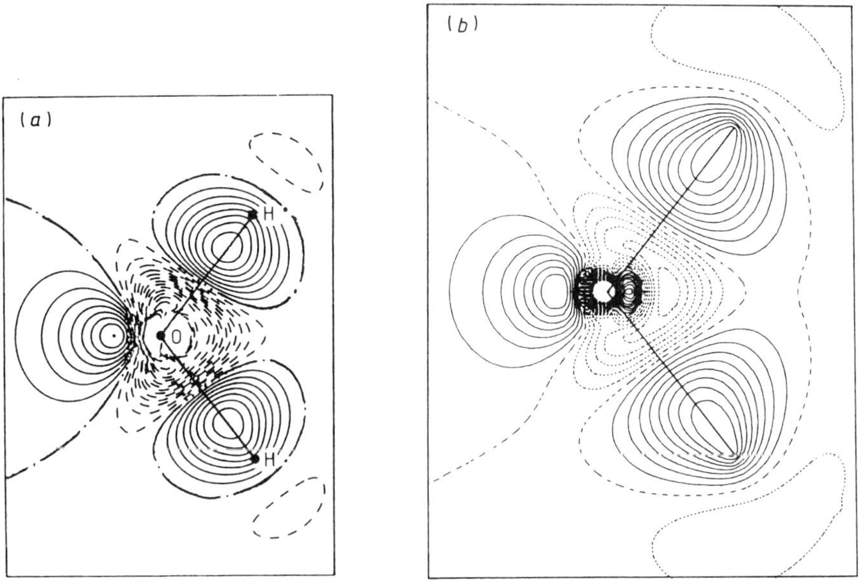

Figure 2.13 Deformation ED in the plane of a water molecule: (*a*) HF method, 4−31G+BF basis set (Breitenstein *et al* 1983); (*b*) X_α LCAO discrete variational method, extended STO basis set (Krijn and Feil 1986). The contour line interval is 0.015 au.

crystal. This is, perhaps, a realistic mean estimation of the accuracy that theory has reached so far.

2.4 Quantum mechanics and topology of the electron density

2.4.1 *The properties of electron density*

Let us consider now properties of ED and its derivatives. Some of these properties have already been discussed above and, therefore, we first recall here some principal points briefly. The ED, $\rho(\mathbf{r})$, is a smooth, finite, non-negative function, defined in real position space. According to Hohenberg and Kohn (1964), all properties of a ground state of many-electron systems (atoms, molecules, and crystals) are uniquely defined by ρ. The condition of N representability of ρ by the antisymmetric N-electron wave-function is the condition of electron charge conservation in a system: $\int \rho(\mathbf{r})\,\mathrm{d}\mathbf{r} = N$. There exists an infinite number of wave-functions (or one-electron DMs) corresponding to the given ED. Or course, only one of these wave-functions describes the ground state. The other functions can correspond to excited states or be realized in systems with only a non-local outer potential, and so on. According to Kohn and Sham (1965), the true wave-function of the ground state can be selected by using a minimum-kinetic-energy criterion.

The continuous ED function must also satisfy asymptotic conditions which provide its proper local behaviour near the nucleus and at a considerable distance from it. Both accurate and HF wave-functions obey the Kato condition (2.15) on nuclei. For ρ this condition has the form (Bingel 1963, Pack and Brown 1966)

$$\lim_{r\to R_a}\left\{\frac{\partial\bar{\rho}(\boldsymbol{r})}{\partial \boldsymbol{r}}\bigg/\bar{\rho}(\boldsymbol{r})\right\} = -2Z_a \tag{2.51}$$

where $\bar{\rho}$ is the spherical average with respect to the ath nucleus. In the vicinity of nuclei it is sufficient to consider the ED of ns (spherical) states ($n=1, 2, \ldots$) only. Accordingly, in the LCAO approximation the general condition for the total ED (2.51) is reduced to a series of partial conditions for ρ_{1s}, ρ_{2s}, etc (Fraga and Malli 1968). At the points $\boldsymbol{r}$, not including nuclear positions, the relation $\lim_{r\to r'}[(\partial\rho(\boldsymbol{r})/\partial\boldsymbol{r}]/\bar{\rho}(\boldsymbol{r})=0$ is valid. Thus, from expression (2.51) it follows that the ED contains all information about the potential in the Schrödinger equation.

The upper level of ED value at a nucleus Z_a of an atom, $\rho(0)$, is determined by the relation given by Hoffman-Ostenhof *et al* (1978):

$$\rho(0)\leqslant(Z_a/2\pi)\int r^{-2}\rho(r)\,\mathrm{d}r. \tag{2.52}$$

At large distances from nuclei ρ varies according to the exponential law (Morrell *et al* 1975)

$$\rho(r)\sim\exp[-2(-2\mu)^{1/2}r]. \tag{2.53}$$

The chemical potential μ in this expression is a function of ρ; it is related to the first ionization potential I_1 by the inequality $\mu\leqslant -I_1$.

Relations (2.51) and (2.53), which describe the local behaviour of ED, and the results of calculations carried out for over 100 many-atomic molecules, have allowed Bader *et al* (1981) to postulate that the local ED maxima in many-electron systems can take place at nuclei positions only. First of all this conclusion substantiates theoretically the determination of atomic structure of molecules and crystals by means of x-ray and electron diffraction methods. The latter identify the local ED maxima with nuclear positions. Further it allows one (as will be shown later) to obtain a quantum mechanics of the ED which is valid for fragments of molecules and crystals. One should note that Gatti *et al* (1987) and Cao *et al* (1987) have found local ED maxima not related to nuclei in Li and Na clusters as well as in Li_2, Li_2^+, Be_2^+, C_2, C_2^+, and Na_2 molecules, and Zou and Bader (1994) have found non-nuclear ED maxima in Si crystal. The use of high-quality basis sets removes the non-nuclear attractors in some molecules and clusters but in the Li_2 molecule they remain at all levels of calculations, probably due to the mixture of the ground and appropriated excited electronic states resulting from electron–electron interaction (Bersuker *et al* 1993).

The local ED behaviour was found to have a great importance for the description of many-electron systems. This statement has been demonstrated by Srebrenik and Bader (1975), Srebrenik (1975), Bader (1975b, 1991, 1994), Srebrenik *et al* (1978), and Bader *et al* (1978). These authors have used the Schwinger (1951) principle of stationary action for the description of many-electron systems. According to this principle, the quantum action integral

$$W_{12}[\Psi]=\int_{t_1}^{t_2} \mathrm{d}t \int L(\Psi, \nabla\Psi, \Psi, t)\, \mathrm{d}\boldsymbol{r} \tag{2.54}$$

remains unchanged with respect to infinitesimal variations of the state function Ψ within the time interval $t_1 - t_2$; rather, it depends on the action of the infinitesimal generator of transformation at time points t_1 and t_2. Function $L(\Psi, \nabla\Psi, t)$ is the many-particle Lagrangian density function, which depends on the coordinates and momentum of all electrons. If one requires, in addition, that the conditions $\delta\Psi = \delta\Psi^* = 0$ be met at finite points of a time interval, then from one universal dynamical principle, $\delta W_{12} = 0$, one can not only obtain the equations of motion and expected values of observed quantities, but also determine the expected values of various characteristics related to total system fragments.

In the case of ground state many-electron systems, it is sufficient to use the one-particle Lagrangian density, $L(\boldsymbol{r})$. Taking account of the specificity of the description of the kinetic energy in the Lagrangian and the Schrödinger equation (Bader and Nguyen-Dang 1981), $L(\boldsymbol{r})$ has (in atomic units) the form

$$L(\boldsymbol{r}) = -\tfrac{1}{4}\nabla^2\rho(\boldsymbol{r}). \tag{2.55}$$

Considering the virial of the force density, i.e. the quantum mechanical average of a dot product of the density of forces acting on an electron at point $\boldsymbol{r}$ from the side of the other electrons and nuclei, and of the vector $\boldsymbol{r}$, Bader (1980) obtained for the stationary state the expression

$$-L(\boldsymbol{r}) = V(\boldsymbol{r}) + 2G(\boldsymbol{r}). \tag{2.56}$$

Here $V(\boldsymbol{r})$ is the density of electron potential energy at point $\boldsymbol{r}$ or the local potential energy of electrons, and $G(\boldsymbol{r})$ is the density of electron kinetic energy at the same point, the local kinetic energy. Combining relations (2.55) and (2.56), one arrives at the expression

$$V(\boldsymbol{r}) + 2G(\boldsymbol{r}) = \tfrac{1}{4}\nabla^2\rho(\boldsymbol{r}). \tag{2.57}$$

This is the virial theorem in the local form: it establishes that densities of kinetic and potential energies at any point of position space are related via the Laplacian of the ED. Further, we can use the Gauss theorem (Arfken 1985) $\int \nabla^2\rho(\boldsymbol{r})\, \mathrm{d}\boldsymbol{r} = \oint_s \nabla\rho(\boldsymbol{r})\cdot\boldsymbol{n}(\boldsymbol{r})\, \mathrm{d}S(\boldsymbol{r})$ and take into account that for a whole many-electron system the latter integral is zero, since both ρ and $\nabla\rho$ are equal to zero on surface S covering a system. Then, integrating (2.57), one easily obtains the virial theorem for average values of kinetic and potential energies (Epstein 1974):

$$\langle V \rangle + 2\langle G \rangle = 0. \tag{2.58}$$

Taking into account that the electron energy of a system (see (2.1)) is $E = \langle H_e \rangle = \langle G \rangle + \langle V \rangle$, one can obtain the other (equivalent) relations, namely,

$$E = -\langle G \rangle \qquad E = \tfrac{1}{2}\langle V \rangle. \tag{2.59}$$

The virial theorem in the form of (2.58) shows that the lowering of the energy does not still guarantee the best quality of ED for a calculation. So, one can lower the value of system energy even for those wave-functions which are still far from the accurate one and, hence, do not reproduce the ED. For this purpose it is sufficient to apply a similarity (or scaling) transformation (Epstein 1974) by introducing the parameter λ, which influences the wave-function coordinates and does not violate its normalization:

$$\Psi(\{\boldsymbol{r}\}, \{\boldsymbol{R}\}) = \lambda^{(3/2)s}\Psi(\lambda\{\boldsymbol{r}\}, \lambda\{\boldsymbol{R}\}) \tag{2.60}$$

(s is the total number of electron and nuclei coordinates). The optimization yields the value $\lambda' = -\langle V' \rangle / 2\langle G' \rangle$ that provides the fulfilment of the virial theorem; $\langle V' \rangle$ and $\langle G' \rangle$ correspond to a non-scaled wave-function. The energy of the system is lowered; however, the wave-function remains far from the correct one and has to be optimized with respect to other parameters. At the same time, the wave-function after λ scaling has remained within the class of trial functions.

The scaling can also be done over the ED function:

$$\rho(\boldsymbol{r}, \boldsymbol{R}) = \varkappa^3 \rho(\varkappa \boldsymbol{r}, \boldsymbol{R}). \tag{2.61}$$

Parameters $\varkappa$ can be introduced, for example, into the crystal superpositional model of atomic type (Coppens *et al* 1979), that is widely used for interpreting the x-ray diffraction data (see below). Petkov *et al* (1986) have pointed out that the local scaling transformation can also be used for construction of the ED functionals. Levy and Perdew (1985) have found, however, that in DFT, where the electron energy (1.14) must be explicitly dependent on kinetic, potential, and exchange–correlation terms, such coordinate scaling is non-trivial.

Another important theorem that concerns the ED is the Hellmann–Feynman theorem (Epstein 1974). It states that for the accurate or HF wave-function the relationship

$$\frac{\partial E_e}{\partial \alpha} = \int \Psi^* \frac{\partial \hat{H}_e}{\partial \alpha} \Psi \, \mathrm{d}V \tag{2.62}$$

must be satisfied. The α is some real parameter in the Hamiltonian, $\hat{H}_e$ (2.1). If $\alpha = \boldsymbol{R}_a$ characterized positions of the ath nucleus in a system, then the derivative in expression (2.62) becomes $\partial \hat{V}_{Ne}/\partial \boldsymbol{R}_a$, since in the Hamiltonian $\hat{H}_e$ only the term describing the electron–nuclear attraction depends on $\boldsymbol{R}_a$. It follows from this that the change of the potential energy of an electron cloud with respect to R_a gives rise to the force acting on each nucleus. The αth component of this force is

$$(F_\alpha)_a = -\int \frac{\partial V_{Ne}}{\partial \alpha_a} \rho(\boldsymbol{r})\, \mathrm{d}\boldsymbol{r} \qquad \alpha = x, y, z. \tag{2.63}$$

This force, thus, can be calculated from the ED (by adding the contribution from other nuclei). The forces acting on nuclei at equilibrium must be equal to zero: this provides the criterion of quality for the function ρ and the criterion of accuracy of nuclear coordinate determination. More important is that the Hellmann–Feynman theorem allows one to treat atomic interactons in a system in terms of forces related to the charge distribution (the binding approach). One should note that the electron cloud is considered here in a classical manner, though initially the ED should be calculated by a quantum mechanical method or constructed from x-ray diffraction data. We will consider this approach in more detail below, but now we just note that its capabilities should not be overestimated. First, according to the Ernshow theorem (Jeans 1947), a static system of electric charges cannot be at equilibrium. Second, as we have already noted above, it is impossible to describe the chemical binding as proceeding from the local ED only: one must also taken into account kinetic terms related to the gradient of the wave-function (Teller 1962).

The Hellman–Feynman theorem can also be obtained from the virial theorem, since the mean potential energy of electrons can be written in terms of a virial of forces acting on an electron from the other electrons and nuclei.

The combination of virial and Hellman–Feynman theorems leads to the following interesting result (Wilson 1962, Frost 1962): introducing screened nuclei charges, $Z_a \rightarrow \lambda Z_a$, and using (2.62), one can write for a single atom $\partial \hat{H}_e/\partial \lambda = -\hat{V}_{Ne}$. Then for a many-nuclear system, the total energy can be expressed by the relationship

$$E_{tot} = \langle V_{nn} \rangle - \sum_a Z_a \int_{\lambda_0}^{1} \int \frac{\rho(\boldsymbol{r}, \lambda)}{|\boldsymbol{r} - \boldsymbol{R}_a|}\, \mathrm{d}\boldsymbol{r}\, \mathrm{d}\lambda \tag{2.64}$$

in which the parameter λ varies from value λ_0, below which the system ceases to be bounded, up to $\lambda = 1$. Thus, if the $\rho = \rho(\lambda)$ dependence is known, then the system energy can be, in principle, calculated from formula (2.64).

2.4.2 *Quantum topological theory of atoms in molecules and solids*

Combination of the virial theorem in the local form (2.57) and the Gauss theorem, results in the expression

$$V(\boldsymbol{r})+2G(\boldsymbol{r})=\oint_S \nabla\rho(\boldsymbol{r})\cdot\boldsymbol{n}(\boldsymbol{r})\,\mathrm{d}S(\boldsymbol{r}). \tag{2.65}$$

The right-hand side of this expression vanishes in molecules and crystals not only in the case when surface S covers them completely. Bader *et al* (1981) have found that the postulate that local ED maxima are present at nuclei positions only entails some important consequences. Namely, since vector $\nabla\rho$ shows the direction of greatest increase of ED, the ED areas covering nuclei are separated in many-nuclear systems by the surfaces on which vector $\nabla\rho$ changes its direction (figure 2.14). In other words, the ED gradient vector flux through these surfaces is zero. This is written in a compact form as

$$\nabla_i\rho(\boldsymbol{r}_i,\boldsymbol{R})\cdot\boldsymbol{n}_i(\boldsymbol{r}_i)=0 \qquad \forall\boldsymbol{r}_i\in S_i. \tag{2.66}$$

Here $\boldsymbol{n}_i(\boldsymbol{r}_i)$ is a unit vector normal to surface S_i at point r_i. Surfaces S_i satisfying condition (2.66) are determined uniquely; they are called zero-flux surfaces.

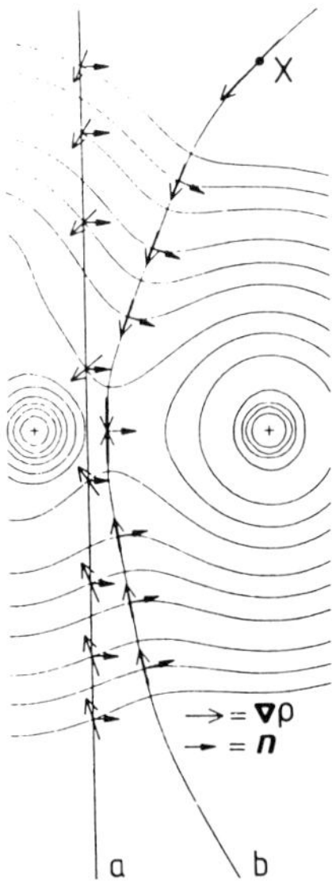

Figure 2.14 The contour map of an ED showing its intersection with two partitioning surfaces. The line of intersection for surface *a* is crossed by the vectors $\nabla\rho$. The vectors $\nabla\rho$ are tangent to the line of intersection of surface *b* at each point on the surface. The two directions on surface *b* meet at a point where $\nabla\rho=0$, a critical point. (Reprinted with permission from Bader 1985. © 1985 American Chemical Society.)

Note that though for terminal atoms these surfaces are, generally speaking, unclosed (going out into infinite), condition (2.66) is met in this case as well, since $\nabla\rho(\infty, \boldsymbol{R})=0$.

Now, if one denotes by Ω_i the volume covered by surface S_i, then from expression (2.65), taking account of (2.66), one can obtain the virial theorem for pseudoatomic fragments

$$V(\Omega_i)+2G(\Omega_i)=0. \tag{2.67}$$

This expression looks formally like the virial theorem for the whole system (2.58). The pseudoatomic virial theorem was postulated by Bader and Beddall (1972) and proved later by Bader (1980) from the stationary action principle. It is important to note that, in spite of the fact that the local kinetic energy of electrons can be determined, in principle, in an infinite number of ways, the integration over volume Ω_i yields the same result regardless of the way the kinetic energy is defined (Cohen 1979). In addition, the Hohenberg Kohn theorem is valid for pseudoatomic fragments (Bader and Becker 1988).

The validity of the virial theorem for pseudoatomic fragments implies that the total energy of a system can be expressed as a sum of atomic energies, each having a physical sense. It was shown by Bader *et al* (1981), for instance, that in LiX molecules (where X = F, Cl, O, H) the energy of the Li pseudoatom is higher than that of a free atom, since the valence electrons of lithium are shifted to a neighbour; at the same time this energy is lower than that of the Li^+ ion.

Thus, the analysis of many-electron many-nuclear systems from the ED topology point of view allows one to separate the ED uniquely into non-overlapping, pseudoatomic subsystems containing a single nucleus and the sum of which fills entirely the whole space. Such a separation is not associated with any separation in the Hamiltonian, as it is in the pseudopotential method, for instance, and does not violate a quantum mechanical requirement of electron indistinguishability. A pseudoatom, determined in accordance with the zero-flux condition (2.66), is an example of an open quantum system: it is a subsystem of a whole many-electron many-nuclear system, with which it freely exchanges charge and momentum, without violating the stationary action principle. It is remarkable that this principle describes in the same manner both the whole system and its fragments. This fact has some important implications. The open system cannot be described by means of a wave-function without speaking about the ED: the density matrix formalism must be used in this case. At the same time, knowing the ED and separating it into pseudoatoms according to (2.66), the expected value of any observed quantity described by a Hermitian operator $\hat{A}$ for the whole system can be presented as a sum of pseudoatomic contributions

$$\langle\hat{A}\rangle=\sum_i \hat{A}(\Omega_i) \tag{2.68}$$

regardless of whether $\hat{A}$ is a one-electron or a many-electron operator. This strong and rather strict result was obtained by Bader with colleagues, who have developed a quantum mechanics for atoms in molecules based on the ED analysis. This result provides a basis for understanding the nature of additivity of molecule and crystal properties, that is well known in experimental chemistry.

The important role of the ED gradient, $\nabla\rho(\boldsymbol{r}, \boldsymbol{R})$, and ED Laplacian, $\nabla^2\rho(\boldsymbol{r}, \boldsymbol{R})$, led Bader (1990) to a thorough analysis of topological properties of the ED. The principal features of Bader's theory are as follows. The topological properties of scalar function ρ are revealed in analysis of the corresponding vector field $\nabla\rho$. The important features of function ρ are described by the number and type of critical points $\boldsymbol{r}_c$ at which $\nabla\rho(\boldsymbol{r}_c)=0$. The type of a critical point is determined by its rank, λ, and signature, σ. Rank λ is equal to the number of non-zero eigenvalues of a curvature matrix, Hessian, with components

$$H_{ij}=(\partial^2\rho/\partial r_i\,\partial r_j)_{\boldsymbol{r}=\boldsymbol{r}_c}. \tag{2.69}$$

Signature σ represents the difference between the numbers of positive and negative eigenvalues of this matrix.

Critical points (λ, σ), for which $\lambda<3$, are degenerate (unstable): at any change of function ρ, for example, caused by nuclear displacement, these points either disappear or break into some number of non-degenerate (stable) critical points. Such points are seldom encountered but play an important part, because they indicate the possibility of structural changes in a system.

The non-degenerate critical points, for which $\lambda=3$, may belong to one of four types: $(3, -3)$, maxima; $(3, -1)$ and $(3, +1)$, saddle points; and $(3, +3)$, minima. Strictly speaking, according to the Kato condition (2.51), the maxima of function ρ corresponding to nuclear positions are not true critical points, since function $\nabla\rho$ discontinues at the nuclear positions. If, however, one smoothes singularities inherent in these maxima by modifying the close vicinity (i.e. by moving to some homeomorphic function), then the cusps can be described by critical points of type $(3, -3)$. Such a smoothing actually takes place in molecules and crystals due to thermal vibrational motion of nuclei.

The universal property of function ρ is a compulsory presence of saddle points of $(3, -1)$ type on lines connecting neighbouring pseudoatoms. The existence of these 'bonding' points is inseparable from the existence of zero-flux surfaces which divide the system into pseudoatomic fragments. The value of ED at these points and its other characteristics can be associated in stable equilibrium systems with the chemical bond character. This matter will be considered in more detail below.

The comprehensive description of the vector field $\nabla\rho$ is provided by gradient paths governed by integral equations. The gradient path passing through a point $\boldsymbol{r}_0$ is given by the equation

$$\boldsymbol{r}(s)=\boldsymbol{r}_0+\int_0^{s} \nabla\rho[\boldsymbol{r}(t), \boldsymbol{R}]\,\mathrm{d}t. \tag{2.70}$$

Such gradient paths always originate in and terminate at critical points, thus reflecting a spatial organization of function ρ (figure 2.15). The beginning α and the end ω of a gradient path are determined by equations $\alpha(\boldsymbol{r}_0)=\lim_{s\to-\infty}\boldsymbol{r}(s)$, $\omega(\boldsymbol{r}_0)=\lim_{s\to+\infty}\boldsymbol{r}(s)$. A critical point of (3, −3) type represents a set of ω-limits of gradient paths. For this reason these points,

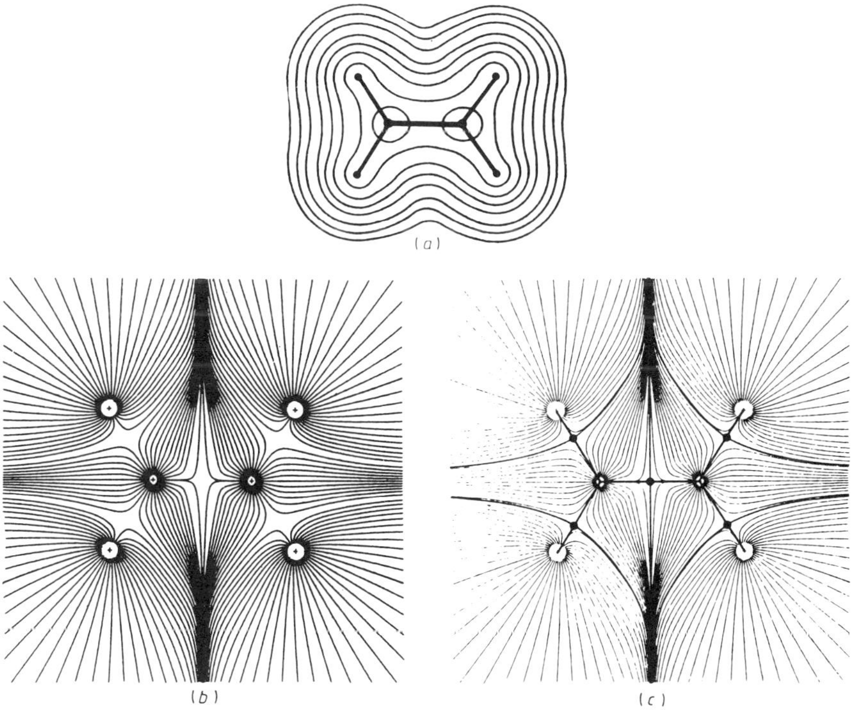

Figure 2.15 ED ρ and topological characteristics of ρ for the ethylene molecule, C_2H_4, in the plane containing the nuclei. (*a*) ED: the contour lines are 2×10^{-3} (outer line), 4×10^{-2}, 8×10^{-1} au etc. (*b*) The gradient paths traced out by the vectors $\nabla\rho$. The paths have been arbitrarily terminated at the boundary of a small circle centred on each nucleus. Only trajectories that terminate at a nucleus are shown. This diagram illustrates the partitioning of space into atomic basins. (*c*) As (*b*) plus the pairs of trajectories which in this plane terminate at each (3, −1) critical point in ρ (denoted by a black dot). Each such pair denotes the intersection of an interatomic surface with this plane. Also shown are the unique pairs of trajectories which originate at (3, −1) critical points and terminate at the neighbouring nuclei. Each pair defines a bond path and collectively they define the molecular graph. (Reprinted with permission from Bader 1985. © 1985 American Chemical Society.)

i.e. the nuclei of atoms, are regarded as the attractions of the vector field $\nabla\rho$. Bader *et al* (1981) have introduced the following definitions. The region inside a closed surface of constant level of ρ values (surface A) is an attractor of field $\nabla\rho$ if: (1) the flux through surface A has constant value; (2) each gradient path, whose α-limit is concluded in the A region, is completely contained in A; (3) there exists, generally speaking, some unclosed neighbourhood B of A such that all gradient paths originating in B terminate in region A. The maximum region B_{max} satisfying these conditions is called the 'basin' of attractor A.

The charge distribution in a molecule or in a crystal is separated into basins, each of which includes only one attractor. This allows one to define a bonded atom (a pseudoatom) as the unity of the attractor and the related basin. An important property of thus-defined atoms is the following one: if two atoms existing in different systems have the same ED distribution within the limits of their basins, then they make equal contributions to the total energy of a system. Atomic contributions to the other properties of the two systems also coincide.

The bonded atoms defined in such a manner are separated by surfaces S_i, each of which contains all gradient paths which terminate at an appropriate critical point of (3, −1) type and satisfy boundary condition (2.66). Thus, the topological definition of an atom coincides with that followed from the stationary action principle modified by Bader.

Of special significance are the pairs of gradient paths issuing from the same saddle point of (3, −1) type and terminated at two neighbouring attractors. Each such pair of gradient paths is determined by its eigenvector, corresponding to a single positive eigenvalue of the Hessian at this point, and forms a line connecting two nuclei (see figure 2.15(*c*)). Along this line, the bond path, function ρ takes values decreasing for any lateral displacements. The saddle point on this line is called the bond critical point. The set of bond paths yields a molecular graph which, thus, has a quite definite physical basis. The graphs constructed in the above manner usually coincide with structural formulae used in a classical chemical molecular structure theory (see e.g. King 1983).

Critical points of types (3, +1) and (3, +3) for an isolated molecule arise only in the presence of cycles (or rings) and fragments called the cages. In the case of a point of (3, +1) type the eigenvectors, associated with two positive Hessian eigenvalues, give rise to an infinite number of gradient paths issuing from this point and forming a ring surface. The third eigenvector yields a pair of gradient paths which are terminated at this point and form a unique axis perpendicular to a cycle surface. This axis is not a straight line, generally speaking; the only requirement is that it should be perpendicular to a cycle surface at the (3, +1) critical point. For example, in a cyclopropane molecule, C_3H_6 (figure 2.16), the point of (3, +1) type is situated at the cycle centre, and the unique axis is a line of intersection of

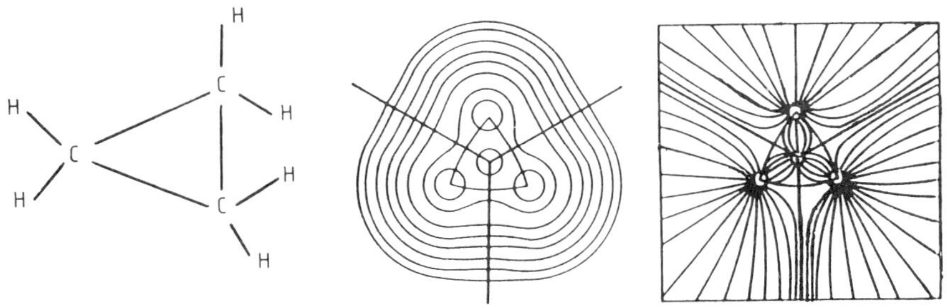

Figure 2.16 Structure (*a*), ED, ρ (*b*), and the gradient vector field $\nabla\rho$ (*c*) for the cyclopropane molecule, C_3H_6, in the plane containing the three carbon nuclei. There is a (3, + 1) critical point at the centre of the bonded ring of carbon atoms. The value $\rho(\boldsymbol{r})$ at this point is 0.182 au, the value of $\rho(\boldsymbol{r})$ at each (3, − 1) critical point is 0.242 au. The boundaries of the three carbon atoms meet at the (3, + 1) critical point (Bader *et al* 1981).

the surface which confines the basins of atoms included in a cycle. It is remarkable that in this molecule the lines of the C–C bond are essentially curved; this implies that the system is stressed.

The eigenvectors associated with a critical point of (3, +3) type, which represents a local minimum of function ρ, generate an infinite number of gradient paths issuing from this point and terminate within a specific region of space. Such a region is confined by two or several cycles situated in an appropriate manner. A typical example is polyhedrane molecules. A significant number of points of (3, +3) type always exists in a crystal as well.

Important information about the character of ED distribution both in bonding and in non-bonding regions is provided by analysing the scalar field $\nabla^2\rho(\boldsymbol{r}, \boldsymbol{R})$ (Bader and Essen 1984). No hypothetical reference state is needed in this case, in contrast to a deformation ED (1.15) when such a state in the form of a set of free non-interacting atoms is used. The Laplacian of the ED, $\nabla^2\rho$, characterizes a curvature of function ρ, i.e. the difference between the ρ value at a given point and its mean value at neighbouring points. In the one-dimensional case one can write

$$\lim_{\Delta x\to 0}\left\{\rho(x)-\frac{1}{2}[\rho(x+\Delta x)+\rho(x-\Delta x)]\right\}$$

$$=-\frac{1}{2}\lim_{\Delta x\to 0}\{[\rho(x+\Delta x)-\rho(x)]-[\rho(x)-\rho(x-\Delta x)]\}$$

$$=-\frac{1}{2}\frac{d^2\rho}{dx^2}(dx)^2. \tag{2.71}$$

For $(d^2\rho/dx^2)<0$ (the negative curvature) the ρ value at point x is larger than the mean ρ value at points $(x+dx)$ and $(x-dx)$ that correspond to concentration of electrons. In the three-dimensional case

$$\nabla^2\rho(r)=\partial^2\rho/\partial z^2+\partial^2\rho/\partial y^2+\partial^2\rho/\partial z^2 \tag{2.72}$$

i.e. the Laplacian of ρ characterizes the ρ curvature in three mutually perpendicular directions. The sign of function $\nabla^2\rho$, that is important for ED analysis, depends on the relation between these three items. Since for a bonded system $E=\langle H_e\rangle<0$, it follows from the virial theorem in a local form (2.62) that in the electron concentration regions where $\nabla^2\rho<0$, the potential energy of electrons is dominant. The region where $\nabla^2\rho>0$, is related to depletion of electrons, though the kinetic energy density $G(\boldsymbol{r})>0$ is not necessarily dominant in this case.

Bader and Beddall (1972) have introduced the local electron energy of a bonded system:

$$H_e(\boldsymbol{r})=G(\boldsymbol{r})+\mathrm{V}(\boldsymbol{r}). \tag{2.73}$$

The sign of $H_e(\boldsymbol{r})$ uniquely shows whether kinetic or potential energy is dominant at the given point of space. If $V(\boldsymbol{r})$ dominates in the internuclear region, $H_e<0$ and the accumulation of electrons is stabilizing here.

Using (2.57) we rewrite expression (2.73) as follows:

$$H_e(\boldsymbol{r})=\tfrac{1}{2}[V(\boldsymbol{r})+\tfrac{1}{4}\nabla^2\rho(\boldsymbol{r})]. \tag{2.74}$$

This expression is remarkable due to the fact that the electron energy density does not depend explicitly on the wave-function derivative that determines the kinetic energy: it is completely defined by the ED, its Laplacian and nuclear (outer) potential. In addition, this expression shows that the condition $\nabla^2\rho>0$ in the internuclear space is still insufficient for the system to become unbounded, since the energy $H_e(\boldsymbol{r})$ in this region can be negative in this case: all depends on the potential energy density value. Thus, the local energy of electrons, $H_e(\boldsymbol{r})$, is an important characteristic of bonding in many-electron and many-nuclear systems.

The topological characteristics are useful in recognition of the electronic shell structure of atoms, which results from the quantum mechanical explanation of the periodical law. These shells are filled by electrons with growing atomic number according to well known rules; one distinguishes K $(n=1)$, L $(n=2)$, M $(n=3)$, N $(n=4)$, etc shells. However, the ED of atoms in the ground state is a monotonically decreasing function of the distance from a nucleus and does not show a shell structure of atoms. To observe electron shells one often uses the one-dimensional radial distribution function

$$D(r)=4\pi r^2\bar{\rho}(r) \tag{2.75}$$

that includes a spherically averaged ED $\bar{\rho}$ of a free atom and possesses minima and maxima. Their number depends on the highest principal quantum number of the atom. The actual existence of these extrema was demonstrated by Bartell and Brockway (1953) by using electrons diffraction. Each of the electronic shells can be determined as the ED described by a set of orbitals with the same principal quantum number; the minima of function $D(r)$ can be treated as boundaries between the shells, and the positions of maxima can be identified with shell's 'radii' (Politzer and Parr 1976). However, Boyd (1977a) has found that the positions of extrema of the $D(r)$ function cannot be distinguished for heavy atoms (see figure 2.17). In particular, beginning with the N shell, the maxima of this function in outer shells of atoms are not revealed, and the number of distinguishable electron shells is one or two less than follows from the analysis of the electron configuration of the atom ground state.

The Laplacian of ED measures the degree of electron concentration, rather than the ED value itself. Bader and Essen (1984) have proposed that the shell structure of atoms results in alternating maxima and minima of function $\nabla^2\rho$, which thus must appear as pairs. The number of pairs must correspond to the number of shells. Shi and Boyd (1988) and Sagar *et al* (1988) have found, however, that this supposition is valid for K and L shells only. Beginning with 3d elements the in-pair appearance of minima and maxima of $\nabla^2\rho$ becomes irregular, and in some atoms the shells (for example the M and N shells in atoms from Sc to Ge and the N and O shells in atoms from Y to Tl) cannot be distinguished using the $\nabla^2\rho$ function (figure 2.18). At the same time, the number of atoms in which the outer shells remain unresolved is lower in the $\nabla^2\rho$ analysis than in the $D(r)$ analysis. In those cases, however,

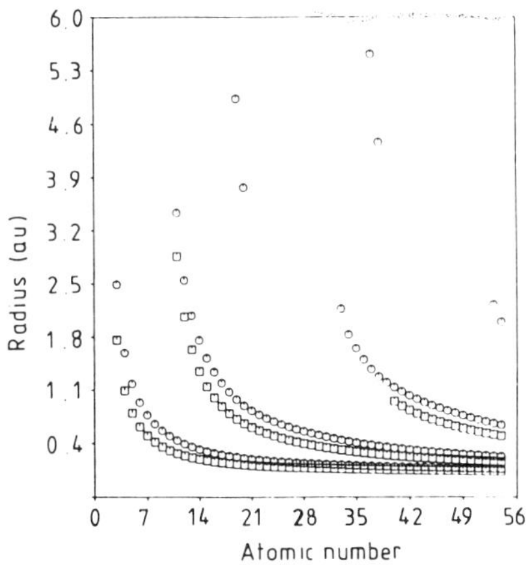

Figure 2.17 Radii (○) of minima in $\nabla^2\rho$ as a function of atomic number. For comparison radii (□) obtained from the partitioning of the radial density, D, are plotted as a function of the atomic number. (Reprinted with permission from Shi and Boyd 1988.)

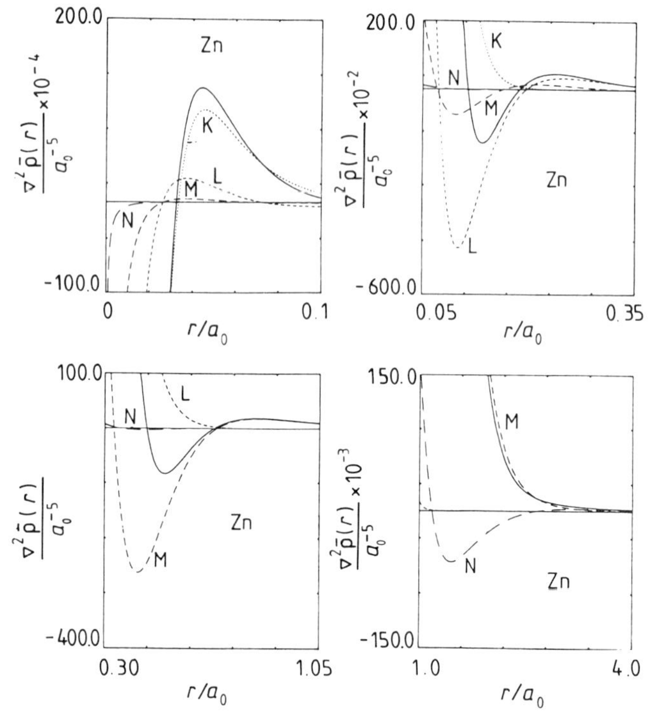

Figure 2.18 Shell (dotted lines) and total (solid line) contributions to the Laplacian of the spherically averaged charge density for the zinc atom ($Z=30$), computed by Sagar *et al* (1988) from the non-relativisitic SCF wave-functions of Clementi and Roetti (1974). Note that in this figure each plot represents different regions of r/a_0, all drawn to different scales. (Reprinted with permission from Sagar *et al* 1988.)

when both methods allow one to reveal the shell structure of atoms, their results are in good agreement with each other.

Kohout *et al* (1991), using the electronic Schrödinger equation modified by Hunter (1986), have tried to study the lack of electronic shell structure in heavy atoms by analysis of the $-|\nabla\rho|/\rho$ value which appears in the equation mentioned. Unfortunately, for the atoms of the fourth and higher rows such an approach is also unable to resolve the valence shells.

In order to better understand the relation between the Laplacian of ED and a shell structure of free atoms let us consider the Laplacian of ED in spherical coordinates:

$$\nabla^2\bar{\rho}(r)=\partial^2\bar{\rho}(r)/\partial r^2+(2/r)\,\partial\bar{\rho}(r)/\partial r. \tag{2.76}$$

In this expression the first term describes the curvature of the function ρ that is related to concentration and depletion of electrons; the sign of this term changes in various regions with increasing r. The second term is always negative, since the function ρ monotonically decreases as r grows. Thus, $\nabla^2\bar{\rho}$ is always positive in the regions of electron depletion, but it can have either positive or negative sign in the regions of electron concentration. Due to this fact the $\nabla^2\rho$ minima do not satisfactorily indicate the positions of maximal concentration of electrons. Sagar *et al* (1988) and Schmider *et al* (1991) have also found that only for each inner shell of first- and second-row atoms does there exist a region of positive or negative values where the behaviour of $\nabla^2\rho$ is determined by the contribution from this shell (see figure 2.17). Unfortunately, for outer shells, beginning with 3d elements, this condition is not obligatory. Moreover, the appearance of the $\nabla^2\rho$ maxima is always a result of domination of positive contributions from individual shells, whereas the $\nabla^2\rho$ minima are formed with different signs of shell contributions. The $\nabla^2\rho$ minima are thus less associated with the shell structure of atoms.

In such a situation the other critical points of the Laplacian of ED turn out to be more informative from the electronic shell study point of view, namely, the points where $\nabla^2\bar{\rho}=0$. As a consequence of conditions (2.51) and (2.53), the $\nabla^2\rho$ function in atoms is negative in the neighbourhood of the point $r=0$ and it changes its sign several times as r is increased. The number of zeros of the $\nabla^2\rho$ depends on the kind of atom. The electrons are accumulated in the region of negative $\nabla^2\rho$ values until this function changes its sign; then the depletion of electrons begins. Thus, the electrons are maximally concentrated at odd zeros, rather than at the $\nabla^2\rho$ minima. Accordingly, the electron charge is most depleted at even zeros of function $\nabla^2\rho$, which can be considered as a topological characteristic of the regions separating atomic electron shells. The odd zeros of the $\nabla^2\rho$ can then be considered as characteristics of shell radii. However, in spite of the fact that corresponding values of r correlate well with orbital radii in the Bohr theory of atoms, the Laplacian of ρ gives, probably, the qualitative pattern of atomic shell structure only.

The analysis of the Laplacian of ED allows one also to reveal some implicit features of ED in valence shells of molecules, which are important for chemical bond and reactivity study. Consider, for instance, ρ and $\nabla^2\rho$ in the H_2O molecule (Bader *et al* 1984). Only one maximum corresponding to the oxygen atom is seen in the ρ distribution in a plane, passing through the HOH angle bisector, perpendicular to the molecular plane (figure 2.19(*a*)). The same section of the $-\nabla^2\rho$ function exhibits two local maxima and two saddle points. They indicate the presence of two regions of high ED concentration, two lone pairs, in the valence shell of the O atom. In the molecular plane (figure 2.19(*b*)) the function ρ has three maxima corresponding to the O and H atoms. Function $-\nabla^2\rho$ again exhibits a more detailed picture: here are revealed the regions of high ED concentration at critical points of $(3,-1)$ type on the O–H bonds and the local maximum representing a saddle point between the maxima of lone pairs of electrons, which are seen in figure 2.19(*a*).

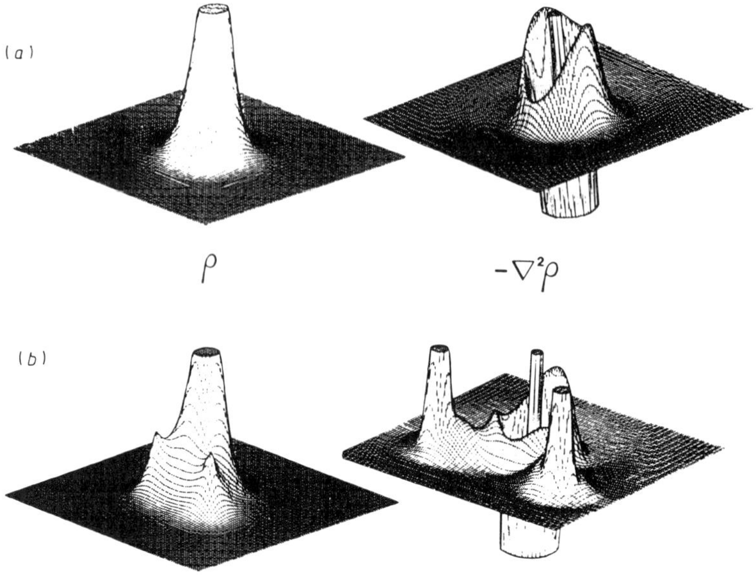

Figure 2.19 The values of ρ and $\nabla^2\rho$ plotted on the third axis for the two symmetry planes of the water molecule. In the non-bonded plane (a), ρ exhibits a single local maximum at the position of the oxygen nucleus. In this plane the valence shell charge concentration exhibits two local minima (two (3, −3) critical points in $\nabla^2\rho$ of charge concentration and two saddle points in $\nabla^2\rho$. In the plane of the nuclei (b), ρ exhibits a local minimum at each of the nuclei, as does $-\nabla^2\rho$. In addition, the valence shell charge concentration of the oxygen atom also exhibits two local maxima, one along each of the bond paths linking the protons to oxygen. What appears as a third maximum in this shell is not a local maximum, but another view of the (3, −1) saddle point between the two non-bonded maxima. The value of $-\nabla^2\rho$ at this saddle point is less than its value at the non-bonded maxima which makes clear the general observation that bonded charge concentrations are smaller than non-bonded concentrations. This illustrates that the uniform sphere of charge concentration of an O atom is transformed into a shell which exhibits four local maxima and four intervening saddle points. Each saddle point possesses at least one axis in the shell along which $-\nabla^2\rho$ is a local minimum. (Reprinted with permission from Bader *et al* 1984. © 1984 American Chemical Society.)

Thus, the $\nabla^2\rho$ function allows us, like a deformation ED (see figure 2.13), to find in the ED of the H_2O molecule four local maxima corresponding to two valence and two lone electron pairs. The first two and the O atom nucleus form an angle of 103.1°, the latter two an angle of 183.3°, that well agrees with Gillespie's (1992) concept of electron pairs. Analogous analysis is applicable for intermolecular and secondary interactions in crystals (Tsirelson *et al* 1995).

Thus, we were convinced the ED is really a fundamental characteristic of many-electron systems. The ED determines all ground state properties and the spatial distribution of local kinetic and potential energies and forces acting in a system. The combination of quantum mechanics and topological analyses allows us to subdivide the many-electron system uniquely into pseudoatomic fragments, all systems properties being an additive sum of properties of these fragments regardless of whether the corresponding operator is one electron or two electron. The topological analysis of ED allows one also to reveal the features related to electron shell structure and to atomic interactions in crystals. All these properties provide a basis for developing the chemical bond theory in terms of ED.

2.5 Electric field characteristics of molecules and crystals

2.5.1 *The internal electric field*

The analysis of charge distribution in molecules and crystals results inevitably in the necessity of considering the electric field related to this charge. The electric field is governed by the vector potential $\boldsymbol{A}(\boldsymbol{r}, t)$ and the scalar potential $\varphi(\boldsymbol{r}, t)$ (Jeans 1947). In a system with motionless nuclei the ED function ρ can be considered to be time independent, and the resulting field, produced by the total charge density

$$\sigma(\boldsymbol{r}) = \sum_a Z_a \, \delta(\boldsymbol{r} - \boldsymbol{R}_a) - \rho(\boldsymbol{r}) \tag{2.77}$$

is electrostatic. The strength of this field, $\boldsymbol{E}$, is related to the scalar potential by the relationship

$$\boldsymbol{E}(\boldsymbol{r}) = -\nabla\varphi(\boldsymbol{r}). \tag{2.78}$$

φ is called the electrostatic potential (EP) in this case. The EP represents the energy of Coulomb interaction of a positive unit charge, situated at point r, with a non-disturbed system of charges. Using one of Maxwell's equations, namely, $\varepsilon_0 \nabla \boldsymbol{E}(\boldsymbol{r}) = \sigma(\boldsymbol{r})$ (the dielectric susceptibility ε is assumed to be unity), one easily obtains Poisson's equation

$$\nabla^2\varphi(\boldsymbol{r}) = (1/\varepsilon_0)\sigma(\boldsymbol{r}). \tag{2.79}$$

Both vector potential $\boldsymbol{A}$ and scalar potential φ are determined non-uniquely. The usual choice of φ potential form corresponds to the vector potential for which $\nabla \boldsymbol{A} = 0$; this case is called Coulomb gauge. There exist also other gauges, which are suitable in electrodynamics, for example (Flygare 1978).

Since the choice of φ is non-unique, one states sometimes that only the electric field strength $\boldsymbol{E}$ and its derivatives have physical meaning. Nevertheless, one can eliminate the non-uniqueness in a potential by counting it, for example, from a zero value at infinity or by taking into account the electrical neutrality of a crystal unit cell and zeroing its dipole and quadrupole moments (Avery *et al* 1984, Becker and Coppens 1990). Then the EP can be assumed to be a quantity experimentally measured by means of diffraction of electrons with energies of 20–50 keV, if one neglects the exchange of electrons between a beam and a target and the target charge polarization (Vainstein 1960, Bonham and Fink 1974). The relationship between ED and electronic EP due to the Poisson equation allows one also to measure this part of the total potential by means of x-ray diffraction (Stewart 1979, Varnek *et al* 1981).

The EP is the major part of a potential in the one-particle effective Hamiltonian in the HF (2.6) and Kohn–Sham (2.50) methods. It plays a key part in the Thomas–Fermi (2.35) method. It also determines the features of molecular crystal packing (Berkovitch-Yellin and Leiserowitz 1980), enters various semiempirical expressions for many-electron systems' energy (Politzer 1980) and allows one to describe the early stages of chemical reactions, when the ED polarization, exchange, and charge transfer are small (Scrocco and Tomasi 1978, Politzer and Truhlar 1981). In addition, the value of potential on nuclei can be related to the chemical shift of energy of K electrons in electron spectroscopy (Schwarz 1970) and to the screening constant in nuclear magnetic resonance (Basch 1970). Therefore, EP can be considered, like ED, to be one of the most important characteristics of atoms, molecules, and crystals.

The insertion of the total charge density of a many-electron, many-nuclear system (2.77) into the Poisson equation yields the expressions for the nuclear

$$\varphi_N(\boldsymbol{r})=\sum_a \frac{Z_a}{4\pi\varepsilon_0|\boldsymbol{r}-\boldsymbol{R}_a|} \tag{2.80}$$

and electronic

$$\varphi_1(\boldsymbol{r})=-\int\left\{\frac{\rho(\boldsymbol{r}')}{4\pi\varepsilon_0|\boldsymbol{r}-\boldsymbol{r}'|}\right\}\mathrm{d}\boldsymbol{r} \tag{2.81}$$

components of total electrostatic potential $\varphi=\varphi_N+\varphi_e$. In contrast to earlier convention (see (1.10)) we have taken into account the negative sign of ED explicitly. The $\varphi_N(\boldsymbol{r})$ component is the local nuclear potential, $v(\boldsymbol{r})$, that, according to the Hohenberg–Kohn theorem, is uniquely related to ED and appears in the energy functional expression (1.14). The function φ_N is positive throughout space for atoms, molecules, and crystals and has maxima at nuclei positions. Tal *et al* (1980) have considered the gradient vector field $\nabla\varphi_N$, compared it with the vector field $\nabla\rho$ and concluded that, though ρ and φ_N do not possess, generally speaking, the same critical points, these scalar functions are homeomorphic, i.e. they can be mapped into each other by means of a continuous transformation that has an inverse transformation. Thus, the unique relation between ED and outer local potential, that was postulated by Hohenberg and Kohn, manifests itself also in topological properties of these functions. Gadre and Bendale (1986) have found that in atoms and molecules there exists also some approximate similarity between the ED, the total EP φ, and its nuclear component φ_N. This circumstance can apparently be used for developing various semiempirical models; for example, it is used in the Thomas–Fermi theory—see (2.35), One should, however, emphasize an essential difference between ED and nuclear EP (Politzer and Zillis 1984). If one writes the difference functions

$$\delta\varphi_N = \varphi_N^{mol} - \sum_{at} \varphi_N^{at} \tag{2.82}$$

then the identity $\delta\varphi(\boldsymbol{r}) \equiv 0$ is valid, for all $\boldsymbol{r}$ values, whereas the deformation ED $\delta\rho(\boldsymbol{r})$ (1.15) takes, for various $\boldsymbol{r}$, positive, zero, or negative values. Thus, the nuclear EP does not reflect any chemical bond effects.

An important rôle in developing methods for calculating many-electron system energy belongs to the EP on nuclei $\varphi(0)$. As follows from the Hellman–Feynman theorem, the potential on the nucleus of a single atom, $\varphi_a(0)$, is related to the average energy of electron–nuclear interaction $\langle V_{Ne} \rangle$ as follows:

$$\varphi_a(0) = -\int \frac{\rho(\boldsymbol{r})}{|\boldsymbol{r}|} \mathrm{d}\boldsymbol{r} = -\frac{1}{Z_a} \langle \hat{V}_{Ne} \rangle. \tag{2.83}$$

Politzer and Parr (1974) have generalized this relationship for the case of many-nuclear systems and obtained an equation that relates their total energy to the electrostatic potential, provided that the number of electrons is constant:

$$E_{tot} = \frac{1}{2} \sum_a \left\{ Z_a \varphi_a(0) - \int_a^{Z_a} \left[Z'_a \left(\frac{\partial \varphi_a(0)}{\partial Z'} \right) - \varphi_a(0) \right]_N \mathrm{d}Z'_a \right\}. \tag{2.84}$$

A remarkable feature of this expression is that it does not include any cross-terms and represents a system's energy as the sum of energies of 'pseudoatoms'. Politzer (1976), using the Thomas–Fermi theory, in which for atoms the relationship $\varphi(0) \sim -Z^{4/3}$ is valid, has shown the energy of each pseudoatom that appears as a term in expression (2.84) is equal to

$$E_{tot}^{at} \cong \tfrac{3}{7} Z \varphi(0) = \tfrac{3}{7} \langle \hat{V}_{Ne} \rangle \tag{2.85}$$

Accordingly, the energy functional for a many-nuclear system has the form

$$E_{tot} \cong \frac{3}{7} \sum_a Z_a \varphi_a(0). \tag{2.86}$$

This relationship has been thoroughly studied in various versions by Politzer (1980, 1987), who has shown that if the electron part of quantity $\varphi_a(0)$ is calculated at HF level, then the accuracy of determination of a molecule's energy using (2.86) is about 1%. Since the potential is an integral characteristic of a charge distribution, then such an estimation is valid when even moderately limited basis sets are used in calculating $\varphi_e(0)$.

The Hellman–Feynman theorem allows one to relate the change of an EP, that occurs in system formation or destruction, with the change of energy (Srebrenik *et al* 1974). Indeed, the force acting on a nucleus shifted from the equilibrium configuration from the side of other nuclei and electrons of the system is proportional to the gradient of the total EP, φ, created by a system on this nucleus:

$$F_a(\boldsymbol{R}) = -Z_a[\nabla'\varphi(\{\boldsymbol{R}\}, \boldsymbol{R}'_a)]_{\boldsymbol{R}'_a=\boldsymbol{R}_a} \tag{2.87}$$

(the gradient operator acts on primed coordinates only). The corresponding change in molecular energy can be defined as the work done in transferring nuclei from configuration $\{\boldsymbol{R}\}_1$ to configuration $\{\boldsymbol{R}\}_2$ in the electrostatic field. The latter, in turn, depends on the nuclear configuration, therefore

$$\Delta E = -\sum_a Z_a \int_{\boldsymbol{R}_{a_1}}^{\boldsymbol{R}_{a_2}} [\nabla'\varphi(\{\boldsymbol{R}\}, \boldsymbol{R}'_a)]_{\boldsymbol{R}'_a=\boldsymbol{R}_a}. \tag{2.88}$$

This relation reveals the role of the electrostatic interactions in the chemical bond formation.

Using formulae of classical electrostatics (Jeans 1947), one can obtain from the potential φ the relations for the electrostatic field strength $\boldsymbol{E}(\boldsymbol{r})$, for field gradient tensor components $\nabla \boldsymbol{E}_{ij} = \partial^2\varphi/\partial x_i\,\partial x_j$, and for other characteristics of an inner electric field at each point in the system. Some of these quantities can be plotted graphically. Besides, tensor components $\nabla \boldsymbol{E}_{ij}$ are associated with characteristics of spectra in nuclear quadrupolar resonance and in the gamma resonance spectroscopy methods (Flygare 1978). Thus, the development of approaches based on the electrostatic potential distribution in a system may prove to be useful for interpreting spectroscopic results.

2.5.2 *Electrostatic interaction*

The energy of the electrostatic interaction of two non-penetrating systems with total charge densities $\sigma_A(\boldsymbol{r})$ and $\sigma_B(\boldsymbol{r})$, creating potentials $\varphi_A(\boldsymbol{r})$ and $\varphi_B(\boldsymbol{r})$, can be written as follows:

$$E_{es} = \frac{1}{2}\int \varphi_A(\boldsymbol{r})\sigma_B(\boldsymbol{r})\,\mathrm{d}\boldsymbol{r} = \frac{1}{2}\int \varphi_B(\boldsymbol{r})\sigma_A(\boldsymbol{r})\,\mathrm{d}\boldsymbol{r}. \tag{2.89}$$

Energy E_{es} is only the first term of the expression for total interaction energy W in the Rayleigh–Schrödinger perturbation theory, where the intermolecular or interionic interaction potential is a perturbation (see e.g. Kaplan 1985). Therefore, E_{es} describes the interaction of hypothetical 'rigid' systems, i.e. the interaction that takes place without any change of the system's electron densities. The total interaction energy W is a sum of the terms taking into account the mutual polarization of ED (E_{pol}), charge transfer (E_{ct}), and the exchange (E_{ex}) and dispersion (E_{disp}) effects (Morokuma 1971):

$$W = E_{es} + E_{pol} + E_{ct} + E_{ex} + E_{disp}. \tag{2.90}$$

Such an energy separation into contributions best corresponds to quantum mechanical analysis of the interaction energy. The E_{es} value usually is 90–95% of the total energy value and in some cases it is sufficient to analyse the EP distribution in order to understand the main features of the interaction. This is the case, for example, when the interaction of a positive ion with a molecule or when the interaction of molecules at large distances is considered. This has stimulated a wide application of the EP for studying the reactivity of molecules (see e.g. Politzer and Truhlar 1981, Politzer and Daiker 1981). Let us consider the interaction of positive ions of alkali metals with crown ethers. The latter are organic molecules whose structure is characterized by the presence of a non-planar macrocyclic fragment (a 'crown') formed by alternating O and C atoms. The different sorts of crown ether bind different ions which are localized inside the cavity of a macrocycle. The study of selective properties of crown ethers is an important part of extraction chemistry; however, the selectivity mechanisms have been only scalarly studied at the microscopic level so far. Meanwhile, some information about this mechanism can be obtained in the EP language. As was found by Glebov (1990), in the dibenzo-18-crown-6, $C_{20}H_{28}O_6$, molecule there exists a negative-potential region inside the cavity, which has the shape of a two-sided 'funnel' (figure 2.20(*a*)). Zero-potential lines separate, in space, the regions of hydrophilic and hydrophobic parts of the molecule. A positive ion falling into the 'funnel' region moves perpendicular to the equipotential surfaces towards the cavity centre where the potential is minimal. The introduction of electron-acceptor NO_2 substituents into benzene rings results in the ED outflow from a macrocyclic ring at the molecule periphery. As a result, the EP in a cavity becomes lower in magnitude, and the 'funnel' is closed by a zero-potential line (figure 2.20(*b*)). This suggests a tendency to increase the activation barrier value in the interaction between dinitrodibenzo-18-crown-6, $C_{20}H_{26}N_6C_{10}$, molecules and a positive complex-producing ion as compared to the initial $C_{20}H_{28}O_6$ molecule, which is just what is observed in extraction chemistry experiments. Such a consideration gives, of course, only qualitative, but simple and clear, pictures of the interaction.

The EP is an integral characteristic of charge density and for a qualitative analysis it is often sufficient to perform a non-empirical potential calculation in a minimal basis set or even a semiempirical calculation. Special numerical experiments (Varnek 1985, Luque *et al* 1990) have shown that the general topographic features of the EP of molecules are reproduced in any basis set and using any method. On the other hand, when the numerical values of

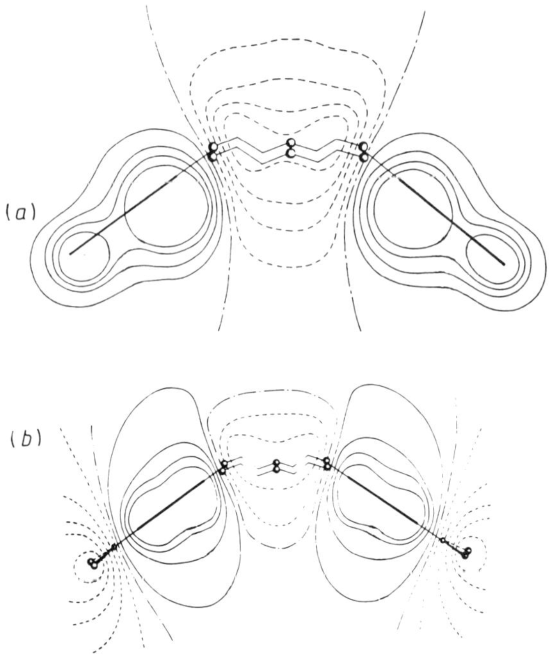

Figure 2.20 EP distribution in crown ether molecules $C_{20}H_{28}C_6$ (*a*) and $C_{20}H_{26}N_2O_{10}$ (*b*). The section plane is perpendicular to the macrocycle fragment. The negative potential values are dashed (Glebov 1990).

extrema are needed, Luque *et al* (1990) recommend using the (DZ+DP) basis set.

The topological analysis of the EP allows one to find in a molecule or crystal the neutral fragments or the fragments possessing some specific charge (Srebrenik *et al* 1975). The neutral fragments are inside the volume Ω_i for which the integral $\int_{\Omega_i} \nabla^2\varphi(\boldsymbol{r})\,\mathrm{d}\boldsymbol{r} = \int_S \nabla\varphi(\boldsymbol{r})\cdot\boldsymbol{n}(\boldsymbol{r})\,\mathrm{d}S(\boldsymbol{r})$ is zero: due to the Poisson equation (2.79) this is equivalent to $\int_{\Omega_i} \sigma(\boldsymbol{r})\,\mathrm{d}\boldsymbol{r} = 0$. The establishment of a Ω_i value for atom *i* allows one to find how the atomic volume varies when the atom enters a bonded state. In the same manner one can determine the volume occupied by a molecule in a crystal. For charged fragments (ions) the integral $\int_{\Omega_i} \nabla^2\varphi(\boldsymbol{r})\,\mathrm{d}\boldsymbol{r}$ must be equal to the fragment's charge (Gadze and Shrivastava 1991).

By analogy with the deformation ED (1.15), one can introduce the deformation electronic EP (Vainstein 1960, Avery 1979, Stewart 1979):

$$\delta\varphi(\boldsymbol{r}) = \varphi(\boldsymbol{r}) - \tilde{\varphi}(\boldsymbol{r}). \tag{2.91}$$

Here $\tilde{\varphi} = \sum_{at} \varphi_{at}$ is the potential of a set of spherical atoms (promolecule or procrystal). The $\delta\varphi$ distribution describes variations in the electronic part of the EP caused by atoms entering into a chemical bond. The potentials of atoms can easily be calculated from atomic wave-function parameters tabulated in the literature. The most complete calculations of such a kind, performed by Varnek (1985) and Sommer-Larsen *et al* (1990), are presented and discussed in appendix D. Potential $\delta\varphi$ takes both positive and negative values, does not depend on the choice of origin, and depends on the deformation ED only due to compensation of the nuclear component. If $\delta\rho$ is known accurately enough, then by combining the relations (2.84)–(2.86), the energy of a homonuclear crystals' bond can be expressed in terms of $\delta\varphi$. Bentley (1979) and Varnek *et al* (1981) have found, however, that the numerical value of a bond energy is obtained in this case with large error due to excessive simplification of the corresponding energy functionals.

Avery *et al* (1981) have indicated that the deformation EP concept can simplify the solution of the Schrödinger equation. Indeed, the main part of the potential corresponding to a procrystal is a sum of easily tabulated atomic contributions. The other part, $\delta\varphi$, can be obtained, for instance, from the electron or x-ray diffraction experiment.

The electric field $\delta\boldsymbol{E}$, related according to (2.78) to EP $\delta\varphi$, allows one to analyse the electrostatic aspect of a chemical bond. Hirshfeld and Rzotkiewitz (1974) have considered the Hellman–Feynman electric field acting on nucleus A in diatomic molecules of AH and A_2 type (here A denotes atoms from Li to F). This field can be represented as a sum

$$\boldsymbol{E}_A = \boldsymbol{E}_{p,A} + \delta\boldsymbol{E}_A \tag{2.92}$$

in which $\boldsymbol{E}_{p,A}$ is the field on nucleus A in a procrystal and $\delta\boldsymbol{E}_A$ is the field on nucleus A produced by the migration of electrons caused by chemical bonding. The migration field, $\delta\boldsymbol{E}_A$, is determined by the deformation ED. As seen from figure 2.21, in diatomic molecules the field $E_{p,A} = |\boldsymbol{E}_{p,A}| = q_B(R_{AB})/R_{AB}^2$ ($q_B = \int \sigma_B(\boldsymbol{r})\,\mathrm{d}\boldsymbol{r} = \int \rho_B(\boldsymbol{r})\,\mathrm{d}\boldsymbol{r}$) depends on the internuclear distance R_{AB} and has a repulsive character. This field is known as the penetrating

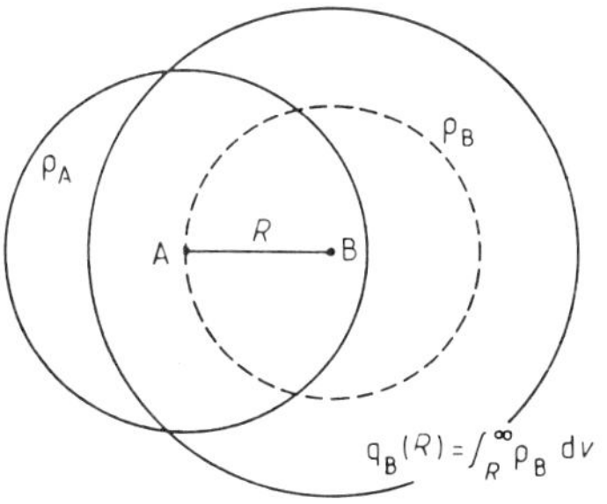

Figure 2.21 The Hellman–Feynman electric field in diatomic promolecule AB. The field at nucleus A is repulsive since it arises from the penetration of the A nucleus inside the electronic cloud of its neighbour (Hirshfeld and Rzotkiewicz 1974).

field since nucleus A seems to be penetrating inside atom B and the nucleus of the latter turns out to be incompletely screened by its own electron shell.

According to the Hellman–Feynman theorem, for an equilibrium molecule $\boldsymbol{E}_A = 0$, and the migration field

$$\delta \boldsymbol{E}_A = -\boldsymbol{E}_{p,A} \tag{2.93}$$

is attractive. Thus, the ED redistribution takes place during chemical bonding in such a manner that the stability of a nuclear equilibrium configuration is provided. The corresponding electric field and the force $F_A = Z_A E_A$ acting on nucleus A is characterized by field δE_A or, which is the same, by field $E_{p,A}$. Therefore, quantity $E_{p,A}$ must provide an approximate estimation of the force of the bond between the atoms. Indeed, in diatomics there exists a fine correlation between the hypothetical energy $E_{p,A}$ and the experimentally measured dissociation energy (figure 2.22). In many-nuclear molecules this correlation will not be seen so clearly but the general approach can be useful in this case as well.

The electric field gradient $\nabla \boldsymbol{E}$ can also be represented as a sum of the contributions from a procrystal and a deformation part:

$$\nabla \boldsymbol{E} = \nabla \boldsymbol{E}_{(procryst)} + \nabla\, \delta \boldsymbol{E}. \tag{2.94}$$

Schwarzenbach and Thong (1979) and Thong and Schwarzenbach (1979) have shown that such a decomposion is suitable for calculating the term $\nabla\, \delta \boldsymbol{E}$ from x-ray diffraction data. We shall consider this approach in detail in chapter 7.

The division of the potential into the parts associated with a procrystal and ED redistribution due to the chemical bond leads to the possibility of energy decomposition as well. The crystal energy E_{cryst} can be represented as a sum

$$E_{cryst} = \sum_{at} E_{at} + E_{(procryst)es} + \Delta E. \tag{2.95}$$

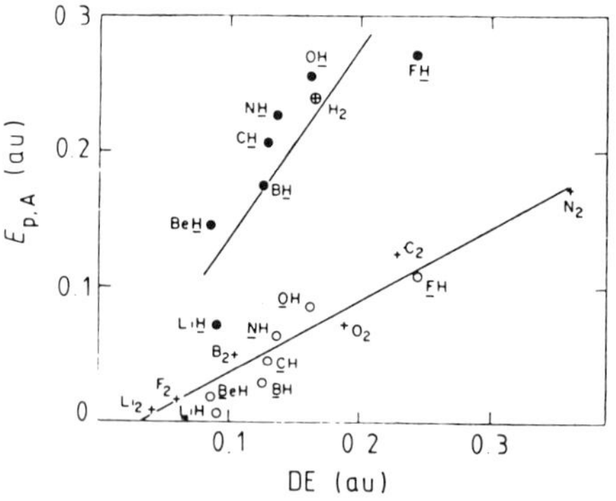

Figure 2.22 Penetration electric field, $E_{p,A}$, versus dissociation energies in A_2 (+) and AH (●) (○, $E_{p,A}$ and ⊕, $-E_{p,H}$) molecules (Hirshfeld and Rzotkiewicz 1974).

The first term on the right-hand side is a sum of the ground state atomic energies E_{at}, the second one is the energy of electrostatic interactions of atoms, and the third term, ΔE, includes the non-classical part of atomic interactions and electrostatic effects related to the deformation ED. Such an energy decomposition is most useful when the ED, rather than a wave-function, underlies the analysis to be performed.

If one considers the energy $E_{(procryst)es}$ from this point of view, then it is clear that its value is composed of a sum of pair energies of electrostatic interactions of atoms, which are easily calculated. Hirshfeld and Rzotkiewicz (1974) have shown that the energy of electrostatic interaction of a pair of spherical atoms with ED distribution is negative (with the exception of H atoms in the H_2 molecule). Thus, a procrystal cannot be described by an antisymmetric wave-function and it does not obey the virial theorem; nevertheless, the forces acting on nuclei in a procrystal are binding. Hence, the procrystal can be applied as an energy model precursor of a stable many-electron, many-atom system formation. Spackman and Maslen (1986)

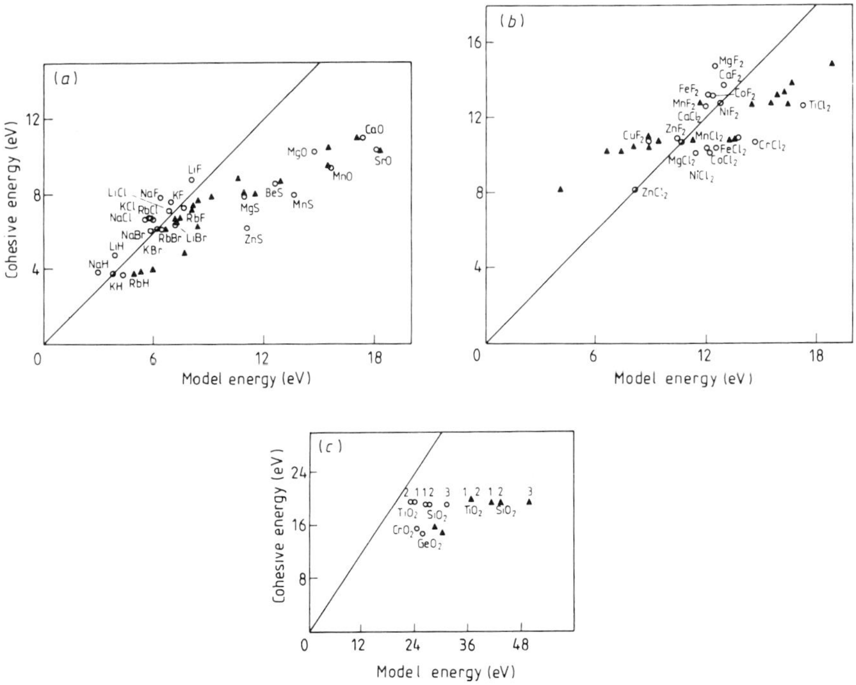

Figure 2.23 The cohesive energy versus model Madelung (triangles) and electrostatic procrystal (circles), E_{es}, energies for RX-type compounds (*a*), difluorides and dichlorides (*b*), and dioxides (*c*) (SiO_2: 1, quartz; 2, cristobalite; 3, tridimite; TiO_2: 1, rutile; 2, anatase) (Trefry *et al* 1987).

have found in diatomic molecules and Trefry *et al* (1987) in crystals with ionic chemical bond character that $E_{(procryst)es}$ well correlates with the binding energy E_{bind} and cohesion energy E_{coh}, respectively. The case of crystals is of special interest. Usually, for evaluating the Coulomb part of cohesion energy one applies the Madelung energy (Ashcroft and Mermin 1976): the electrostatic energy of an ensemble of ions supposed to be point charges. The estimation of the value of those charges in complex compounds with a partially covalent character of the chemical bond causes a major difficulty in calculating the Madelung energy. Trefry *et al* (1987) have found that in crystals of alkali halide compounds, dichlorides, difluorides, and some silicates the $E_{(procryst)es}$ energy, calculated for neutral atoms, will provide a better estimation of cohesion energy than the Madelung energy. This fact is illustrated by figure 2.23. Thus, the procrystal model, which partially reproduces the bonding, represents a good initial point for studying the energy properties of crystals. Besides, it is grounded on clear physical approximations and is applicable to a greater number of compounds than the Madelung model is.

3

X-RAY DIFFRACTION EXPERIMENT

There is another approach to the determination of the ED and the electric field characteristics of crystals. This approach is based on an indirect measurement of EC by using the diffraction of electromagnetic radiation with wavelength $\lambda \leqslant 10^{-10}$ m. More correctly, the method deals with the ED reconstruction from accurate diffraction data obtained with automatic diffractometers. The information about the reflection phase is usually lost in such measurements. Besides, the measured intensities also contain, along with a coherent component, contributions from various subsidiary diffraction effects. Therefore, the measurement results are supplemented and corrected by means of special procedures which need some additional theoretical or experimental information.

X-ray radiation is mostly used in experiments of this kind. Therefore, our considerations in this chapter will be based on this radiation source. The methods which use electrons, γ-rays, and synchrotron radiation are described in chapter 4.

3.1 Physical principles of the electron density reconstruction

The reconstruction of the ED and characteristics of the thermal motion of nuclei in crystals from the x-ray scattering data presents the specific case of an inverse problem of diffraction theory. Before proceeding with this problem, we shall consider the direct problem of x-ray diffraction in a single crystal. We shall follow here Feil (1977a), Stewart (1977a), Born and Huang (1954), Reisland (1973), Tsirelson (1989, 1993), and Tsirelson *et al* (1992), paying special attention to the approximations made in the diffraction description and the related uncertainty in the results.

3.1.1 *X-ray scattering by mean-electron systems with fixed nuclei*

Let a small single crystal ($r \sim 0.1$ mm) be 'bathed' in a strictly monochromatic x-ray plane wave with wave-vector $\boldsymbol{k}_0$ ($\omega = c|\boldsymbol{k}_0| \sim 10^{19}\ \mathrm{s}^{-1}$). The x-ray quanta $\hbar\omega \sim 10^4$ eV greatly exceeds the binding energy of electrons to nuclei (~ 10 eV). The radiation reflected from a specimen is recorded at different angles by a quantum counter (figure 3.1). An individual scattering act of duration $\omega^{-1} = 10^{-19}$ s can be considered as absorption of a photon from an incident beam and emission of a photon in the direction $\boldsymbol{k}_1$. To describe

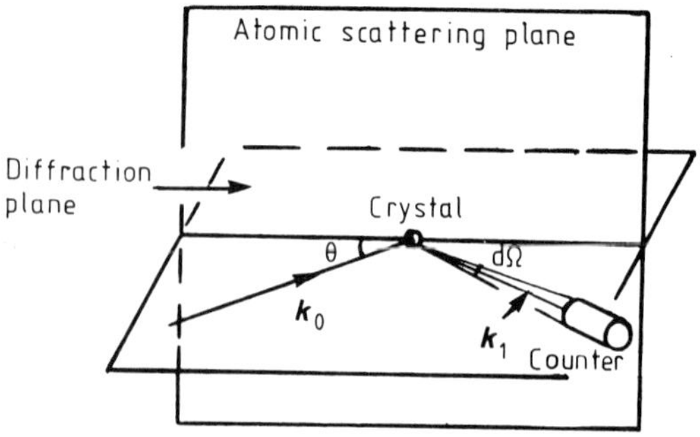

Figure 3.1 The arrangement of the direction of the incident, $\boldsymbol{k}_0$, and diffracted, $\boldsymbol{k}_1$, x-ray beams. The diffraction and scattering planes are also shown.

the scattering process the probability of this event should be related to the scattered radiation intensity measured. The x-ray scattering cross-section on electrons is about 10^7 times larger than that on nuclei; therefore, we shall consider the photon–electron interaction only.

Consider first the scattering on a one-electron motionless atom. When the x-ray photon interacts with an electron of an atom, the energy and momentum conservation laws should be obeyed. Before interaction the Hamiltonian of the 'atom + radiation' system is a sum of the atomic Hamiltonian, $\hat{H}_{at}$, and the radiation Hamiltonian, $\hat{H}_{rad}$. The wave-functions of stationary states of a system before interaction, Ψ_n^0, are the products of non-perturbed wave-functions of components: the electron wave-functions $\varphi_n^0(\boldsymbol{r})\exp(-\mathrm{i}\omega_n t)$ and the radiation state vectors $|\mathrm{Rad}(n_{\boldsymbol{k}_l})\rangle = |n_{\boldsymbol{k}_0}, 0_{\boldsymbol{k}_1}\ \ldots\rangle$ $n_{\boldsymbol{k}_0}$ is the number of x-ray photons moving along the $\boldsymbol{k}_0$ direction). An additional term $\hat{H}'$ in the total Hamiltonian, $\hat{H} = \hat{H}_{at} + \hat{H}_{rad} + \hat{H}'$, is responsible for interaction. The system's wave-function, $\Psi(t)$, which obeys the non-stationary Schrödinger equation, is represented after interaction as an expansion over the functions which are the products of atomic and radiation field wave-functions:

$$\Psi(t) = \sum_n \sum_{\boldsymbol{k}_l} c_n(\boldsymbol{k}_1, t)|\varphi_n^0(\boldsymbol{r})\exp(-\mathrm{i}\omega_n t)\rangle|\mathrm{Rad}(n_{\boldsymbol{k}_l})\rangle. \tag{3.1}$$

The terms with wave-vectors different from $\boldsymbol{k}_0$ appear in the radiation state vector $|\mathrm{Rad}(n_{\boldsymbol{k}_l})\rangle$ after interaction, that is, scattering, takes place.

The squares of the expansion coefficients, c_n, characterize the photon scattering probability in certain directions. These coefficients can be found from non-stationary perturbation theory (Messiah 1966), which yields for them the expression known as the Born perturbation series:

$$c_n(\boldsymbol{k}_l, t) = -\frac{\mathrm{i}}{\hbar}\sum_m \sum_{\boldsymbol{k}_q} \int_0^t \mathrm{d}t' \exp(\mathrm{i}\omega_{nm}t')H'_{nm}(\boldsymbol{k}_l, \boldsymbol{k}_q, t')c_m(\boldsymbol{k}_q, t'). \tag{3.2}$$

Here $H'_{nm}(\boldsymbol{k}_l, \boldsymbol{k}_q, t') = \langle n_j \boldsymbol{k}_l|\hat{H}'(\boldsymbol{r}, t)|m_j \boldsymbol{k}_q\rangle$ is the matrix element of the perturbation Hamiltonian, $\hat{H}'$. Each term of the series describes the system's transition from some state m into state n with a simultaneous emission (or absorption) of one or several x-ray photons, the initial and final states may be the same. The scattering in a certain chosen direction may be carried out in one step (onefold scattering), in two steps, etc. Since the interaction between radiation and atom is weak, the onefold scattering is most probable, and it is sufficient to consider this type of scattering only (the first Born approximation). Then for a one-electron atom, which was in the ground state, φ_0^0, before scattering and transferred into state φ_n^0 under the action of radiation during time t, one has for scattering in the $\boldsymbol{k}_1$ direction:

$$c_n(\boldsymbol{k}_0, \rightarrow \boldsymbol{k}_1, t) \cong -\frac{\mathrm{i}}{\hbar}\int_0^t \mathrm{d}t' \exp(\mathrm{i}\omega_{n0}t')H'_{n0}(\boldsymbol{k}_0, \boldsymbol{k}_1, t'). \tag{3.3}$$

The frequency $\omega_{n0}=(E_n-E_0)/\hbar$ characterizes the electron energy change under the action of radiation.

The explicit form of the non-relativistic interaction Hamiltonian is

$$H'(\boldsymbol{r}, t) = -(e/m)\hat{A}(\boldsymbol{r}, t)\cdot\hat{p} + (e^2/2m)\hat{A}^2(\boldsymbol{r}, t). \tag{3.4}$$

Here e, m, and $\hat{p}$ are the charge, mass, and electron momentum respectively; $\hat{A}(\boldsymbol{r}, t)$ is the vector potential of the radiation field. Since during the field–atom interaction photon absorption or emission takes place as well as the change of atomic electronic states, the $\hat{A}(\boldsymbol{r}, t)$ potential should be considered as an operator acting on the corresponding state vectors (see appendix E). Considering the radiation field in some volume Ω_0 comprising a crystal, the $\hat{A}(\boldsymbol{r}, t)$ potential is conveniently written in the form of the Fourier expansion over plane waves corresponding to x-ray photons with certain wave-vectors (Loudon 1973):

$$\hat{A}(\boldsymbol{r}, t) = \left(\frac{\hbar}{2\varepsilon_0\Omega_0}\right)^{1/2} \sum_{\boldsymbol{k}_l,\mu} \omega_l^{-1/2}\{\hat{a}(\boldsymbol{k}_l, \mu)\exp[\mathrm{i}(\boldsymbol{k}_l\cdot\boldsymbol{r}-\omega_1 t)]$$
$$+\hat{a}^+(\boldsymbol{k}_l, \mu)\exp[-\mathrm{i}(\boldsymbol{k}_l\cdot\boldsymbol{r}-\omega_l t)]\}\boldsymbol{e}(\boldsymbol{k}_l, \mu). \tag{3.5}$$

Here $\boldsymbol{e}(\boldsymbol{k}_l, \mu)$ is the photon polarization vector; $\hat{a}$ and $\hat{a}^+$ are the operators of photon annihilation and creation in the radiation field.

The scattering is associated with the second term of equation (3.4). Substituting (3.5) into (3.4) and using the results from appendix E, one can calculate the matrix element $\hat{H}'_{n0}$ which appears in (3.3). Let us assume initially that only one photon is scattered, i.e. only one plane wave is present in expression (3.5). Then

$$H'_{n0}(\boldsymbol{k}_0 \rightarrow \boldsymbol{k}_1, t) = [2\pi\hbar c^2 r_0/\Omega_0(\omega_0\omega_1)^{1/2}]f_{n0}(\boldsymbol{S})\exp[-\mathrm{i}(\omega_0-\omega_1)t]$$
$$\times[\boldsymbol{e}(\boldsymbol{k}_0, \mu)\cdot\boldsymbol{e}(\boldsymbol{k}_1, \nu)]. \tag{3.6}$$

Here $r_0=e^2/4\pi\varepsilon_0 mc^2$ is a classical electron radius, $\boldsymbol{S}=\boldsymbol{k}_1-\boldsymbol{k}_0$ ($|\boldsymbol{S}|=4\pi\sin\theta/\lambda$).

Scattering amplitude f_{n0} is given by

$$f_{n0}(\boldsymbol{S}) = \int \varphi_n^0(\boldsymbol{r}) \exp(-\mathrm{i}\boldsymbol{S}\cdot\boldsymbol{r})\varphi_0^0(\boldsymbol{r})\,\mathrm{d}\boldsymbol{r}. \tag{3.7}$$

Now expression (3.3) can be written in the form

$$c_n(\boldsymbol{S}, t) = -\mathrm{i}\frac{2\pi r_0 c^2}{\Omega_0(\omega_0\omega_1)^{1/2}} f_{n0}(\boldsymbol{S})[\boldsymbol{e}(\boldsymbol{k}_0, \mu)\cdot\boldsymbol{e}(\boldsymbol{k}_1, \nu)] \times \int_0^t \exp\{\mathrm{i}[\omega_{n0} - (\omega_0 - \omega_1)]t'\}\,\mathrm{d}t'. \tag{3.8}$$

The total probability of photon scattering in the $\boldsymbol{k}_1$ direction during time t is proportional to $\sum c_n^2(\boldsymbol{S}, t)$, provided the interference between the waves is absent and the relativistic Breit–Dirac correction, which describes the small x-ray photon frequency shift in scattering (James 1967), is neglected. This is the Debye–Waller scattering case. One of the terms in this sum, c_0, corresponds to the situation when $\omega_{n0} = 0$ and $\omega_0 = \omega_1$ (in this case the kinetic energy of the electron after interaction remains lower than the electron binding energy). This term includes the quantity

$$f_{00}^2(\boldsymbol{S}) = \left\{\int [\varphi_0^0(\boldsymbol{r})]^2 \exp(-\mathrm{i}\boldsymbol{S}\cdot\boldsymbol{r})\,\mathrm{d}\boldsymbol{r}\right\}^2 \tag{3.9}$$

which describes the coherent x-ray scattering on the ground state atomic electron density, $[\varphi_0^0(\boldsymbol{r})]^2$. The remaining coefficients c_n ($n \neq 0$) describe the incoherent scattering associated with electron transitions into excited states.

In order to generalize this result for the scattering of any number of photons in a many-electron system, operator $\hat{H}'$ should be replaced by $\Sigma_i \hat{H}_i'$ and operator $\hat{A}$ by $\Sigma_i \hat{A}(\boldsymbol{r}_i, t)$ (summing over electrons). Then all plane waves in (3.5) should be taken into account. Each of these waves corresponds to the interaction of one photon with one electron. A sum of exponents appears in expression (3.7) and the many-electron wave-function $\Psi_0(\boldsymbol{x}_1, \ldots, \boldsymbol{x}_N; \boldsymbol{R}_0)$ appears instead of one-electron wave-function φ^0. Using the electron local density operator $e\,\Sigma_i\,\delta(\boldsymbol{r} - \boldsymbol{r}_i)$, one can write (omitting subscripts at a scattering amplitude)

$$\begin{aligned} f(\boldsymbol{S}) &= \sum_i^N \int_{r_1} \cdots \int_{s_1} \cdots \Psi_0^*(\boldsymbol{x}_1, \ldots, \boldsymbol{x}_N; \boldsymbol{R}_0)\delta(\boldsymbol{r} - \boldsymbol{r}_i)\exp(-\mathrm{i}\boldsymbol{S}\cdot\boldsymbol{r}_i) \\ &\quad \times \Psi_0(\boldsymbol{x}_1, \ldots, \boldsymbol{x}_N; \boldsymbol{R}_0)\,\mathrm{d}\boldsymbol{x}_1 \cdots \mathrm{d}\boldsymbol{x}_N \\ &= N\int \cdots \int \cdots \Psi_0^*(\boldsymbol{x}_1, \ldots, \boldsymbol{x}_N; \boldsymbol{R}_0)\delta(\boldsymbol{r} - \boldsymbol{r}_i)\exp(-\mathrm{i}\boldsymbol{S}\cdot\boldsymbol{r}_i) \\ &\quad \times \Psi_0(\boldsymbol{x}_1, \ldots, \boldsymbol{x}_N; \boldsymbol{R}_0)\,\mathrm{d}\boldsymbol{x}_1 \cdots \mathrm{d}\boldsymbol{x}_N. \end{aligned} \tag{3.10}$$

Comparing (3.10) with the expression for the ground state ED (1.10), one can see that

$$f(\boldsymbol{S}) = \int \rho(\boldsymbol{r}, \boldsymbol{R}_0) \exp(-\mathrm{i}\boldsymbol{S}\cdot\boldsymbol{r})\, \mathrm{d}\boldsymbol{r}. \tag{3.11}$$

Taking into account the crystal periodicity, the integration in the latter expression is carried out over the unit cell volume V, and then the result is summed over the unit cells. The radiation reflected from a crystal differs from zero only in the directions defined by the Bragg diffraction condition, $\boldsymbol{S} = 2\pi\boldsymbol{H}$ ($\boldsymbol{H}$ is the reciprocal lattice vector—see appendix F). Function $\rho(\boldsymbol{r}, \boldsymbol{R}_0)$ in expression (3.11) is in this case the static ED of a unit cell which corresponds to the equilibrium nuclear configuration, $\boldsymbol{R}_0$. Its Fourier transform is called the static structure amplitude of coherent scattering and is denoted as $F(\boldsymbol{H}, \boldsymbol{R}_0)$. We will ignore the crystal periodicity for a time and the notation $F(\boldsymbol{S}, \boldsymbol{R}_0)$ will be used.

Thus, the coherent structure amplitude and the ED of scattering system are mutually related by the Fourier transformation. Hence, the separation of a coherent part from the total x-ray scattering is necessary for ED analysis.

We have so far considered an extremely idealized diffraction situation. In real diffractometric experiments, however, the incident radiation is not polarized and not strictly monochromatic: the x-ray line half width is 1–5 eV. The counter, which has finite size, records all photons with any polarization state scattered within some solid angle $\mathrm{d}\Omega$ and with wave-vectors close to $\boldsymbol{k}_1$. The wave-vector interval is small, and the summation over $\boldsymbol{k}$ can be replaced by integration:

$$\sum_{\boldsymbol{k}} \to \frac{\Omega_0}{(2\pi)^3} \int k^2\, \mathrm{d}k\, \mathrm{d}\Omega = \frac{\Omega_0}{(2\pi c)^3} \int \omega^2\, \mathrm{d}\omega\, \mathrm{d}\Omega.$$

Then the total number of photons coherently scattered during 1 s into a unit solid angle $\mathrm{d}\Omega = 1/B_0^2$ (B_0 is the distance from a crystal to a counter) in the direction close to $\boldsymbol{k}_1$ is determined by the photon scattering probability. It is given by

$$N = \frac{c r_0^2}{\Omega_0 B_0^2} |F(\boldsymbol{S}, \boldsymbol{R}_0)|^2 \int \frac{\omega^2}{\omega_0 \omega_1} \delta(\omega_0 - \omega_1)\, \mathrm{d}\omega \sum_{\mu,\nu} [\boldsymbol{e}(\boldsymbol{k}_0, \mu)\cdot\boldsymbol{e}(\boldsymbol{k}_1, \nu)]^2$$

$$= \frac{c r_0^2 p}{\Omega_0 B_0^2} |F(\boldsymbol{S}, \boldsymbol{R}_0)|^2. \tag{3.12}$$

The polarization factor

$$p = \sum_{\mu,\nu} [\boldsymbol{e}(\boldsymbol{k}_0, \mu)\cdot\boldsymbol{e}(\boldsymbol{k}_1, \nu)]^2 = (1 + \cos^2 2\theta)/2 \tag{3.13}$$

appears as a result of summation over all possible polarization states (2θ is the angle between incident and scattered beams). The Dirac delta function, which appears in (3.12), expresses the energy conservation law during coherent scattering. It arises from the integral of the exponent in (3.8) after extension of the limit of integration to $t=\infty$, when the measurement time is large enough (Loudon 1973).

The incident radiation intensity, I_{inc}, is the radiation energy flux through a unit area per second, $I_{inc}=\hbar\omega c/\Omega_0$. The number of photons transferred through unit area per second is $I_{inc}/\hbar\omega=c/\Omega_0$. The fraction of scattered photons is $N\Omega_0/c$, therefore the intensity of coherently scattered radiation is

$$\mathfrak{I}_0(\boldsymbol{S})=I_{inc}N\Omega_0/c=I_{inc}(r_0^2p/R_0^2)|B(\boldsymbol{S}, \boldsymbol{R}_0)|^2. \tag{3.14}$$

Thus, we have related the coherently scattered radiation intensity with the Fourier transform of the ED. The expression (3.14) will slightly underestimate the intensity value actually measured. First, the measurement result includes additionally the incoherent scattering contribution. This is practically constant within the measurement interval of $\sim 1^0$ (Pohler and Hanson 1965) and can be included into the background and eliminated at the experimental data processing stage. Second, the relativistic effects were not taken into account in (3.14). According to Grelland (1985), the Fourier transformation of the total atomic scattering amplitude in the Pauli approximation has the form $\{\rho(r)+(5N/6m^2c^2)r^{-2}\,\mathrm{d}[(r^2\,\mathrm{d}/\mathrm{d}r)\gamma(\boldsymbol{r}, \boldsymbol{r}')]/\mathrm{d}r|_{r=r'}\}$ and along with ρ contains the part depending on the spinless one-electron density matrix $\gamma(\boldsymbol{r}, \boldsymbol{r}')$. One should note that the relativity affects ED mainly near nuclei and is prominent for heavy atoms only. This effect is taken into account in the structural crystal model by using the relativistic atomic scattering functions at the time of the experimental data processing.

Let us return now to expression (3.4). This expression contains, along with a scattering-related term, a term linear in vector potential $\hat{A}(\boldsymbol{r}, t)$ that includes also the electron momentum operator $\hat{p}$. This term describes the effects accompanying changes in the radiation state vector (photons are absorbed or created) and in the electronic wave-function (the electronic states are changed). Depending on whether the final electronic state at absorption relates to a discrete or continuous energy spectrum, these effects result in atomic ionization or excitation (Heitler 1954). The ionization is accompanied by an internal photoeffect that leads to the so-called true absorption. The excitation results in incoherent processes of photon re-emisssion: a phosphorescence (the multiplicity of the electronic state changes; the lifetime τ is $\sim 10^{-4}$ s) or a fluorescence (the multiplicity does not change; τ is $\sim 10^{-8}$ s). There also exist some other secondary effects (James 1967). All these phenomena decrease the probability of a scattered photon's appearance in the $\boldsymbol{k}_1$ direction. The true absorption is taken into account by introducing a multiplicative correction to the intensities measured (see section 3.3.1). The remaining effects are accompanied by the emission of a photon with a continuous wave-vector spectrum. These effects influence the background and are corrected during the experimental data processing.

There are some cases which require inclusion in consideration of the second term of the Born series of perturbation theory (3.2). The corresponding coefficient of this series has the form

$$c_n(\boldsymbol{k}_0, \rightarrow \boldsymbol{k}_1, t) = -\left(\frac{\mathrm{i}}{\hbar}\right)\int_0^t \mathrm{d}t' \exp(\mathrm{i}\omega_{n0}t')H'_{n0}(\boldsymbol{k}_0, \boldsymbol{k}_1, t')$$
$$+\left(\frac{\mathrm{i}}{\hbar}\right)^2 \int_0^t \mathrm{d}t' \int \mathrm{d}t'' \sum_m \exp(\mathrm{i}\omega_{nm}t')H'_{nm}(\boldsymbol{k}_0, \boldsymbol{k}_1, t)$$
$$\times \exp(\mathrm{i}\omega_{m0}t'')H'_{m0}(\boldsymbol{k}_0, \boldsymbol{k}_1, t''). \qquad (3.15)$$

The second-order matrix elements, which contain the square of a radiation field vector potential, describe the phenomena associated with a simultaneous participation of four quanta in the interaction. The probability of these phenomena is usually small. However, the matrix elements which include the term linear in vector potential from the interaction Hamiltonian (3.4) at radiation frequencies close to that of resonance electron frequencies (but not equal to it) make a contribution which is comparable in magnitude with the term describing the scattering in the first order of the perturbation theory. If the initial and final states are both ground states ($n=0$) and the intermediate state is excited ($m \neq 0$), then the second term in (3.15) describes the resonance fluorescence (Heitler 1954), This phenomenon is known in crystallography as anomalous scattering. Its physical nature is as follows. The electrons of atoms placed in a radiation field can be considered to be virtual oscillators. The excited electron energy levels have a finite width and, hence, a finite lifetime, τ, during which the electrons, excited by the radiation field, emit the absorbed photons. However, the energy of an excited level (or the amplitude of the wave package emitted by an oscillating electron) exponentially attenuates, and the energy conservation law is no longer satisfied. Therefore, in the exponent in the integrand of (3.8) the additional real term $\varepsilon(\tau)t'$ appears, appropriately transforming the δ-function in expression (3.12). As a result, the factor $[(\omega_{n0}-\omega)^2+\varepsilon^2(\tau)]^{-1}$ appears in the expression for a photon emission probability and the factor $[\omega_{n0}-\omega-\mathrm{i}\varepsilon(\tau)]^{-1}$ in the scattering amplitude. The latter becomes a complex quantity. Thus, the anomalous scattering changes the phase of oscillations and it should be taken into account in the experimental data processing.

Thus, in the case of small (or corrected) absorption, insignificant relativistic effects, and a weak wave interference in a crystal specimen, the intensity of coherently scattered x-ray radiation can be very simply connected with a spatial ED distribution. The above-listed description composes a physical basis for a kinematic diffraction theory that is widely used in x-ray and neutron crystallography (Cowley 1975). A specific feature of this theory is the fact that when it is used for diffraction description, the information on reflection phases is completely lost. This is a consequence of the uncertainty relation, which includes in the given case a pair of non-commuting variables 'the number of photons–the wave phase' (Loudon 1973).

3.1.2 *X-ray scattering by many-electron systems with vibrating nuclei*

Now we shall take one more step towards approaching a diffraction description closer to reality: we shall taken into consideration the thermal motion of atoms in a crystal, which is anharmonic in the general case.

A crystal is an ordered equilibrium dynamical system of electrons and nuclei. The measured intensity can be considered as an average over all nuclear configurations, each of which is characterized by a multidimensional vector $\boldsymbol{R}_a$. Describing the motion of nuclei by a total set of eigenfunctions of the nuclear Hamiltonian, $\chi_a \equiv |a\rangle \equiv \chi_{0k}$ (see (1.5)), one can write this intensity in electronic units as follows:

$$I(\boldsymbol{S}) = (\mathfrak{J}_0(\boldsymbol{S})/I_{inc})\frac{B_0^2}{r_0^2 p} = \sum_a W_a \int \chi_a^*(\boldsymbol{R}_a)|F(\boldsymbol{S}, \boldsymbol{R}_a)|^2 \chi_a(\boldsymbol{R}_a)\, \mathrm{d}\boldsymbol{R}_a$$

$$= \sum_a W_a |\langle a|F(\boldsymbol{S}, \boldsymbol{R}_a)|a\rangle|^2 + \sum_a W_a \sum_{b \neq a} |\langle a|F(\boldsymbol{S}, \boldsymbol{R}_a)|b\rangle|^2. \tag{3.16}$$

The first term in this expression describes the elastic scattering on (dynamic) ED, the second, the non-elastic thermal diffuse scattering. The Boltzmann weight factors $W_a = \mathrm{e}^{-E_a\beta}/\sum_n \mathrm{e}^{-E_n\beta}$ ($\beta = 1/kT$) describe the probability of a system staying at some vibrational state a with energy $E_a \equiv E_{0k}$. The $\chi_a(\boldsymbol{R}_a)$ and E_a are solutions of the Schrödinger equation for nuclear motion (1.5), which depends on potential energy operator $\hat{\Phi}(\boldsymbol{R}) \equiv \hat{\Phi}_m(\boldsymbol{R})$. We are interested in the consideration of the inverse problem of the diffraction, therefore, the solution of this Schrödinger equation is superfluous. Consider the potential energy expansion over powers of operator $\hat{u}$ describing the small nuclear displacement from the equilibrium positions (see (1.7)):

$$\hat{\Phi} = \hat{\Phi}_0 + \hat{\Phi}_1 + \hat{\Phi}_2 + \hat{\Phi}_3 + \hat{\Phi}_4 + \cdots = \Phi(\cdots \boldsymbol{R}_{n\mu} \cdots) + \sum_{n\mu i} \Phi_i^{\mu}{}^{n} \hat{u}_{n\mu i}$$

$$+ \frac{1}{2!} \sum_{\substack{n\mu i \\ m\nu j}} \Phi_{ijk}^{\mu\nu\varkappa}{}^{nmp} \hat{u}_{n\mu i}\hat{u}_{m\nu j} + \frac{1}{3!} \sum_{\substack{n\mu i \\ m\nu j \\ p\varkappa k}} \Phi_{ijk}^{\mu\nu\varkappa}{}^{nmp} \hat{u}_{n\mu i}\hat{u}_{m\nu j}\hat{u}_{p\varkappa k}$$

$$+ \frac{1}{4!} \sum_{\substack{n\mu i \\ m\nu j \\ p\varkappa k \\ r\lambda l}} \Phi_{ijkl}^{\mu\nu\varkappa}{}^{nmp} \hat{u}_{n\mu i}\, \hat{u}_{m\nu j}\hat{u}_{p\varkappa k}\hat{u}_{r\lambda l} + \cdots. \tag{3.17}$$

Here $\boldsymbol{R}_{n\mu} = \boldsymbol{R}_n + \boldsymbol{r}_\mu$ is the equilibrium position of the μth atom in the nth unit cell; $u_{n\mu i}$ is the component of the displacement operator of this atom from the equilibrium position. Quantities $\Phi_{ij\ldots}^{\mu\nu\ldots}{}^{nm\ldots}$ are coupling parameters of the kth order (Born and Huang 1954): these are tensors of kth rank, whose components are the kth derivatives of potential energy with respect to displacements. $\overset{n}{\Phi}{}_i^{\mu}$ is the ith component of force acting on the μth atom in the nth unit cell, when the remaining atoms occupy their equilibrium positions; $\overset{nm}{\Phi}{}_{ij}^{\mu\nu}\hat{u}_{m\nu j}$ is the ith component of force acting on the μth atom in the nth unit cell at a small displacement $u_{m\nu j}$ of atom ν in the mth cell in the j direction, etc.

The constant term $\hat{\Phi}_0$ is the crystal energy at the equilibrium thermodynamic state, the static crystal energy. This term can be zeroed by the appropriate choice of reference of potential energy. The linear term $\hat{\Phi}_1=0$ due to the expansion of $\hat{\Phi}$ in a series is carried out about the equilibrium atomic positions.

In the harmonic approximation the terms of order higher than the second are neglected in the expansion (3.17). In this case the expression for the potential energy operator $\hat{\Phi}_2$ can be essentially simplified by eliminating the cross-terms. For this purpose one transforms to normal coordinates—complex quantities vibrating independently of each other according to the harmonic law (see e.g. Willis and Pryor 1975). The corresponding independent vibrations are normal vibrations or modes, and the quanta of their energies are phonons. The polarization states of the modes characterize the type of atomic displacement transferring the energy. In normal coordinates the displacements of atoms in the potential field can be represented in the form

$$\hat{u}_{n\mu}=\left(\frac{\hbar}{2sNm_{\mu}}\right)^{1/2}\sum_{\alpha}(\omega_{\alpha})^{-1/2}\boldsymbol{e}_{\alpha}^{\mu}\exp(\mathrm{i}\boldsymbol{q}_{\alpha}\cdot\boldsymbol{R}_{n\mu})\hat{B}_{\alpha}. \tag{3.18}$$

Here m_{μ} is the mass of the μth atom in a unit cell, $\boldsymbol{e}_{\alpha}^{\mu}$ is the unit vector of polarization of vibrational mode α with frequency $\omega_{\alpha}(\boldsymbol{q}_{\alpha})$ and wave-vector $\boldsymbol{q}_{\alpha}$ for an atom of type μ ($\alpha=1, 2,\ldots 3sN$, $3s$ is the number of frequency spectrum branches, N is the lattice symmetry-allowed values of $\boldsymbol{q}_{\alpha}$ vector in the first Brillouin zone). Operator $\hat{B}_{\alpha}$ is determined in terms of operators of creation and annihilation of phonons (Reisland 1973), the action of which on a nuclear weave-function transfers a crystal into the other vibrational state. The expression for the crystal's potential energy operator in the harmonic approximation takes the following form with using $\hat{B}_{\alpha}$ operators:

$$\hat{\Phi}=\hat{\Phi}_2=-\frac{1}{2!}\frac{\hbar^2}{2}\sum_{\alpha}\omega_{\alpha}\hat{B}_{\alpha}\hat{B}_{\alpha}. \tag{3.19}$$

The total energy operator $\hat{H}_{N,0}=\hat{T}_N+\hat{\Phi}_2$ also has a simple diagonal form in the harmonic approximation (see appendices B and E). As a result, the vibrational energy of a crystal is presented as a sum of energies of non-interacting harmonic oscillators corresponding to the normal vibrations (see expression (B.4) in appendix B) and the variables in the harmonic nuclear wave-function χ_a^{harm} are separated: it can be written now as a product of wave-functions of harmonic oscillators: $\chi_{a_\alpha}(\xi_\alpha)=e^{-1/2\xi_\alpha^2}H_{a_\alpha}(\xi_\alpha)$. The $H_{a_\alpha}(\xi_\alpha)$ are normalized Hermite polynomials of the a_αth order, ξ_α is the αth normal coordinate.

If the interrelated harmonic atom vibrations of a crystal are described by a set of normal modes, the equilibrium positions of atoms are determined by the potential energy minima, the mode frequencies are independent of temperature, and the thermal expansion in a crystal is not taken into account. Obviously, this approximation is not valid and the anharmonicity should be taken into account. In this case the Hamiltonian of a system of vibration is represented as

$$\hat{H}_N=\hat{H}_{N,0}+\hat{H}'_N \qquad \hat{H}_{N,0}=\hat{T}_N+\Phi_2 \qquad \hat{H}'_N=\hat{\Phi}_3+\Phi_4+\cdots. \tag{3.20}$$

This does not allow the exact diagonalization of the Hamiltonian; therefore, the corresponding wave-function cannot be found accurately, and the anharmonic crystal energy cannot be presented as a sum of energies of normal vibrations. In other words, the normal mode language is, strictly speaking, inapplicable here. Fortunately, the anharmonic wave-functions, χ_a, and weight factors, W_a (3.16), can be found by approximate methods. The anharmonic effects are weaker than the harmonic ones, and to take these effects into account it is convenient to apply the time-independent perturbation theory (Hahn and Ludwig 1961, Krivoglaz and Techonova 1961, Maradudin and Flinn 1963, Kashiwase 1965, 1973, Cowley 1963, Maiz 1980, Shukla and Hübschle 1989). This approach allows one to express the perturbative Hamiltonian $\hat{H}'_N$ in (3.20) in terms of displacement vectors written in normal coordinates (3.18). The corresponding $\hat{H}'_N$ components will depend on cross-terms which take into account the interaction between phonons. In such an approximation one has

$$\begin{aligned}\hat{\Phi}_3&=\frac{1}{3!}\frac{1}{\sqrt{sN}}\left(\frac{\hbar}{2}\right)^{3/2}\sum_{\alpha\beta\gamma}\frac{\Phi_{\alpha\beta\gamma}}{(\omega_\alpha\omega_\beta\omega_\gamma)^{1/2}}\hat{B}_\alpha\hat{B}_\beta\hat{B}_\gamma\\ \hat{\Phi}_4&=\frac{1}{4!}\frac{1}{\sqrt{sN}}\left(\frac{\hbar}{2}\right)^{2}\sum_{\alpha\beta\gamma\delta}\frac{\Phi_{\alpha\beta\gamma\delta}}{(\omega_\alpha\omega_\beta\omega_\gamma\omega_\delta)^{1/2}}\hat{B}_\alpha\hat{B}_\beta\hat{B}_\gamma\hat{B}_\delta.\end{aligned} \tag{3.21}$$

Functions $\Phi_{\alpha\beta\gamma}$ and $\Phi_{\alpha\beta\gamma\delta}$, anharmonic parameters, are Fourier-transformed coupling parameters of the third and fourth order, respectively, from (3.17).

In order to calculate the thermal-averaged x-ray scattering intensity (3.16) in a form which allows one to study the characteristics of the ED and thermal motion in a crystal, it is necessary to specify the electron dynamic structural model of a crystal. Following Born and Huang (1954), the crystal can be considered as an equilibrium thermodynamic ensemble of atoms which occupy all possible vibrational states during the measurement time. Each atom, which vibrates in the force field produced by its environment, is a part of a crystal and should be described by means of a density matrix (see appendix B). Therefore, the approximations for crystal ED and for potentials in which the nuclei are vibrating should be made and the character of the relationship between them should be defined. For this reason a superpositional crystal structural model is usually used, in which the static ED in a crystal unit cell is separated into the sum of pseudoatoms:

$$\rho(\boldsymbol{r}, \{\boldsymbol{u}_\mu\}) = \sum_\mu \rho_\mu(\boldsymbol{r} - \boldsymbol{r}_\mu - \boldsymbol{u}_\mu). \tag{3.22}$$

Further, following Born (1942–1943), one supposes that each pseudoatom 'rigidly' follows the motion of 'its own' nucleus in the adiabatic potential, which is anharmonic in general. This means that each atom is assigned, in this case, its own probability density function, PDF $\tilde{p}_\mu(\boldsymbol{u}_\mu)$, which describes the distribution of atomic positions over displacements from an equilibrium position. The Fourier transformation of this PDF

$$T(\boldsymbol{S}) = (2\pi)^3 \int \tilde{p}_\mu(\boldsymbol{u}_\mu) \exp(\mathrm{i}\boldsymbol{S}\cdot\boldsymbol{u}_\mu)\, \mathrm{d}\boldsymbol{u}_\mu \tag{3.23}$$

is the atomic temperature factor. By means of the atomic PDF, the average (over nuclear displacements or, as it results from the ergodic hypothesis, over time) dynamic ED is represented as

$$\rho(\boldsymbol{r}) = \sum_{n\mu} \int \rho_\mu(\boldsymbol{r} - \boldsymbol{R}_{n\mu} - \boldsymbol{u}_\mu)\tilde{p}_\mu(\boldsymbol{u}_\mu)\, \mathrm{d}\boldsymbol{u}_\mu. \tag{3.24}$$

Now, using the convolution theorem (Cowley 1975) and expressions (3.11) and (3.23), one can easily obtain the structure amplitude in the form

$$F(\boldsymbol{S}) = \sum_\mu f_\mu(\boldsymbol{S}) T_\mu(\boldsymbol{S}) \exp(\mathrm{i}\boldsymbol{S}\cdot\boldsymbol{r}_\mu). \tag{3.25}$$

In this expression f_μ is, generally speaking, a complex scattering amplitude for the ED pseudoatom associated with a vibrating nucleus μ. Epstein and Stewart (1979) have shown that for diatomic molecules, containing elements with $Z \leqslant 9$, the correction to scattered x-ray intensity due to the 'non-rigidity' of the pseudoatomic electron densities does not exceed 0.13%. For heavier elements this correction will be smaller. Thus, at the modern accuracy of the x-ray diffraction method, the convolution approximation used is valid.

One should emphasize that the approximation of the superpositional crystal structural model, accepted for describing the structure amplitude, is not equivalent to the approximation of independently vibrating atoms. The $\tilde{p}_\mu(\boldsymbol{u}_\mu)$ functions are marginal (see appendix H) with respect to many-particle PDF $P\{\boldsymbol{u}_\mu\}$) that takes into account the correlations in the thermal motion of all atoms (Scheringer 1977) and corresponds to the crystal potential energy

$\Phi_{\mu\mu}$ (3.17). The marginality of $\tilde{p}_\mu(\boldsymbol{u}_\mu)$ functions implies that the above-mentioned correlations in the thermal motion of atoms are taken into account by a set of these functions on average only. In the general case, however, $P(\{\boldsymbol{u}_\mu\}) \neq \prod_\mu \tilde{p}_\mu(\boldsymbol{u}_\mu)$ and $\Phi(\{\boldsymbol{u}_\mu\}) \neq \sum_\mu \Phi_\mu(\boldsymbol{u})$. The additive scheme for Φ_μ will be valid in the harmonic approximation only, due to the properties of Hermite polynomials appearing in the expression for wave-functions of harmonic oscillators. Nevertheless, the additivity can be conserved in the anharmonic approximation as well if the real potential (3.17) is replaced by a sum of effective atomic potentials $\sum_\mu V_\mu(\boldsymbol{u})$ and the marginal atomic PDFs are replaced by effective ones: $\tilde{p}_\mu(\boldsymbol{u}) \to p_\mu(\boldsymbol{u})$. It is important (Willis 1969) that the dependences of functions $\tilde{p}_\mu$ and p_μ on scattering vector $\boldsymbol{S}$ and on temperature turn out to be the same (some possible exceptions were discussed by Scheringer (1977)), and that the above replacements do not have any effect on the determination of parameters of these functions from the x-ray diffraction experiment.

Now, retaining the perturbation theory scheme, one replaces the PDF $\tilde{p}_\mu$ in (3.23) by p_μ and, considering the expansion parameters (3.17) as effective parameters one rewrites expression (3.16) in the form

$$I(\boldsymbol{S}) = \sum_{\mu\mu'} f_\mu^*(\boldsymbol{S}) f_{\mu'}(\boldsymbol{S}) \sum_{nn'} \langle \mathrm{e}^{\mathrm{i}\boldsymbol{S}(\boldsymbol{u}_{n'\mu'} - \boldsymbol{u}_{n\mu})} \rangle \, \mathrm{e}^{\mathrm{i}\boldsymbol{S}(\boldsymbol{R}_{n'\mu'} - \boldsymbol{R}_{n\mu})}. \tag{3.26}$$

Angle brackets designate here the averaging over a canonical ensemble of vibrating atoms. This averaging by means of a DM is performed as

$$\langle \mathrm{e}^{\mathrm{i}\boldsymbol{S}(\hat{u}_{n'\mu'} - \hat{u}_{n\mu})} \rangle = \frac{\mathrm{Tr}\{\mathrm{e}^{\mathrm{i}\boldsymbol{S}(\hat{u}_{n'\mu'} - \hat{u}_{n\mu}) - \beta\hat{H}_N}\}}{\mathrm{Tr}\{\mathrm{e}^{-\beta\hat{H}_N}\}}. \tag{3.27}$$

The Hamiltonian of a nuclear system of an anharmonic crystal, $\hat{H}_N$, is considered here as effective. Within the perturbation theory frame the numerator and denominator of this expression are determined by the energy of the nuclear system, vibrating harmonically, and by matrix elements of operators $\exp\{\mathrm{i}\boldsymbol{S}(\hat{u}_{n'\mu'} - \hat{u}_{n\mu})\}$ and $\hat{H}_{N'}$ between harmonic nuclear vibrational wave-functions. Separating the terms depending and not depending upon $\boldsymbol{R}_n$ and $\boldsymbol{R}_{n'}$, and among the latter the terms related to vibrations of one (μ or μ') and two (μ and μ') atoms in the unit cell, we arrive at the generalized Debye–Waller factor

$$\langle \exp\{\mathrm{i}\boldsymbol{S}(\hat{u}_{n'\mu'} - \hat{u}_{n\mu})\} \rangle = T_\mu^*(\boldsymbol{S}) T_{\mu'}(\boldsymbol{S}) \exp\{-\Delta M_{\mu\mu'}(\boldsymbol{S})\} \exp\{D_{\mu\mu'}^{nn'}(\boldsymbol{S})\}. \tag{3.28}$$

The one-particle temperature factor $T_\mu(\boldsymbol{S})$ and two-particle terms $\Delta M_{\mu\mu'}(\boldsymbol{S})$ and $D_{\mu\mu'}^{nn'}(\boldsymbol{S})$ contain the contributions corresponding to the various orders of the perturbation theory. The temperature factor, for an atom in a general position which accounts for the terms up to second order, can be represented as follows:

$$T_\mu(\boldsymbol{S}) = \exp[-M(\boldsymbol{S})] = \exp[-M_\mu^a(\boldsymbol{S}) - M_\mu^b(\boldsymbol{S}) - M_\mu^c(\boldsymbol{S}) - M_\mu^d(\boldsymbol{S}) - M_\mu^e(\boldsymbol{S}) - M_\mu^f(\boldsymbol{S}) - M_\mu^g(\boldsymbol{S}) - M_\mu^h(\boldsymbol{S})]. \tag{3.29}$$

The term M_μ^a, corresponding to zero-order perturbation, in the high-temperature limit, when the classical approximation for filling of photon states is valid, has the form

$$M_\mu^\alpha(\boldsymbol{S}) = \frac{kT}{2sN} \sum \frac{|\boldsymbol{S} \cdot \boldsymbol{e}_\alpha^\mu|^2}{m_\mu \omega_\alpha^2}. \tag{3.30}$$

It contains the temperature-dependent phonon mode frequency ω_α, i.e. it corresponds to a quasiharmonic approximation, though its formal appearance does not differ from the appearance of a similar term in the harmonic approximation (Willis and Pryor 1975). This term determines the decrease in intensity of elastic scattering of x-rays by a crystal due to the non-elastic scattering associated with the change of phonon system states in the radiation field. The creation and annihilation of phonons without their interaction takes place in this case.

The remaining terms in (3.29) describe the purely anharmonic effects. The explicit forms of all these terms for arbitrary temperature and for arbitrary positions of atoms was given by Tsirelson (1993), who also discussed their physical meaning and compared them to one another. The imaginary terms M_μ^b and M_μ^c depend on the scattering vector $\boldsymbol{S}$ and on the temperature in a different manner. They relate to the three-phonon scattering processes and are responsible for the appearance of 'forbidden' x-ray (and neutron) reflections of anharmonic nature; for atoms occupying centrosymmetric positions in a unit cell they are zeros. The real M_μ^d, M_μ^g, and M_μ^h terms are associated with four-phonon processes and with pairs of three-phonon processes; they depend upon $\boldsymbol{S}$ and T in a different manner as well. The contributions M_μ^f and M_μ^l approximately compensate each other. Note that term M_μ^d, caused by anharmonicity, has the same dependence on the scattering vector square $|\boldsymbol{S}|^2$ as the quasiharmonic term M_μ^a (3.30) does. These terms, however, can be distinguished by analysing their temperature dependence, which is linear for M_μ^a and quadratic for M_μ^d.

Regrouping the M_μ terms having the same dependence on $\boldsymbol{S}$, one can find appropriate numerical coefficients, even at a single temperature, by optimizing the crystal structural model. It is usually realized in practice using the least-squares method and minimizing the deviation between measured and model intensities or structure amplitudes. This is the general way of solving the inverse problem of diffraction by x-ray structure analysis.

Expression (3.28) includes the cross-term $\exp\{-\Delta M_{\mu\mu'}(T^3, \boldsymbol{S})\}$ which is related to anharmonic vibrations of atoms μ and μ' in the same unit cell. Tsirelson (1993) has shown that this non-elastic term arises due to the presence of four-phonon and pairs of three-phonon interactions in a crystal and can be both positive and negative. Components of $\Delta M_{\mu\mu'}$ have the same dependences on the scattering vector and the temperature, as similar dependences for terms $M_\mu^g(\boldsymbol{S})$ and $M_\mu^h(\boldsymbol{S})$ in the temperature factor (3.29). Since all these terms have the same order in a perturbation, this means that, strictly speaking, when the anharmonicity is taken into account up to second order, it is impossible to separate the scattering power of a crystal's unit cell into the product of one-particle terms, each of which belongs to the atom of a single sort only. Hence, by optimizing the dynamic structural model of a crystal with the x-ray diffraction data, it is impossible to determine correctly thermal atomic parameters, having the $|\boldsymbol{S}|^4$ dependence, without preliminary calculation of the anharmonic cross-term. This can be done, for example, by using the data on a temperature dependence of heat capacity. This is not the case in practice and these parameters and parameters of ED may be distorted in an uncontrollable manner. This, of course, impedes their interpretation, but since the one-particle potentials are considered in the crystal structural model as effective potential, the neglect of an anharmonic cross-term can be assumed to be an acceptable approximation. Hereafter we shall suppose that $\exp\{-\Delta M_{ij}(\boldsymbol{S})\}=1$.

Expression (3.28) also includes the positive-definite matrix $D_{\mu\mu}^{nn'}(\boldsymbol{S})$, which is related to the Born scattering matrix (Born 1942–1943). It describes pair correlations in the thermal motion of atoms and determines the thermal diffuse scattering (TDS) of the x-rays. The eigenvalues of this matrix are much less than unity, and $\exp\{D_{\mu\mu}^{nn'}(\boldsymbol{S})\}$ can be expanded into a series, $\exp\{D_{\mu\mu}^{nn'}(\boldsymbol{S})\}=1+D_{\mu\mu}^{nn'}(\boldsymbol{S})+\frac{1}{2}[D_{\mu\mu}^{nn'}(\boldsymbol{S})]^2+\cdots$. Consequently, the expression for average intensity (3.26) can be written as

$$I(\boldsymbol{S})=\sum_{\mu\mu} f_\mu^*(\boldsymbol{S})f_\mu(\boldsymbol{S})T_\mu^*(\boldsymbol{S})T_\mu(\boldsymbol{S})\{1+D_{\mu\mu}^{nn'}(\boldsymbol{S})+\tfrac{1}{2}[D_{\mu\mu}^{nn'}(\boldsymbol{S})]^2+\cdots\}. \tag{3.31}$$

Now one can separate the elastic I_0 (zero-phonon) and non-elastic $I_1, I_2, \ldots$ (one-, two-phonon, ...) scattering contributions and write

$$I(\boldsymbol{S})=I_0(\boldsymbol{S})+\sum_{k\neq 0} I_k(\boldsymbol{S}). \tag{3.32}$$

Term I_0 during the diffraction on a crystal has sharp maxima at reciprocal lattice points (the Bragg diffraction). The diffraction condition in this case is $\boldsymbol{S}=2\pi\boldsymbol{H}$ and, using (3.25), one can write the Bragg intensity in the form

$$I_0(\boldsymbol{H})=N^2\sum_{\mu\mu'} f_\mu^*(\boldsymbol{H})f_{\mu'}(\boldsymbol{H})T_\mu^*(\boldsymbol{H})T_{\mu'}(\boldsymbol{H})\exp[\mathrm{i}2\pi\boldsymbol{H}(\boldsymbol{r}_\mu-\boldsymbol{r}_{\mu'})=N^2|F(\boldsymbol{H})|^2. \tag{3.33}$$

(N is the number of unit cells in a crystal). The diffraction condition for the remaining term in (3.32) is $\boldsymbol{S}=2\pi\boldsymbol{H}+\sum_r \boldsymbol{q}_r$ where subscript $r=1, 2, \ldots$ relates to one-, two-, etc photon scattering. The permitted values of wave-vectors of vibrational modes $\boldsymbol{q}_r$ lie in the first Brillouin zone; therefore, intensities I_k

are diffusely distributed in reciprocal space. Near the Bragg reflection, the one-phonon TDS intensity reaches a maximum value, and in some cases the two-phonon TDS should also be taken into account. For reasons which will be discussed later, the integral intensity is measured experimentally, i.e. the summation over all values of vector $\boldsymbol{S}$ inside some small scan volume is carried out. If one neglects the shift of frequency in diffuse terms, which is of the order of $10^{12}\ \mathrm{s}^{-1}$ (the corresponding shift of energy is lower than 0.1 eV and lies within the limits of x-ray spectral line width, $\sim$1–5 eV), then the scattering vector $\boldsymbol{S}$ in expressions for intensities I_k can be replaced by $2\pi\boldsymbol{H}$, and summation over $\boldsymbol{S}$ should be replaced by summation over all values of the wave-vector of the phonon mode. As a result, the expression for intensity I_k can be written as follows:

$$I_k(\boldsymbol{H}) = N^2 \sum_{\mu\mu'} f_\mu^*(\boldsymbol{H}) f_{\mu'}(\boldsymbol{H}) T_\mu^*(\boldsymbol{H}) T_{\mu'}(\boldsymbol{H}) \exp\{i2\pi \boldsymbol{H}(\boldsymbol{r}_\mu - \boldsymbol{r}_{\mu'})\} I_{\mu\mu'}^{(k)} \quad (3.34)$$

(the anharmonic cross-term $\Delta M_{\mu\mu'}$ is neglected here). The contributions to intensities of one-phonon $I_{\mu\mu'}^{(1)}$ and two-phonon $I_{\mu\mu'}^{(2)}$ scattering and their physical nature are considered in appendix G. As can be seen from formulae (G.4)–(G.10), these contributions have the scattering vector dependences of $|\boldsymbol{S}|^2$, $|\boldsymbol{S}|^3$, $|\boldsymbol{S}|^4$, i.e. just the same as that of atomic thermal parameters M_μ^i which appear in the temperature factor (3.29). Hence, the optimization of an electron dynamic structural crystal model with the x-ray diffraction data must be preceded by correction of the experimental data for TDS. Moving further to integral intensity, $T(\boldsymbol{H})$, measured experimentally, we can write, using (3.31)–(3.34) and the results obtained in appendix G,

$$T(\boldsymbol{H}) = T_0(\boldsymbol{H})\left(1 + \sum_{k>0} \alpha_k\right). \quad (3.35)$$

Such a representation of intensity is suitable, because the TDS corrections of kth order, α_k, can be determined, in general, from elastic properties of a crystal, its heat capacity at various temperatures, and from the thermal expansion coefficient (a reciprocal space volume scanned in the x-ray diffraction experiments is assumed to be known). The way of calculating the TDS corrections is described in appendix G.

Now one is appropriately comparing the initial expression for a scattered intensity (3.16) with its approximate expression (3.31). One can easily notice that for $T > 0$ K the Bragg intensity I_0 is not equal to the average over a canonical thermodynamic ensemble of the coherent elastic part of a measured intensity (the first item in (3.16)), since I_0 describes the x-ray scattering in a time-averaged dynamic electron density ρ. Fortunately, during the time of measurement for each reflection, lasting at least 1–2 s, the majority of canonical ensemble members have time enough to be in all vibrational states, and intensity I_0 is close to the ensemble average. According to Stewart and Feil (1980), even for a microcrystal with 10^6 normal modes of vibrations with frequencies ω_q, the relative value of the above difference is 10^{-6}, when $kT \approx \hbar\omega_q$. This is beyond the diffraction measurement precision limit.

Thus, the most concentrated information on ED and thermal motion of atoms in a crystal (excluding correlated atomic displacements) is contained in the coherent, elastic (kinematic) part of the x-ray diffraction intensities. Now we are able to clarify in what manner and form this information can be extracted from the x-ray diffraction data. In order to reduce the problem to the case described by a kinematic diffraction theory, the intensities measured should be corrected for a thermal diffuse and anomalous scattering, absorption and extinction, in addition to experimental factors mentioned above. Further, in order to generate the structure amplitudes, additional information on reflection phases is needed. This can be obtained from multiple scattering experiments or within some theoretical model framework. Then, one can reconstruct the electron density in a crystal unit cell in the form of a Fourier series. This will be a dynamic ED. The parametrization, by any method, of the scattering amplitudes f_μ in (3.33) allows one to obtain, by optimization of the crystal structural model, the numerical ED parameters having a pseudostatic sense. The parameters of thermal vibrations of atoms, obtained in the same manner, will have effective character only. The mathematical problems which arise during crystal structural model optimization should be, of course, properly taken into account, in order that the meaning of the model parameters be refined not distorted.

Epstein and Stewart (1979) have studied the accuracy of the convolution approximation in the frame of the superpositional crystal structure model. Their estimation of corresponding uncertainty in the intensities is equal to $\sim 0.1\%$. This can be considered as a model uncertainty of the theory presented above.

Now it remains to evaluate the accuracy of the measurement of scattered x-ray intensities which will allow us to study the chemical-bond-related effects and anharmonicity. The non-empirical calculations allow one to obtain an answer to this question. So, analysis of table 2.4, which presents the x-ray coherent static structure amplitudes for an Si crystal calculated by different methods, shows that change in structure amplitudes caused by ED redistribution during formation of this homoatomic covalent crystal from atoms are 0.5–2%. In an ionic LiF crystal these changes reach 4% (Zunger and Freeman 1977b) and in BN they reach 14% with average relative deviation 4% (Lichanot *et al* 1995). In urea corresponding changes due to intermolecular interaction also reach 4% (Dovesi *et al* 1990). Therefore, the accuracy of measurement of reflection intensities, equal to 1–2% (0.5–1% for structure amplitudes), can be considered to be *a priori* sufficient for their further recalculation into kinematic structure amplitudes used for calculating the deformation EDs. The same accuracy is required for studying the anharmonicity of a thermal motion of atoms in crystals far from phase transition points (Tsirelson 1993).

3.2 Precise measurements of the diffracted x-ray beam intensities

There are a few books and reviews which deal with different sides of precise measurements of x-ray intensities (Rees 1977a, Feil 1977b, Dunitz 1979, Lehmann 1980a, b, Sayre 1985, Seiler 1987, Blessing 1987, Aslanov *et al* 1989). That is why we will restrict ourselves to only modern aspects of intensity measurement procedure and to estimation of its precision.

Kinematical diffraction theory is valid for crystals consisting of perfect small mosaic blocks. However, real crystals contain impurity atoms and defects. This results in the appearance of static diffuse scattering which distorts the diffraction picture (Krivoglaz 1983). There are growth defects, mechanical deformation defects, thermal defects, non-homogeneity defects, etc. Many of them can be removed by thermal treatment, but the rest are inherent. As a result, the crystal sample is usually a set of ideal disoriented blocks, 10^{-7} m in size (figure 3.2).

The direct method to look at defects is electron high-resolution ($\sim 2 \times 10^{-10}$ m) microscopy (Cowley and Smith 1987, Spence 1993). The x-ray diffraction picture is also applicable to measure the concentration of defects and their distribution. For example, analysis of the static diffuse scattering contribution to the Debye–Waller factor allows us to determine the atomic static displacements (Metzger *et al* 1981). The strong influence of dislocations on the scattered radiation polarization can also be used (Olechnovich 1973).

The sample under study should be stable in an x-ray radiation field, untwined, uniformly dense, and mosaic. If its size is ~ 0.2 mm or less, the crystal can be placed in the uniform area of an x-ray beam. The spherical form of the sample allows one to reduce the error due to the x-ray beam non-uniformity. The real sample size is determined by crystal mechanical properties, extinction, absorption, and reflectivity, especially at high scattering angles. Of course, the sample should be typically representative of the substance investigated. Detailed consideration of the sample preparation is given by Wilson (1992).

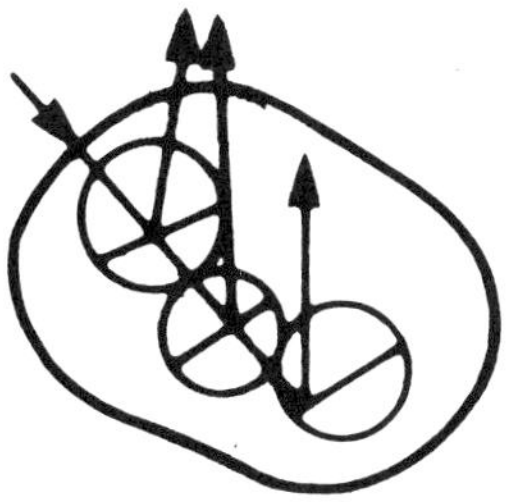

Figure 3.2 An illustration of the scattering on a mosaic block crystal. The x-ray beam intensity that reaches the second block is diminished due to scattering from the first block, oriented in nearly the same way.

The choice of a sample is usually realized experimentally (Wilson 1992). Among the criteria for such a choice are the symmetric form of diffraction peak profiles and the agreement between intensities of the symmetry-equivalent x-ray reflections: the spread of their values should not exceed 2%.

3.2.1 *X-ray diffractometry*

The basic device for accurate measurements is the computer-controlled four-circle x-ray diffractometer. X-ray tubes with Mo or Ag anodes, which have a characteristic line spectrum, serve as the x-ray sources. The K series (α_1–α_2) doublet is most often used ($\lambda_{Mo\,K\alpha_1}=0.7093\times10^{-10}$ m, $\lambda_{Ag\,K\alpha_2}=0.5594\times10^{-10}$ m). The sample under examination is placed on a goniometer head in the centre of the goniometer (figure 3.3) to 'bathe' in the incident x-ray beam and is turned by on-line computer into the reflection position (to fulfil the Bragg condition) by ω, φ, and χ angle variations. The counter is moved in the horizontal plane fixing the angle 2θ. The uncertainty of angle setting for the most modern diffractometers is $\pm0.01°$, and the discrepancy of goniometer centre position and beam axis is ~20 μm. The presence of four variable angles exceeds the experimental necessity and allows one to use low- (Luger 1993, Boese 1994, Larsen 1995, Zobel *et al* 1993) and high- (Tuinstra and Fraase Storm (1978) temperature attachments. In these cases samples are overblown by a steam of cold or warm gas with a temperature stability of 1–2 K.

The incident x-ray beam is often monochromated by a crystal monochromator, the quality of which should provide the homogeneity of the beam used. For example, the monochromator mosaicity in SYNTEX diffractometers is $\sim0.5°$ and the subsequent beam non-uniformity in a

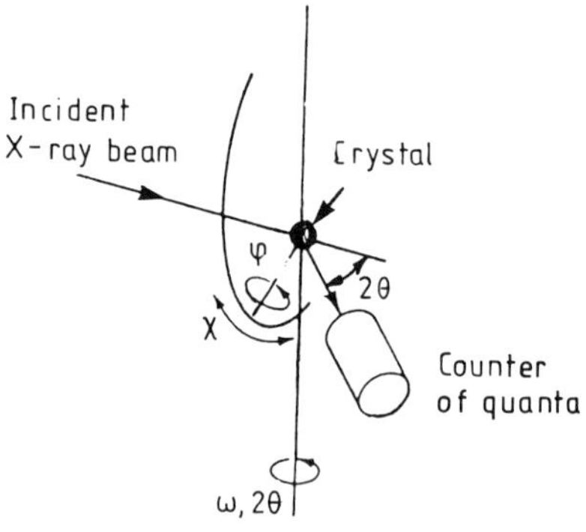

Figure 3.3 The scheme of measurements with a four-circle (four-axis) goniometer with Eulerian (SYNTEX–Nicolet) geometry: φ is the angle of the crystal rotation around the goniometer head axis; χ is the angle of φ axis inclination; ω is the angle of the rotation of the plane in which the χ circle is lying; 2θ is the angle of the quantum counter rotation around the vertical axis.

section of 0.3×0.3 mm^2 can reach 3%. The monochromator changes the incident beam polarization. That is why the value of polarization factor, p (3.13), depends on the degree of monochromator perfection and on its orientation. For a graphite monochromator and for Mo radiation corresponding deviations do not exceed 1% (Rees 1977a). The β-filter is sometimes preferred (Dunitz 1979).

The cross rate of the reflection sphere by a reciprocal point and, hence, the intensity value measured depends in a four-circle diffractometer only upon scattering angle θ. When both incident and reflected beams are in the equatorial plane (figure 3.1) this dependence is accounted for by an integral Lorentz factor, $L = 1/\sin 2\theta$. If there is more than one reciprocal point on the Ewald sphere the reflected beam is affected by simultaneous scattering (see appendix F). This phenomenon can distort the intensity values measured drastically, in the cases of weak reflections and of large unit cells especially. For example, false reflections can appear in the positions of crystal-symmetry-forbidden ones. Crystal rotation around the scattering vector $\boldsymbol{H}$, which is equivalent to the additional Ψ angle scan, allows one to reveal simultaneous scattering; the intensity of onefold reflection is not dependent on such rotation provided possible fluctuations in extinction, absorption, and incident beam inhomogeneity are neglected (figure 3.4). The examination of simultaneous scattering can also be done by diffraction picture analyses (Tanaka and Saito 1975, Aslanov *et al* 1989).

Because of the natural width of the x-ray characteristic line, mosaicities of the crystal monochromator and sample, and the non-zero counter collimation system the intensity is distributed in the reciprocal space (figure 3.5). Therefore, in order to measure the reflection intensity it is necessary to integrate the intensity within the distribution area or along some direction. The interval of integration depends on the primary beam divergence and its wave dispersion, on crystal mosaic block disorientation and on the scattering angle. The $\omega/2\theta$ scan of x-ray diffraction peak profiles is usually used in accurate x-ray experiments when the crystal is rotated with the speed $\dot{\omega}$ and the counter is rotated with the speed $2\dot{\theta}$, the speed ratio being 1:2. In this case the peak width depends mainly on radiation wave dispersion, and a fixed counter window width can be used. The scan interval is presented as $\Delta\omega = \omega_0 + B \tan\theta$, where $\omega_0(\theta) \cong 1°$ and $B(\text{Mo K}\alpha) = 0.345°$ provided only the $(\alpha_1 - \alpha_2)$ doublet splitting is taken into account, and $B = 0.8$–$0.9°$ when the doublet line broadening is accounted for (Mathieson 1984). Denne (1977) and Destro and March (1987) found that in the case of an $\omega/2\theta$ scan the value of integral intensity, obtained after background subtraction, depends on scattering angle value very much: the intensities of high-angle reflections are underestimated relative to low-angle ones. Mathieson (1989) stated later that a better approximation for the scan interval which improves the situation is the expression

$$\Delta\omega = \omega_0'' + B'' \tan\theta + [(\omega_0')^2 + (B' \tan\theta)^2]^{1/2}. \tag{3.36}$$

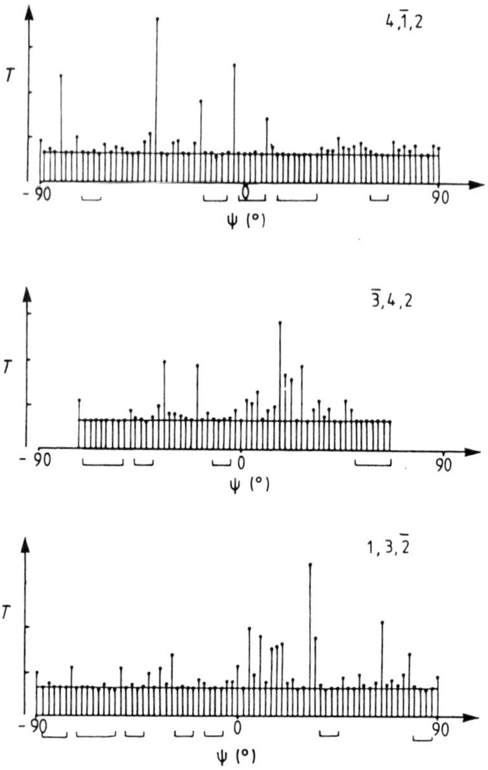

Figure 3.4 The azimuthal intensity profile for some low-order angle reflections from the Li_2BeF_4 crystal at 81 K (Seiler 1987). The intensity due to simultaneous scattering occurs as sharp spikes.

Because of the large difference in intensity values, the scan speed varies depending on the counting rate. This decreases the statistical non-equiprecision in the measured intensities.

Scintillation counters, supplied by modern electronics, are more often used as radiation detectors. This kind of counter is characterized by a dead time of $\tau_d = \sim 10^{-6}$ s; this is why the number of quanta passing through a counter can exceed their registered number. To take the difference into consideration an expression $T = T_m \exp(T\tau_d)$ is usually used, where T and T_m are the 'true' and measured intensities, respectively (Chipman 1969). This formula is approximated by a parabolic expression $T = T_m + T^2\tau_d$ (it is fulfilled well at $T_m \leqslant 5 \times 10^4$ quanta s^{-1}). If a continuous scan is used the rates of quantum counts in the different points of the diffraction peak due to the existence of dead time are unknown and introduction of dead time correction is impossible. This is why, beginning with 5×10^4 quanta s^{-1}, an attenuator should be used in this case. A continuous scan also makes impossible the

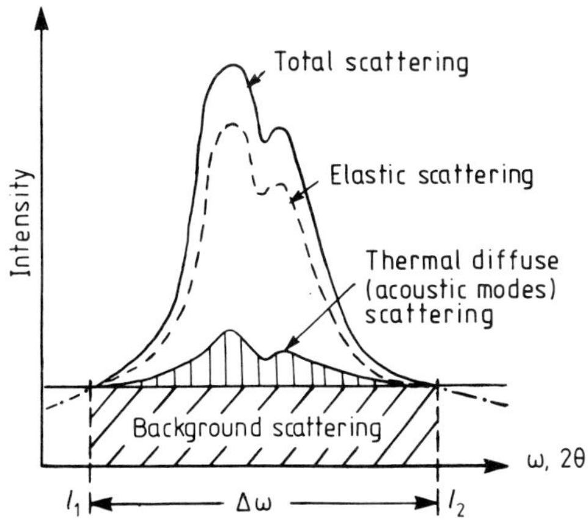

Figure 3.5 A schematic picture of typical features of the profile of x-ray reflection, showing the various component contributions to the total scattering near the Bragg position. The two-humped profile results from α_1–α_2 splitting of the x-ray line. Background scattering due to Compton, air, and instrumental scattering and thermal diffuse (optic model) scattering is assumed constant over the $\Delta\omega$-range.

correct evaluation of the reflection variances. As a result, the discontinuous step-by-step scan, which allows one to also analyse the measurement results *post factum*, is preferable. The reflection profile is divided into equal intervals and quantum rates are recorded in every interval. An ith scan step in the intensity scattered is presented by the expression

$$T_i = (t_i - b_i)/L_i p_i \tag{3.37}$$

where

$$t_i = \dot{\omega} N_i/(1 - n_i \tau_d) \tag{3.38}$$

with scan speed, $\dot{\omega}$, number of quanta measured, N_i, and quantum count rate value, n_i, being taking into account.

There are many factors which give rise to the background intensity under a Bragg peak, b_i: instrumental factors, Compton and air scattering, static and thermal diffuse scattering, etc. The background value is measured at angles outside the Bragg peak position. Generally speaking, a diffraction peak has infinitely decreasing wings (see figure 3.5) and its borders, l_1 and l_2, should be chosen in such a manner that the scan truncation error will be less than intensity statistical error. The search for l_1 and l_2 point positions is called reflection profile analysis.

Strictly speaking, infinite peak wings do not allow an accurate intensity determination in a limited scan interval. However, intensities are measured on a relative scale, therefore the uniform truncation of peak wings for all reflections does not destroy their relative values. The subtraction of background leads to the 'net' intensity set after summing T_i values (3.37) in the l_2–l_1 range.

The overestimation of the peak size leads to an overestimation of its statistical uncertainty; the underestimation introduces systematical error in net intensity. Therefore, profile analysis is an important and non-trivial problem. As was shown by van der Wal *et al* (1979), different methods of profile analysis lead to change of positional parameters within 1–2 estimated standard deviations and to a scale factor deviation of 1%. This can distort the thermal atomic parameters obtained and the deformation ED.

There are two main approaches in profile analysis, both oriented on intensities of nearly equal precision. In the first, the sequence of counts at every scan step is analysed and l_1 and l_2 values are determined for each reflection. So, in the Lehmann and Larsen (1974) method modified by Blessing *et al* (1974) l_1 and l_2 values are suggested to be determined from the condition of the minimum relative intensity error. In the Grant and Gabe (1978) method the points i are searched for where the condition $t_{i+1} = t_i \pm 0.67\sigma(t_i)$ begins to be fulfilled. These points are identified as peak borders. Rigoult (1979) has used the fact that background intensities at neighbouring points are random values and at the same time the reflection intensity profile points are correlated ones. Therefore, l_1 and l_2 can be found as points where the full measured intensity self-convolution is minimal.

The peak borders undergo statistical fluctuations and the squence analysed is the result of the onefold counting rate measurement. Therefore, the statistical estimate of counting rate variance, $\sigma^2(t_i)$, in the methods described is not correct. These methods cannot be applied to weak reflections because of their sensitivity to signal-to-noise ratio. Besides, the Lehmann–Larsen method underestimates the scan interval, and the Grant–Gabe and Rigoult methods overestimate the peak intensities. These shortcomings were partly eliminated by Gerr (1985) who used preliminary short-time background measurements at points 1.3–1.5 times further than at l_1 and l_2. The peak asymmetry is accounted for here automatically. Then on the peak slope, beginning from the border of the scan interval, points are considered which satisfy condition $t_{i+1} \geqslant (1.0 + 1.5)\sigma(\tilde{b}) + \tilde{b}$, where $\tilde{b}$ is the background intensity, normalized to the single point measurement time. The peak size of weak and smoothed reflections is determined by using averaged results of the preceding analysis. The time for additional computations is small and weak reflection intensities are more properly measured. The last is important for ED studies.

Another approach to profile analysis is based on the approximation of each experimental reflection by a model function. The last can be given as some analytical function (Hanson *et al* 1979), as a convolution of functions, which are associated with experiment (Chulichkov *et al* 1987), or can be experimentally measured. In order to adjust a model function to a diffraction peak some complementary knowledge on peak shape, on the local behaviour of the background, and on the particular reflection position in reciprocal space is needed. Therefore, such profile treatments are tedious and, besides, are applicable only with difficulty to reflections with the unresolved (α_1–α_2) doublet.

To avoid the drawbacks of such methods while preserving their advantages, Streltsov and Zavodnik (1989) suggested the following algorithm, which is applied after the experiment is over. At the first stage a set of 'strong' reflections is formed from the whole data set. The criteria are the ratio of scan speed of a given reflection to a minimal one and $T/\sigma(T)$ ratio. After this the values l_1 and l_2 are determined, for example by the Lehmann and Larsen (1974) method for these reflections. Then components of the tensor which describes l_1 and l_2 dependence on the reciprocal lattice reflection coordinates are calculated by least squares. When an $\omega/2\theta$ scan is used and there are no abnormal anisotropic effects, these tensors are nearly isotropic, and l_i values can be considered constant for all reflections from the total set. After such a treatment, normalized peak profiles with background subtracted are collected and averaged over intervals $\Delta(2\theta)=5°$. The result gives the dependence of model peak form on 2θ angle, which is used for determination of T and $\sigma(T)$ values of weak reflections. Background measurement is not needed in this case and the correct evaluation of intensity variances is possible.

The number of reflections measured over the whole Ewald sphere can reach several thousand. Measurements can last a few weeks and long-term variations can influence the intensity because of drift or fluctuations in diffractometer operation, the changes in conditions of x-ray beam transmission in air, the damage of the sample by influence of x-rays, etc. In order to take long-term variations into account some standard reference reflections are measured periodically and interpolative correction of data is made. Abrahams (1973) has suggested the use of several strong, medium, and weak reflections spread uniformly over all reciprocal space. However, in practice one is restricted to a small number of reference reflections. The long-term variations can be anisotropic and there is an effective algorithm which takes this fact into account (Streltsov and Zavodnik 1990).

The expression for the net integral intensity of any reflection can be finally written as

$$T=\kappa(\tau)\sum_{l_1}^{l_2}(t_i-b_i)/(L_i p_i). \tag{3.39}$$

The scale factor $\kappa(\tau)$ takes into consideration the long-term intensity variations which depend on time τ.

3.2.2 *Precision of measured x-ray intensities*

The statistical error analysis (appendix H) provides the evaluation of the intensity (3.39) measurement precision. The number of quanta registered obeys the Poisson statistics (Loudon 1973)

$$P_\mu(N)=(\mu^N/N!)\exp(-\mu). \tag{3.40}$$

Here N is the number of counts in a given time interval and μ is the expected averaged number of counts in the same interval while replicating counting procedure. The property of the Poisson distribution is the fact that its variance approaches the average count number: $\sigma_N^2 \to \mu$. That is why $\sigma^2=N$ is the best estimation at a single measurement. Starting with expression (3.40) and accounting for the error distribution law, one can estimate the uncertainties in diffraction data at each stage of the treatment process. For example, the variance of intensity measured at each scan step (3.38) can be described as

$$\sigma^2(t_i)\cong\dot{\omega}^2 N_i/(1-n_i\tau_d)^4 \tag{3.41}$$

provided the errors in angular velocity $\dot{\omega}$ and counter dead time τ_d are significantly less than the statistical error in the number of counts. For estimation of the background variance the denominator in (3.41) should be taken as equal to unity. Considering the values in equation (3.39) as independent and using equation (H.10) from appendix H, one can arrive at the expression for estimation of variance of the net integral intensity of any reflection:

$$\sigma^2(T)=T^2\frac{\sigma^2[\kappa(\tau)]}{\kappa^2(\tau)}+\sum_{i=l_1}^{l_2} t_i^2\left\{\frac{[\sigma^2(t_i)+\sigma^2(b_i)]}{(t_i-b_i)^2}+\frac{\sigma^2(L_i)}{L_i^2}+\frac{\sigma^2(p_i)}{p_i^2}\right\}. \tag{3.42}$$

Blessing (1987) evaluated contributions to (3.42) from different sources. He came to the conclusion that uncertainty of intensity resulting from long-term scale drift is about 0.5%. As was stressed by McCandlish *et al* (1975) this uncertainty is increasing in time even if $\kappa(\tau)$ remains constant ('zero position drift'). The error in integral Lorentz factor maximizes at very small and very high scattering angles where $\sigma(L_i)/L_i \leqslant 0.01$. The polarization factor is nearly θ independent, $\sigma(P_i)/P_i \leqslant 2.5\times10^{-4}$. Therefore, the total uncertainty in net integral intensity depends mainly on counting statistics, on diffractometer operation stability, and on crystal stability to the x-ray radiation.

If the supposition on statistical independence of errors caused by different sources and errors of successive measurements of the same value is not valid, the terms which account for correlations between corresponding values in (3.42) should be added (Blessing 1987).

The most complete set of x-ray intensities can be obtained by measuring all reflections from a single crystal under examination over the whole Ewald reflection sphere. Instead of repeating measurements, which would increase the statistical precision, one usually uses the Laue crystal symmetry and the symmetry-related reflections are measured. Averaging of these reflections is useful from the point of view of elimination of crystal anisotropy and inhomogeneity effects. Supposing the statistical independence of each measurement and numbering each intensity which belong to the same group of the Laue equivalents by a j-index we can write the averaged weighted intensity as

$$\bar{T}(\boldsymbol{H})=\left[\sum_j T_j(\boldsymbol{H})\sigma^{-2}(T_j(\boldsymbol{H}))\right]\Big/\sum_j \sigma^{-2}(T_j(\boldsymbol{H})). \tag{3.43}$$

The variance of the averaged weighted intensity is

$$\sigma^{-2}(\bar{T}(\boldsymbol{H}))=1\Big/\sum_j \sigma^{-2}(T_j(\boldsymbol{H})) \tag{3.44}$$

and their sample variance is

$$S^2(\bar{T}(\boldsymbol{H}))=\frac{n}{(n-1)}\sum_j^n (\bar{T}(\boldsymbol{H})-T_j(\boldsymbol{H}))^2\sigma^{-2}(T_j(\boldsymbol{H}))\Big/\sum_j^n \sigma^{-2}(T_j(\boldsymbol{H})) \tag{3.45}$$

(n is the number of symmetry-equivalent reflections). Those measurement results which differ significantly from the averaged values are defined according to statistical criteria as missing and removed from the set. The final set of diffraction data consists of averaged and symmetry-independent values of intensities (3.43) and estimates of their variances. Finally, the largest values of σ^2 (3.44) or S^2 (3.45) are usually chosen. In order to judge the data set quality the 'internal' discrepancy factor for each group of equivalent reflections is calculated as

$$R_{int}=\sum_j^n \frac{(\bar{T}-T_j)}{n\bar{T}}. \tag{3.46}$$

Figure 3.6 presents the distribution in reciprocal space of R_{int} values which were calculated for the spinel crystal, $MgAl_2O_4$ (Tsirelson *et al* 1986). The

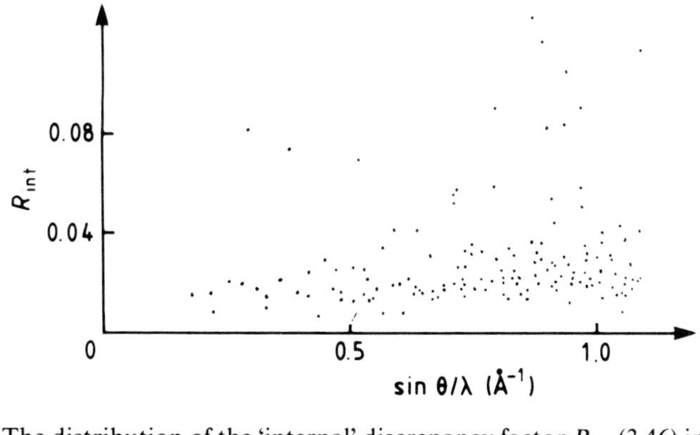

Figure 3.6 The distribution of the 'internal' discrepancy factor R_{int} (3.46) in reciprocal space for the $MgAl_2O_4$ crystal: such data are useful for assessment of quality of experimental results and for remeasurements, if needed.

picture presented is typical: the dispersion of equivalent reflection values increases with scattering angle because of change for the worse of quantum counting statistics. The R_{int} value increases correspondingly. The R_{int} distribution points out which reflections should be remeasured to provide the intensity statistical precision needed.

The uncertainty of measurements often appears to be underestimated because the correlations of errors from different sources are not taken into account. Table 3.1 contains a collection of the experimental factors which influence the experimental $|F_0|^2$ values. The ranges of correlation coefficients between different measurement pairs are also presented. These coefficients

Table 3.1 Errors affecting F_0^2 values classified according to the possibility of their detection and estimation by repeated measurements (Rees 1977, Blessing 1987). (Reprinted with permission from Blessing 1987.)

Group	Detection	Error source	Contribution to $\sigma(\|F_0\|)$[a]	Range of the correlation coefficient
1	Repeated measurement	1.1 Counting statistics	$N^{1/2}$	0
		1.2 Short-term fluctuations	$pNt^{-1/2}$	0
2	Periodic remeasurement of the same reflection	2.1 Long-term fluctuations	$\sim I$	0, +1
		2.2 Crystal decomposition	$f(I, t)$	0, +1
		2.3 Scan-angle setting error (step scans)	$\sim I$	0
		2.4 Scan motion irregularities (continuous scans)	$\sim I$	0, +1
		2.5 Temperature fluctuations	$f(I, 2\theta)$	0, +1
3	Comparison of measurement reflections or ψ-scans	3.1 Beam inhomogeneity	$\sim I$	0, +1
		3.2 Crystal misalignment	$\sim I$	0, +1
		3.3 Crystal dimensions (absorption anisotropy)	$\sim I$	0, +1
		3.4 Extinction anisotropy		+1
		3.5 Multiple reflection		−1, +1
4	Undetectable by repeated measurements	4.1 Non-monochromatic beam		+1
		4.2 Counter dead time, τ	$N^2\sigma(\tau)$	+1
		4.3 Attenuator coefficient	$\sim I$	+1
		4.4 Polarization factor	$\sim I$; $f(\tau)$	+1
		4.5 Absorption coefficient	$\sim I$	+1
		4.6 Extinction factor, y	$\sim yI^2$	+1
		4.7 Thermal diffuse scattering	$\sim I$; $f(2\theta)$	+1
		4.8 Scale factor	$\sim I$	1

[a] N is the count registered during time t, p is the instrumental instability parameter (McCandlish *et al* 1975).

usually have the same sign; therefore, it is hopeless to expect that errors will compensate each other. This is why it is important to reveal the presence of error correlation dependences. Statistical methods are useful for this purpose (Blessing 1987).

What is the real reproducibility of the measured integral intensities? The answer to this question can be obtained from the investigation by Mackenzie and Maslen (1968) of the same CaF_2 single crystal by different diffractometers and from Denne's (1972) study of six different crystals of α-glycin, $C_2H_5NO_2$, of different shape. In the first work the distribution of intensities was 2–3%, in the second, ~0.5%. In both cases the crystal was recognized as the main source of the discrepancies. Hence, choosing the sample to be investigated carefully and examining the measurement process attentively, one is able to provide a precision of intensity measurement of about 1%.

3.3 From intensities to kinematic structure amplitudes

After the primary treatment of the x-ray diffraction data we have the set of symmetrically independent (averaged) intensities $\bar{T}(\boldsymbol{H})$ together with their variances $\sigma^2(\bar{T})$. The first step on the path to ED reconstruction from these data is the extraction of the elastic coherent contribution to the intensities, $|F_0(\boldsymbol{H})|^2$. Then the kinematic structure amplitude moduli, $|F_0(\boldsymbol{H})|$, can be determined. The path from intensities to structure amplitudes is not trivial. It is realized by the introduction of a number of corrections which account for diffraction effects distorting the kinematic scattering picture. Obviously, these corrections depend very much on the level of understanding of these physical effects to be taken into account and on the adequacy of methods of correction determination.

The estimator for the value of the net integral intensity of each independent reflection (3.43) measured on the same relative scale is usually described by the expression

$$\bar{T}(\boldsymbol{H})=Y(\boldsymbol{H})\left[1+\sum_{k\geq 1}\alpha_k(\boldsymbol{H})\right]A(\boldsymbol{H},\mu) \tag{3.47}$$

in which

$$Y(\boldsymbol{H})=y(\boldsymbol{H})k^2|F(\boldsymbol{H})|^2 \tag{3.48}$$

$$k^2=I_{inc}(r_0 v/V)^2\lambda^3. \tag{3.49}$$

Here α_k are the corrections for one-phonon ($k=1$), two-phonon ($k=2$), etc TDS; $A(\mu)$ is the transmission factor which depends on the linear absorption coefficient μ; y is the correction for extinction; I_{inc} is the intensity of the incident radiation with wavelength λ which falls on a unit square area; r_0 is the classical electron radius; and v and V are crystal and unit cell volumes;

the factor $(\lambda^3 v/V)$ appears as the result of Laue interference function integration over the reciprocal volume scanned. Hence, all factors which influence the results of measurements are taken into account in these expressions.

The angle-independent values are concentrated in expression (3.49). The coefficient k, the scale factor, transfers the intensity values into an absolute scale. Measurements can be directly carried out in the absolute scale (Stevens and Coppens 1975) however in this case it is difficult to evaluate precisely the incident beam intensity and the very small volume of crystal sample studied. As a result, the $\sigma(k_{exp})/k_{exp}$ value is, in reality, about 3–5%. Therefore, the scale factor, as a rule, is determined later during the crystal structural model refinement.

Let us consider successively the factors which distort the kinematical picture of scattering and, hence, should be accounted for.

3.3.1 *Absorption*

In section 3.1 describing the x-ray scattering by a crystal we noted that a term $(-e/m)(\hat{A}\cdot\hat{p})$ in the perturbation Hamiltonian (3.4) is related to the absorption of quanta. The direct calculation of beam intensity lowering is too troublesome. Instead, a more simple and sufficiently precise method is usually used, which preserves the formalism of section 3.1: the transmission factor, $A(\boldsymbol{H}, \mu)$ is introduced into the intensity expression (3.47). For this purpose the crystal is represented by a collection of separate atoms and the linear absorption coefficient, μ, is calculated according to an additive scheme using the data measured for all elements by means of special methods (Wilson 1992). The possible anisotropy of absorption is ignored. Then, the crystal is divided into elementary volumes for which the path lengths of incident, $t_1(\boldsymbol{H})$, and scattered, $t_2(\boldsymbol{H})$, x-ray beams (figure 3.7) as well as the value $\exp[-\mu(t_1+t_2)]$ are found for each direction. The intensity weakening for

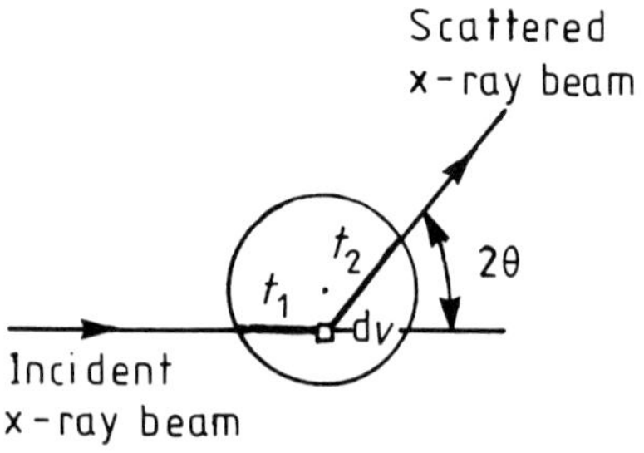

Figure 3.7 Path lengths t_1 and t_2 for some reflection before and after scattering on an element volume dv in a crystal. As may be seen, $t = t(2\theta)$.

a sample is calculated from

$$A(\boldsymbol{H}, \mu)=\frac{1}{V}\int_V \exp[-\mu(t_1+t_2)]\,\mathrm{d}V \tag{3.50}$$

for every reflection. The precision of μ and (t_1+t_2) value determination plays the most important role in $A(\boldsymbol{H}, \mu)$ estimation. The path length depends on the crystal shape and dimensions; usually $\sigma(t)/t \sim 0.01$. The calculation of integral (3.50) for a spherical or regular sample shape is very simple. The shape of other crystals is described by some arbitrary polyhedron and integral (3.50) is calculated by numerical integration (Harkema *et al* 1980, Chidambaran 1981, Flack 1984, Aslanov *et al* 1989). The x-ray absorption coefficients are usually 10^3–10^5 m^{-1} and their uncertainty is usually not more than 2% (Creagh and Hubbell 1987); only in heavy atoms can it reach ~5%. This leads to the absorption correction uncertainty $\sigma(A)/A \sim 1$–2%, mostly important in the case of absolute measurements (Rees 1977b). Hewat (1979) noted that the transmission factor can be represented as the product of fictitious 'scale' and fictitious 'temperature' factors: both multipliers will deviate from their true values when $A(\boldsymbol{H}, \mu)$ is not correct. It is obvious that the error in absorption correction will influence the ED maps most in the vicinity of the nuclei.

Empirical absorption correction is made sometimes by symmetry-related reflection examination although non-corrected anisotropic extinction and anisotropic thermal diffuse scattering distort the result. Therefore, the theoretical calculation of the transmission factor is preferable.

3.3.2 *Thermal diffuse scattering*

Thermal diffuse scattering is the inelastic scattering caused by correlated atomic thermal movements. The separation of TDS and elastic scattering in x-ray diffraction experiments is impossible because the energy shift due to TDS, ~0.1 eV, is much smaller than the natural x-ray line width, 1–5 eV. At the same time, this shift is much more than the width of the resonance emission line of Mossbauer γ-ray radiation, $\sim 10^{-8}$ eV. Therefore, elastic and inelastic effects can be separated with the help of γ-diffractometry (Crow *et al* 1987). Unfortunately, such experiments are done in a few cases only.

The contributions to TDS, described by expressions (G.4–G.10) from appendix G, depend on the scattering vector as $|\boldsymbol{S}|^2$, $|\boldsymbol{S}|^3$, $|\boldsymbol{S}|^4$. The same dependences are found for terms in the atomic temperature factor expression (3.29). Therefore, the atomic temperature parameters are suffering from the TDS effect first of all. The TDS has both harmonic and anharmonic components which are related to manifold phonon inelastic processes. Neglecting contributions higher than second order the correction for TDS

(3.47) can be presented in the form

$$\alpha_1+\alpha_2 \cong (1+\alpha_T)\alpha_1^{qh}+\alpha_2^{qh}. \tag{3.51}$$

Values α_T (G.14), α_1^{qh} (G.18) and α_2^{qh} (G.20) describe anharmonic additions to one-phonon TDS in the Leibfried and Ludwig (1961) approximation and one- and two-phonon TDS in the quasiharmonic approximation, respectively.

What does happen if one ignores TDS correction while treating x-ray diffraction data? In order to answer this question we will follow Stevenson and Harada (1983) and will write

$$\begin{aligned} T'(\boldsymbol{H}) &\sim \left|\sum_\mu f_\mu \exp(-M'_\mu)\exp(\mathrm{i}2\pi \boldsymbol{H}\cdot\boldsymbol{r}'_\mu)\right|^2 \\ T(\boldsymbol{H}) &\sim \left|\sum_\mu f_\mu \exp(-M_\mu)\exp(\mathrm{i}2\pi \boldsymbol{H}\cdot\boldsymbol{r}_\mu)\right|^2. \end{aligned} \tag{3.52}$$

Here T' is the intensity of scattered x-ray radiation with TDS contribution; T is the elastic intensity. Correspondingly, M'_μ and M_μ, r'_μ and r_μ describe thermal and positional atomic parameters obtained by using the same x-ray diffraction data without and with TDS correction. Let us consider TDS correction to be small. This is not always the case: nevertheless further consideration, having general character, will be valid. As long as

$$(1+\alpha_1+\alpha_2)^{-1} \approx \exp[-(\alpha_1+\alpha_2)] \tag{3.53}$$

then (3.47) can be rewritten as

$$T \sim \left|\sum_\mu f_\mu \exp[-M'_\mu-(\alpha_1+\alpha_2)/2]\exp(\mathrm{i}2\pi \boldsymbol{H}\cdot\boldsymbol{r}'_\mu)\right|^2. \tag{3.54}$$

Comparing (3.52) and (3.54) one obtains

$$M'_\mu \cong M_\mu-(\alpha_1+\alpha_2)/2 \qquad \boldsymbol{r}'_\mu \cong \boldsymbol{r}_\mu. \tag{3.55}$$

Some of the M_μ terms are real; the others are imaginary. Because the TDS correction is a real value, TDS does not influence imaginary anharmonic thermal parameters. The neglect of TDS is equivalent to the artificial decrease of the real thermal parameters of all atoms. Simultaneously, it results in the change of direction of the main axes of thermal vibration ellipsoids which obey the local symmetry of the atomic position whereas the artificial decrease does not. Consequently, it leads to distortion of scale factor and extinction parameters if they are refined along with other parameters of the crystal structural model, because in an interval 0.3–0.7 Å^{-1} thermal motion and extinction phenomena overlap.

In the case of isotropic thermal motion, when $M_\mu=B_\mu(\sin\theta/\lambda)^2$, the neglect of TDS leads to the change of the B_μ value, ΔB_μ. Experimental evaluation shows (Lehmann 1980b) that ΔB is 0.10–0.25 Å^{-1}; this is equivalent to an error of 10–20% in B for inorganic crystals.

If TDS is neglected, the ED maximum heights are changed without displacement of their positions and their slopes are distorted. As a result, false peaks can appear in deformation ED maps at atomic positions (Helmholdt and Vos 1977).

First-principles calculation of the harmonic and anharmonic components of TDS requires knowledge of the polarization vectors, the normal mode frequencies, and anharmonic properties of a crystal. Such an approach is too complicated for practical realization. Therefore, the TDS correction is calculated now using approximate schemes, which include the additional experimental information. The general approach and corresponding formulae for the calculation of α_1^{qh} α_2^{qh}, and α_T values are given in appendix G. The concrete method of the calculation and the level of approximations as well as the amount of additional information used can be different. The more exhausting computational scheme (Tsarkov and Tsirelson 1991) requires knowledge of the total matrix of crystal elastic constants, the heat capacity, and their temperature dependences, as well as characteristics of thermal expansion of the crystal. These values can be measured precisely enough by ultrasound and calorimetric methods (Mason and Thurston 1964), by the thermal diffuse scattering of x-rays and neutrons (Wooster 1962), by inelastic neutron scattering (Willis and Pryor 1975), by time-of-flight neutron experiments (Schofield and Willis 1987, Carlile and Willis 1989), and by Brillouin scattering spectroscopy (Fabelinskii 1968). As a rule, crystals of large size ($\sim 10^{-6}$ m^3) are needed for such experiments; their orientation is supposed to be known. This is not usually available, especially for the new compounds. Therefore, taking account of anharmonical contributions to the TDS is a very troublesome task (see appendix G). The importance of these contributions depends on the properties of the crystal examined, the problem studied, and the precision of the diffraction data. As was found by Tsarkov and Tsirelson (1991), the α_T value at 300 K for the NaCl crystal is -0.086, for NaI $+0.049$, for CsCl -0.081, and for $KZnF_3$ $+0.08$. For CsCl, for example, this means that the anharmonicity contribution to the total x-ray scattering reaches 4% in the high-angle region and should be taken into account. The data listed lead also to an interesting conclusion: the anharmonicity of atomic motion can result in both the increase and the decrease of inelastic scattering intensity.

The quasiharmonic TDS corrections, α_1^{qh} and α_2^{qh}, need only additional single-temperature experimental data. The realizations of the quasiharmonic approximation by Stevens (1974), Sakata and Harada (1976), Merisalo and Kurittu (1978), Helmholdt *et al* (1983), and Dudka *et al* (1989) differ only in the manner of (G.18) and (G.20) integral evaluation. Independent of computational method, the TDS correction depends on crystal size, on scattering vector value, and on experimental conditions such as divergence and non-monochromaticity of the x-ray beam, initial scan interval, detector aperture, etc. These facts should be taken into account when the tactics of the x-ray diffraction experiment are chosen. For example, it follows from TDS theory that, by decreasing the detector aperture and initial scan interval width, it is possible to decrease the TDS contribution to the scattering. This point was experimentally confirmed by Göttlicher (1968). He measured, with two detector aperture dimensions, the x-ray reflection intensities for the same NaCl single crystal. The reciprocal space volumes scanned and, consequently, the TDS contributions in the scattering were different in the two experiments; the TDS corrections should be different as well. This difference was actually observed. This can be seen in table 3.2. After TDS correction the structure amplitude values appear to be nearly equal for the two experiments.

The values α_1^{qh}, $\alpha_1^{qh}(1+\alpha_T)$, and $\alpha_1^{qh}(1+\alpha_T)+\alpha_2^{qh}$, as functions of scattering angle for the high-temperature superconductor $YBa_2Cu_3O_{6.85}$ far from the temperature of phase transition, are presented in figure 3.8. The angular

Table 3.2 TDS correction for some structure amplitudes of NaCl corresponding to two different sizes of detector aperture (Göttlicher 1968).

	Large aperture		Small aperture	
hkl	$\sqrt{1+\alpha}F_0$	F_0	$\sqrt{1+\alpha}F_0$	F_0
444	20.90	19.40	20.35	19.25
860	7.97	6.74	7.50	6.70
11 1 1	2.95	2.41	2.72	2.43

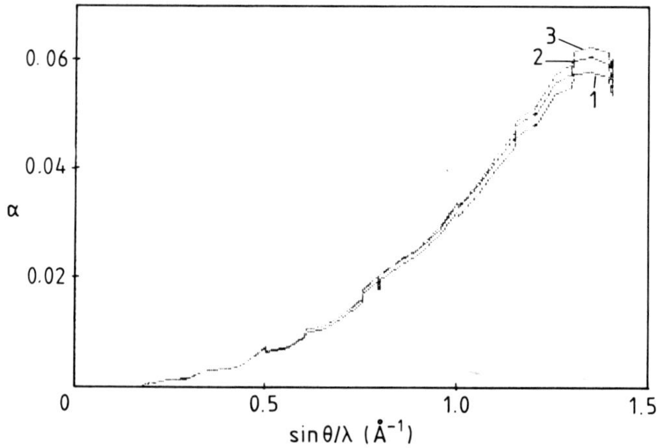

Figure 3.8 Values of thermal diffuse anisotropic scattering corrections for a $YBa_2Cu_3O_{6.85}$ quasitetragonal crystal: 1, one-phonon quasiharmonic correction α_1^{qh}; 2, one-phonon anharmonic correction $\alpha_1^{qh}(1+\alpha_T)$; 3, total correction $\alpha_1^{qh}(1+\alpha_T)+\alpha_2^{qh}$.

dependence of TDS corrections is not monotonic due to the anisotropy of TDS and has a maximum at some $\sin\theta/\lambda$ value, in agreement with TDS theory. The TDS contribution in total scattering reaches 6% and, therefore, it should be taken into account in order to obtain the exact values of atomic thermal parameters, which are of great interest for the physics of superconductivity. At the same time, the two-phonon scattering and anharmonicity have a very small influence on the TDS in the crystal in question: the α_2^{qh} and α_T values are only 0.002 and 0.04, respectively.

The data presented are typical. They show that the one-phonon quasiharmonic scattering of the x-rays on acoustic phonons gives the main contribution to TDS. The two-phonon contribution and anharmonicity should be taken into account in some cases only. However, at temperatures not too far from the phase transition point the α_T value can be of great value.

The TDS anisotropy can be very important, especially for acoustooptical crystals or fast-ionic conductors. However, if the anisotropy is not large, rather simple approximations for calculation of the TDS correction can be used. For example, scan volume can be approximated by a sphere (Pryor 1966) and for calculation of α_1^{qh} the average longitudinal and transverse velocities of sound can be used (Willis 1969, Jennings 1970). If there are no data on elastic properties of the crystal studied, the TDS correction can be calculated by the Born–Karman model with approximation of the pair interactions by atom–atom potentials (Criado *et al* 1985). If the reciprocal space scan volume and Brillouin zone of the crystal are approximated by spheres, then the TDS correction depends only on the sphere radius ratio (Stevenson and Harada 1983). The empirical TDS correction of the thermal parameters derived by least squares can be made in this case without knowledge of the elastic constants of the crystal. This method has been checked on some compounds; the results are listed in table 3.3. They demonstrate that, in spite of the very approximate character of the correction, satisfactory compensation of the TDS effect on thermal parameters can be achieved. Another empirical approach to the elimination of the TDS has been suggested by Blessing (1987). He has approximated the TDS peak under

Table 3.3 Corrections $\Delta B = 8\pi^2 \Delta\bar{u}^2$ (Å^2) to atomic thermal parameters for some crystals (Stevenson and Harada 1983).

Crystal	ΔB calculated from elastic constants	ΔB calculated without elastic constants
KCl	0.16	0.14
CdS	0.08	0.11
CdSe	0.11	0.12
BaF_2	0.05	0.05

the total reflection intensity profile by triangle, exponential, Gaussian, or Lorentzian functions. The numerical characteristics of the function chosen are determined by its fitting to the total intensity profile in the regions of peak profile wings. Zavodnik and Stash (1996) have modified this approach. They used low-angle (TDS-free) reflections in order to obtain an average normalized 'reference' profile of the Bragg peak which takes into account instrumental and other functions. Such profiles with addition of TDS profiles (approximated by a parabola) and accounting for spectral broadening are calculated for all reflections and are adjusted to experimental peaks. As a result, an estimate of TDS correction is obtained. The first applications of this method to urea and sodium chloride (Zavodnik *et al* 1994) have shown that it results in TDS correction values which are in good agreement with calculated quantities.

There is one more effect, called the critical scattering, which arises in the vicinity of a phase transition (Landau 1937). It gives a nonlinear contribution, which looks like an addition to TDS and is more prominent for weak reflections. This scattering results from the correlated displacements of some atoms in a crystal due to long-range density fluctuations while approaching the phase transition. The modulation of atomic thermal parameters is a result of this phenomenon, if it is not corrected. There is no general procedure to account for critical scattering. Harada (1988) has shown that if this scattering is caused by a soft phonon mode then a corresponding correction can be treated as an addition to the TDS correction. That addition value depends on the closeness to the phase transition temperature, on the density fluctuation character, and on the phonon structure factor components along the main reciprocal lattice directions.

3.3.3 *Extinction*

Extinction appears when the x-ray radiation passes through the crystal and there is energy redistribution between the primary and reflected beams. The kinematic theory of diffraction supposes that the primary x-ray beam has equal access to any point into a crystal. However, absorption, crystal structure defects, and wave interference destroy this supposition. Intensities of strong, mainly low-angle, reflections became lower than expected. Therefore, in order to make the kinematical theory valid the experimental results need to be corrected. For this purpose, a model is proposed (Darwin 1922) in which a crystal is presented as a mosaic set of perfect crystalline blocks disoriented with respect to each other (figure 3.2). The secondary reflections from different blocks, which are in the same reflecting position, appear to be weaker (secondary extinction), even though true absorption has not taken place at all. The rescattering probability is proportional to the mosaic block orientation distribution and to the crystal size. Besides, in the case of large blocks the interference of beams coherently scattered by different parts of the same mosaic block takes place (primary extinction).

To account for these effects the factor $y(\boldsymbol{H}) = y_1(\boldsymbol{H}) \cdot y_2(\boldsymbol{H})$ is introduced into (3.48), where y_1 and y_2 are the corrections for primary and secondary extinctions. These corrections account for the decrease in intensity by interference in the same and in different blocks, correspondingly. In order to take extinction into account both corrections should be expressed as a function of scattering angle, block size, crystal size, and degree of mosaic block disorientation. The block size and disorientation characteristics can be measured by synchrotron radiation rocking curves (Hoche *et al* 1987). Because of the small angular divergence of the synchrotron beam ($\sim 20''$) the rocking curve half-width is nearly equal to the crystal mosaicity. Another possibility is to analyse the diffraction peak measured with a four-circle diffractometer in $\Delta\omega$–$2\,\Delta\theta$ coordinates (Mathieson and Stevenson 1986): the block disorientation can be estimated in this way. High-resolution electron microscopy (Cowley and Smith 1987) can also be used for this. However, the extinction correction is usually determined by least squares together with crystal structural parameters.

The neglect of extinction leads to the distortion of the scale factor and deviation of the atomic thermal parameters, i.e. leads to systematical errors in x-ray diffraction experiments. Moreover, physically meaningful anisotropic thermal parameters cannot be obtained in some cases without taking extinction into account: the diagonal elements of thermal motion tensors can appear to be negative.

The example of the Be crystal demonstrates how extinction influences the deformation ED. The set of 27 structure amplitudes for the plate-shaped and 1 mm thick crystal was measured by Brown (1972) with Ag Kα radiation. Brown had arrived at results which she accepted as indication that extinction in the sample was negligible. However, Suortti (1982) and Larsen and Hansen (1984) have shown that the first three reflections (100), (002), and (101) from Brown's structure amplitude set were reduced by 6–12%. Besides, Larsen *et al* (1980) established that Brown's data set as a whole is lower than absolute values. In order to correct the Brown set, Tsirelson *et al* (1987a) have calculated the theoretical structure amplitudes using the Larsen *et al* (1980) neutron thermal parameters and atomic scattering functions and have determined the scale factor using high-order reflections with $\sin\theta/\lambda > 0.66$ Å^{-1} only. In fact, Brown's data set appeared to be lowered by 7%. Then, the structure amplitudes were rescaled and the $|F_0|$ values for (100), (002), and (101) reflections were substituted by those measured by Suortti (1982) in an absolute scale with experimental correction for primary and secondary extinctions. The new structure amplitudes appeared to coincide within estimated standard deviation with those measured by Larsen and Hansen (1984) with γ-rays.

The deformation ED in the internuclear space of the Be crystal is not changed qualitatively (figure 3.9) after the procedure described. However,

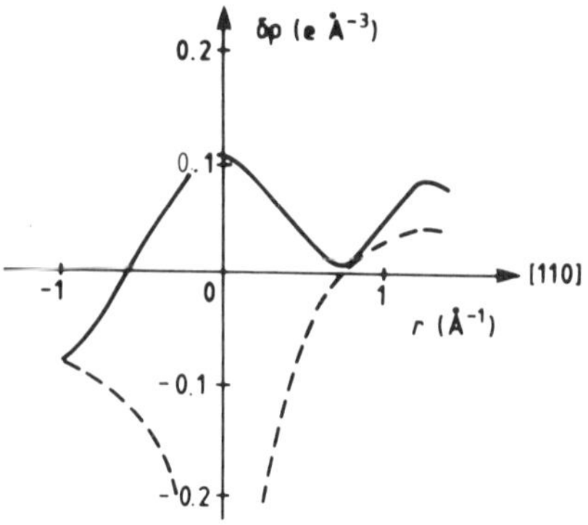

Figure 3.9 Extinction effects on the deformation ED, $\delta\rho$, of the Be crystal: dashed line, $\delta\rho$, calculated from Brown's original (1972) data; solid line, $\delta\rho$, calculated with Brown's data placed in an absolute scale, after substitution of (100), (002), and (101) reflections by Sourtti's extinction-free (1982) reflections.

significant changes appear near nuclei. The $\delta\rho$ function changes its sign here and reaches the value 0.11 e Å^{-3}, whereas the change of scale factor only led to the value 0.03 e Å^{-3}. Therefore, the $\delta\rho$ increase observed can be mainly attributed to correction of the x-ray data set for extinction. Note that the sign of the deformation EP at nucleus positions is also changed.

The deformation ED distortions in electron lone-pair regions as a result of extinction neglect or inadequate correction procedure were also observed in different crystals by Tanaka (1979), van der Wal *et al* (1987), and Tsirelson (1989). This may influence the confusion about the valence state of an atom in a crystal. Thus, extinction correction is an important stage of experimental ED study.

Let us consider now the modern aspects of extinction treatment in accurate x-ray diffraction studies. The extinction theory usually suggests that absorption, anomalous scattering, and TDS have already been taken into account. Becker and Coppens (1974a, b, 1985) gave a widely used, sufficiently general pseudokinematical way to calculate y correction for a mosaic crystal. They took into account that the successive x-ray scatterings from different mosaic blocks are incoherent and so the beam interaction can be described via beam intensities, not amplitudes. The interference term which depends on wave phases can be averaged before consideration of wave interaction in this case. The intensity balance is given by the Hamilton (1957) equations

$$\begin{aligned} \partial I_1/\partial t_1 &= -\sigma(I_2 - I_2) \\ \partial I_2/\partial t_2 &= -\sigma(I_2 - I_1) \\ \partial I_1/\partial t_1 &+ \partial I_2/\partial t_2 = 0. \end{aligned} \tag{3.56}$$

Here I_1 and I_2 are the incident and reflected beam intensities, σ is the scattering cross-section reduced to unit volume, and t_1 and t_2 are coordinates of the scattering point along corresponding beam paths (figure 3.7). If the absorption is preliminarily accounted for, weakening of beam intensity is described by introduction of the absorption-averaged path length value $\bar{T}_\mu$. The extinction correction factor y can then be calculated by integration of (3.56) over all beam paths in a crystal sample. y depends on unit kinematic scattering intensity $Q_0=(r_0C^2|F_0|/V)^2\lambda^3/\sin 2\theta$ (C is a polarization factor which depends on mutual monochromator and crystal measured orientation—see Blessing 1987), on crystal size, and on mosaic orientation spread. The secondary extinction, y_2, depends also on the scattering cross-section within a single block and, therefore, on y_1. However, at first approximation the primary and secondary extinction corrections can be considered as independent and the approximate expression $y=y_1y_2$ is valid. Becker and Coppens (1974a, b) modified the integral form of expression (3.56) by supposing that mosaic blocks are spherical and the distribution of their orientation is uniform. They arrived at an expression for y with the block radius as a parameter. If the averaged block size $t<\Lambda=[\lambda/(Q_0 \sin 2\theta)]^{1/2}$, $\Lambda\sim 10^{-6}$ m is the extinction length (Cowley 1975), then the primary extinction can be neglected and the secondary extinction can be calculated in the following way. The σ value in (3.56) depends on the angle between incident beam and ideal Bragg scattering direction, ε. The angular block disorientation is described by function $W(\Delta)$ which is of the Gauss, Lorentz, or Fresnel type. Then the effective cross-section, $\sigma(\varepsilon)$, averaged over a sample which consists of many blocks ($\sim 10^6$–10^8), is presented as convolution over two functions: $\bar{\sigma}(\varepsilon)=\int \sigma(\varepsilon-\Delta)W(\Delta)\, d\Delta$. If block disorientation predominates, the crystal is designated type I (Zachariasen 1967). In the case of a diffuse W function, the block size predominates in the convolution integral and the crystal is designated type II. The latter case is always found with small scattering angles; however, it is impossible to recognize which factor is predominant in the general crystal. Parameters of block disorientation and effective block size are usually included in the list of parameters of a crystal structural model and are determined at the stage of model refinement by least squares after TDS and absorption corrections.

The expressions which Becker and Coppens used in order to approximate the correction factors y_1 and y_2 formally have the same form. If the sample and the mosaic blocks are spherical and the block orientation distribution is isotropic, the corresponding expression has the form

$$y_{1,2}=[1+D_{1,2}x+A_{1,2}(\theta)x^2/(1+B_{1,2}(\theta)x)]^{-1/2} \qquad x=\tfrac{2}{3}Q_0\alpha_{G(\mathfrak{F},L)}\bar{T}_\mu \tag{3.57}$$

which is valid if $\sin\theta/\lambda<0.9$ Å^{-1} and $\mu r<4$ (r is the radius of the mosaic block). The coefficient $D_{1,2}$ and functions $A_{1,2}$, $B_{1,2}(\theta)$, and $\alpha_{G(\mathfrak{F},L)}$ are different for primary and secondary extinction. Sabine (1988) recomends using as block disorientation function a Gaussian (G) for secondary extinction and a Fresnelian ($\mathfrak{F}$) or Lorentzian (L) for primary extinction.

Extinction is often corrected according to the Zachariasen (1967) simplified treatment which starts from the same equation (3.57) as the general Becker and Coppens approach. In the Zachariasen approach the angular dependence of mosaic block thickness and the primary extinction are neglected. This excludes crystals with $\bar{t} \gg \Lambda$ from consideration. The corresponding expression for extinction correction resulting from suppositions of spherical blocks with radius r and a Gaussian orientation distribution with the parameter g is

$$y=(1+2x)^{-1/2} \qquad x=Q_0\bar{T}_\mu r^*/\lambda. \tag{3.58}$$

Here $r^*=r[1+(r/\lambda g)^2]^{-1}$ is the generalized extinction parameter which is included in the crystal structural model. In general, r^* has no clear physical meaning. However, for a type I crystal with $(r/\lambda)\gg g$ and $r^*\approx\lambda g$, the mosaic block disorientation can be evaluated by this parameter. In a type II crystal with $(r/\lambda)\ll g$ and $r^*\approx r$, r^* is a measure of block size.

Comparing equations (3.57) and (3.58), one can see that the additional θ dependence of the extinction parameters, which is present in Becker–Coppens theory, is absent in the Zachariasen approach. This leads to a difference in correction values at high and small scattering angles in the case of significant extinction in irregularly shaped crystals. Besides, the coordinate system used by Zachariasen differs from that which was used by Hamilton (1957). Fortunately, this has not practically influenced numerical results, as was shown by Brown and Fatemi (1974). Further, Becker and Coppens (1974a) tried to take interference inside a single block into account. However, their approach, as was noticed by Willis (1984), is not able to describe the beam energy redistribution at diffraction by a perfect crystal: the dynamical wave theory (Cowley 1975), not the differential equations for intensities (3.56), should be applied for this purpose.

It is important to note that extinction parameters are refined in practice by least squares together with other crystal structural model parameters. This makes the extinction correction model dependent. The model structure amplitudes are often calculated at this stage with the atomic scattering amplitudes for free spherical atoms or ions, i.e. the electron redistribution in a crystal due to chemical bond formation is neglected. As a result, the θ dependence of the y correction (3.57) in the Becker and Coppens theory can be severely distorted because of the inadequacy of the structural model used. For this reason, Cooper (1979) has doubted the advantage of the Becker and Coppens approach from a practical point of view. The same model defect does not permit one to consider the Zachariasen r^* parameter as the averaged mosaic block size even in an ideal mosaic crystal. If the more flexible crystal structural models are used, this drawback is much less pronounced. However, the dependence of the y correction value on the least-squares weight scheme is retained (Streltsov and Maslen 1992).

Borovskaya *et al* (1989) have studied the influence of different extinction models on deformation ED. The extinction corrections were determined in the frame of an atomic-type structural model. It was found that in H_3BNF_3, where absorption is low ($\mu r = 0.043$), deformation ED maps are the same, within one estimated standard deviation (± 0.05 e Å^{-3}) if extinction corrections in Zachariasen and Becker–Coppens approximations are found (figure 3.10). All important chemical ED pecularities, such as bonding electron peaks on the internuclear lines and electron lone-pair peaks, near positions of N and F atoms are clearly seen on both maps. If extinction correction is neglected, these details are not so evident and corresponding deformation ED values are distorted.

It was noted that the changes of the deformation ED maps resulting from the extinction model used only slightly correlate with the reliability (or discrepancy) factor, R, which characterizes the crystal structural model adjustment (see expression (H.19) in appendix H). Besides, as can be seen in table 3.4, R-factor analysis gives no indication to decide which extinction correction model is favourable in comparison with full extinction neglect.

Borovskaya *et al* (1989) arrived at the same result with crystal $CdCr_2Se_4$, where absorption is high. The lowest R-factor has been achieved using the Becker–Coppens approach for a type II crystal with a Lorentzian block orientation function (see table 3.4). The deformation ED maps in this compound appear to be the same in the 0.2 e Å^{-3} limit for all extinction models.

Thus, Sabine's (1988) prediction was experimentally confirmed: the different extinction theories based on the same equation (3.56) and using the same constant σ in this equation should lead to close results.

It is unreasonable to expect that the mosaic block orientations are distributed isotropically in a real crystal. Hence, variations in azimuthal intensity dependence reflections could appear and symmetry-related reflections will be distinct also. Extinction anisotropy is more pronounced for strong reflections. It can be distinguished experimentally, provided a temperature gradient which changes the deformation distribution in the crystal is created (Seiler and Dunitz 1978). In order to take extinction anisotropy into account, the mosaic blocks are considered as ellipsoids anisotropically distributed in orientation. This model was realized in different variants by Coppens and Hamilton (1970), Thornley and Nelmes (1974), and Becker and Coppens (1975). The last model was mofidied by Popa (1987) who considered an ellipsoidal crystal with mosaic ellipsoidal blocks. A matrix of 3×3 dimension describes the block's size. The orientation distribution is described with Gauss or Lorentz anisotropic prabability laws, which depend on another matrix of the same dimension. Thus, the number of independent parameters which should be found increases from two to 12 when anisotropy

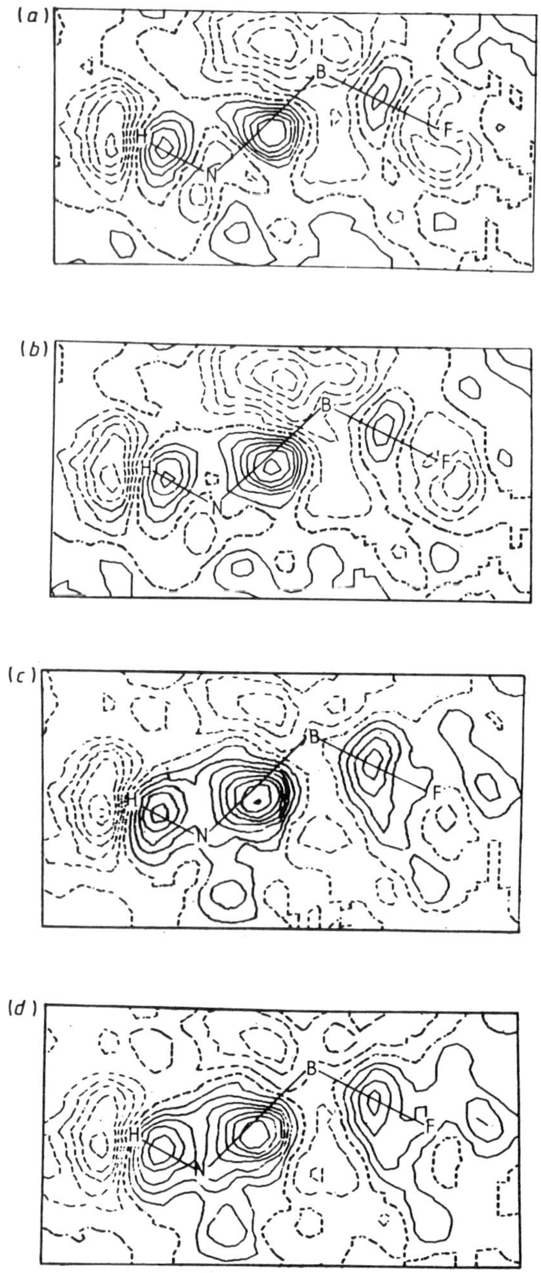

Figure 3.10 The deformation ED maps for the H_3NBF_3 crystal (Borovskaya *et al* 1989) with and without extinction corrections. (*a*) Without correction, (*b*)–(*d*) corrections according to (*b*) Kato, (*c*) Zachariasen, and (*d*) Becker and Coppens. Line intervals are 0.05 e Å^{-3}.

Table 3.4 Reliability factors as optimization indicators of rigid superposition crystal models for H_3NBF_3 and $CdCr_2Se_4$ with different manners of extinction consideration.

	R	
Method of extinction consideration	H_3NBF_3	$CdCr_2Se_4$
Without extinction correction	0.0300	0.0273
Zachariasen (1967) algorithm	0.0316	0.0158
Becker–Coppens (1974a) algorithm	0.0302	0.0135
Kato algorithm (Kawamura and Kato 1983)	0.0299	—

of extinction is taken into consideration. The probability of correlation between parameters to be refined by least squares increases also, especially that of correlation between extinction and thermal anisotropic parameters. All these circumstances diminish the advantage of this more rigorous approach. Therefore, Lewis and Schwarzenbach (1981a, 1982) are doubtful of the effectiveness of the anisotropic extinction correction, although it is in principle preferable.

The division of extinction into primary and secondary is somewhat artificial. The mosaic block model is formal and does not reflect the real crystal structure. Therefore, the extinction theory, which is free from suppositions on the degree of reflected beam coherence and on the distribution of defects in a crystal, is preferable. Such a theory was developed by Kato (1976, 1979, 1980, 1991, 1994) and Kawamura and Kato (1983) and modified by Becker and Al Haddad (1989). The dynamical Takagi–Taupin equations (Takagi 1969) were used:

$$\begin{aligned} \partial D_1/\partial t_1 &= ik_H \exp[i2\pi \boldsymbol{H}\cdot\boldsymbol{u}(\boldsymbol{r})]D_2 \\ \partial D_2/\partial t_2 &= ik_H \exp[-2\pi \boldsymbol{H}\cdot\boldsymbol{u}(\boldsymbol{r})]D_1 \end{aligned} \tag{3.59}$$

in which D_1 and D_2 are amplitudes of incident and scattered waves, k_H are imaginary values, $|k_H| = \Lambda^{-1}$, and $\boldsymbol{u}(\boldsymbol{r})$ is a function describing atomic displacements from their positions in a perfect crystal. The wave amplitude at a crystal point is consisidered as a sum of waves repeatedly scattered by other (nodal) crystal points. Suggesting that these waves are planar and that the resulting intensities can be presented as averaged over a statistical ensemble of nodal points, Kato introduced the correlation function

$$f(z) = \langle \exp i2\pi \boldsymbol{H}[\boldsymbol{u}(\boldsymbol{r}+z) - \boldsymbol{u}(\boldsymbol{r})] \rangle \tag{3.60}$$

which is characterized by the coherence length $\tau = \int_0^\infty f(z)\,dz$. The crystal defect structure is described here statistically and the wave interference is accounted for by the initial equations. As the distance between scattering points is increased the degree of scattered wave coherence is decreased. If the defect distribution is uniform and isotropic, then one can arrive at the Hamilton equations (3.56) provided $\tau < \Lambda$ (Kawamura and Kato 1983).

Solving equations (3.59) Kato described extinction correction for a spherical crystal by a single coefficient. Kawamura and Kato (1983) suggested for this a simple analytical approximation which is valid for $\mu r \leqslant 3.0$ and $\theta \leqslant 30°$. They supposed that for moderate ($y > 0.8$) extinction this correction value should be close to that calculated according to Becker and Coppens. However, Borovskaya *et al* (1989) obtained the opposite result. They found that in a low-extinction H_3BNF_3 single crystal corrected according to Kato the deformation ED map appeared to be different from all others (figure 3.10). Moreover, these maps were found to be closer to those where extinction was neglected. The electron lone-pair regions for N and F atoms suffered mainly. At the same time, the *R*-factor for the Kato algorithm is the smallest (see table 3.4). Hence, statistical dynamical theory in the initial version cannot be considered reliable. The theory of Becker and Al Haddad (1989), where correlations between waves scattered by two centres were taken into account, gives much better agreement with experiment, as has recently been found by Schneider *et al* (1992) and Takama and Harima (1994).

In some special cases dealing with large perfect mosaic blocks, the stationary waves, arising from the interference of incident and reflected radiation, can be created. The extremes of these waves can coincide with the crystal atomic planes or can occur in between them. When large atomic absorption takes place and the antinodes of the stationary wave coincide with the positions of atoms, a large absorption develops. In the opposite case, when the nodes of stationary waves coincide with interatomic space (free from absorbers), the absorption is less pronounced and can vanish completely. The absorption is connected with electron excitation in atoms and this is a local phenomenon. Therefore, there are some directions in a crystal where an abnormal radiation transmission can be observed. This dynamical phenomenon leads to the remarkable reduction of absorption and is called the Borrmann effect (Cowley 1975). In particular, it is pronounced in inorganic crystals and can influence extinction very much. Zachariasen (1968) tried to account for the Borrmann effect with his mosaic crystal model; however this approach cannot be considered satisfactory: it is necessary to apply wave theory in order to take the Borrmann effect into account properly.

Thus, the problem of extinction consideration has not been resolved properly until now. Some attempts to work out its solution can be mentioned. Olechnovich and Markovich (1980) have obtained exact solutions of the Takagi–Taupin dynamical equations (3.59) for crystals of cylinder and parallelepiped forms and suggested a simple one-parameter approximation for primary extinction in a spherical sample. This approach can be combined with the secondary-extinction treatment with the energy transfer equations (3.56) and leads to results coinciding with those after the Becker and Coppens extinction treatment, if $y > 0.5$ (Aslanov *et al* 1989). In the same way, Kulda (1987, 1988a, b) has described blocks inside which the dynamical theory is valid as a stochastic sequence of fragments with elastically deformed borders. The first application of this theory to an SrF_2 crystal showed that closer agreement of experimental and calculated structure amplitude as well as more realistic thermal atomic parameters has been obtained with this approach. The minimal-model-dependent approach to extinction has also been suggested by Maslen and Spadaccini (1992). The extinction correction, applied to irregularly shaped samples, is evaluated by optimization of parameters in any extinction formula over intensities of equivalent reflections with different path lengths. The correlation between this correction and other parameters of the crystal structural model is absent in this case. However, non-uniformity of the incident x-ray beam and anisotropic diffraction phenomena, TDS for example, can distort the extinction estimate if they are not taken into account properly.

3.3.4 *Anomalous scattering*

The kinematic theory of diffraction connects the intensities of reflections and the squares of coherent scattering amplitudes. This theory is based on the first term of the Born perturbation theory series. If there are anomalous scatterers in a many-electron system then the Debye–Waller theory is no longer valid and the second term of the perturbation theory series should be considered. Precisely speaking, the anomalous scattering should be introduced into consideration at the stage of calculation of the x-ray quantum scattering probability in the $\mathbf{k}_1$ direction. However, this would complicate the intensity expression very much. Moreover, it is not clear how this model should be parametrized in order to make correct calculation possible. Therefore, an approximate method is applied to correct intensities for anomalous scattering. For this purpose the atomic scattering amplitude, $f(s)$, $s = \sin\theta/\lambda$, is presented in the form $f(s) = f_0(s) + \Delta f'(s) + \mathrm{i}\,\Delta f''(s)$, where f_0 is described by the expression (3.9) and $\Delta f'$ and $\Delta f''$ are the corrections for real and imaginary parts of anomalous atomic scattering. In such a form f is introduced into the expression for structure amplitude, F_{model}, which describes the refined crystal structural model.

For light atoms at the usual wavelengths, the values $\Delta f'$ and $\Delta f''$ are small, but they increase non-monotonically with atomic number and for 4d elements can reach 10–15% of the f_0 value at high scattering angles. This can lead to significant errors in deformation ED maps, especially near atomic nuclei due to errors in the thermal parameters and scale factor.

Values $\Delta f'$ and $\Delta f''$ for free atoms can be calculated. For crystals they can be measured by special methods such as interferometry and attenuation methods (see e.g. Creagh 1985). The anomalous scattering appears as the result of inner electron excitation and the relativistic effects are essential here. Therefore, relativistic methods in the electric dipole (Cromer and Liberman 1970, 1981, 1983) and quadrupole (Creagh 1988, Kissel and Pratt 1990) approximations, as well as DM formalism (Kissel *et al* 1980) are used for $\Delta f'$ and $\Delta f''$ calculations. Creagh (1988) found that the two last methods give good consistent results which are not far from experimental data. Their precision is estimated as $\sim 5\%$.

The uncertainty in structure amplitudes due to non-accuracy in $\Delta f'$ and $\Delta f''$ values is usually less than 1% (far from resonance frequencies). However, the atomic resonance frequency depends on the atomic valence state and may be shifted for the same atom in different compounds causing change in $\Delta f'$ and $\Delta f''$ values (Jensen *et al* 1994). In practice this effect is completely neglected, as well as possible anisotropy of anomalous scattering (Templeton and Templeton 1988). Tables of $\Delta f'$ and $\Delta f''$ values calculated for all atoms in the relativistic electric multipole approximation have been recently published by Wilson (1992).

In order to correct the coefficients in ED Fourier series for anomalous dispersion the moduli of experimental structure amplitudes are usually recalculated by

$$|F_0| = \{|F_{model}|/|F^{an}_{model}|\}|F^{an}_0|. \tag{3.61}$$

The subscript 'an' indicates that the anomalous scattering contribution is included in the structure amplitude. Another method was suggested by de Titta as cited by Takazawa *et al* (1988). The corrected complex structure amplitude is presented as

$$F = \frac{|F^{an}_0|}{|F^{an}_{model}|}[A^{an}_{model} + \mathrm{i}B^{an}_{model}] - [(A^{an}_{model} - A_{model}) + \mathrm{i}(B^{an}_{model} - B_{model})] \tag{3.62}$$

(the notation $F = A + \mathrm{i}B$ is used here). From a general point of view such a method for the correction is more favourable because the uncertainties in phases resulting from anomalous scattering in non-centrosymmetric structures, especially for weak reflections, are minimal. Takazawa *et al* (1988) and Zhurova *et al* (1991) noted that the noise level on ED maps is less if expression (3.62) is used, though all chemically significant details of maps are the same for both correction approaches.

3.3.5 *Moduli of kinematic structure amplitudes and their variances*

The coherent elastic x-ray scattering contribution in the intensity $\bar{T}(H)$ is described by the expression

$$|F_0(\boldsymbol{H})|^2 = \bar{T}(\boldsymbol{H}) \bigg/ \left[\left(1 + \sum_{k \geqslant 1} \alpha_k(\boldsymbol{H})\right) A(\boldsymbol{H}, \mu) y(\boldsymbol{H}) k^2\right]. \tag{3.63}$$

Now we can move from $|F_0(\boldsymbol{H})|^2$ to moduli of structure amplitudes. This procedure is not trivial because of the statistical fluctuations in experimental data. The distribution of $|F_0|^2$ values about the 'true' values is symmetrical but this is not the case for $|F_0|$ (Rees 1977a, Wilson 1979). Therefore, the transition from $|F_0|^2$ to $|F_0|$ is not a linear procedure and can be most correctly realized by expansion of $|F_0|^2$ values into a series over small fluctuations around their average values, $|F|$. The average value of fluctuation is zero and the average value of fluctuation squared is the variance $\sigma^2(|F_0|^2)$. If $|F_0|^2 \gg \sigma(|F_0|^2)$, then, retaining the second-order term in the Taylor series, one can write the average value of kinematic structure amplitude in the form

$$|F| \cong |F_0| + \tfrac{1}{8}\sigma^2(|F_0|^2)/|F_0|^3. \tag{3.64}$$

It can be seen that the $|F(H)|$ values depend on the precision of $|F_0(H)|^2$ determination. The addition to $|F_0|$ is statistically significant for weak reflections. It is easy to show that this addition is worthwhile if $3\sigma(|F_0|^2) \leqslant 5\sigma(|F_0|^2)$. The expansion (3.64) is not valid for the weak reflections with $|F_0|^2 < 3\sigma|F_0|^2)$. Rees (1977a, b) and Blessing (1987) considered the behaviour of the average value of $|F_0|^2/\sigma(|F_0|^2)$ in comparison with the 'true' distribution $|F|^2/\sigma(|F|^2)$ and concluded that an approximation $|F| = c|F_0|$ with $c = 1$ if $|F_0|^2 < 0.3\sigma(|F_0|^2)$ is valid for very weak reflections. Table 3.5 should be used to chose a c value if $0.3\sigma|F_0|^2) \leqslant |F_0|^2 \leqslant 5\sigma(|F_0|^2)$.

If all corrections in (3.63) are properly introduced, the errors of contributions in $|F_0(\boldsymbol{H})|^2$ can be considered as random and treated with statistical error analysis (appendix H). In the case of independent errors, the variance of $|F_0|^2$ is given by the expression

$$\sigma^2[|F_0(\boldsymbol{H})|^2] = |F_0(\boldsymbol{H})|^4 \left\{ \frac{\sigma^2[\bar{T}(\boldsymbol{H})]}{\bar{T}^2(\boldsymbol{H})} + \frac{\sum_{k\geqslant 1}\sigma^2[\alpha_k(\boldsymbol{H})]}{[1+\sum_{k\geqslant 1}\alpha_k(\boldsymbol{H})]^2} + \frac{\sigma^2[A(\boldsymbol{H},\mu)]}{A^2(\boldsymbol{H},\mu)} \right.$$
$$\left. + \frac{\sigma^2[y(\boldsymbol{H})]}{y^2(\boldsymbol{H})} + 4\frac{\sigma^2(k)}{k^2} \right\}. \tag{3.65}$$

The value of the first term in the curly brackets is approximately $(1\text{–}2) \times 10^4$, provided all corrections have been adequately introduced. The uncertainty of TDS correction depends on the precision of elastic constant determination ($\sim 5\%$); for quasiharmonic correction the corresponding term in (3.65) is $\sim 10^{-4}$. The uncertainty of the absorption correction depends on the precision of the external sample form determination and on the precision of the linear absorption coefficient values. The average estimate of the third term in (3.65) is $\sim 4 \times 10^{-4}$. Other terms are very difficult to estimate. The main contribution to variances gives the uncertainty of the models used. For this reason, the extinction and scale factor variance estimates, as calculated from results of least-squares refinement, cannot be reliable. They are usually strongly underestimated. Our own experience shows that the estimate of the precision of both values is $\sim 1\%$ in the best case.

Table 3.5 Correction factors for the bias in $(F_0^2)^{1/2}$ toward values for F_0.

$F_0^2/\sigma(F_0^2)$	$F/(F_0^2)^{1/2}$	$F_0^2/\sigma(F_0^2)$	$F/(F_0^2)^{1/2}$
$\leqslant 0.3$	1.000	1.5	1.065
0.4	1.040	2.0	1.040
0.5	1.080	2.5	1.025
0.6	1.095	3.0	1.015
0.7	1.100	3.5	1.010
0.8	1.100	4.0	1.008
0.9	1.095	4.5	1.007
1.0	1.090	5.0	1.005
1.1	1.085	>5.0	$1+\frac{1}{8}[\sigma(F_0^2)/F_0^2]^2$
1.2	1.080		

For reflections with $|F_0|^2 > \sigma(|F_0|^2)$ the expression

$$\sigma^2(|F|^2) \cong \sigma^2(|F_0|) \cong \sigma^2(|F_0|^2)/4|F_0|^2 \tag{3.66}$$

is usually used for estimate of variances of the experimental structure amplitudes. Rees (1976) recommends the use of the estimate

$$\sigma^2(|F|) \cong \sigma(|F_0|^2)/4 \tag{3.67}$$

for very weak reflections with $|F_0|^2 \leqslant \sigma(|F_0|^2)$.

3.4 Crystallographic structural models

3.4.1 *Crystal structural models and phases of the structure amplitudes*

The set of structure amplitude moduli, $|F|$, and estimates of their variances, $\sigma^2(|F|)$, are the result of the x-ray data treatment procedure described. The problem of reconstruction of the ED from these data closely resembles the inverse problem of scattering. However, in this case it is not an interaction law that should be found but rather the density of scattering matter, the ED. Unfortunately, the x-ray diffraction experiment as formulated above does not provide sufficient information to solve this problem. First of all, the information concerning the phases of kinematic structure amplitudes is lost: the measurements with diffractometers and kinematic theory of diffraction give only absolute values of the structure amplitudes. This information allows us to calculate the unit cell parameters (with an accuracy to 10^{-14} m) and to determine the crystal space group, but not the ED itself: in order to calculate ED in the form of a Fourier series the phases of the reflections are needed. These phases are determined by the choice of unit cell origin; however, the transfer of origin changes phases of all reflections simultaneously and the ED of the crystal appears to be independent of the coordinate system.

In principle, the manifold scattering (see appendix F) or anomalous scattering of x-rays or γ-rays can be applied to measure phases of reflections. However, other well developed methods are usually used for this purpose, the so-called 'direct methods' (see e.g. Ladd and Palmer 1980). The direct methods are based on the analysis of trigonometrical and statistical relationships between structure amplitude moduli which result from the crystal symmetry and ED properties. In particular, these concern the ED sign constancy in position space and the fact that ED maxima coincide with nuclear positions (see (2.51) and (2.52)). As a result, the approximate positions of atoms in a unit cell can be found by direct methods. Since the EDs of atoms are monotonically decreasing functions (see (2.53)), phases of their Fourier transforms are caused by the geometric parts of structure amplitudes, e.g. nuclear positions. This fact solves, in principle, the phase problem for reflections whose contribution to the scattering is described by the sum of atomic EDs.

In order to refine the nuclear positions in the unit cell, the electron dynamic structural model of the crystal, parameters of which give the best agreement with x-ray data, is usually used. This model is an estimator (see appendix H) of the structural parameters of the crystal. It also determines the accuracy of calculation of the reflection phases. That is why the problem of adequate crystal structural model choice is very important for ED reconstruction with a Fourier series. Moreover, the parametrization of the model mentioned can be performed in such a fashion that characteristics of the crystal electron structure can be included in it. Let us consider the problem of describing a crystal by different structural models in detail (Tsirelson *et al* 1991).

We should first mention the fact that all structural models used in crystallography are of the superpositional type. They approximate a crystal by superposition of pseudoatomic fragments: electron basins with nuclei inside. Therefore, the numerical ED characteristics which can be calculated with such models are the characteristics of pseudoatoms. This distinguishes the crystallographic structural models from quantum chemical ones. In the latter the crystal electron structure is described by wave-functions or by orbital decomposition of the DM without division of a crystal into pseudoatomic pieces. It should be noted that for weak reflections and those which are associated with ED deviations from the superpositional model the phases calculated with crystallographic models can be distorted.

Generally, the EDs of pseudoatoms, $\rho_\mu(\boldsymbol{r})$ in (3.22), are not spherical and their vibrations, described by the PDF, $p_\mu(\boldsymbol{u})$ in (3.24), are anharmonic. This means that the Fourier transforms of both functions, $f_\mu(\boldsymbol{H})$ and $T_\mu(\boldsymbol{H})$, have a complex form in reciprocal space. Let us present, after Dawson (1967a), ρ_μ and p_μ functions as the sums of centrosymmetric (c) and antisymmetric (a) parts. Then the model structure amplitudes can be written in general form as

$$F_{model}(\boldsymbol{H})=\sum_{\mu}\{[f_{\mu}^{c}(\boldsymbol{H})T_{\mu}^{c}(\boldsymbol{H})-f_{\mu}^{a}(\boldsymbol{H})T_{\mu}^{a}(\boldsymbol{H})]$$
$$+\mathrm{i}[f_{\mu}^{c}(\boldsymbol{H})T_{\mu}^{a}(\boldsymbol{H})+f_{\mu}^{a}(\boldsymbol{H})T_{\mu}^{c}(\boldsymbol{H})]\}\exp(2\pi\mathrm{i}\boldsymbol{H}\cdot\boldsymbol{r}_{\mu}). \tag{3.68}$$

The antisymmetric atomic scattering functions, f_{μ}^{a}, differ from zero only in the multipole ED model, and T_{μ}^{a} components appear when anharmonicity is taken into account (they are linked with terms which contain the scattering vector of odd orders). The parametrization of functions f_{μ}^{i} and T_{μ}^{i} allow one to determine electron and thermal vibration characteristics of pseudoatoms by minimizing the functional

$$D_{w}=\sum_{\boldsymbol{H}}w(\boldsymbol{H})[|\mathfrak{F}(\boldsymbol{H})|-y^{1/2}k^{-1}|F_{model}(\boldsymbol{H})|]^{2}. \tag{3.69}$$

Here $|\mathfrak{F}|$ are structure amplitude moduli measured in an arbitrary scale, k is the scale factor ($k|\mathfrak{F}|=|F|_{0}$), y is the extinction correction, and $w(\boldsymbol{H})$ are statistical weights of measurements. In the majority of structure investigations the least squares (LS) are used to minimize the functional (3.69). The indicators of the degree of model optimization are considered in appendix H. It is clear that model defects would directly influence the values of all parameters obtained even if atomic positions are fixed, say, due to crystal symmetry restrictions. That is why the adequate structural modelling of a crystal and the determination of the parameters of models are very important for an ED study.

3.4.2 *Description of the pseudoatom thermal vibrations*

The crystalline potential near atomic nuclei, which determines their vibrations, is usually not known, and therefore some approximations are used for its description. As has been shown in section 3.1, if the superpositional, atomic-type crystal structural model is used, the vibrations of each pseudoatom can be considered as taking place in the effective potential produced by its neighbours. This is equivalent to considering the crystal potential energy (3.17) as a sum of one-particle atomic-like terms $\Phi(\{\boldsymbol{u}\})=\sum_{\mu}V_{\mu}(\boldsymbol{u})$ and to introducing, for every atom, the temperature factor $T_{\mu}(\boldsymbol{H})$ (3.29) in reciprocal space. The temperature factor and effective PDF, describing the atomic displacement $\boldsymbol{u}$ from its equilibrium position, $p_{\mu}(\boldsymbol{u})$, are connected by Fourier transformation

$$T_{\mu}(\boldsymbol{H})=\int_{-\infty}^{\infty}p_{\mu}(\boldsymbol{u})\exp(2\mathrm{i}\boldsymbol{H}\cdot\boldsymbol{u}_{\mu})\,\mathrm{d}\boldsymbol{u}_{\mu}. \tag{3.70}$$

In its turn, the PDF is determined by the effective potential near the equilibrium position:

$$p_{\mu}(\boldsymbol{u})=(1/z_{\mu})\exp\{-\beta V_{\mu}(\boldsymbol{u})\}. \tag{3.71}$$

The normalizing factor is $z_\mu = \iiint_{-\infty}^{+\infty} \exp\{-V_\mu\}(\boldsymbol{u})\, d\boldsymbol{u}_\mu$, $\beta = (kT)^{-1}$. The expressions (3.29) and (3.70) are the initial ones used in diffraction structure analysis for description of atomic thermal vibrations. The main idea consists of approximation of the terms depending on different orders of vector $\boldsymbol{H}$. This can be done in different ways. It is possible to expand V_μ or p_μ functions for all atoms into a Taylor series over small atomic displacements from equilibrium positions and then to parametrize the temperature factors properly. Another way is to expand the temperature factor over cumulants considering the T_μ as characteristic functions of the atomic PDF. In all cases, a crystal structural model, apart from positional parameters, will contain parameters describing the atomic thermal vibrations. These parameters are determined by minimization of functional (3.69). Although both approaches are interrelated to each other by equations (3.70) and (3.71), only decomposition of the potential V_μ provides the direct physical interpretation of the parameters of the thermal motion. At the same time, the T_μ decomposition can be used in the description of statistiscally disordered structures (see Kuhs 1988, 1992).

The harmonic (or, more exactly, quasiharmonic) approximation is often used in practice for the temperature factor. In this case $T_\mu^a = 0$ in (3.69) and the temperature factor is independent of initial suppositions accepted as thermal motion descriptions (see table 3.7). The thermal parameters refined are components of the second-order tensor $U_{ij}^\mu = u_b^\mu u_j^\mu$ which describes the motion of atom μ in terms of mean squares of the vibration amplitudes (the component dimension is m^2). However, in many cases a more adequate anharmonic approximation is needed when $T_\mu^a \neq 0$. Therefore, we will not restrict ourselves to the harmonic case.

Consider first the description that is based on the Taylor expansion of the one-particle potentials V_μ (Willis 1969). If higher-order terms are preserved in addition to the quadratic ones, this expansion has the form

$$V_\mu(\boldsymbol{u}) = V_\mu(0) + \sum_i \left[\frac{\partial V_\mu(\boldsymbol{u})}{\partial u_i}\right]_0 u_i + \frac{1}{2!}\sum_{ij}\left[\frac{\partial^2 V_\mu(\boldsymbol{u})}{\partial u_i\, \partial u_j}\right]_0 u_i u_j + \frac{1}{3!}\sum_{ijk}\left[\frac{\partial^3 V_\mu(\boldsymbol{u})}{\partial u_i\, \partial u_j\, \partial u_k}\right]_0 u_i u_j u_k + \frac{1}{4!}\sum_{ijkl}\left[\frac{\partial^4 V_\mu(\boldsymbol{u})}{\partial u_i\, \partial u_j\, \partial u_k\, \partial u_l}\right]_0 u_i u_j u_k u_l + \cdots. \quad (3.72)$$

The derivative values are parameters of the potential. They are taken at potential energy minimum positions, usually identifying with equilibrium atomic positions; therefore, the second term in (3.72) vanishes. The origin is usually accepted as the point where $V_\mu(0) = 0$. Tanaka and Marumo (1983) adapted the expansion (3.72) for atomic positions with any symmetry. The temperature factor, which corresponds to the potential (3.72), contains quadratic, cubic, and quartic terms (U_{ij}^μ, C_{ijk}^μ, D_{ijkl}^μ, respectively). They are multipliers of terms which include the corresponding degrees of components of the reciprocal lattice vector $\boldsymbol{H}$. The transition to the crystallographic coordinate system, related to the crystal cell, allows one to obtain easily programmed expressions for T_μ. Its harmonic part has the usual form (see table 3.6) whereas the anharmonic part includes the components of the third- and fourth-order tensors, which are multipliers of the corresponding degrees of the $\boldsymbol{H}$ vector.

The point symmetry of atomic positions puts some restrictions on the possible independent values of the U_{ij}, C_{ijk}, and D_{ijkl} tensor components and significantly decreases their number. These restrictions follow from the condition of invariance of the potential energy (3.17) relative to symmetry transforms. The list of non-zero components of U_{ij}, C_{ijk}, and D_{ijkl} for different point symmetries can be found in the works of Ibers and Hamilton (1974) and Tanaka and Marumo (1983).

The possible divergence of the series (3.72) at some parameter values is a serious drawback of the potential decomposition. In this case, the PDF will not be an integrated function in the limits from $-\infty$ to $+\infty$. For example, for an atom in a cubic crystal field, the potential can be presented as

$$V_\mu(\boldsymbol{u}) = V_\mu(0) + \tfrac{1}{2}\alpha_\mu r^2 + \beta_\mu u_1 u_2 u_3 + \gamma_\mu r^4 + \delta_\mu(u_1^4 + u_2^4 + u_3^4 - \tfrac{3}{5}r^4) \qquad (3.73)$$

where $r^2 = u_1^2 + u_2^2 + u_3^2$. If the decomposition is restricted to the cubic term or the coefficient $\gamma_\mu < 0$, then $V_\mu(u) \to \infty$ at $u \to \infty$. The Boltzman distribution converges only when the last term in the expansion has even degree and the coefficient in front of it is positive.

Kurki-Suonio *et al* (1979) proposed taking anharmonicity into account by the decomposition of $V_\mu(\boldsymbol{u})$ over Hermite polynomials (eigenfunctions of the harmonic oscillator). It permits one to obtain Fourier-invariant expressions for PDFs and temperature factors (table 3.6), which make the interpretation of the atomic thermal parameters obtained much easier.

Let us now consider the parametrization of the PDF. According to Wilson (1992) the PDF can be presented as a series over increasing degrees of derivatives:

$$p_\mu(\boldsymbol{u}) = [1 - a^p\hat{D}_p + \tfrac{1}{2}b^{pq}\hat{D}_p\hat{D}_q - (1/3!)c^{pqr}\hat{D}_p\hat{D}_q\hat{D}_r + (1/4!)d^{pqrs}\hat{D}_p\hat{D}_q\hat{D}_r\hat{D}_s - \cdots]\,p_\mu^{harm}(\boldsymbol{u}) \qquad (3.74)$$

($\hat{D} = \partial/\partial u_p$ is the partial derivative operator). If the mean value and variance of the PDF are defined by the harmonic part, $p_\mu^{harm}(\boldsymbol{u})$, i.e. equilibrium atomic positions coincide with potential minima, the terms with components a^p and b^{pq} in this expression can be omitted. The atomic position symmetry restrictions for the c^{pqr} and d^{pqrs} coefficients are presented by Ibers and Hamilton (1974) and Kuhs (1984). Using the definition of multidimensional Hermite polynomials (Kendall and Stuart 1969), $H(\boldsymbol{u})_{pq\ldots w}p_\mu^{harm}(\boldsymbol{u}) = (-1)^w\hat{D}_p\hat{D}_q\ldots\hat{D}_w p_\mu^{harm}(\boldsymbol{u})$, and accepting that $a^p = b^{pq} = 0$, one can arrive at the Gram–Charlier expansion. Approximating the PDF (3.74) by the three-dimensional Edgeworth (1905) series, $p_\mu^E(\boldsymbol{u})$, decomposition, in which the coefficients k^{pqr} are included instead of tensor components $c^{pqr\dots}$ in the Gram–Charlier expansion, can be obtained (Johnson 1969). The coefficients in both decompositions are interrelated: $c^{pqr}=k^{pqr}$, $d^{pqrs}=k^{pqrs}$, $f^{pqrst}=k^{pqrst}$, $g^{pqrstv}=k^{pqrstv}+\frac{1}{72}k^{pqr}k^{stv}$. The extended Edgeworth decomposition, $p_\mu^{ext}(\boldsymbol{u})$, also can be used in which the mean value and variance of the PDF are split into harmonic and anharmonic parts. The formulae for all expansions mentioned are presented in table 3.6. The Fourier transforms of $p_\mu^E(\boldsymbol{u})$ and $p_\mu^{ext}(\boldsymbol{u})$ decompositions coincide, therefore numerical values of both functions should be the same as well. Zucker and Schulz (1982) have shown, however, that this is not always the case. Moreover, the mean value and variance of PDF are not always correctly determined from their harmonic part only, especially in the case of large anharmonicity, when an average atomic position is shifted to the site with lower symmetry.

The Gram–Charlier and Edgeworth decompositions over infinite number of terms are identical. However, a limited number of terms is really used and the results obtained with different models for the same crystal can be different. For example, Zucker and Schulz (1982) found, studying the Li_3N crystal at 888 K, that the occupancy of one of two crystallographically independent positions of Li atoms is 0.93(1) for the Edgeworth and 0.98(1) for the Gram–Charlier model (series up to fourth order were used). The discrepancy factors were 0.027 and 0.015, correspondingly. This result is typical. It permits one to recommend the Gram–Charlier decomposition in practice as it provides the better model optimization. This decomposition allows also the accurate Fourier transform of the PDF, whereas the Edgeworth temperature factor is always divergent in the reciprocal space due to its exponential form. Besides, the truncated Gram–Charlier decomposition formally does not suffer from indeterminancy of the phase of structure amplitudes as occurs for odd-order terms of the Edgeworth series for acentric space groups (Nelmes and Tun 1987, 1988, Hansen 1988). At the same time, as has been noticed by Kuhs (1992), the difference between the two approaches is the same as that between $\exp(-x)$ and $1-x$ and the refinement of any anharmonic structural model is, as a rule, impossible without fixing certain anharmonic terms (Hazell and Willis 1978).

The significant drawback of all PDF expansions and the corresponding temperature factors is the strong correlation between even- and odd-termed parameters separately which arises at the stage of structural model parameter refinement. Johnson (1969, 1980) proposed lowering this correlation by introduction of the quasi-orthogonalizing α parameter (table 3.6) into terms of third and fourth orders. However, Zucker and Schulz (1982) have found that this lowering can be insignificant.

A formal statistical approach to the description of atomic thermal motion is also possible. It uses the decomposition of the temperature factor over cumulants k^i, k^{ij}, ... which can be defined as follows (Johnson 1969):

Table 3.6 PDFs and temperature factors. $\boldsymbol{u}$ is the displacement vector of an atom from its equilibrium position, $\boldsymbol{H}$ is the scattering vector with components H_i, $(h_1, h_2, h_3)=(h, k, l)$, $H_{hqrz}(x)$ is the Hermite polynomial of zth order, $p, q, r, \ldots = 1, 2, 3, \ldots, n_1, n_2, n_3$ and $\lambda, \mu, v=0, 1, 2, \ldots, \beta^{pq}, B_i, c^{pqr}, k^{pqr}, c_{n_1n_2n_3}, d^{pqrs}$, and k^{pqrs} are components of tensors of second, third, fourth, etc rank correspondingly, a_p^* and a_q^* are elements of the reciprocal lattice vector, $\hat{P}_n$ is the permutation operator which indicates that the term in angle brackets must be averaged by all permutations of indices which give different terms. The sum sign in all expressions is omitted. The coordinate system for the Fourier-invariant expansion is coincident with the main axes of the thermal ellipsoid.

Expansion	PDF	Temperature factor
The harmonic approximation	$p_\mu^{harm}(\boldsymbol{u})=\{(\det[g])^{1/2}/(2\pi)^{3/2}\}\exp\{-1/2g_{pq}(u^p-u_0^p)(u^q-u_0^q)\}$ $[g]=2\pi^2[\beta]^{-1}$ $\beta^{pq}=2\pi^2a_p^*a_q^*u^pu^q$	$T_\mu^{harm}(\boldsymbol{H})=\exp\{-\beta^{pq}h_ph_q\}$
Gram–Charlier	$p_\mu^{GC}(\boldsymbol{u})=p_\mu^{harm}(\boldsymbol{u})\{1+(1/3!)c^{pqr}H_{pqr}(\boldsymbol{u})+(1/4!)d^{pqrs}H_{pqrs}(\boldsymbol{u})$ $+(1/5!)f^{pqrst}H_{pqrst}(\boldsymbol{u})+(1/6!)g^{pqrstv}H_{pqrstv}(\boldsymbol{u})+\cdots\}$	$T_\mu^{GC}(\boldsymbol{H})=T_\mu^{harm}(\boldsymbol{H})\{1+[(2\pi i)^3/3!]c^{pqr}h_ph_qh_r$ $+[(2\pi i)^4/4!]d^{pqrs}h_ph_qh_rh_s+[(2\pi i)^5/5!]f^{pqrst}h_ph_qh_rh_sh_t$ $+[(2\pi i)^6/6!]g^{pqrstv}h_ph_qh_rh_sh_th_v+\cdots\}$
Edgeworth	$p_\mu^E(\boldsymbol{u})=p_\mu^{harm}(\boldsymbol{u})\{1+(1/3!)k^{pqr}H_{pqr}(\boldsymbol{u})+(1/4!)k^{pqrs}H_{pqrs}(\boldsymbol{u})$ $+(1/3\times 4!)k^{pqr}k^{stv}H_{pqrstv}(\boldsymbol{u})+\cdots\}$	$T_\mu^E(\boldsymbol{H})=T_\mu^{harm}(\boldsymbol{H})\exp\{[(2\pi i)^3/3!]k^{pqr}h_ph_qh_r$ $+[(2\pi i)^4/4!]k^{pqrs}h_ph_qh_rh_s+\cdots\}$

Extended Edgeworth	$p_\mu^{ext}(\boldsymbol{u}) = p_\mu^{harm}(\boldsymbol{u})\{1 + k^p H_p(\boldsymbol{u}) + (1/3!)k^{pqr}H_{pqr}(\boldsymbol{u})$ $+ [(\tfrac{1}{2})k^{pq}H_{pq}(\boldsymbol{u}) + \tfrac{1}{2}k^p k^q H_{pq}(\boldsymbol{u}) + (1/3!)k^p k^{qrs} H_{pqrs}(\boldsymbol{u})$ $+ (1/4!)k^{pqrs}H_{pqrs}(\boldsymbol{u}) + (1/3\times 4!)k^{pqr}k^{stv}H_{pqrstv}(\boldsymbol{u}) + \cdots]\}$ $k^p = u^p_{anharm} - u^p_{harm} \qquad k^{pq} = \beta^{pq}_{anharm} - \beta^{pq}_{harm}$	$T_\mu^E(\boldsymbol{H}) = T_\mu^{harm}(\boldsymbol{H})\exp\{[(2\pi i)^3/3!]k^{pqr}h_p h_q h_r$ $+ [(2\pi i)^4/4!]k^{pqrs}h_p h_q h_r h_s + \cdots\}$
Cumulant	$p_\mu(\boldsymbol{u}) = p_\mu^E(\boldsymbol{u})$	$T_\mu(\boldsymbol{H}) = \exp\{2\pi i k^p h_p + [(2\pi i)^2/2!]k^{pq}h_p h_q$ $+ [(2\pi i)^3/3!]k^{pqr}h_p h_p h_r$ $+ [(2\pi i)^4/4!]k^{pqrs}h_p h_q h_r h_s + \cdots\}$
α-formalism	$p_\mu^\alpha(\boldsymbol{u}) = [p_\mu^{harm}(\boldsymbol{u})/M][1 + T_3(\boldsymbol{u}) + T_4(\boldsymbol{u}) + \cdots]$ $T_3(\boldsymbol{u}) = (c^{pqr}/3!)\{H_{pqr}(\boldsymbol{u}) + \alpha_\mu \hat{P}_3\langle H_p(\boldsymbol{u})g_{rs}\rangle\}$ $T_4(\boldsymbol{u}) = (d^{pqrs}/4!)\{H_{pqrs}(\boldsymbol{u}) + \alpha_\mu \hat{P}_{10}\langle H_{pq}(\boldsymbol{u})g_{rs}$ $+ \alpha_\mu^2 g_p \hat{P}_3\langle {}_q g_{rs}\rangle\}$ $M = 1 + (\alpha_\mu^2/4!)g_p \hat{P}_3\langle {}_q g_{rs}\rangle d^{pqrs}$	$T_\mu^\alpha(\boldsymbol{H}) = T_\mu^{harm}(\boldsymbol{H})[1 + [(2\pi i)^3/3!]G_{pqr}(\boldsymbol{H}, \alpha_\mu)c^{pqr}$ $+ [(2\pi i)^4/4!]G_{pqrs}(\boldsymbol{H}, \alpha_\mu)d^{pqrs} + \cdots]$ $G_{pqr}(\boldsymbol{H}, \alpha_\mu) = h_p h_q h_r - (\alpha_\mu/2)\hat{P}_3\langle[\beta]^{-1}_{pq}h_r\rangle$ $G_{pqrs}(\boldsymbol{H}, \alpha_\mu) = h_p h_q h_r h_s - (\alpha_\mu/2)\hat{P}_{10}\langle h_p h_q [\beta]^{-1}_{rs}\rangle$ $+ \alpha_\mu^2/4[\beta]_p^{-1}\hat{P}_3\langle {}_q[\beta]^{-1}_{rs}\rangle$
Fourier invariant	$p_\mu^{FIV}(\boldsymbol{u}) = 1/N\exp\{-\tfrac{1}{2}\sum B_i^2 u_i^2\}$ $\times[1 - \sum_{n_1}\sum_{n_2}\sum_{n_3} C_{n_1 n_2 n_3}H_{n_1}(B_1 u_1)H_{n_2}(B_2 u_2)H_{n_3}(B_3 u_3) + \cdots]$ $N = (2\pi)^{3/2}/(B_1 B_2 B_3)\left[-\sum_{\lambda\mu\nu}\frac{(2\lambda)!\,(2\mu)!\,(2\nu)!}{\lambda!\,\mu!\,\nu!}C_{2\lambda 2\mu 2\nu}\right]$	$T_\mu^{FIV}(\boldsymbol{H}) = [(2\pi)^{3/2}/(NB_1 B_2 B_3)]\exp\{-\tfrac{1}{2}\sum b_i^2 q_i^2\}$ $\times\{1 - \sum_{n_1}\sum_{n_2}\sum_{n_3} i^{(n_1+n_2+n_3)}C_{n_1 n_2 n_3}H_{n_1}(b_1 q_1)$ $\times H_{n_2}(b_2 q_2)H_{n_3}(b_3 q_3) + \cdots\}$ $b_i = 2\pi/B_i$

$k^i = \langle u^i \rangle$, $k^{ij} = \langle u^i u^j \rangle - \langle u^i \rangle \langle u^j \rangle$, $k^{ijk} = \langle u^i u^j u^k \rangle - \langle u^i \rangle \langle u^j u^k \rangle - \langle u^j \rangle \langle u^i u^k \rangle - \langle u^k \rangle \langle u^i u^j \rangle + 2 \langle u^i \rangle \langle u^j u^k \rangle$, etc. There is no exact Fourier transformation of the cumulant temperature factor. To obtain the PDF in this case, the Edgeworth approximation is usually used. This is why both approaches are the same, in practice.

It is impossible to say *a priori* which approach to describing it is better. Kuhs (1988) presented an example of study of anharmonicity in the ZnS crystal at two different temperatures with different formalisms. The discrepancy factors and goodness-of-fit indices for different anharmonic models were nearly the same for each temperature. Therefore, the confident choice of a model could not be made only with these, 'internal', statistical criteria: some additional considerations are necessary.

The use of expressions (3.70) and (3.71) is valid only in the high-temperature (classic) limit. At low temperatures the quantum statistics, as described in appendix B, should be used (Mair and Wilkins 1976, Kara and Merisalo 1982). If anharmonicity is such that the condition $[V_{anh}(\boldsymbol{u})] \ll kT$ is not fulfilled, the use of formal temperature factor decomposition over cumulants can be recommended.

The PDF must be positive everywhere in position space. This PDF property allows one to check the physical significance of the atomic thermal parameters achieved. Scheringer (1988) has evaluated the size of negative regions arising in a few crystals for different descriptions of the atomic thermal vibrations and has found that the negative PDF regions are not large, as a rule. If the experimental situation is within the validity of the approximation of the model used, these regions approach zero. At the same time, the cumulant PDF decomposition always has negative regions. Scheringer (1988) described also a mathematical procedure which provides a positive PDF over all space.

3.4.3 *Rigid-atom-type superpositional models*

The simplest approximation for the ED distribution in a crystal is the superpositional structural model consisting of spherical particles (atoms or ions) which vibrate in the same way as their nuclei. If these particles are considered as unchanged during model optimization (i.e. the changes of their electronic shell are not taken into account), we will call such a model a rigid-atom-type superpositional model. In this model, because of spherical averaging of atomic ED, $f^a_\mu = 0$ and $f_\mu = f^c_\mu$, the latter depending only on the scattering vector absolute value, $|\boldsymbol{S}|$, i.e. only on $\sin\theta/\lambda$ and not on $\boldsymbol{S}$ vector direction. The temperature factor and, consequently, the form of structure amplitude, F_{model} (3.68), depends on the type of approximation which is used for the atomic thermal vibration description, harmonic or anharmonic.

The atomic (ionic) x-ray scattering functions, f_μ, calculated for free particles by the relativistic HF method are usually used in (3.68). They are tabulated in numerical and parametric forms for all atoms and ions in their ground states for $0 \leqslant (\sin\theta/\lambda) \leqslant 2$ Å^{-1} by Ibers and Hamilton (1974) and for $2 \leqslant (\sin\theta/\lambda) \leqslant 6$ Å^{-1} in exponential polynomial form by Fox *et al* (1989). The calculation precision is 0.2–0.5% ($\sim$1% for elements with large atomic numbers). The drawback of these functions, the neglect of electron correlation, has been overcome recently by Su and Coppens (1994) and atomic scattering functions calculated taking account of electron correlation and relativistic effects are now available.

The mentioned uncertainty estimate characterizes the calculation precision but not the goodness of the approximation of the crystal ED by the rigid-atom superpositional structural model. Chemical bonds destroy the sphericity of atoms and f_μ functions depend both on the length and on the direction of scattering vector. The spherical atom model does not take ED deformation into account, therefore the structural and thermal parameters obtained will be influenced by model inadequacies. For example, a bonded oxygen atom is aspherical because of lone-pair electrons. This is so pronounced that oxygen positional parameters, as found with the rigid superpositional structural model with a total set of measured x-ray structure amplitudes included in the refinement, differ from those found by neutron diffraction by $(0.7–1.3) \times 10^{-2}$ Å (Coppens 1978). Hence, the problem is to find more precise parameters in the framework of a rather rough structural model of a crystal.

The simplest and most effective way to adapt the model discussed for obtaining more accurate structural parameters was suggested by Stewart (1968) on the basis of the analysis of the x-ray scattering by different atomic electron subshells. The electrons of a valence shell, which are responsible for the chemical bond formation, scatter the x-rays only in the low-angle region of reciprocal space. This can be seen in figure 3.11 which presents the partial atomic scattering amplitudes for different electronic subshells of the Si atom. The core electron distribution in a crystal is nearly the same as in free atoms, but the valence electrons are distributed in a crystal unlike those in the superpositional model. Hence, spherical scattering amplitudes are a poor approximation for valence electrons in a crystal. Stewart suggested the refinement of the atomic position, thermal parameters, and scale factor using high-angle reflection only, if the rigid superpositional structural model is used. This procedure has been examined many times and is now standard. Position and thermal atomic parameters derived with the superpositional model in the high-angle approximation coincide well with those determined from neutron scattering. Correspondingly, interatomic distances are determined more precisely from high-angle data (figure 3.12).

Where should one draw the border between low- and high-angle regions of the reciprocal space? Analysing the light- (C, N, O etc) atom spherical x-ray scattering functions, Stewart concluded that this border lies in the range 0.6–0.8 Å^{-1}. However, some pecularities of atomic valence subshell asphericity, especially electron lone pairs, can give a noticeable contribution

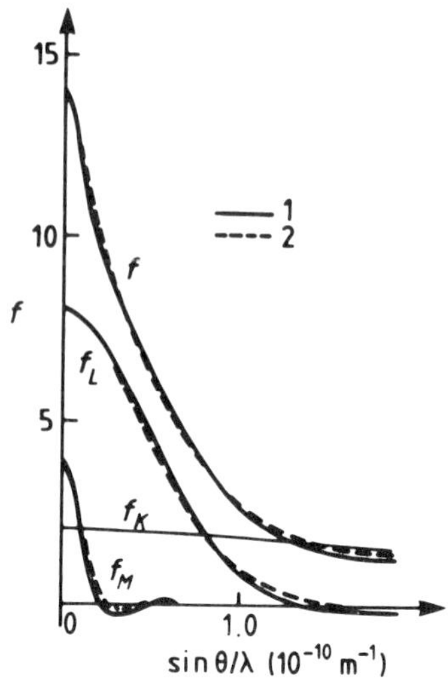

Figure 3.11 Partial x-ray scattering amplitudes for K, L, and M electron shells of the Si atom calculated from STOs (1) and non-relativistic HF AOs of Clementi and Roetti (1974) (2).

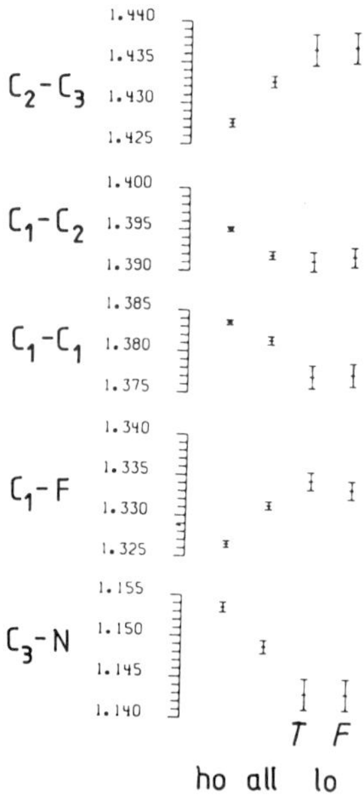

Figure 3.12 An intercomparison of interatomic distances (Å) for tetrafluoroterephthalonitrile $C_8F_4N_2$ at 98 K obtained by high-angle or high-order (ho) refinement with $\sin\theta/\lambda > 0.84$ Å^{-1}, all-reflection (all) refinement with $0 < \sin\theta < 1.15$ Å^{-1} and low-order (lo) refinement with $\sin\theta/\lambda < 0.65$ Å^{-1}. In the latter case the refinement on intensities T and on structure amplitudes F was undertaken (Seiler *et al* 1984).

the scattering up to 1.0 Å^{-1} (Coppens 1974). The core electron shell contraction can contribute even further in reciprocal space (Lobanov *et al* 1984). Moreover, light atoms give a small contribution in that region of reciprocal space where the heavy-atom asphericity is still pronounced. This is why the structure amplitudes which contain the information on the non-spherical ED distribution can be incorrectly included in the refinement of the position and thermal parameters instead of in the calculation of the deformation ED. This will result in biased parameter values and in distortion of the corresponding Fourier series.

In order to find the regions in reciprocal space where the spherical particle approximation is valid, the following procedure can be recommended (Lobanov *et al* 1989). Let the refinement of parameters of the rigid-atom superpositional structural model be previously performed using high-angle reflections chosen in any way. The mean weighted value

$$\langle \Delta F(\boldsymbol{H}) \rangle_w = \sum_{\boldsymbol{H}} w(\boldsymbol{H})\, \Delta F(\boldsymbol{H}) \Big/ \sum_{\boldsymbol{H}} w(\boldsymbol{H}) \tag{3.75}$$

for the high-angle region is then calculated and the reflections for which the inequality

$$|\Delta F(\boldsymbol{H}) - \langle \Delta F(\boldsymbol{H}) \rangle_w| \geqslant \beta \sigma \langle \Delta F(\boldsymbol{H}) \rangle \tag{3.76}$$

is fulfilled are excluded from the structure amplitude set used. In these expressions $\Delta F(\boldsymbol{H}) = |F(\boldsymbol{H})| - |F_{model}(\boldsymbol{H})|$, $\sigma \langle \Delta F \rangle_w = \{\sum_i w_i(\boldsymbol{H})\}^{-1/2}$. The remaining reflections are subdivided into $k = 4$–8 groups, and for each of them the value $\langle \Delta F(\boldsymbol{H}) \rangle_w^g$, analogous to (3.75), is calculated. The number of groups depends on the whole number of reflections in the high-angle region: the number of reflections in each group should be large enough to make a statistical procedure valid. From groups mentioned, those for which the condition

$$|\langle \Delta F(\boldsymbol{H}) \rangle_w^g|_1 \approx |\langle \Delta F(\boldsymbol{H}) \rangle_w^g|_2 \approx \cdots |\langle \Delta F(\boldsymbol{H}) \rangle_w^g|_k \approx \varepsilon \tag{3.77}$$

is fulfilled are chosen. ε is a small value in comparison with statistical error. The reflections in each group which do not obey condition (3.76) are excluded and the rest of the reflections are used in the rigid superpositional structural model refinement. For $\beta = 1.5$–2.0 in (3.76) only those reflections for which the model discussed is valid with 87–95% probability and higher are included in the refinement. If necessary the procedure is repeated and self-consistency is usually reached after one or two iterations. As a result, the reflections influenced by electron redistribution in inner electronic subshells of atoms are excluded from model refinement and can be further used in ED reconstruction together with low-angle scattering data.

The procedure described above was tested by Lobanov *et al* (1988, 1989) in the refinements of structural models of the crystals with garnet-type structure. The reciprocal space distribution of $\langle \Delta F(\boldsymbol{H})| \rangle_w^g$ values for

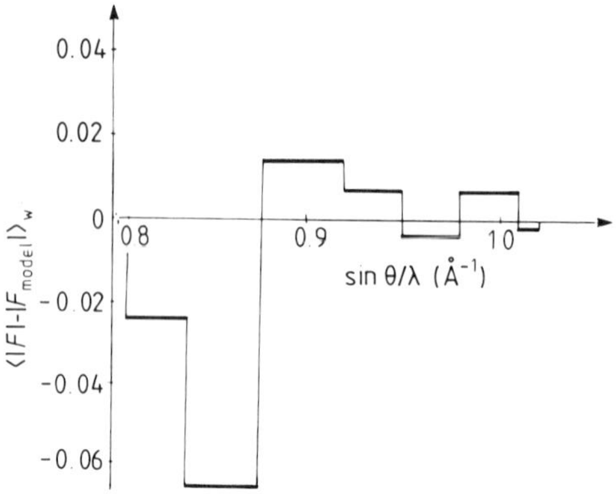

Figure 3.13 Comment on the search of the region in reciprocal space for the refinement of the superpositional atomic structural model in the high-angle approximation: distribution of $\langle F - F_{model} \rangle_w$ values for garnet, $Na_3Sc_2V_3O_{12}$.

$Na_3Sc_2V_3O_{12}$, after preliminary refinement, are presented in figure 3.13. The 171 reflections with $(\sin\theta/\lambda) > 0.88$ Å^{-1} satisfy condition (3.76). If all reflections with $(\sin\theta/\lambda) \geqslant 0.8$ Å^{-1} were included in the refinement, the mean value of $\langle \Delta F(\boldsymbol{H}) \rangle_w$ over the range (0.8–1.022) Å^{-1} was equal to -0.062; this is equivalent to the artificial enlargement of scale factor and to the shift of values of other model parameters.

The procedure described gives the best optimization of the rigid-atom superpositional structural model (see table 3.7). The statistical significance of the results obtained can easily be checked by Hamilton's test (Ibers and Hamilton 1974). Note that the refinement results are stable to small changes of the weighting scheme used.

3.4.4 *Flexible-atom-type superpositional models*

The rigid-atom-type superpositional structural model ignores the asphericity of the valence electron subshells of atoms and electron transfer from one atom to another in a crystal. These changes can be taken into account in the first approximation if the limitation of the rigidity of a model is removed. For this purpose the atomic electron cloud is divided into the spherical core and valence parts; the electron occupancy of the latter are varied. Stewart (1970), who suggested this method and applied it to some organic compounds, called the approach the L shell projection method. The ED in the unit cell is presented as

$$\rho(\boldsymbol{r}) = \sum_{\mu} [\rho_{\mu}^{core}(\boldsymbol{r}) + P_{\mu}\rho_{\mu}^{val}(\boldsymbol{r})] \tag{3.78}$$

Table 3.7 High-order refinement indices for rigid superpositional models of garnet-type crystals with different selection of the refinement region in the reflection set.

	$Na_3Sc_2V_3O_{12}$		$Na_{0.9}Ca_{2.38}Mb_{1.72}V_3O_{12}$		$Ca_{3.02}Nb_{1.68}Ga_{3.2}O_{12}$	
Refinement region in reciprocal space	0.8–1.022	0.88–1.022	0.8–1.024	0.847–1.024	0.8–1.027	0.814–1.07
Number of reflections[a]	271	185(15)	135	110(4)	98	93(7)
R	0.118	0.0085	0.0065	0.0058	0.0130	0.0109
R_w	0.0091	0.0071	0.0067	0.0049	0.0109	0.0099
s	1.1482	1.0060	1.2317	1.0790	3.0696	1.1182

[a] The numbers of reflections rejected according to (3.76) are given in brackets.

and the x-ray scattering amplitude of each atom as

$$f_\mu(H) = f_\mu^{core}(H) + P_\mu f_\mu^{val}(H) \tag{3.79}$$

ρ_μ^{val} and f^{val} being normalized to one electron. The rigid cores with fixed integer electron occupancies are supposed to scatter in the high-angle part of reciprocal space as spherical particles. The corresponding scattering amplitudes can be calculated with atomic wave-functions and be used in high-angle refinement. The atom-in-crystal valence subshell occupancies, p_μ^{val}, are accepted as model parameters and can be then refined using only low-angle reflections, when high-angle position and temperature parameters are fixed. The partial scattering amplitudes for subshells of many atoms have been calculated and tabulated by Stewart (1970), Fukamachi (1971), Ibers and Hamilton (1974), Tsirelson *et al* (1980) and Su and Coppens (1995). The calculational procedure is described in appendix I.

It is important that correlation between electron occupancies of valence subshells of atoms and other structural parameters of such models does not destroy the physical meaning of the parameters during their optimization. This has been demonstrated by Sasaki *et al* (1980) who have studied the $D_w = \sum_i w_i[|F| - |F_{model}|]_i^2$ value for different ionization states of Si and O atoms in Mn_2SiO_4. The curves of D_w dependence on the O atom charge at fixed Si atom charge, while the charges on Mn(I) and Mn(II) atoms were refined independently at each step, are presented in figure 3.14. The resulting

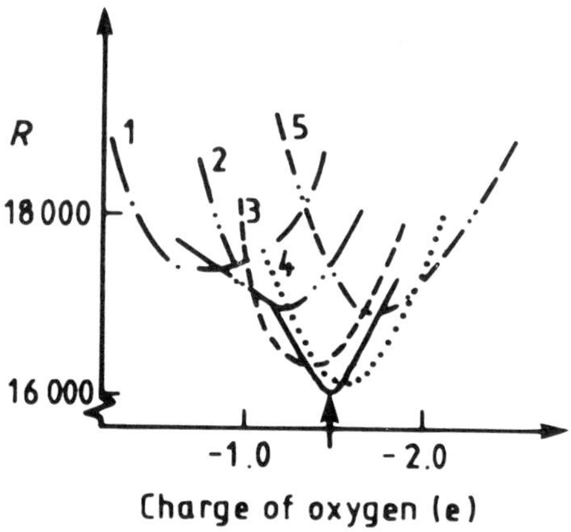

Figure 3.14 Convergence of atomic charge refinement in tephroite, Mn_2SiO_4 (Sasaki *et al* 1980). Each non-solid curve shows the variation of residual factors $R = \sum_i w[|F| - |F_{model}|]_i^2$ when the charges of the Mn atoms are varied with that of the Si atom fixed (1, neutral atom Si; 2, Si^+; 3, Si^{2+}; 4, Si^{3+}; 5, Si^{4+}) and the charge of the O atoms fixed step by step. The solid line indicates the variation of converged residual factors when varying the charge of Si. The arrow indicates the convergence point when all atomic charges are simultaneously refined.

curve has a minimum which corresponds to the minimum of the functional refined. The simultaneous refinement of all atomic charges results in the same minimum of the functional and in the same atomic charge values. It can be concluded that the atomic charge variation is not a mathematical artefact but reflects the crystal electronic structure peculiarities.

Coppens *et al* (1979) suggested a more flexible modification of the L shell projection method. They noticed that the changes of the atomic valence states because of chemical bond formation lead not only to occupancy variations but also to the deformation of valence electron subshells. This deformation can be accounted for by the introduction of a scale factor $\varkappa_\mu$ into the radial part of the atomic ED which describes the spherical contraction or extension of valence electron subshells in comparison with free atoms. This approach is called the $\varkappa$-technique. The unit cell ED is presented in this case as an expression

$$\rho(r)=\sum_\mu [\rho_\mu^{core}(r)+P_\mu \kappa^3 \rho_\mu^{val}(\varkappa_\mu r)] \tag{3.80}$$

and, consequently, the scattering amplitude for each atom in a crystal is

$$f_\mu(H)=f_\mu^{core}(H)+P_\mu f_\mu^{val}(H/2\varkappa_\mu). \tag{3.81}$$

The p_μ and $\varkappa_\mu$ values are included into the refined model parameters; the optimization of the structural model can be performed with the total structure amplitude set.

The $\varkappa$-model is very attractive. Retaining the simplicity of the approximation of spherical particles, it makes possible evaluation of the main changes in electron subshells of atoms due to the chemical bond. The case with $\varkappa > 1$ corresponds to contraction of the atomic valence electron subshell, whereas $\varkappa < 1$ means that the atom is expanded in the crystal. It is important that the $\varkappa$-technique is directly connected with the quantum mechanical scale transform (2.66). Besides, the $\varkappa$-parameter values obtained are consistent with well known Slater rules, which are used in quantum chemistry for chosing exponential screening parameters in radial parts of atomic orbitals (Flygare 1978). For example, the N atomic charges in some nitrogen containing molecular crystals, produced by Coppens *et al* (1979) with the $\varkappa$-technique, linearly correlate with $\varkappa$ values for the same atom (figure 3.15). This dependence agrees very well with the dependence of the screening constant in the valence AO radial part on the charge of the same atom, which follows from Slater rules. Thus, the results obtained with the $\varkappa$-technique are very useful for finding the data which are needed for quantum chemical calculations with a minimal basis set.

The $\varkappa$-model is less useful for atoms with a small number of valence electrons (say one or two) or, in contrast, with nearly fully occupied valence subshells. For example, the electron occupancies of s subshells of single-charge cations, such as Li^+, Na^+, etc, are difficult to establish because they scatter

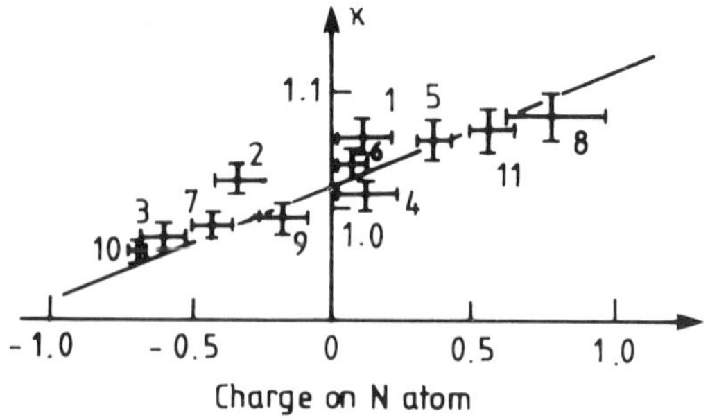

Figure 3.15 The relation between $\varkappa$ and charge on N atoms in a number of compounds as found by Coppens (1982): 1 and 2, glycylglycine; 3, formamide; 4 and 5, *p*-nitropyridine N-oxide; 6, sulphamic acid; 7 and 8, NH_4SCN; 9, NaSCN; 10 and 11, KN_3. Bars indicate estimated standard deviation. The full line demonstrates the relation predicted by Slater's rule for AO exponents.

the x-rays in a very low-angle region of reciprocal space. At the same time, the charge variations in negative ions such as F^- and Cl^- are not small; therefore, the refinement of the structural model containing these ions will not be stable. To avoid this difficulty Sasaki *et al* (1980) has suggested the use of the atomic scattering amplitude in the form

$$f_\mu(H) = f_\mu^{core}(H) + f_{\mu m}^{val}(H/2\varkappa_\mu) + p_\mu f_{\mu n}^{val}(H/2\varkappa_\mu) \qquad (3.82)$$

instead of (3.81). Indices m and n indicate that scattering functions relate to different ionic valence states. Parameter p_μ shows the difference in electron occupancies of valence shells of these states. Though the $\varkappa$-parameter in this modification of the method sometimes has no clear physical meaning as before, the computational advantages are obvious.

The optimization of structural crystal models in x-ray structure analysis is usually done by LS. In principle the realistic physical structural model should provide the correct solution of the problem of parameter determination. However, because of the limited accuracy of experimental data and of the approximations within the models used, instead of a precise system of normal LS equations, one has to deal with only an approximate one. The increase in the number of parameters to be determined and the complication of the structural models make the corresponding LS equations ill conditioned, i.e. their solution is unstable relative to small errors in the data. Such tasks are called ill posed in mathematics.

The general approach to the solution of perturbed systems of linear algebraic equations was described by Tichonov and Arsenin (1986) in the framework of regularization theory. This theory provides stable approximate solutions when approximate structural models are refined, the parameters are correlated, and the experimental data suffers from statistical errors. Tichonov and Tschedrin (1986) and Streltsov *et al* (1988a) described some aspects of the application of this theory to the problem of crystal structure investigation by x-ray diffraction. The application of the regularized LS to optimization of the $\varkappa$-model can be demonstrated with fluorite, CaF_2 (Streltsov *et al* 1988a); the mathematical details of the method are given in appendix J.

The Ca^{2+} ions in the CaF_2 structure (space group $Fm3m$) occupy the positions with local symmetry O_h ($m3m$). They are cubically surrounded by eight anions. Half of the fluorine polyhedra are occupied by cations; the rest are empty. The point symmetry of the F^- ion position is T_d ($\bar{4}3m$). Therefore, there is a non-zero component of the temperature factor $T^a(\boldsymbol{H})$ which includes a single anharmonic parameter describing the anion motion along the [111] direction. The description in terms of the one-particle effective potential gives for the anharmonic part of the F atom temperature factor the expression

$$T^a_F(s) = -\exp[-B_F s^2](B_F/4\pi a) b_F hkl. \qquad (3.83)$$

Here $B_F = 8\pi^2 \langle u \rangle^2$ is the anion isotropic thermal parameter, b_F is the anion anharmonic cubic parameter, a is the parameter of the cubic unit cell, $s = \sin\theta/\lambda$. Streltsov *et al* (1988a) described the CaF_2 crystal using the $\varkappa$-model (3.82) with the anharmonicity accounted for in the form of (3.83). Other structural models were also examined: the harmonic $\varkappa$-model and the harmonic and anharmonic rigid-ion superpositional models. This allows us to compare the results which were obtained with different structural models. f_m in (3.82) were valence electron scattering amplitudes of neutral F and Ca atoms and f_n were the scattering amplitudes of 2p (F) and 4s (Ca) electrons, normalized to one electron. The results of refinements of different models are presented in table 3.8. Let us look at these results more closely.

First, introduction of the single additional parameter b_F into a rigid superpositional model increases the conditional number of the LS normal equation matrix. However, the refinement process remains stable and the harmonic temperature parameter values are the same in both cases. The transition to the flexible superpositional structural model and simultaneous refinement of all parameters of the model lead to the appearance of nearly linear dependences between parameters. The conditional number of the LS normal equation matrix increases significantly and the usual LS does not allow one to refine the model: the process of refinement is divergent. The use of the regularized LS allows one to overcome this difficulty, though at the cost of increasing the discrepancy factor, R, and change for the worse of the goodness-of-fit, S, indices.

Second, introduction of the anharmonic thermal parameter of the F atom into both rigid and flexible superpositional models only weakly influences the values of isotropic thermal parameters. The electronic parameters (atomic charges and $\varkappa$-parameters) of Ca and F atoms also remain stable, when anharmonicity is taken into account. This means that effects of thermal vibrations and modification of atomic ED in a crystal are separated when

Table 3.8 Refinements of CaF_2 crystal structure with different superpositional models.

Model characteristics	Rigid model		Flexible ($\varkappa$) model		Neutron diffraction (Cooper and Rouse 1971)	Theoretical calculation (Elcombe and Pryor 1970)
	A[a]	B	A	B		
Refinement region ($Å^{-1}$)	0.7–1.1	0.7–1.1	0–1.1	0–1.1	—	—
B_{Ca} (10^{20} m^2)	0.522(1)	0.522(1)	0.495(1)	0.502(2)	0.507	0.518
B_F (10^{20} m^2)	0.772(4)	0.722(9)	0.750(1)	0.750(4)	0.697	0.711
b_F (10^{11} J m^{-3})	—	−2.8(3)	—	−2.9(3)	−5.7(2)	—
Anion charge	—	—	−0.336(2)	−0.319(1)	—	—
$\varkappa_{Ca}$	—	—	1.0047(3)	1.0040(4)	—	—
$\varkappa_F$	—	—	0.963(1)	0.964(1)	—	—
R	0.0073	0.0066	0.0157	0.0159	—	—
R_W	0.0082	0.0074	0.0107	0.0083	—	—
S	0.781	0.709	1.440	1.370	—	—
Conditional number (before refinement)	107	625	803	845	—	—
Filter parameter β	0	0	1.5	1	—	—

[a] A is the harmonic approximation and B is the anharmonic one for atomic thermal vibration.

the regularized LS is used. Therefore, in this case the physical meaning of $\varkappa$-model parameters is not distorted by defects of the computational procedure. The thermal parameters from high-angle refinement of the rigid model appear to be the same within 3–5% as those obtained with the sophisticated electron-dynamic crystal structural model. This means that the high-angle approximation can be considered reliable for determination of the precise atomic position and thermal parameters.

Further, isotropic thermal parameters coincide within 3% for Ca and 7% for F atoms with the corresponding values from theoretical calculations of Elcombe and Pryor (1970) and from neutron diffraction (Cooper and Rouse 1971) which are also presented in table 3.8. The value of anharmonic thermal parameter, b_F, at room temperature proved to be small, less than the neutron value. This discrepancy can be attributed to lower accuracy of the Cooper and Rouse neutron data in comparison with the x-ray diffraction data discussed.

The averaged values of atomic charges in CaF_2 obtained with the $\varkappa$-technique, Ca, $+0.66\,e$ and F, $-0.33\,e$, differ remarkably from the formal oxidation numbers of $+2$ and -1, respectively. This result is typical for a large number of studies of a wide range of compounds, including alkali halides with nearly ideal ionic bond compounds (Coppens *et al* 1979, Sasaki *et al* 1980, Streltsov *et al* 1988a, Zavodnik 1994). This means that the ionic model with formal charge values does not correspond to the real distribution of the ED in ionic crystals. It is impossible to properly improve this model by atomic charge variations. The data of table 3.8 support this statement. In particular, this leads to the conclusion that in theoretical modelling of CaF_2 properties of the cation can be approximated by a modified free-ion wave-function, but the anion, referring to the $\varkappa_F$-value, needs stricter (self-consistent) treatment. The $\varkappa$-model parameters give a good initial approximation for this purpose.

3.4.5 *Multipole models*

The most flexible crystallographic ED model which has some common elements with the LCAO approximation is the multipole model. In this model crystal ED is presented as a sum of aspherical pseudoatoms, $\rho_\mu(\boldsymbol{r})$. In order to take asphericity into account, each $\rho_\mu(\boldsymbol{r})$ function is decomposed into a convergent Laplace series (Arfken 1985) i.e. it is presented as an infinite expansion over spherical harmonics, $Y_{lm}(\theta, \varphi)$ or over their real combinations, $y_{lm\pm}$, which satisfy the symmetry of the local environment. The general form of the Laplace series for ED is

$$\rho_\mu(\boldsymbol{r}) = \sum_{lm} C_{\mu lm} R_{\mu l}(r) y_{lm\pm}(\theta, \varphi). \tag{3.84}$$

Here $C_{\mu lm}$ are the occupancies (weights) of the terms and $R_{\mu l}$ are radial functions. The terms in the series (3.84) are called multipoles like the corresponding terms in the potential decomposition in electrostatics. The integral over the whole space for the multipole with $l=0$ gives the pseudoatom electric charge; for multipoles with $l>0$ it is equal to zero. This means that multipoles with $l>0$ describe the electron redistribution in the pseudoatomic volume. The dipole ($l=1$) describes the displacement of the electron centre of gravity with respect to the nucleus position; the higher-order multipoles describe more complicated features of electron shell deformation (figure 3.16). The ED in a unit cell is presented in the form of a multicentre sum $\rho_M(\boldsymbol{r})=\sum_\mu \rho_\mu(\boldsymbol{r}-\boldsymbol{r}_\mu)$ each term of which is described by a convergent series (3.84). As a result, the multipole model can take into account fine details of ED in a crystal.

The expansion (3.84) is very useful for analysis of x-ray diffraction data,

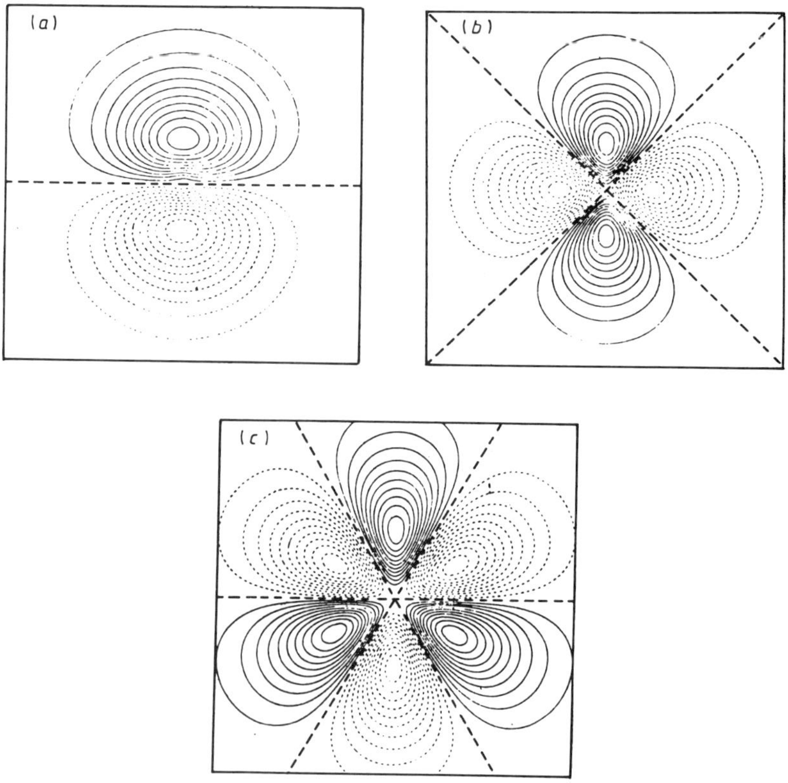

Figure 3.16 ED plots of (*a*) dipole ($xr\exp(-\alpha r)$ at $z=0$, $\alpha=3.9$ au, (*b*) quadrupole ($(x^2-y^2)\exp(-\alpha r)$ at $z=0$, $\alpha=3.9$ au), (*c*) octupole ($(x^2-3y^2)x\exp(-\alpha r)$ at $z=0$, $\alpha=3.9$ au). Contours are proportional to $e\,a_0^{-3}$ (Stewart 1976).

since it can be rewritten in reciprocal space as

$$f_\mu(\boldsymbol{H}) = \sum_{lm} C_{\mu lm} f_{\mu l}(H) y_{\mu lm\pm}(\theta_H, \varphi_H) \tag{3.85}$$

where spherical harmonics have the same form as in expression (3.84). The $f_{\mu l}$ function is the Fourier–Bessel transform of radial functions of multipoles (Arfken 1985)

$$f_\mu(H) = 4\pi \mathrm{i} \int_0^\infty R_{\mu l}(r) j_{\mu l}(2\pi Hr) r^2 \, \mathrm{d}r \tag{3.86}$$

($j_l(x)$ is the spherical Bessel function). The details of $f_{\mu l}$ calculation are given in appendix I.

As in simpler structural models, the electron cloud of each pseudoatom is assumed to tightly follow the vibrations of its nucleus. This allows one to combine the multipole model with any description of atomic vibrations. The convolution approximation is usually used here; therefore, the multipole expansion parameters can be considered to be static (pseudostatic) ones.

A limited number of terms in the expansion (3.84) is used in practice, of course. There is an important theorem stated by Stewart *et al* (1975): independent of radial function form, the maximum order of the multipole expansion term used determines the maximum order of the static electric properties of pseudoatoms which can be calculated exactly from the multipole model. This theorem allows one to limit the length of series (3.84) in a way which preserves the information required for the properties to be calculated. Indeed, due to the completeness of the Laplace series, each multipole with $l > 0$ describes the ED deformation of a definite symmetry with respect to the nucleus, independent of the number of terms included in (3.84). That is why even the use of a limited number of terms in the multipole expansion allows one to calculate a property of a system which depends on the corresponding deformation of atomic ED in a crystal. For example, the electric field gradient at the nucleus position arises from the quadrupole deformation of ED of its 'own' pseudoatom, provided that the influence of other pseudoatoms is small. To reduce the number of terms in the expansion (3.84) the local coordinate axis can be oriented in accordance with the local pseudoatom symmetry.

In practice some analytical approximation for radial functions must be chosen. The form of these functions should obey some requirements which will be considered below.

There are several variants of multipole model; the first one was, as far as we know, suggested by Konobeevskij (1951). In the simplest parameterless models the radial functions were either approximated by analytical functions with fixed coefficients (Dawson 1967b) or determined from the experimental structure amplitudes by the inverse Fourier transform (Kurki-Suonio 1968). Such models are useful for simple ionic crystals of NaCl type (Vidal-Valat *et al* 1978, Kurki-Suonio and Sälke 1984) or covalent crystals of diamond type (Dawson 1967b).

Hirshfeld (1971) and Harel and Hirshfeld (1974) suggested for study of organic molecular crystals a parametric multipole model in which each pseudoatom is presented as

$$\rho_\mu(r) = \rho_\mu^0(r_\mu) + \sum_l C_{\mu l}\, \delta\rho_{\mu l}(r_\mu, n_l). \tag{3.87}$$

Here ρ_μ^0 is the ED of the free spherical atom which is calculated from HF atomic wave-functions. Atomic deformation functions are taken in the form $\delta\rho_{\mu l}(r_\mu, n_l) = N_{\mu l} r_\mu^{n_l} \exp(-\alpha_\mu r_\mu) \cos^{n_l} \theta_k$. The n_l values are integer numbers: $0 \leqslant n_l \leqslant 4$. For $n_l = 0$ the radial function is of s type, $n_l = 1$ corresponds to three functions of p type, etc. θ_k is the angle between a radius vector $\boldsymbol{r}$ and polar axis k; $N_{n_l} = (n_l + 1)\alpha_{\mu l}^{n_l + 3}/4\pi(n_l + 2)!$ is the normalization coefficient. For even n_l values the integral of each multipole term over the whole space is equal to one electronic charge. For odd n_l the same integral is zero: corresponding multipoles describe the intra-atomic redistribution of ED. The model parameters, multipole occupations, C_l, and exponential factors in radial functions, α_μ, are refined by minimizing of functional (3.69) with LS.

Stewart (1976) suggested another multipole model, in which the valence part of ED of the pseudoatom is decomposed over the real spherical functions:

$$\rho_\mu(\boldsymbol{r}) = \rho_\mu^{core}(\boldsymbol{r}) + \sum_{l=0} \sum_{m=0}^{l} C_{\mu lm} R_{\mu l}(r) y_{\mu lm}(\theta, \varphi). \tag{3.88}$$

The atomic cores in a crystal are considered to be undisturbed by a chemical bond and their EDs are calculated from atomic wave-functions. The radial functions have the form

$$R_\mu(r) = (1/4\pi)\{\alpha_\mu^{n_l + 3/(n_l + 2)!}\} r_\mu^{n_l} \exp(-\alpha_\mu r_\mu) \tag{3.89}$$

and the structure amplitude is presented as

$$F(\boldsymbol{H}) = \sum_\mu \left\{ f_\mu^{core}(\boldsymbol{H}) + \sum_{lm} C_{\mu lm} f_{\mu l}(H) y_{\mu lm\pm}(\theta_H, \varphi_H) \right\} T_\mu(\boldsymbol{H}) \exp(i2\pi \boldsymbol{H} \boldsymbol{r}_\mu). \tag{3.90}$$

Coefficients $C_{\mu lm}$ and exponential factors α_μ are determined by the LS adjustment of these amplitudes to experimental values.

Price and Maslen (1978) added to the Stewart model the deformation monopole composed of $C_i r^{n_i} \exp(-\xi_i r)$ terms in order to better take into account the spherical deformation of a pseudoatom. Parameters C_i and ξ_i are chosen to ensure the true asymptotical behaviour of radial functions when $r \Rightarrow 0$, as results from the Kato condition (2.51).

Another multipole model was suggested by Hansen and Coppens (1978). They used the $\varkappa$-technique for variation of exponential parameters of the radial functions of multipoles. The ED for every pseudoatom in this model is expressed as

$$\rho_\mu(\boldsymbol{r}) = \rho_\mu^{core}(\boldsymbol{r}) + C_\mu^{val} \varkappa_{\mu 1}^3 \rho_\mu^{val}(\varkappa_{\mu 1} r) + \sum_{l=0}^{4} \varkappa_\mu^3 R_{\mu l}(\varkappa_\mu r) \sum_{m=-l}^{l} C_{\mu l m} y_{\mu l m \pm}(\theta, \varphi). \tag{3.91}$$

The ρ^{core} and ρ^{val} densities are calculated using the HF wave-functions of free atoms. The radial functions have the form of (3.89). The structure amplitude in the Hansen–Coppens model is calculated by

$$F(\boldsymbol{H}) = \sum_\mu \Bigg\{ f_\mu^{core}(\boldsymbol{H}) + C_\mu^{val} f_\mu^{val}(H/2\varkappa_{\mu 1}) + \sum_{l=0}^{4} f_{\mu l}(H/2\varkappa_\mu) \sum_{m=-l}^{l} C_{\mu l m} y_{\mu l m \pm}(\theta_H, \varphi_H) \Bigg\} T_\mu(\boldsymbol{H}) \exp(\mathrm{i}2\pi \boldsymbol{H}\cdot\boldsymbol{r}_\mu). \tag{3.92}$$

The real spherical harmonics are normalized as follows: $\iint |y_{lm\pm}| \sin\theta \, d\theta \, d\varphi = 2 - \delta_{0l}$. This means that at $C_{lm} = 1$ one electron moves from a negative multipole lobe to a positive one, if $l > 0$. The list of normalizing coefficients was given by Hansen and Coppens (1978). The electron occupancies of multipoles and $\varkappa$-parameters are refined by LS.

Different compounds were investigated by many authors using the multipole models described: ionic, covalent, molecular compounds, etc. The results can be summarized as follows.

There are many arguments in favour of the use of the multipole model. First, all types of multipole expansion have high flexibility, as was found by Baert *et al* (1982), Delaplane *et al* (1985), and Parini *et al* (1985). They equally well describe the magnitude and character of atomic deformations in a crystal. As a result, the ED of complicated heterodesmic compounds can be readily described by multipole models (figure 3.17). The details of ED, which are not described by the model, have values approximately equal to statistical error. The multipole model possesses a filtering property regarding ED: the random details ('noise') are removed by the model and the noise level of the deformation ED maps decreases (see figure 3.18). In addition, the phases of structure amplitudes can be calculated with higher accuracy. This is important in the case of non-centrosymmetrical crystal structures, especially for weak reflections†.

Second, the spherical harmonics form the basis of irreducible representation of the full rotation group. Therefore the pseudoatoms are invariant towards rotations and their EDs are independent of the local coordinate systems

†The remark in subsection 3.4.1 concerning the phase indeterminacy of structure amplitudes, including odd terms of PDF decomposition, which take place in acentric space groups, is applicable to odd terms in the multipole model as well (Hansen 1988).

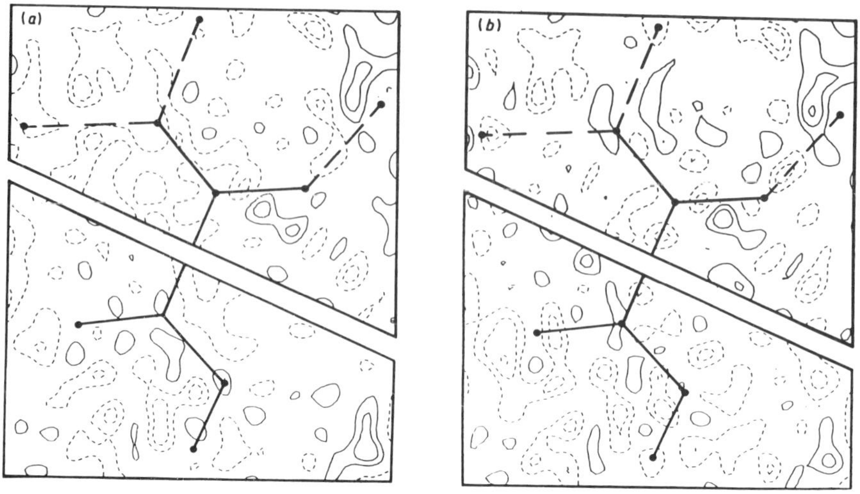

Figure 3.17 The residual densities $\rho_{exp} - \rho_{multipole}$ for $HC_2O_4^{2-}$ in $NaHC_2O_4 \cdot H_2O$: (*a*) Hirshfeld model; (*b*) Hansen and Coppens model. The line intervals are 0.05 e Å^{-3} (Delaplane *et al* 1985).

chosen at atomic centres. This is why pseudoatoms can easily be adapted to the local environment symmetry: the rules for choosing the local coordinate system were given by Kurki-Suonio (1977a) for different point groups. In order to transfer from one local system to another it is necessary to replace each coefficient in the multipole expansion by a linear combination of coefficients with the same l value.

Extending the multipole series, one can arrive at an accurate description of ED at each unit cell point. In quantum chemistry the same result is reached by extension of the set of basis functions. Finally, the analytical representation of ED enables one to simplify the task of crystal electronic properties calculation (Stewart 1972, Moss and Coppens 1981).

Naturally the multipole model is not free of shortcomings. The analytical form of radial functions used influences the model parameters. This influence is clearly seen in figure 3.11, where the x-ray scattering amplitudes for spherical K, L, and M subshells of an Si atom which have been calculated with HF atomic wave-functions and with one-exponential STOs, orthogonalized according to Lowdin (1970), are depicted. In different regions of reciprocal space the partial scattering amplitudes deviate in different directions. Therefore, the different radial functions can result in different multipole parameter values as found by LS. Correspondingly, the values of electric field characteristics of a pseudoatom also depend on the radial function form. The deviation in pseudoatom thermal parameters derived by LS will be another consequence of difference in the radial functions. In order to decrease this influence one should vary the exponential parameters, α, in the radial

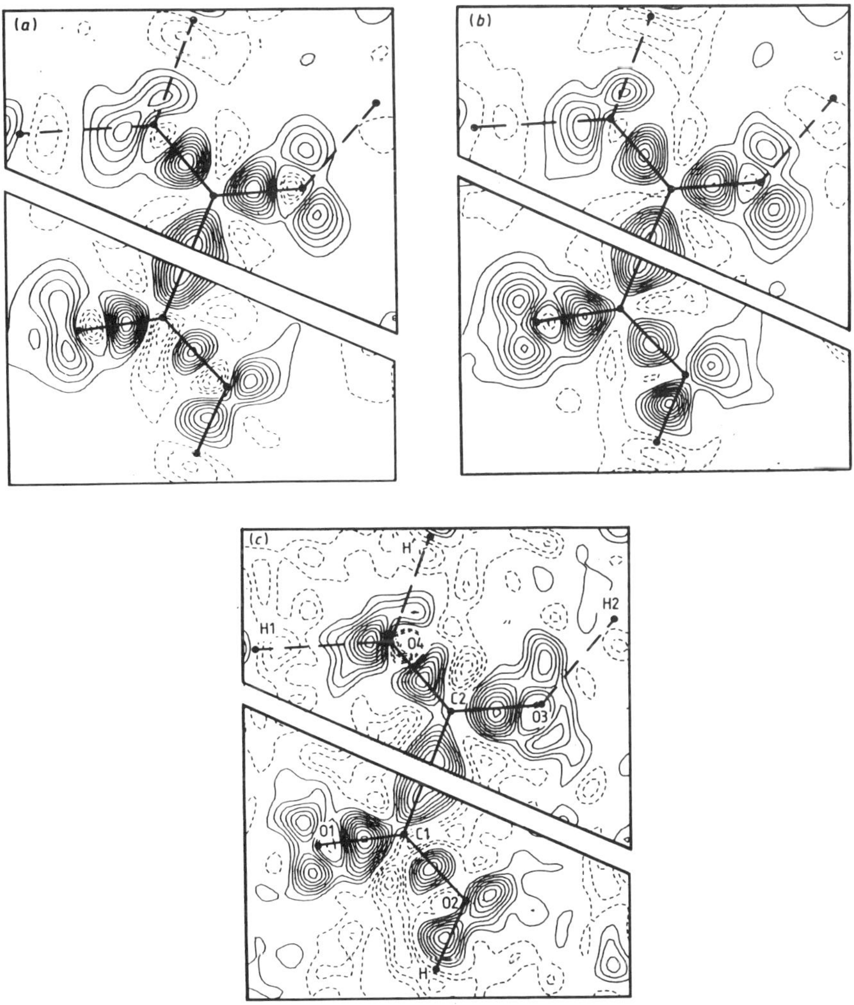

Figure 3.18 A comparison of deformation dynamic multipole EDs for $HC_2O_4^{2-}$ in $NaHC_2O_4 \cdot H_2O$ (120 K): (*a*) Hirshfeld model; (*b*) Hansen and Coppens model. The deformation ED as calculated by Fourier series is also shown (*c*). The line intervals are 0.05 e Å^{-3} (Delaplane *et al* 1985).

parts of multipole functions (Bentley and Stewart 1976, Chandler and Spackman 1982). It is necessary here to keep in mind that the parameters varied should agree with the quantum mechanical laws which regulate the behaviour of electrons. In particular, the Kato asymptotical condition (2.56) should be fulfilled. Further, the exponential parameters derived from multipole model refinement and from the minimum-system-energy requirement (see appendix C) differ in their physical meaning. The first correspond to the best model ED fit to x-ray scattering and depend to a greater degree on the ED of the valence subshell, the second are more sensitive to core electrons whose energies are higher. Moreover, that part of ED which is associated with AO overlap is ignored in the multipole model which consists of atom-like terms. Bentley and Stewart (1976) in their analysis of diatomic molecules have found that multipole radial function parameters are closer to the similar atomic values than to standard molecular ones (table 3.9). It was also found that the multipole-reconstructed valence electron subshells of B, C, and N atoms are more diffuse than valence subshells calculated with parameters derived by the variation principle from the minimum-energy condition. It is noted as well that the multipole exponential parameters of the same atom differ only slightly in different bonding situations. This circumstance can be used in the investigation of compounds with the same fragments.

The structure amplitude derivatives with respect to orbital exponential parameters, which are needed for LS refinement, can be found from the recurrent expression (Epstein and Stewart 1977) $(\partial/\partial\alpha_\mu)f_{\mu n_l l}(\boldsymbol{H}, \alpha_\mu) = (f_{\mu n_l, l}(\boldsymbol{H}, \alpha_\mu) - (n_l + 3)f_{\mu n_l + 1, l}(\boldsymbol{H}, \alpha_\mu))/\alpha_\mu$, which is valid for radial functions (3.89). It is seen from this expression that undesirable nearly linear dependences can appear in the matrix of the LS normal equation and this equation will be ill conditioned. The diffuse nature of the multipole radial functions enhances the uncertainty of the model parameters as well. For example, Stewart (1973a) has found that even in diamond, where there is no interatomic charge transfer, the ED value at the C atom due to neighbouring atom hexadecapoles is 18% of the general ED value. The corresponding octopole and monopole contributions are equal to 7 and 2%, respectively. This increases the probability of correlation between parameters of the model, of course. Besides, though the ED remains invariant, the information about pseudoatomic deformations of ED of different orders, which is extracted from the x-ray diffraction data, does in fact depend upon both the length of the multipole expansion and the analytical radial function form. In diamond, where uncertainties in structure amplitudes range from 1 to 2%, the precision of octopole occupancy determination is about 15% and that of the

Table 3.9 Valence exponents α for first-row atoms (Bentley and Stewart 1976).

α	B	C	N	O	F
α_A from multipole analysis of AH molecules	2.39	3.05	3.75	4.46	5.52
α_A from multipole analysis of AB molecules	2.34	3.04	3.72	4.43	5.10
α from HF atom	2.36	3.05	3.77	4.47	5.16
Standard molecular value (see appendix C)	3.00	3.44	3.90	4.50	5.10

hexadecapole is 28% (Stewart 1973a). Thus, the error of the multipole parameter determination increases with l enhancement.

The results of investigation of diatomics of AB and AH type (A, B = B, C, N, O, F), based on multipole decomposition of ED, calculated from very accurate wave-functions, show that the quadrupole level of the expansion proved the molecular ED reconstruction with a precision of 0.2% and the hexadecapole level of expansion gives 0.025% precision (Bentley and Stewart 1976, Chandler and Spackman 1982). The precision of electric property calculations increases when the terms with $l = 3, 4$ are introduced into the multipole series. Therefore, the terms with $l \leqslant 4$ are usually included in expansion (3.84).

The thermal vibrations and the ED asphericities can mimic each other under optimization of the crystal structural model. Therefore, the tactics of separate refinement of corresponding model parameters are often used. The study of bis(pyridine)(mesotetraphenylporphinate)Fe(II), $FeC_{54}H_{38}N_6$ (space group, $P\bar{1}$; 8497 independent reflections; $\sin\theta/\lambda \leqslant 1.15$ Å^{-1}; $T = 110$ K) by Mallinson *et al* (1988) allows one to estimate the effectiveness of that empirical approach. The simultaneous refinement of the parameters of the Hansen–Coppens multipole model and of the anharmonic Gram–Charlier model for the Fe atom (local symmetry D_{4h}) by using all measured sets of reflections diverged due to correlations between parameters. After this the authors returned from the multipole to a rigid-atomic superpositional model and refined their position and thermal parameters including 15 anharmonic parameters of the Fe atom up to fourth order, in the high-angle approximation (table 3.10). The discrepancy factor R_w was equal to 0.0363. Then the anharmonic thermal parameters of the Fe atom were fixed and the multipole, position, and anisotropic harmonic thermal parameters of all atoms were refined with all reflections. In this case the refinement converged to $R_w = 0.0286$, which appeared to be very near to $R_w = 0.0285$ in the multipole harmonic model. The anharmonic thermal parameters of the Fe atom were then included in the model and their optimization over all reflections resulted in the lowering of the R_w factor to 0.0277. However, the refinement process appeared to be unstable, in spite of the fact that the ratio of number of reflections to number of parameters was nearly 17. The largest positive correlation coefficients (~ 0.9) were observed between anisotropic harmonic and some anharmonic thermal parameters of the Fe atom. A remarkable negative correlation between some harmonic thermal and quadrupole multipole parameters of the Fe atom was also found.

The results of the various refinements are presented in table 3.10. The values of the harmonic vibration tensor, U_{ij}, for a multipole anharmonic model are approximately 1.5 times higher than those for a multipole harmonic model. At the same time, they are lower than for an atomic superposition anharmonic model; the latter are more than twice as high as harmonic thermal parameters obtained when the multipole model is used. The change

Table 3.10 The temperature parameters of the Fe atom in $FeC_{54}H_{38}N_6$ (Mallnison *et al* 1988).

Parameter	I	II	III
U_{11} (Å^2)	0.005 01(10)	0.011 55(27)	0.007 52(42)
U_{22}	0.007 72(10)	0.016 58(31)	0.012 03(76)
U_{33}	0.007 74(10)	0.016 44(31)	0.011 97(32)
U_{12}	0.002 44(8)	0.005 72(22)	0.003 59(17)
U_{13}	0.002 87(8)	0.006 09(22)	0.004 27(18)
U_{23}	0.002 93(10)	0.005 82(23)	0.004 18(48)
C_{1111} (10^2 Å^4)		0.070(5)	0.018(4)
C_{2222}		0.116(7)	0.048(9)
C_{3333}		0.121(7)	0.049(5)
C_{1112}		0.032(3)	0.008(2)
C_{1222}		0.048(4)	0.014(3)
C_{1113}		0.047(3)	0.014(2)
C_{1333}		0.045(4)	0.018(2)
C_{2223}		0.032(4)	0.013(5)
C_{2333}		0.045(4)	0.016(3)
C_{1122}		0.044(4)	0.013(2)
C_{1133}		0.053(3)	0.019(2)
C_{2233}		0.050(3)	0.020(3)
C_{1123}		0.027(2)	0.008(1)
C_{1223}		0.027(2)	0.009(2)
C_{1233}		0.033(2)	0.010(1)

in harmonic thermal parameter values when anharmonicity is taken into account contradicts the requirement of their stability in the frame of the same ED model (see table 3.8). At the same time, the structural model which takes into account the pseudoatom ED asphericity and anharmonicity of its atomic thermal vibrations had high statistical significance according to statistical tests. Thus, the changes in thermal parameters observed are caused, perhaps, by the differences in electron-dynamic crystal structural models and by inter-influence of parameters refined. Hence, the adequacy of the complicated crystal structural model and the tactics of their parameter refinement are very important for obtaining reliable and physically meaningful information from the x-ray diffraction data. The general recipe to overcome this problem is clear; however, the solution is sought empirically every time. It should be noted that the correlation between thermal and multipole parameters is decreased if atomic vibrations are small and if the experimental structure amplitude set is spread over very high values of scattering angle (Stewart 1976).

In order to increase the reliability of parameters obtained with a multipole model, some restrictions upon these parameters can be posed. These restrictions can follow both from theory and from independent physical measurement data. Hirshfeld (1984a, b) has proposed to use, during optimization of the model, the consequence of the Hellmann–Feynman theorem (2.63) which states that the electric field value at the nucleus positions of a stable many-electron system should be zero. Dipole terms with variable electron occupancies and exponential coefficients were added to the multipole model (3.87) to take this restriction into consideration. However, the application of this approach to the tetrafluoroterephthalonitrile, $C_6N_2F_4$, did not change either the discrepancy factor or the deformation ED. At the same time, strong correlation between parameters of the model took place. The contributions to the electric field at nucleus positions from both standard multipole and additional dipole-type terms are given in table 3.11. The latter contributions do not change the electric field values even within one estimated standard deviation. Probably, this restriction will be more useful when very accurate neutron values of atomic coordinates are known.

Schwarzenbach and Thong (1979) have proposed to use as a restriction in the Hirshfeld model the value of the electric field gradient at nucleus positions, measured separately by Mössbauer spectroscopy. NQR, or NMR methods. The study of $AlPO_4$, Li_3N, and α-Al_2O_3 crystals (Schwarzenbach and Lewis 1982) has shown that the convergence of LS refinement is better in this case; however, a significant change in deformation ED was observed in the vicinity of the nucleus only.

Crystal electroneutrality is now the general restriction. Symmetry restrictions can also be imposed on the chemically equivalent parts of molecules, as well as restriction on thermal vibration tensor components in the case of rigid bonds (Hirshfeld 1976). All these restrictions are introduced in the LS equation and lead to the improvement of the physical meaning of the

Table 3.11 The electric field (e Å^{-2}) on nuclei in the $C_6N_2F_4$ crystal: the contributions from the promolecule E_p, deformation ED $E_{\delta\rho}$ and additional dipole terms E_{dip} which ensure Hellmann–Feynman[a] theorem fulfilment (Hirshfeld 1984b).

		Refinement without restrictions	Refinement with additional dipole terms	
Atom	E_p	$E_{\delta\rho}$	$E_{\delta\rho}$	E_{dip}
C(1)	0.143	−0.546(73)	−0.539(69)	0.396
C(2)	0.031	−0.693(172)	−0.703(169)	0.672
C(3)	−0.240	0.776(139)	0.773(145)	−0.533
F	0.314	1.588(327)	1.532(319)	−1.845
N	0.640	0.613(421)	0.613(431)	−1.252

[a]According to the Hellmann–Feynman theorem $E_p + E'_{\delta\rho} + E_{dip} = 0$; $\delta(E_{dip}) = \delta(E'_{\delta\rho})$.

parameters of the structural model. This is especially important in the nuclear vicinity, where the ED resolution is poor if usual wavelengths of x-radiation are used.

The advantages and shortcomings of the multipole model force us to look attentively at their physical foundation. There are several points here which deserve analysis (Stewart 1977b, Kurki-Suonio 1977b, Price and Maslen 1978, Parini *et al* 1985). The cusp condition (2.56) should be taken into account during model parameter refinement for the correct (in quantum mechanical terms) ED description near a nucleus. Only in this case thermal parameters and electric properties of multipoles near a nucleus (e.g. the electric field gradient) can be determined for certain (Stewart 1973b). However, the cusp condition is taken into account only in the Hirshfeld multipole model (3.87). Since the STOs of ns type do not satisfy this condition for $n > 1$, the valence monopole in the Stewart model badly describes the s electron behaviour near a nucleus.

In order to satisfy the Poisson equation (2.79), it is necessary to have the principal quantum number n_l of a radial function not less than the harmonic order l. This was demonstrated by Stewart (1977b) who considered the asymptotical properties of the integral form of the Poisson equation at $r \Rightarrow 0$ for multipoles with radial STO-type functions. He also found that at $n \geqslant l$ there was no logarithmic singularity at $r \Rightarrow 0$ for the atom with its nucleus removed. Besides, for $n \geqslant l$ the radial part of a pseudoatom in the STO form satisfies the non-divergence condition for the integrals which describe the electrical field and its derivatives at nuclear position. This is important for crystal properties calculated with the multipole model.

The orthogonality of multipole density functions should decrease the danger of producing near-linear dependences in the LS equations. For different poles of a valence subshell of the same pseudoatom the orthogonality is fulfilled because spherical harmonics are themselves orthogonal. Besides, valence subshell poles should also be orthogonal to the functions with the same angle dependence describing the core ED. The last requirement is not fulfilled in the Stewart and Hansen–Coppens models in which the valence subshells are modified: the overlap integral between the density functions $S_{1s2s}(\varkappa) \neq 0$ at $\varkappa \neq 1$. In the Hirshfeld model the density overlap between initial and deformation monopoles also exists and results in the distortion of the multipole occupancy parameters, as determined by LS. The multipoles on different pseudoatoms are not orthogonal in any case; however, their overlap integrals have smaller values.

An attempt to provide the orthogonality of core and valence radial functions with $\varkappa$-parameter variation was undertaken by Parini *et al* (1985) in the study of diamond. An s-type valence monopole function was orthogonalized by the Schmidt algorithm with respect to s-type core function. However, this method was unreasonable from the theoretical point of view because at some $\varkappa$ values the 2s electron cusp condition at the nucleus position is not fulfilled. An attempt to choose the valence monopole s function in a sophisticated form which does not suffer from this shortcoming was then made, but a change of multipole parameter values was not found.

Part of the ED of atomic valence s subshells in a crystal is localized near the nuclei. The deformation of ED in this region has a dipole character (Hirshfeld and Rzotkiewicz 1974); its influence on structure amplitudes is not larger than 1% (Bentley and Stewart 1974), and seems inaccessible to x-rays due to limited resolution of the Fourier series. Nevertheless, this part of the valence ED should be taken into account in a multipole model. As was found by model calculations (Bentley and Stewart 1976, Chandler and Spackman 1982), the variation of electronic occupancies of the atomic cores improves the calculated electronic characteristics of the pseudoatoms, making them closer to quantum mechanical ones. The effective values of occupancy parameters increase with the atomic number and exceed two for all first-row atoms. This result, which seems meaningless from the orbital theory point of view, shows the necessity of taking into account the monopole component of ED near a nucleus. For this purpose the use of the two-exponential radial functions is very useful (Chandler and Spackman 1982). Unfortunately, this recommendation is difficult to fulfil in practice because of the significant increase of the number of crystal model parameters refined.

In order to make the multipole model applicable to complicated crystals with heavy atoms, it is important to take into account the deformations of the inner electron subshells of bonded atoms. There are many pieces of evidence concerning this deformation, especially from x-ray and photoelectron core ionization spectroscopy (Saethre *et al* 1991). The x-ray diffraction experiment can also 'feel' this effect, so Tsirelson *et al* (1988b) have found that in V metal the positive region of deformation ED, which exists near the nucleus position, is surrounded by a negative-deformation-ED region. A correlation between the depth of minima and the arrangement of atoms in first and second coordination spheres was found. The introduction of additional deformation atomic monopoles can account for all these features without violating the cusp condition.

Parini *et al* (1985) have tried to improve some shortcomings of the multipole model. They presented the ED of each pseudoatom in the form

$$\rho_\mu(\boldsymbol{r}) = \rho_{\mu 0}(\boldsymbol{r}) + \delta\rho_{\mu 0}(\boldsymbol{r}) + \delta\rho_\mu(\boldsymbol{r}). \tag{3.93}$$

The multipole term $\rho_{\mu 0}(\boldsymbol{r}) = \rho_{\mu at}(\boldsymbol{r}) - (C_0/n_{ion})[\rho_{\mu ion}(\boldsymbol{r}) - \rho_{\mu at}(\boldsymbol{r})]$ is constructed from HF wave-functions for free spherical atoms (*at*) and ions (*ion*), n_{ion} being the formal ionic charge. The refined parameter C_0 characterizes the charge transfer; for monatomic crystals $C_0 = 0$. This type of monopole includes, naturally, the spherical atom ED which satisfies cusp condition (2.56). The deformation monopole has the form

$$\delta\rho_{\mu 0}(\boldsymbol{r})=\frac{1}{4\pi}\Bigg\{C_{\mu 1}[N_0+(3-6z/\gamma_\mu)N_1 r+(6z/\gamma_\mu-4)N_2 r_\mu^2]$$
$$+\sum_{i\geqslant 1}\sum_{j\geqslant 1} C_{\mu ij}(-N_i r_\mu^i+N_j r_\mu^j)\Bigg\}\exp(-\gamma_\mu r_\mu) \tag{3.94}$$

where $C_{\mu 1}$, $C_{\mu ij}$, and γ_μ are parameters to be refined. The term in curly brackets describes the spherical deformation of ED at distance 0.3–0.5 Å away from a nucleus. It takes into account condition (2.56) and the charge conservation law $\int \delta\rho_{\mu 0}(\boldsymbol{r})\,\mathrm{d}\boldsymbol{r}=0$. The remaining terms in (3.94) describe a spherical valence subshell deformation and do not influence the cusp condition.

The aspherical pseudoatom deformation is described by the term

$$\delta\rho_\mu(r)=\sum_{l>0}\sum_m C_{\mu lm}\frac{\alpha_\mu^{n_l+3}}{4\pi(n_l+2)!}r^{n_l}\exp(-\alpha_\mu r)y_{\mu lm\pm}(\theta,\varphi). \tag{3.95}$$

The electron occupancies $C_{\mu lm}$ and exponential factor α_μ are to be refined. The multipole model described permits one to take into account both ED changes in pseudoatom valence shells, and the changes in cores. The model provides good asymptotic behaviour of ED near a nucleus. The application of this model to hexamethylenetetramine, $C_6H_{12}N_4$, was described by Parini (1988). The compound mentioned crystallizes in the cubic non-centrosymmetric space group $I\bar{4}3m$ with two molecules in the unit cell (figure 3.19(*a*)). The C atoms are connected to two H and two N atoms, the latter being connected to three C atoms. The local symmetry of the C atom is *mm*2, the N atom 3*m*, and the H atom *m*. The x-ray structure amplitudes of Stevens and Hope (1975) (room temperature; $\sin\theta/\lambda\leqslant 1.1$ Å^{-1}) corrected for TDS were used in the calculation. The position and thermal parameters of H atoms were taken from neutron diffraction work of Duckworth *et al* (1970).

Due to acentricity of the space group, the refinement of the multipole anharmonic structural model of hexamethylenetetramine provides a worst-case example of refinement of complicated structural models. As was mentioned above, the strong correlation between odd terms in anharmonic temperature factors as well as odd terms in multipole expansion is observed in such structures. The refinement procedure used by Parini (1988) was as follows. First, position and thermal vibration parameters, including the Edgeworth third order of anharmonicity for C and N atoms as well as the scale factor, were refined with high-angle reflections. Significant positive correlations between some position and anharmonic parameters of each pseudoatom with correlation coefficient close to unity were found. To avoid correlation the third-order anharmonic parameter k^{123} for the C atom was fixed at zero. After this stage of refinement the harmonic thermal parameters proved to be close to those calculated by Willis and Howard (1975)

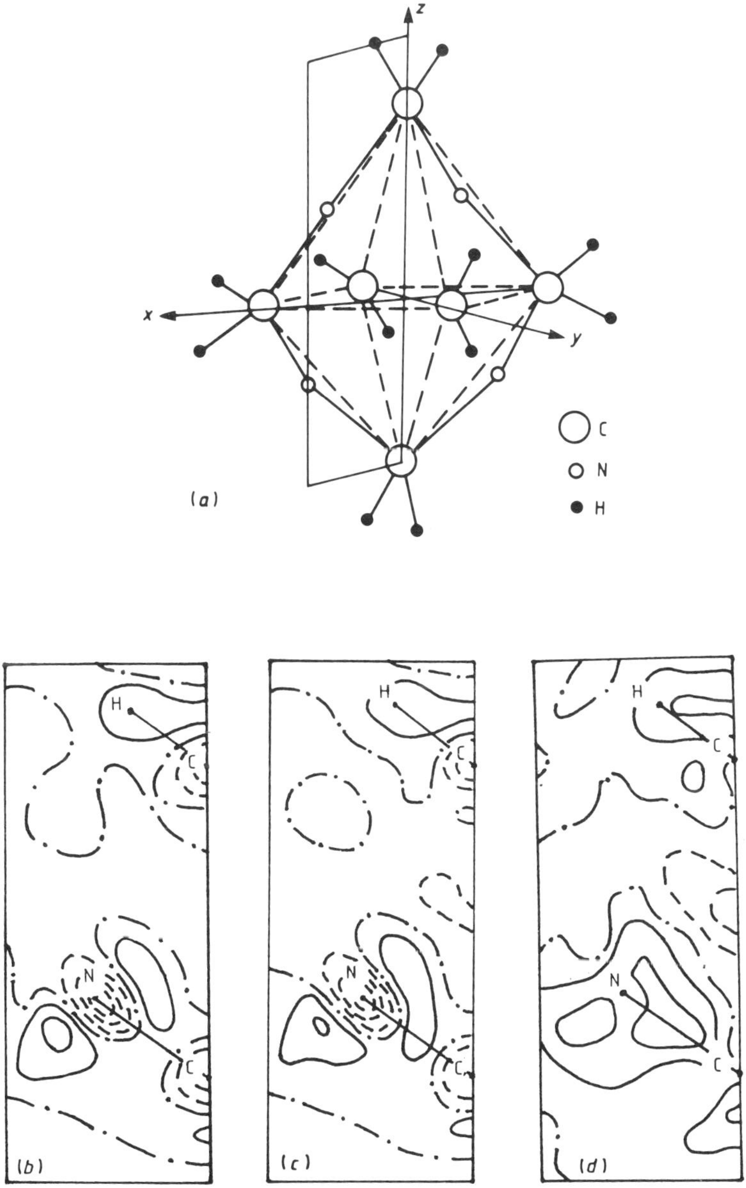

Figure 3.19 Hexamethylenetetramine, $C_6H_{12}N_4$: (*a*) the atom arrangement and the (110) plane containing the crystallographic *z* axis; (*b*) and (*c*) dynamic multipole deformation electron density in the (110) plane, $l=2$, and $l=4$, correspondingly (Parini 1988); (*d*) dynamic deformation ED in the same plane calculated by Fourier series (Stevens and Hope 1975). The line intervals are 0.1 e Å^{-3}.

according to the Born theory. Further, the electronic parameters of all multipoles and the scale factor were refined with the total set of structure amplitudes. Standard molecular values presented in table C.1 were taken as multipole exponential parameters. The deformation monopole, $\delta\rho_0$, parameters of C and N atoms were refined subsequently and, regardless of the multipole expansion length, the positive C_1 and negative C_{24} coefficients appeared to be statistically significant for both atoms. Hence, the ED in the 'middle' part of C and N atoms is decreased because of the chemical bonding effect. The H pseudoatom is contracted along the C–H bond as often observed in organic compounds (Tsirelson 1993).

The quadrupole level of the multipole model ($l \leqslant 2$) gave the R-factor value 0.0270. The set charges of C, N, and H atoms, as calculated from monopole electron occupancies, were $+0.28(16)$, $-0.45(14)$, and $+0.01(14)$ e, respectively, in agreement with results of semiempirical calculations of free hexamethylenetetramine molecules, in the approximation of neglect of diatomic differential overlap: $+0.20$, -0.20, and -0.03 e. In spite of this, it was found that the dynamic multipole ED of the hexamethylenetetramine on the quadrupole level of the model used (figure 3.19(b)) is not sufficient for describing the N pseudoatom ED. In particular, the value of the model deformation ED, $\delta\rho_M$, on the C–N bond is only 0.08 e Å^{-3}. The N atom is in the apex of the solid angle which is formed by three carbon atoms, and is in the sp^3 hydridization valence state. In order to produce all the ED features, the multipole expansion (3.95) was extended with octupole ($l=3$) and hexadecapole ($l=4$) terms for C and N pseudoatoms included in the model. Significant correlation between some dipole and octupole multipole terms was observed; however, a better ED description was achieved in this case (figure 3.19(c)): the $\delta\rho_M$ peak on the C–N bond increased up to 0.14 e Å^{-3}. The level of the residual ED, which was not described by the multipole model, was only 0.05 e Å^{-3}. The new $\delta\rho_M$ map is very close to the similar Fourier map (figure 3.19(d)). At the same time, the filtration of random ED peaks by the multipole model can be seen on the multipole $\delta\rho_M$ map.

In spite of parameter correlation, the electron occupancies of octupole and hexadecapole terms appeared to be two to three times larger than their estimated standard deviations. The dipole and quadrupole parameters did not change by more than $(0.2\text{–}1.0)\sigma$ after the extension of the multipole expansion. As a result, the components of tensors of second moment and electric field gradient on the nucleus, calculated for the N pseudoatom from quadrupole parameters of the multipole model discussed (table 3.12) are only slightly dependent on the length of the expansion. We will consider the methods of calculation and treatment of pseudoatomic moments in detail below (see sections 6.2 and 7.1). One can conclude that multipole parameters obtained are physically meaningful.

The electron occupancies of the multipoles do not have such clear sense as similar parameters in orbital theories of the chemical bond. The diagonal

Table 3.12 The components of electric field gradient (eq_{ij}) and quadrupole moment (eQ_{ij}) of N pseudoatoms in hexamethylenetetramine, $C_6H_{12}N_4$, at different lengths of multipole expansion (all values in local coordinate system).

Property	Pseudoatom N	
	$l\leqslant 2$	$l\leqslant 4$
eq_{xx}	0.48(15)	0.60(10)
$eq_{yy}\times 10^{-22}$ V m^{-2}	0.48(15)	0.60(10)
eq_{zz}	−1.00(30)	−1.00(20)
eQ_{xx}	0.19(5)	0.21(3)
$eQ_{yy}\times 10^{39}$ C m^2	0.19(5)	0.21(3)
eQ_{zz}	−0.39(10)	−0.42(7)

elements of the matrix $p_{\mu\nu}$ (2.26), which represents the one-electron DM in the atomic orbital basis set, can be considered as electron orbital changes, and the non-diagonal elements as bond order values (Mestechkin 1977). The multipoles describe the ED in a form which does not allow, generally speaking, division on the orbital contributions. However, there is the important case when the relation between orbital and multipole electron occupancies can be established. This is the case of d subshell populations of transition-metal atoms. If the d orbital is described by a single STO and the overlap of transition-metal atom and ligand orbitals is small, then the electron multipole populations can be recalculated to d orbital occupancies. This has been shown by Holladay *et al* (1983) on the basis of point-group-specific relation analysis. The density of the d electrons can be written in terms of the AO, d_i, as

$$\rho_d=\sum_{i=1}^{5} P_i d_i^2+\sum_{i=1}^{5}\sum_{j>1}^{5} P_{ij}d_id_j \tag{3.96}$$

with $d_i=R(r)y_{lm\pm}$. The cross-term d_id_j do not occur in the case of an isolated atom for which the ED is diagonal in orbitals. In the molecule or crystal the cross-terms occur only between orbitals belonging to the same symmetry representation, which appear together in the MO LCAO approximation. Thus, equation (3.96), in which all cross-terms are represented, corresponds to the point group symmetry $\bar{1}$ (C_i). In this case all d orbitals transform according to the representation a_g and the mixing is symmetry allowed. In higher-symmetry point groups, this does not occur. For example, in the square-planar point group, $4/mmm$ (D_{4h}), the d orbitals are of b_{1g}, a_{1g}, b_{2g}, and e_g symmetry, so only the diagonal terms in (3.96) can occur. The corresponding ED expression is

$$\rho_{3d}=P_1(d_{x^2-y^2})^2+P_2(d_{z^2})^2+P_3(d_{xy})^2+\tfrac{1}{2}P_4(d_{xz}^2+d_{yz}^2). \tag{3.97}$$

If the Hansen and Coppens multipole model is used, the 3d pseudoatom ED is

$$\rho_d = C_{val}\varkappa_1^3\rho_{val}(\varkappa_1 r) + \sum_{l=0}^{4} \varkappa^3 R_1(\varkappa r) \sum_{m=0}^{l} C_{lm\pm} y_{lm\pm}(\theta, \varphi). \qquad (3.98)$$

The spherical harmonics compose a complete set of functions; therefore, the product of two spherical harmonics is a linear combination of member set. The product $y_{lm\pm} y_{l'm'\pm}$ contains terms with $l''=|l-l'|$, $|l-l'+2|$, ..., $l+l'$ and with $m''=|m-m'|_{\pm}$ and $|m+m'|_{\pm}$. Similar rules are valid for products $y_{lm} y_{l'm'}$, except $l''=0$ and $m''=0$ which do not occur in this case. With the coefficients listed by Biedenharn and Louk (1981), expression (3.96) can be rewritten as a sum of spherical harmonics by analogy with the multipole expansion (3.98).

To equate the two expressions, the difference in normalization between AOs and density functions (see above) should be taken into account. The relations between orbital coefficients P_i and multipole populations $C_{lm\pm}$ are then obtained: $C_{lm\pm} = MP_i$. P_i is the 15-element vector of coefficients of (3.96), $C_{lm\pm}$ is the vector containing coefficients of 15 spherical harmonic functions generated by the products of d orbitals, for which $l''=0$, 2, or 4. The d orbital occupancies are derived from the experimental multipole populations by the use of the equation

$$P_i = M^{-1} C_{lm\pm}. \qquad (3.99)$$

The general case of inverse matrix M^{-1} is presented in table 3.13. The 'index picking' rules for spherical harmonic functions under all crystallographic site symmetries were given by Kurki-Suonio (1977a). Due to the fact that $l=0$, 2, or 4, the nine different cases can be distinguished, as summarized in table 3.14. The matrix M^{-1} for specific point groups is obtained from table 3.13 by eliminating the columns corresponding to non-allowed multipoles. This leads to rows consisting only of zeros which represent non-symmetry-allowed cross-terms in (3.96).

The example of d orbital occupancy analysis via the multipole model is chromium hexacarbonyl, $Cr(CO)_6$, investigated by Holladay *et al* (1983). The accurate x-ray diffraction data for this octahedral complex were reported by Rees and Mitscher (1976). Only C_{00}, C_{40}, and C_{44+} terms are allowed for the octahedral local symmetry of the Cr atom position. The separate refinement of the two latter terms, which is possible in this case because of the lower crystallographic symmetry, gave a ratio very close to that required in the octahedral field. The electron occupancy of the Cr atom 4s subshell could not be refined with certainty because of the small contribution of 4s scattering to the x-ray diffraction intensities. Therefore, the refinements were done with and without taking account of 4s electrons.

The results listed in table 3.15 are almost independent of the 4s electron treatment. The experimental values of orbital occupancies deviate appreciably

Table 3.13 The matrix M^{-1} (Holladay *et al* 1983).

d orbital populations	Multipole populations					
	c_{00}	c_{20}	c_{22}	c_{40}	c_{42}	c_{44}
p_{z^2}	0.200	1.04	0.00	1.40	0.00	0.00
p_{xz}	0.200	0.520	0.943	−0.931	1.11	0.00
p_{yz}	0.200	0.520	0.943	−0.931	1.11	0.00
$p_{x^2-y^2}$	0.200	−1.04	0.00	0.233	0.00	1.57
p_{xy}	0.200	−1.04	0.00	0.233	0.00	−1.57

	Mixing terms										
	c_{21}	c_{21-}	c_{22}	c_{22-}	c_{41}	c_{41-}	c_{42}	c_{42-}	c_{43}	c_{43-}	c_{44-}
$p_{z^2/xz}$	1.09	0.00	0.00	0.00	3.68	0.00	0.00	0.00	0.00	0.00	0.00
$p_{z^2/yz}$	0.00	1.09	0.00	0.00	0.00	3.68	0.00	0.00	0.00	0.00	0.00
p_{z^2/x^2-y^2}	0.00	0.00	−2.18	0.00	0.00	0.00	1.92	0.00	0.00	0.00	0.00
$p_{z^2/xy}$	0.00	0.00	0.00	−1.85	0.00	0.00	0.00	2.30	0.00	0.00	0.00
$p_{xz/yz}$	0.00	0.00	0.00	1.60	0.00	0.00	0.00	1.88	0.00	0.00	0.00
p_{xz/x^2-y^2}	1.88	0.00	0.00	0.00	−1.60	0.00	0.00	0.00	2.10	0.00	0.00
$p_{xz/xy}$	0.00	1.88	0.00	0.00	0.00	−1.60	0.00	0.00	0.00	2,10	0.00
p_{yz/x^2-y^2}	0.00	−1.80	0.00	0.00	0.00	1.60	0.00	0.00	0.00	2.10	0.00
$p_{zy/xy}$	1.88	0.00	0.00	0.00	−1.60	0.00	0.00	0.00	−2.10	0.00	0.00
$p_{x^2-y^2/xy}$	0.00	0.00	0.00	0.00	0.00	0.00	0.00	0.00	0.00	0.00	3.14

Table 3.14 Allowed multipole functions describing d orbital density (Holladay *et al* 1983).

Point group		Allowed values of l, $m\pm$[a]	Dimension of M
I	$1, \bar{1}$	$l=0, 2, 4$, all m	15×15
II	$2, m, 2/m$	00, 20, 22+, 22−, 40, 42+, 42−, 44+, 44−	9×9
III	$222, m2m, mmm$	00, 20, 22+, 40, 42+, 44+	6×6
IV	$4, 4/m, \bar{4}$	00, 20, 40, 44+, 44−	$5\times 5 (4\times 4)$[b]
V	$422, 42m, 4mm, 4/mmm$	00, 20, 40, 44+	4×4
VI	$3, \bar{3}$	00, 20, 40, 43+, 43−	$5\times 5 (4\times 4)$[c]
VII	$32, 3m, \bar{3}m$	00, 20, 40, 43+	4×4
VIII	$6, \bar{6}, 6/m, 622, 6mm, \bar{6}m2, 6/mmm$	00, 20, 40	3×3
IX	$23, m3, 432, \bar{4}3m, m3m$	00, 40+ (0.7403)44+	2×2

[a]The principal symmetry axis is the z axis.
[b]The dimension can be reduced by rotation of the coordinate system.
[c]This function is usually described as the cubic harmonic K_4.

Table 3.15 3d orbital populations for $Cr(CO)_6$ (Holladay *et al* 1983).

	With 4s	Without 4s	Theoretical (Heijser 1979)	Spherical atom
e_g	1.42(5)	1.37(7)	1.12	2
t_{2g}	3.40(5)	3.32(8)	3.26	3
Total d	4.82(9)	4.69(9)	4.38	5
4s	1	—	−0.085	1
$4p_x$			−0.011	
$4p_y$			−0.011	
$4p_z$			−0.011	

from the spherical-atom configuration and are in very reasonable agreement with theoretical values calculated by Heijser (1979) with the X_α method.

Holladay *et al* (1983) and Streltsov *et al* (1993) have studied some other compounds with 3d elements and concluded that atomic net charge values obtained only weakly depend on the inclusion of the 4s subshell in the multipole model. The 4s electrons have a diffuse distribution which peaks near neighbouring ligand atoms, and the scattering of these electrons can be accounted for about equally well by ligand-centred multipoles. Thus, the asphericity of ED is a much better defined property than the net atomic charge. The 3d orbital occupancies, recalculated from experimental multipole paramerters, are in satisfactory agreement with the results of theoretical LCAO calculations and allow one to study the transition-metal states in coordination compounds. The discrepancies between experimental and theoretical populations reflect a covalency effect ignored in the multipole model.

3.5 Quantum chemical models

All superposition structural models considered above describe a crystal as a set of pseudoatomic fragments. The latter can be considered as 'rigid' particles or can be subdivided into a rigid core and variable valence part. The electron clouds of pseudoatoms can penetrate into each other, however, the ED related to pseudoatom overlap is disregarded. This circumstance leads to the fact that chemical bond effects related to a quantum phenomenon, for example, to interference of atomic orbitals, are lost when even a very flexible model is used. Indeed, following Ruedenberg (1951), the ED related to the overlap of two atomic orbitals at different centres can be approximately written as $\varphi_\mu^*\varphi_\nu \approx cS_{\mu\nu}\{\varphi_\mu^*\varphi_\mu + \varphi_\nu^*\varphi_\nu\}$, where $S_{\mu\nu} = \int_{-\infty}^{+\infty} \varphi_\mu^*\varphi_\nu \, dV$ is the overlap integral. Since constant c is never known precisely, the approximation of ED in a crystal unit cell by a sum of one-centre terms always contains some uncertainty. At the same time, the overlap density is responsible for the appearance of forbidden reflections.

There are quantum chemical structural models of a crystal, which take into account the orbital overlap effect and whose parameters can be determined from the x-ray diffraction data. One can distinguish two types of such models. One of them takes into account explicitly the shell structure and uses wave-functions or one-electron DMs for describing crystal ED. The other utilizes the bond charges concept. These models are very attractive, because they provide a unified description of the same quantity, the ED of a crystal, determined from completely different origins: from the solution of quantum chemical equations and from treatment of the x-ray diffraction measurement results. Such a coupling of theory with experiment arises rarely and deserves special attention.

3.5.1 *Wave-function-based models*

Let the wave-function Ψ of some system of N electrons be approximated by a single determinant composed of N orthonormalized orbitals $\varphi_i(\mathbf{r})$. This corresponds to the HF method in quantum chemistry. Each of these orbitals can be presented as a superposition of atomic orbitals Φ_μ (MO LCAO approximation) and the ED of a system can be presented as

$$\rho(\boldsymbol{r})=\sum_i |\varphi_i(\boldsymbol{r})|^2=\sum_i \sum_{\mu\nu} c^*_{i\mu}c_{i\nu}\Phi^*_\mu(\boldsymbol{r})\Phi_\nu(\boldsymbol{r}). \tag{3.100}$$

The Fourier transform of $\rho(\boldsymbol{r})$ gives the model structure amplitude in the form

$$F_{model}(\boldsymbol{H})=\sum_i \sum_{\mu\nu} c^*_{i\mu}c_{i\nu}f_{\mu\nu}(\boldsymbol{H}) \tag{3.101}$$

where $f_{\mu\nu}$ is an amplitude which describes the x-ray scattering of the ED arising from $\Phi^*_\mu\Phi_\nu$ orbital overlap. The methods of calculation of amplitudes, $f_{\mu\nu}$, are given in appendix I. The expression (3.101) permits, in principle, calculation of coefficients $c_{i\mu}$ of the MO expansion over AOs from the x-ray diffraction data. To do this, the values $F_{model}(\boldsymbol{H})$ should be introduced into the functional (3.69). Then the coefficients $c_{i\mu}$ can be determined with, say, LS, using the orthonormalization condition (2.4) as a constraint. The Lagrange unknown multiplier method is most suitable for this purpose.

The use of the orthonormalization condition as a restriction during determination of coefficients $c_{i\mu}$ allows one to interpret the latter as parameters of orbitals similar to those in the HF method. Besides, there are linear or near-linear dependences between matrix elements $f_{\mu\nu}$ in (3.101). This is clearly seen if, for example, the radial dependences of 2s and 2p AOs are described by a single STO, when $R_{2s}=R_{2p}$. Then function $f_{\mu\nu}$ describing the x-ray scattering by the 2s electron subshell will be indistinguishable from the spherical part of scattering by the sum of $2p_x2p_x$, $2p_y2p_y$, and $2p_zp_z$ densities. The derivatives of $F_{model}(\boldsymbol{H})$ (3.101) with respect to parameters $c_{i\mu}$ and $c_{i\nu}$, which are necessary for LS, will differ from each other only by a numerical factor in front of the same function $f_{\mu\nu}$. As a result, the LS normal equation will either have no solution or will be ill conditioned. The use of the additional condition (2.4), which is nonlinear in parameters $c_{i\mu}$, breaks down linear dependences in the LS equations, and the determination of these coefficients becomes possible, though the problem of correlation between refined parameters still remains, of course.

The method of determination of $c_{i\mu}$ coefficients was proposed in this form by Tanaka (1988a). In spite of the fact that the atomic position symmetry can reduce the number of coefficients $c_{i\mu}$ (their total number is $M\times m$, where M is the number of molecular orbitals and m is the number of basis functions), their number remains, in the general case, too large to be determined with a limited set of structure amplitudes.

There exists one important special case, however, when this approach is practically useful. This is the case of the determination of parameters of orbitals of transition d-metal ions in coordination compounds if overlap of these orbitals with orbitals of ligands is small and can be ignored. The crystal field removes the degeneracy of the d energy levels. The wave-functions in the general case may be expressed as a linear combination of five real orthonormalized basis functions, d AOs. The specific form of this linear combination depends on a crystal field symmetry. Every combination of AOs is chosen in such a manner that the resulting d orbitals are transformed as fully symmetric representations of the corresponding symmetry group. The number of independent real expansion coefficients $c_{i\mu}$ is determined by symmetry and is small.

The ED of the transition d ion is written in the adapted approximation as

$$\rho_d(\boldsymbol{r}) = \rho_{core}(\boldsymbol{r}) + \sum_i^5 p_i \sum_{\mu\nu} c_{i\mu} c_{i\nu} \Phi_\mu(\boldsymbol{r}) \Phi_\nu(\boldsymbol{r}) \tag{3.102}$$

and the corresponding structure amplitude as

$$f_d(\boldsymbol{H}) = f_{core}(\boldsymbol{H}) + \sum_i p_i \sum_{\mu\nu} c_{i\mu} c_{i\nu} f_{\mu\nu}(\boldsymbol{H}). \tag{3.103}$$

f_{core} describes the x-ray scattering on the inner spherical part of the ED of the d ion and is easily calculated (Ibers and Hamilton 1974). The $f_{i\mu\nu}$ are basis orbital (partial) scattering amplitudes. In the case of d ions, the most suitable tool for calculating these amplitudes is the Weiss and Freeman (1959) method that is described in a general form in appendix I. Quantities p_i are the electron populations of d orbitals to be determined.

There are a few examples of application of this approach to the study of electron structure of crystals: $(C_6H_{12}N_2)Cu(NO_3)_2$ (Tanaka 1988b, 1993), PtP_2 (Tanaka *et al* 1992), and 3d sesquioxides Ti_2O_3, V_2O_3, and α-F_2O_3 with a corundum structure (Tsirelson *et al* 1989, Streltsov *et al* 1990). Let us consider the latter group of compounds. In the crystal structure of corundum (space group $R\bar{3}c$) the O atoms form a hexagonal close packing, in which two-thirds of the octahedral holes are occupied by cations. The anions form along the 3_1 axis trigonally distorted octahedra, each pair of adjacent cation-occupied octahedra being separated by a single non-occupied octahedron. The anions can be subdivided into two symmetrically equivalent groups belonging to a common face of adjacent populated octahedra (O) and the remaining ones (O′). The cation–O distances are always larger than similar distances to the O′ atoms. The cations are slightly shifted with respect to plane (0001) and occupy positions along the *c* axis at heights approximately equal to one-third and two-thirds of the distance between the layers of anions. Each cation has one closest neighbour along the *c* axis and three neighbours in the basal plane. Such a crystal structure is usually explained by the presence of a strong electrostatic interaction between anions, and the chemical bond in such crystals is supposed to be predominantly ionic. Thus, the compounds of this type are suitable for analysis of their electron structure by the method described above.

In sesquioxide crystals the ions of transition 3d metals are in a trigonally distorted cubic crystal field. The octahedral coordination is distorted and the symmetry lowers from O_h to D_{3d}. In this case the energy levels of 3d orbitals are split into a singlet level a and two doubly degenerate e levels. If the c axis of the unit cell is chosen as the quantization axis z and the axis x is in the plane determined by axis z and one of the M–O bonds the basis functions of irreducible representations, a, e^σ, and e^π are conveniently selected as follows:

$$\text{a:}\quad d_{z^2}=t_2^0 \tag{3.104}$$

$$e^\pi:\quad \begin{cases}(\sqrt{2}\,d_{x^2-y^2}+d_{xz})/\sqrt{3}\\(\sqrt{2}\,d_{xy}-d_{yz})/\sqrt{3}\end{cases} \tag{3.105}$$

$$e^\sigma:\quad \begin{cases}(d_{x^2-y^2}-\sqrt{2}\,d_{xz})/\sqrt{3}\\(d_{xy}+\sqrt{2}\,d_{yz})/\sqrt{3}.\end{cases} \tag{3.106}$$

Using the results from appendix I, one can write the expressions for partial scattering amplitudes of 3d electrons in a trigonally distorted octahedral field as

$$f_i(\text{a})=\langle j_0\rangle-\tfrac{5}{7}(3\cos^2\beta-1)\langle j_2\rangle+\tfrac{9}{28}(35\cos^4\beta-30\cos^2\beta+3)\langle j_4\rangle \tag{3.107}$$

$$f_i(e^\pi)=\langle j_0\rangle+\tfrac{5}{14}(3\cos^2\beta-1)\langle j_2\rangle - \{\tfrac{1}{28}(35\cos^4\beta-30\cos^2\beta+3)-(5/\sqrt{2})(\sin^3\beta\cos\beta\cos 3\gamma)\}\langle j_4\rangle \tag{3.108}$$

$$f_i(e^\sigma)=\langle j_0\rangle-\{\tfrac{1}{8}(35\cos^4\beta-30\cos^2\beta+3)+(5/\sqrt{2})(\sin^3\beta\cos\beta\cos 3\gamma)\}\langle j_4\rangle. \tag{3.109}$$

The Fourier–Bessel transforms of the radial parts of the wave-functions (the 'radial integrals') $\langle j_n\rangle$ in these expressions are basis scattering factors for d electrons. The $\cos\beta$ and $\cos\gamma$ are direction cosines of quantization axis z. Note that subscripts μ and ν in these integrals are omitted, since they are the same for the same values of n.

The d electron scattering amplitude in (3.103) contains a linear combination of partial scattering amplitudes $f_i(\text{a})$, $f_i(e^\pi)$, and $f_i(e^\sigma)$ multiplied by electron population coefficients, p_i. The latter can be refined by LS along with the other crystal structural model parameters.

For calculation of model structure amplitudes, F_{model}, and their derivatives appearing in LS equations, the local atomic coordinate systems should be related to the crystallographic system. The detailed derivation was given by Streltsov *et al* (1990). Note that both the form of the resulting expressions and the values of direction cosines depend on the choice of the local coordinate system, though the partial scattering amplitudes are invariant with respect to this choice. For example, in 3d sesquioxides, for a trigonally distorted cubic crystal field, the local coordinate system for the 3d ion can be turned by 60 or 180° around the c axis with respect to the system stated above. In this case the signs in the last terms in expressions for $f_i(e^\pi)$ and $f_i(e^\sigma)$ will be changed and, simultaneously, the direction cosine cos 3γ also changes its sign. As a result, the values of partial scattering amplitudes will remain the same.

The electron populations of 3d orbitals of transition-metal ions in Ti_2O_3, V_2O_3, and α-Fe_2O_3 were found by Streltsov *et al* (1990) with the regularized LS. The normalization condition was not applied explicitly; the total number of 3d electrons was not fixed, although the unit cell electroneutrality was retained. Quite complete and accurate sets of structure amplitudes, measured at room temperature, were used in the refinements. The data of an x-ray diffraction experiment at 153 K were also used for α-Fe_2O_3. In the latter case, the refined electron populations were less distorted by thermal motion of the atoms in the crystal. The results of refinement of 3d populations with radial integrals $\langle j_n \rangle$ calculated by means of atomic (model I) and ionic M^{3+} (model II) wave-functions from Ibers and Hamilton (1974) are presented in table 3.16. The population of 4s orbitals was not included in the model. The statistical analysis based on Fisher's criterion (Hamilton 1964) has shown that the use of atomic wave-functions at the significance level $\alpha = 0.01$ leads to a more optimal orbital model for Ti_2O_3 and V_2O_3. For α-Fe_2O_3 at 153 K models I and II are statistically equivalent, and at 298 K model II has an insignificant statistical advantage.

Table 3.17 presents the results of refinements of electron populations of 3d orbitals for the same crystals accounting for the populations of 4s orbitals. Radial integrals $\langle j_n \rangle$ have been calculated using atomic wave-functions only. EDs of 4s orbitals are very diffuse and stretched towards the ligands. As a result, their populations are not always accessible for analysis. Holladay *et al* (1983), who applied a multipole model for analysis of 3d orbital populations of Cr, Co, and Mn compounds, have found that the inclusion of 4s orbitals in a model only weakly influences the refinement results. In the orbital model under consideration it was found that in Ti_2O_3 and V_2O_3 taking account of 4s orbitals improves the model at the significance level of 0.01. In α-Fe_2O_3 the model was not improved; on the contrary the inclusion of these orbitals resulted in the values of their electron populations not having any physical meaning.

Thus, one can conclude that within the framework of the model under consideration and the accuracy of experimental data the refinement results for α-Fe_2O_3 at 295 K and 153 K are not sensitive to the presence of an additional parameter for the 4s shell of the Fe atom, in agreement with the results of Holladay *et al* (1983). On the other hand, in Ti_2O_3 and V_2O_3 crystals the introduction of additional 4s population parameters noticeably improves the model adjustment. This fact can be associated, apparently, with the presence in Ti and V atoms of a proportion of the 4s electrons which are more localized near the cores, in contrast to Cr, Mn, Co, and Fe atoms with the 3d subshells filled with a greater number of electrons.

Table 3.16 Electron populations of the 3d orbitals in Ti_2O_3, V_2O_3. and α-Fe_2O_3 without taking into account 4s orbital populations.

	Ti_2O_3 295 K		V_2O_3 295 K		α-Fe_2O_3 153 K		α-Fe_2O_3 295 K	
Orbitals	I[a]	II	I	II	I	II	I	II
e^{σ}	1.302(5)	1.130(2)	1.247(5)	1.062(3)	2.700(9)	2.574(9)	2.815(17)	2.685(1)
(%)[b]	41.5	42.6	31.7	30.1	40.0	40.2	40.5	40.7
e^{π}	0.778(5)	0.684(1)	1.874(4)	1.697(2)	2.445(8)	2.343(7)	2.552(13)	2.449(1)
(%)	24.8	25.8	47.7	48.2	36 2	36.6	36.7	37.2
a	1.056(2)	0.846(1)	0.810(2)	0.764(2)	1.604(9)	1.480(8)	1.581(17)	1.455(1)
(%)	33.7	31.6	20.6	21.7	23.8	23.2	22.8	22.1
R (%)	1.214	1.307	1.093	1.232	1.478	1.461	2.007	2.039
R_w (%)	0.930	1.080	1.101	1.233	2.110	2.100	2.602	2.634
S	1.599	1.858	1.615	1.807	1.822	1.814	1.705	1.727

[a] I, II, radial integrals $\langle j_n \rangle$ have been calculated with atomic and ionic wave-functions, respectively.
[b] The value of the relative population as a percentage.

Table 3.17 Electron populations of the 3d orbitals in Ti_2O_3, V_2O_3, and α-Fe_2O_3 taking into account the 4s orbital population.

Orbitals	Ti_2O_3[a] 295 K	V_2O_3 295 K	α-Fe_2O_3 153 K	α-Fe_2O_3 295 K
e^σ	0.790(7)	0.860(5)	2.874(14)	2.745(19)
(%)	37.5	28.6	40.0	40.5
e^π	0.485(6)	1.533(3)	2.618(13)	2.482(20)
(%)	23.1	51.0	36.4	36.7
a	0.830(2)	0.615(2)	1.696(12)	1.545(19)
(%)	39.4	20.4	23.6	22.8
4s	1.628(6)	1.600(4)	−0.715(41)	2.83(56)
R (%)	1.080	1.028	1.445	2.016
R_w (%)	0.785	0.914	2.099	2.600
S	1.351	1.342	1.814	1.706

[a] Electron populations of 3d orbitals in free spherical symmetrical Ti, V, and Fe atoms:

	Ti	V	Fe
e^σ	0.8 (40%)	1.2 (40%)	2.4 (40%)
e^π	0.8 (40%)	1.2 (40%)	2.4 (40%)
a	0.4 (20%)	0.6 (20%)	1.2 (20%).

Analysing the results presented in tables 3.16 and 3.17 for Ti_2O_3, V_2O_3, and α-Fe_2O_3, one can notice that the total refined number of 3d electrons exceeds their number in free atoms. This can be attributed both to the influence of 4s electron scattering and to the ignored hybridization of 3d, 4s, and 4p orbitals and their overlap with orbitals of ligands. Besides, the use of the atomic many-electron wave-function in the form of a single Slater determinant is often insufficient for correct description of an electron structure of transition 3d ions. Nevertheless, the relative values of populations of electron 3d sublevels of transition elements obtained are stable enough for different variants of model for every compound. The results allow one to make some conclusions about the chemical bond character in sesquioxides. So, in Ti_2O_3 the a orbital is more filled by electrons as compared to the e^π orbitals. In V_2O_3, on the other hand, the electron population of the e^π orbitals is considerably higher. This indicates the possibility of V–V bonding in V_2O_3 via the overlap of π orbitals of V atoms in the basal plane and of Ti–Ti bonding in Ti_2O_3 along the c axis of the cell, into which the overlap of a orbitals of the Ti atom gives a contribution as well. Judging by 3d populations in α-Fe_2O_3, the a orbital is filled to a greater extent than each of the e^π orbitals. This indicates the possibility of a direct Fe–Fe interaction along the c axis direction of a cell, which takes place with the participation of d orbitals of Fe atom. These conclusions agree with the scheme of band structure for sesquioxides that was proposed by Goodenough (1971), as well as with deformation ED maps which have been discussed by Tsirelson *et al* (1989).

3.5.2 *Density-matrix-based models*

By analogy with theory (see section 2.1), the problem of determining from x-ray diffraction data the numerical characteristics of the one-electron ground state density matrix, DM1, written in the orbital approximation, can be formulated. In the single-determinant approximation and using the MO LCAO method one can write DM1 in position space in the form

$$\rho(\boldsymbol{r}, \boldsymbol{r}') \equiv \gamma(\boldsymbol{r}, \boldsymbol{r}') = \sum_{\mu\nu} P_{\mu\nu} \Phi_\mu^*(\boldsymbol{r}) \Phi_\nu(\boldsymbol{r}'). \tag{3.110}$$

The charge–bond order matrix with elements $P_{\mu\nu}$ has already been considered in section 2.1. It fully represents DM1 in the atomic orbital basis set, $\{\Phi_\mu\}$, and, using relation (2.19), allows one to calculate the mean values of one-electron quantum mechanical operators. Supposing, implicitly, a continuous dependence of DM1 upon spatial coordinates $\boldsymbol{r}$ and $\boldsymbol{r}'$, Stewart (1969) proposed searching for $P_{\mu\nu}$ coefficients by minimizing the functional (3.69) in which $F_{model}(\boldsymbol{H})$ is the Fourier transform of diagonal elements of DM1 only, i.e. of the electron density:

$$F_{model}(\boldsymbol{H}) = \sum_{\mu\nu} P_{\mu\nu} \int \Phi(\boldsymbol{r}) \Phi_\nu^*(\boldsymbol{r}) \exp(2\pi \mathrm{i} \boldsymbol{H} \cdot \boldsymbol{r}) \, \mathrm{d}\boldsymbol{r} = \sum_{\mu\nu} P_{\mu\nu} f_{\mu\nu}(\boldsymbol{H}). \tag{3.111}$$

Stewart's idea, which is close to the ideology of DFT, has undergone considerable change, but it is still being discussed in the literature (Saenz and Weyrich 1994). Attempts to apply this approach by means of LS were first undertaken by Coppens *et al* (1971a,b) and by Matthews *et al* (1972). The authors studied molecular crystals using non-orthogonal basis sets of STOs or HF AOs. The atomic thermal motion, which is related to the overlap of AOs centred at neighbouring atoms, was described using the approximation of independent vibrations of these atoms (Coppens *et al* 1971a). The displacements of the overlap density represent the root mean square of the neighbouring atom displacements, and the temperature factor has the form $T_{\mu\nu} = (T_\mu + T_\nu)/2$. This is a rather rough approximation, of course. These works have shown that the determination of $P_{\mu\nu}$ coefficients is far from a trivial problem. The redundancy of expansion (3.110) and the indistinguishability of contributions from orbital components of the s^2 and $p_{x^2} + p_{y^2} + p_{z^2}$ type and from some of the two-centre overlap densities to the x-ray scattering give rise to linear dependences in the LS equations. The transition to the statistically uncorrelated set of electronic parameters by means of orthogonal transformation (Jones *et al* 1972) or the decrease of the number of coefficients to be determined by using the hybrid AO basis set (Allen-Williams *et al* 1975) allows us to reduce the correlation between parameters and obtain the solution. However, the eigenvalues of matrix $\hat{P}$ were not equal to zero or unity, as the HF theory requires (Löwdin 1955). Therefore, the $P_{\mu\nu}$ coefficients, as found by means of a formal approximation of ED by expansion (3.110), do not represent DM1 in the basis set of chosen AOs.

Thus, the reconstruction of DM1 from x-ray diffraction data has difficulties which are not computational. The modern point of view on this problem is as follows. It seems at first sight that the practical implementation of the above-considered approach intends the extraction from ED of more information than it contains in reality. This is not the case, however. The Hohenberg–Kohn (1964) theorem states that the knowledge of ED is sufficient to determine the ground state DM1 (but the recipe for DM1 determination is unknown). Bader (1980) has shown that most of the physical information in DM1 is contained in the neighbourhood of its diagonal elements $\boldsymbol{r}=\boldsymbol{r}'$! In addition, the one-electron DM in the single-determinant approximation has a factorized form (2.22) or (3.110) and, hence, the amount of information in it is lower than in an arbitrary DM1. Further, according to the theorem of N representability of the ground state ED (Harriman 1981, Kolmanovich and Reznik 1981), discussed in section 2.2, the infinite number of antisymmetric many-electron wave-functions (and, consequently, DMs) correspond to the same ρ and, hence, to the same set of structure amplitudes $F_{model}(\boldsymbol{H})$ which are interrelated by a linear Fourier transformation with ρ. Some of these DM1 can be attributed to excited states of a many-electron system in the field of some outer potential; the others correspond to systems with the non-local outer potential only. In order to strictly select the proper ground state matrix $\hat{P}$ one should additionally require that $\hat{P}$ provides the kinetic energy minimum (Levy 1979, 1990, Zhao and Parr 1993). This conclusion follows from the Kohn–Sham (1965) approach, according to which the kinetic energy of a gas of non-interacting electrons with a given density, which is described by the system of one-electron functions, can be separated from the total energy. The DM1 mentioned can be found by varying an energy functional of (1.14) type, in which the electronic part depends on ρ only. In other cases the non-uniqueness of the ρ expansion will result in the approach under consideration having many solutions for sets of coefficients $P_{\mu\nu}$.

A similar difficulty is well known in the other more widespread quantum chemical methods. For example, in the HF theory the equation is solved to yield a stationary state energy rather than a minimum energy. Nevertheless, in almost all cases the procedure for equation solution is constructed in such a manner that the result corresponds to the energy minimum (see e.g. Clementi and Roetti 1974, Pisani *et al* 1988, Koelling 1981). This is achieved because the initial approximation used is much closer to the ground state of a system than to any other state, and the minimum can be found by introducing a small correction to this approximation. In theoretical calculations, the atomic superpositional model is usually such an approximation. When one finds the parameters of DM1 which give the best description of the x-ray scattering the same initial approximation can be used. Another variant is to use a properly symmetrized set of AOs to describe the atoms in the system or a Fourier transform of their products.

Another thing is that the specificity of the problem under consideration consists of the attempt to extract all DM1 elements in one representation ($P_{\mu\nu}$) from only diagonal elements of the one-electron DM in another representation ($\rho(\boldsymbol{r}, \boldsymbol{r}')$). Quantities $P_{\mu\nu}$ may be determined as matrix elements of DM1 in the basis set of orthonormalized AOs:

$$P_{\mu\nu} = \int \Phi_\mu^*(\boldsymbol{r})\rho(\boldsymbol{r}, \boldsymbol{r}')\Phi_\nu(\boldsymbol{r}')\,\mathrm{d}\boldsymbol{r}\,\mathrm{d}\boldsymbol{r}'. \tag{3.112}$$

If, however, these quantities are obtained by minimizing the discrepancy between the experimental and model ED, then one cannot obtain an expression for $P_{\mu\nu}$ similar to (3.112). Indeed, if the experimentally measured ED is approximated by the expression $\rho_0(r) = \sum_{\rho\sigma} P_{\rho\sigma}\Phi_\rho(r)\Phi_\sigma^*(r)$ then the integral

$$\int \Phi_\mu^*(\boldsymbol{r})\rho_0(\boldsymbol{r})\Phi_\nu(\boldsymbol{r})\,\mathrm{d}\boldsymbol{r} = \sum_{\rho\sigma} P_{\rho\sigma}\Gamma_{\rho\sigma\mu\nu} = G_{\mu\nu}(P) \tag{3.113}$$

does not reduce to $P_{\mu\nu}$, but represents a linear combination of these quantities. Four-index one-electron 'superoverlap' integrals appear, $\Gamma_{\rho\sigma\mu\nu} = \int \Phi_\mu^*(\boldsymbol{r})\Phi_\nu(\boldsymbol{r})\Phi_\rho(\boldsymbol{r})\Phi_\sigma^*(\boldsymbol{r})\,\mathrm{d}\boldsymbol{r}$, and the $G_{\mu\nu}$ matrix does not obey the normalization (charge conservation) condition

$$\int_{-\infty}^{\infty} \rho(\boldsymbol{r})\,\mathrm{d}r = \mathrm{Tr}\,\hat{P} = N. \tag{3.114}$$

Thus, only linear combinations of all elements of matrix $\hat{P}$ can be determined from the ED without imposing some additional (theoretical or experimental) conditions or without using the properties of the basis set chosen. Fortunately, these elements are not independent in the HF theory: matrix $\hat{P}$, which represents DM1 in the AO basis set, possesses the idempotency property (see (2.23)) that has the form

$$\hat{P}^2 = \hat{P}. \tag{3.115}$$

This matrix has only $M \times (m - M)$ independent parameters, where M is the number of one-electron orbitals, m is the number of basis set functions; $m \geqslant M$. The nonlinear relation (3.115) mixes diagonal and off-diagonal elements of the $\hat{P}$ matrix. This expression presents the necessary and sufficient condition for coefficients $P_{\mu\nu}$ to correspond to a many-electron wave-function which is described by a single Slater determinant. Therefore, if one wishes to find physically meaningful coefficients $P_{\mu\nu}$, one should take into account property (3.115) as a restriction.

The nonlinear character of the idempotency condition violates the linear dependences among coefficients $P_{\mu\nu}$. If their number is not too large, one can hope to construct the solution in such a way that a good approximation to the true DM1 can be found.

Harriman (1986, 1990) has found that by using a basis set which provides a linear independence of a finite number of $\Phi_\mu\Phi_\nu^*$ products, DM1 elements can be uniquely constructed from a given ED even without additional conditions. Levy and Goldstein (1987) have shown that in this case the number of linearly independent products $\Phi_\mu\Phi_\nu^*$ must be not lower than $M\times(m-M)$. They have also found that the condition of unique determination of DM1 from the ED, reconstructed from x-ray diffraction data, is the presence of at least $M\times(m-M)$ measured independent x-ray reflections. However, the model calculations carried out by Morrison (1988) and Schwarz and Miller (1990) have shown that one cannot hope to obtain solutions for real many-electron system by using only basis set properties. Therefore, a combination of quantum chemical calculation schemes, which take into account DM1 and basis set (Hoch and Harriman 1995) properties, and x-ray diffraction information of ED is the preferable method.

Such an approach can be implemented in two ways. The first one consists of minimizing the functional (3.69) with $F_{model}(\boldsymbol{H})$ in the form of (3.111) under the restrictions (3.114) and (3.115). The second consists of constructing the $\hat{P}$ matrix by using any quantum chemical iterative procedure with the additional condition of best reproduction of experimental structure amplitudes with the $P_{\mu\nu}$ parameters obtained. The latter approach was implemented in practice. Clinton *et al* (1969) and Clinton and Massa (1972) suggested constructing the idempotent matrix $\hat{P}$ by applying the McWeeny iteration procedure (McWeeny and Sutcliffe 1976). This is based on variation of the scalar quantity $\mathrm{Tr}(\hat{P}^2-\hat{P})^2$ which is zero for stationary states:

$$\delta\,\mathrm{Tr}(\hat{P}^2-\hat{P})^2=2\,\mathrm{Tr}(2\hat{P}^3-3\hat{P}^2+\hat{P})\,\delta\hat{P}=0. \tag{3.116}$$

This relation yields the equation for $\hat{P}$ with arbitrary $\delta\hat{P}$ variation:

$$2\hat{P}^3-3\hat{P}^2+\hat{P}=0 \tag{3.117}$$

if the idempotency condition is fulfilled. The iterative self-consistent process of finding matrix $\hat{P}$ follows from (3.117):

$$\hat{P}_{n+1}=3\hat{P}_n^2-2\hat{P}_n^3=0. \tag{3.118}$$

Such an algorithm is used in quantum chemistry for finding matrices $\hat{P}$ corresponding to the stationary state with minimal energy. In the approach under consideration, the condition of best reproduction of the experimental x-ray diffraction pattern by means of matrix $\hat{P}$ is used instead. This condition can be met in the LS sense and then the additional restriction has the form of (3.69) or a linear restriction (Clinton *et al* 1973)

$$\sum_{\boldsymbol{H}} |F(\boldsymbol{H}) - \sum_{\mu\nu} P_{\mu\nu} f_{\mu\nu}(\boldsymbol{H})| = \varepsilon \tag{3.119}$$

can be applied. The introduction of the small quantity ε takes into account the presence of errors in the experimental data, because of which expression (3.119) cannot be made equal to zero.

Tsirelson *et al* (1977) have developed another method for determining the matrix $\hat{P}$. Having approximated $F_{model}(\boldsymbol{H})$ in functional (3.69) by expression (3.111), they transformed this functional to the form

$$D = \frac{1}{V} \sum_{\boldsymbol{H}} |F(\boldsymbol{H})|^2 + \mathrm{Tr}\ \hat{P}(\hat{A} + \hat{H}). \tag{3.120}$$

Here $\hat{A} = \hat{H} + \hat{G}(\hat{P})$, where matrix $\hat{H}$ has elements $H_{\mu\nu} = -(1/V)\sum_{\boldsymbol{H}} F(\boldsymbol{H}) f^*_{\mu\nu}(\boldsymbol{H})$ and matrix $G(\hat{P})$ has already been defined above. The minimization of the quantity D with respect to $\hat{P}$ results in an equation of the form

$$\mathrm{Tr}\ \delta\hat{P}\ \hat{A} = 0. \tag{3.121}$$

Due to the idempotency property, the variation $\delta\hat{P}$ is not independent, but satisfies the relation $\delta\hat{P} = \delta\hat{P}\ \hat{P} + P\ \delta\hat{P}$ and, therefore,

$$\hat{P}\ \delta\hat{P}\ \hat{P} = 0 \tag{3.122}$$

which follows from this relation. The property (3.122) can be taken into account in (3.121) by means of the technique of projection operators (Mestechkin 1977). The result is the following equation for determination of $\hat{P}$:

$$\hat{A}\hat{P} = \hat{P}\hat{A}. \tag{3.123}$$

It is remarkable that this relation has the same formal form as the matrix HF equation (2.25), and matrix $\hat{A}$, that contains the x-ray structure amplitudes, depends on matrix $\hat{P}$ in exactly the same manner as the expression for energy in the HF method. Thus, one manages to obtain a formally unified description for the charge–bond order matrix from different principles. It is important that, by using the projection operator formalism, the algorithm for solution of the basic equation of a theory can be constructed in such a manner that it is not necessary to apply the inversion of matrices (Tsirelson and Ozerov 1979, Tsirelson *et al* 1980b). Therefore, the problem of nearly linear dependence, which usually plays a negative role in crystal structural model optimization, is less important in this case.

In spite of the theoretical attractiveness of this approach, the nonlinearity of equation (3.123) with respect to $\hat{P}$ complicates the procedure of its solution. It is possible to construct the solution; however, it strongly depends on the initial approximation which should be close to the true $\hat{P}$ matrix. This is why the linear restriction (3.119) seems preferable from the computational point of view. This restriction, as well as the charge conservation condition (3.114), can be taken into account by means of the unknown Lagrange multipliers method. Then the equation for determination of $\hat{P}$ has the form (Frishberg and Massa 1982, Clinton *et al* 1983)

$$\hat{P}_{n+1} = 3\hat{P}_n^2 + 2\hat{P}^3 + \lambda_\varepsilon \hat{G} + \lambda_N \hat{I}. \tag{3.124}$$

Here λ_ε and λ_N are the unknown Lagrange multipliers and $\hat{I}$ is a unit matrix. The numerical solution of equation (3.124) is carried out by a self-consistent iterative method in which the procedure of idempotent matrix construction is regularly followed by successive reduction in ε (3.119).

Aleksandrov *et al* (1989) took into account the translation periodicity of the crystal lattice and, using crystal orbitals (2.16) as the basis set functions, extended the above approach to diamond-type crystals. Diamond and silicon were studied using the experimental structure amplitude sets reduced to zero temperature measured for diamond by Göttlicher and Wölfel (1959) and compiled for silicon by Scheringer (1980). The minimal valence basis set of s- and p-type orbitals was used. The single Baldereschi (1972) special point $k^* = (2\pi/a_0)$ (0.6223, 0.2953, 0), as found by Srivastava (1982), was used in the calculations. The experimentally measured forbidden reflections of 222 type were not included in calculations in order to check on the method afterwards. In this way all $\hat{P}$ matrix elements were found. The values of ε_{start} and ε_{end} were 1.711 and 0.652 for diamond and 2.806 and 1.115 for silicon, respectively.

The method described provided good adjustment of the model structure amplitudes to experimental values. Figure 3.20 depicts the $\delta_1 = (|F| - |F_{\rm DM1}|)/|F|$ and $\delta_2 = (|F| - |F_{\rm HF}|)/|F|$ values for the first few x-ray

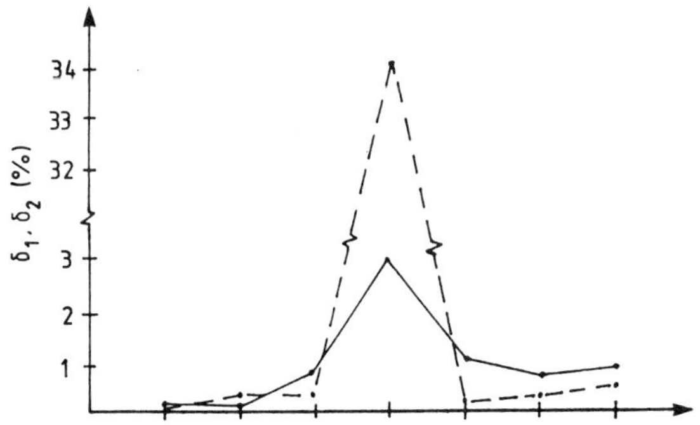

Figure 3.20 The values of $\delta_1 = (F - F_{DM1})/F$ and $\delta_2 = (F - F_{HF})/F$ for some low-angle reflections of silicon (F_{DM1} are structure amplitudes calculated from DM1 which had been obtained from x-ray diffraction experiments; F_{HF} are those calculated from HF wave-functions). δ_1 is represented by the solid curve.

reflections for the Si single crystal (the reflections chosen are mostly affected by valence ED distribution in a crystal). It can be seen that the HF method gave worse agreement with x-ray experiment than the approach discussed. The respective discrepancy factors were $R=0.014$ and 0.023.

The following tendency can also be noticed: neither weight scheme nor filtration have been used in the calculation for diamond and silicon; however, the agreement between structure amplitudes is better for those reflections which had been measured more precisely. This circumstance can be interpreted as follows: the x-ray $P_{\mu\nu}$ values reproduce the ED peculiarities so adequately that they can themselves produce filtration, exclusing from consideration the effect of 'badly' measured reflections. The $P_{\mu\nu}$ parameters better describe the 'forbidden' intensities for diamond which appear mainly because of the ED redistribution after chemical bonding.

Another test of the approach is the estimation of the electron kinetic energy. This quantity in silicon and diamond has been calculated by Aleksandrov *et al* (1989) according to the relation $\langle \hat{T} \rangle = \mathrm{Tr}\, \hat{P}\hat{T}$ (see (2.19)), where $\hat{T}$ is the kinetic energy matrix calculated with the same basis set. The results are given in table 3.18. On the basis of the virial theorem, the absolute magnitude of total energy has been taken as the true 'experimental' value for comparison with T. The 'experimental' T values and those calculated from DM1 in diamond are in good agreement. For silicon the agreement is worse. The difference between x-ray results, the 'experimental' data, and HF values can be attributed partially to the limitation of the minimal basis set used in the DM1 reconstruction.

Aleksandrov *et al* (1989) have also calculated the directed Compton profiles from the x-ray $P_{\mu\nu}$ parameters for both crystals. These values depend on not only the diagonal, but all elements of DM1. The impulse approximation has been used (Eisenberger and Platzman 1970), when the final state of a scattering

Table 3.18 Energy (atomic units) per atom in diamond and silicon crystals.

Crystal	Method	Kinetic energy
Diamond	Experiment[a]	38.145
	X-ray DM calculation	37.860
	HF calculation[a]	37.437
Silicon	Experiment[b]	287.898
	X-ray DM calculation	285.747
	HF calculation[a]	285.639

[a] Dovesi *et al* (1980).
[b] Dovesi *et al* (1981).

system is thought to be a free-electron state. The corresponding wave-function is the plane wave. The directed Compton profile is the one-dimensional projection of the electron momentum $\boldsymbol{p}$ space density in the direction of momentum transfer $\boldsymbol{\mu}$. It can be written as

$$\mathfrak{I}(p_z)=\int \cdots \int \rho(\boldsymbol{r}, \boldsymbol{r}') \exp[\mathrm{i}\boldsymbol{p}(\boldsymbol{r}-\boldsymbol{r}')]\, \mathrm{d}\boldsymbol{r}\, \mathrm{d}\boldsymbol{r}'\, \mathrm{d}p_x\, \mathrm{d}p_y \tag{3.125}$$

where $p_z = p_z(\mu)$. Inserting the DM1 in the orbital representations (3.110) and (3.125), one can calculate the integral via parameters found from x-ray diffraction experiment.

A good general agreement between Compton profiles derived from x-ray DM1, those from HF calculations, and those directly measured has been achieved for silicon, as can be seen in figure 3.21. The greatest difference from direct experimental evidence is seen for small values of momentum deviation from the centre of the Compton line, μ. The same difference is often observed in the theoretical computation and is usually attributed to the simplified basis set used.

Thus, the approach outlined above allows one to describe, approximately, the incoherent diffraction effects in diamond and silicon on the basis of coherent x-ray diffraction data. The fundamental role of the electron density is evident here. At the same time, the problem of the reconstruction of all DM1 elements, corresponding to the kinetic electron energy minimum, from experimental data has not yet been solved rigorously. Schmider and Weyrich (1988) have suggested the reconstruction of the quantum mechanically valid

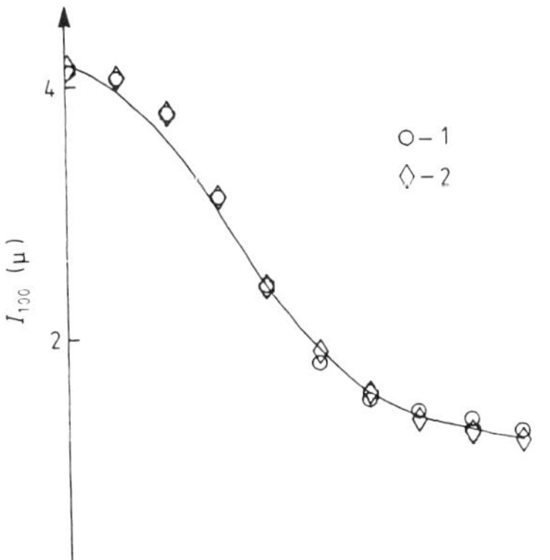

Figure 3.21 The Compton profiles in the [100] direction for silicon: the solid line is the calculation result obtained with DM1 (Aleksandrov *et al* 1989); 1, Reed and Eisenberger's (1972) measurement; 2, Angonoa *et al*'s (1981) HF calculation. (Reprinted with permission from Aleksandrov *et al* 1989.)

DM1 from the condition of the best agreement with the results of various experimental measurements acting as a restriction. Schmider *et al* (1992b) have studied this approach, taking the beryllium atom as a model example, and used for this purpose the sets of structure amplitudes $F(\boldsymbol{H})$ and reciprocal structure amplitudes $B(\boldsymbol{k})$. The latter are Fourier transforms of momentum space electron density and contain information about the off-diagonal part of DM1. Both sets of structure amplitudes were calculated from the same, very exact, correlated wave-functions. It was found that the use of additional complementary information from position and momentum space quantities resulted in a reconstructed DM1 which is much closer to the initial one than it is in the case when each sort of restriction is used separately. Unfortunately, the analysis of kinetic electron energy has not been reported in this study.

Schülke and Mourikis (1986) and Schülke (1988) have shown that the special measurement of the energy distribution of inelastically scattered x-rays under the experimental condition where the initial photon state is represented by a standing wave field allows one to obtain information about off-diagonal elements of the DM1 matrix in momentum space. Parametrizing the DM1 in the orbital approximation and, subsequently, its Fourier transform, and combining the elastic and non-elastic x-ray scattering experiments one can, in principle, reconstruct both diagonal and off-diagonal elements of DM1. This matrix will correspond to the ground electron state of a system and, thus, it should provide the kinetic electron energy minimum. Time will show whether such an approach could be implemented in practice.

3.5.3 *The bond charge model*

We mentioned in section 2.1 that in quantum chemical calculations one can achieve a more flexible description of ED by inclusion in the basis set of polarization functions or bond functions centred in the internuclear space (Breitenstein *et al* 1983). A similar structural model was used in x-ray crystallography by Ewald and Hönl (1936) for explaining the non-zero value of 'forbidden' reflection (222) in diamond. Later, Hellner (1977) modified this model for analytical description of ED in molecular organic crystals in the x-ray diffraction analysis. He presented ED in a unit cell as a sum of three parts:

$$\rho(\boldsymbol{r})=\sum_{\mu}\left[\rho_{\mu}^{core}(\boldsymbol{r})+\rho_{\mu}^{nb}(\boldsymbol{r})\right]+\sum_{\mu\nu}\rho_{\mu\nu}^{b}(\boldsymbol{r}). \qquad (3.126)$$

The first term describes atomic cores with fixed electron populations, the second deals with non-bonding electrons in spherical atomic valence subshells, and the third describes the valence electrons concentrated in the interatomic space (bond charges). Each of these three components vibrates, in the general case, according to its 'own' law and possesses its own centre of gravity. The corresponding structure amplitude has the form

$$F(\boldsymbol{H})=\sum_{\mu}\left[f_{\mu}^{core}(\boldsymbol{H})T_{\mu}^{core}(\boldsymbol{H})\exp(2\pi i\boldsymbol{H}\cdot\boldsymbol{r}_{\mu}^{core})+p_{\mu}^{nb}f_{\mu}^{nb}(\boldsymbol{H})T_{\mu}^{nb}(\boldsymbol{H})\exp(2\pi i\boldsymbol{H}\cdot\boldsymbol{r}_{\mu}^{nb})\right]$$

$$+\sum_{\mu\nu}p_{\mu\nu}^{b}f_{\mu\nu}^{b}(\boldsymbol{H})T_{\mu\nu}^{b}(\boldsymbol{H})\exp(2\pi i\boldsymbol{H}\cdot\boldsymbol{r}_{\mu\nu}^{b}). \qquad (3.127)$$

Thermal and position parameters of the atomic cores are determined by the LS in the high-angle approximation, whereas the electron populations of non-bonding spherical parts of valence subshells p_{μ}^{nb} and the bond charges p_{μ}^{b} are varied, along with respective thermal and position parameters, over reflections from the low-angle region. The scattering functions f_{μ}^{nb} and $f_{\mu\nu}^{b}$, normalized to unity, are chosen as the Gaussian. If necessary, the model can be supplemented by scatterers describing lone electron pairs. Thermal parameters of bond charges are considered as formal parameters.

The model described provides much better values of structure amplitude phases than a rigid-atom superpositional structural model. It allows us also to obtain from the x-ray diffraction data the ED close to similar quantum chemical values. Table 3.19 presents the values of deformation ED maxima on bonds in a thiourea molecule, CH_4N_2S, obtained from x-ray diffraction within the bond charge model and calculated non-empirically with the 4-31G+BF basis set. Significant discrepancies are observed for the C=S bond only; for remaining bonds and for lone electron pairs of the S atom the coincidence of deformation ED values is very good. Similar results were obtained for urea, cyanuric acid, and other molecular crystals. However, though the achieved analytical description of ED is good, a rather strong

Table 3.19 Deformation ED maxima ($e\,Å^{-3}$) for bonds in thiourea, CH_4N_2S, at 123 K (Kutoglu *et al* 1982).

		Bonds			
Molecule view	Method	C=S[a]	C–N	N–H[b]	S atom lone pairs
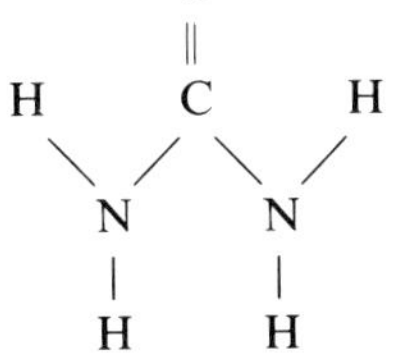	Bond charge model	0.1/0.4	0.4/0.4	0.4, 0.4/0.4, 0.4	0.1/0.1
	Quantum chemical calculation, 4−31G+BF basis set, thermal smearing	0.4	0.5	0.4, 0.4/0.4, 0.4	0.1/0.1

[a]Values to the left or right of the inclined line refer to independent molecules in the asymmetric unit.
[b]Values to the left or right of the comma refer to crystallographically different bonds.

correlation between refined model parameters, whose number is high, distorts the physical meaning attributed to the parameters. Besides, the electron parameters obtained are not interpreted directly in terms of orbital presentations, and the analogy with quantum chemistry turns out to be not so deep as desired.

Pietsch (1980) managed to make the bond charge model, applied for x-ray diffraction data analysis, more physically meaningful. He related additional scattering centres on bonds to the ideas of a dielectric theory of the chemical bond (Phillips 1970). The appearance of a bond charge is a consequence of incomplete screening of an ionic potential by the electron cloud due to a finite value of dielectric constant ε. The value of this charge can be assumed to be equal to Ze/ε in the simplest case of a diamond-type crystal. Accordingly, the x-ray scattering amplitude for spherical parts of atomic valence shells is lowered by a factor of $(1-4/n\varepsilon)$, where $n=4$ if both s and p electrons are screened, and $n=2$ if only s or only p electrons are screened. The total atomic amplitude is as follows:

$$f_\mu(\boldsymbol{H})=f_\mu^{core}(\boldsymbol{H})+(1-4/n\varepsilon)f_\mu^{val}(\boldsymbol{H})+\alpha_H f_{\mu\nu}^b(\boldsymbol{H}). \qquad (3.128)$$

The bond charge form and corresponding scattering amplitude are described by the Gaussian $f_{\mu\nu}^b(H)=(A\pi^{3/2}/a^3)\exp(-\pi^2H^2/a^2)$, where parameters A and a are determined by LS. Parameter α_H is determined from the x-ray reflections, including forbidden ones. The atom and the bond charge part belonging to it are described by the same temperature factor.

As in any model, the bond charge parameters obtained turn out to be dependent on the type of analytical function approximating the ED. So, the maximum of ED values for a bond charge in silicon is changed from 0.13 to 0.18 e Å^{-3} for various functions: simultaneously, the bond charge width varies from 1.82 to 1.88 Å. Moreover, Reid and Pirie (1980) have shown that bond charges oscillate more weakly than the atomic cores, and their description by the same temperature factor turns out to be too rough. Nevertheless, the bond charge model has been quite successfully used for studying static and dynamic properties of semiconductor compounds of A^NB^{8-N} type (Pietsch 1982a), GaAs (Pietsch 1983, 1993), ternary chalcopyrite (Pietsch 1981), and rock-salt structure compounds (Pietsch 1982b). Below we shall return to the physical meaning of results obtained within this model framework (see section 7.3).

3.6 Electron densities via Fourier series

3.6.1 *Electron density and its Laplacian*

When the absolute values of kinematic structure amplitudes, $|F(\boldsymbol{H})|$, are obtained from measured intensities of reflections and their phases are calculated by any crystal structural model, according to (3.11), the problem of crystal ED reconstruction is reduced to the calculation of the Fourier series with complex structure amplitudes $F(\boldsymbol{H})$ as coefficients:

$$\rho(\boldsymbol{r})\frac{1}{V}\sum F(\boldsymbol{H})\exp(-2\pi i\boldsymbol{H}\cdot\boldsymbol{r}) \tag{3.129}$$

(V is the unit cell volume). This function has dynamical character, since it includes thermal motion effects. It is sign constant at infinite summation limits and, with a properly determined scale factor, it satisfies the condition of electroneutrality of the unit cell: $\int_V \rho(\boldsymbol{r})\,d\boldsymbol{r}=N$. As shown in section 2.2, such an ED is N representable, i.e. it can be described by some antisymmetrical many-electron wave-function.

The application of simple linear Fourier transformation to structure amplitudes does not meet any difficulty at first sight. Meanwhile, there are two circumstances which should be considered specially. The first one results from the fact that the number of measured reflections is finite, the second one results from the fact that structure amplitudes contain random errors, due to which the summation of approximate values in (3.129) is unstable. Let us consider both problems in more detail.

The limitation on the number of reflections in the low-angle scattering region is determined by the crystal unit cell size (Dawson 1975). For example, for a cubic unit cell the reciprocal lattice vector length is $|\boldsymbol{H}|=(h^2+k^2+l^2)^{1/2}/a$, where a is the cell dimension. The first reflection of (100) type can be measured for $|\boldsymbol{H}|=1/a$. The low-angle reflections owing to the Fourier transform property are associated with the distribution of valence electrons, which are at a great distance from the nuclei. Therefore, the details of deformation of atomic electron shells in a crystal, which are determined by x-ray diffraction, and cell dimensions are rigidly interrelated. On the other hand, if one disregards x-ray diffractometer capabilities and crystal reflectivity, then the maximal reflection that can be measured, is limited by the value of $(\sin\theta/\lambda)\leqslant 1/\lambda$, where λ is the radiation wavelength used. As Wilson (1979) has noted, with x-ray diffraction we are 'looking' at the ED function through a small 'window' in reciprocal space.

Thus, the Fourier series (3.129) turns out to be incomplete. At radiation wavelengths, applied in precision x-ray structure analysis, the resolution $\sim\lambda/3\approx 0.2$ Å is achieved in ED synthesis according to the Rayleigh criterion (Wilson 1979). This imposes a limitation on the resolution of ED details much lower than 0.2 Å by means of the Fourier series and results in false maxima and minima in the ED, which distort the true picture (figure 3.22(a)). The distortions are especially significant in those regions of space where ρ varies most rapidly, near nuclei for instance. A simple growing up of the Fourier series by adding theoretical values of structure amplitudes would lead to an increase of the ED variance (Rees 1976), which may annihilate the resolution achieved, therefore other methods for weakening ED distortion are needed.

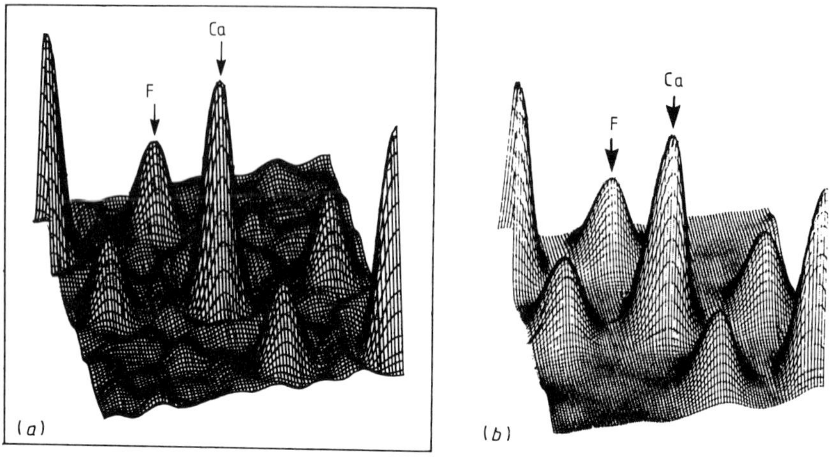

Figure 3.22 The electron density in CaF_2: (*a*) the result of direct Fourier summation; (*b*) the result of Fourier summation using the Lanczoc multipliers $\sin\theta/\lambda \leqslant 1.164$ Å^{-1}.

One of the possible approaches consists in the introduction into the Fourier series of weight factors, the so-called Lanczoc (1957) σ factors. The oscillations in a truncated Fourier series have a period which corresponds to the frequency of the first rejected member of a series (or of the last retained one). If one performs averaging over this period, then one has

$$\sigma_{hkl} = \prod \frac{\sin[\pi h_i/(H_i+1)]}{\pi^3 h_i/(H_i+1)} \tag{3.130}$$

where $h_i = h, k, l$ and $H_i = H, K, L =$ maximal values of h, k, l. Streltsov *et al* (1985) have found that the use of σ factors leads to a satisfactory smoothing of the ED in the interatomic space (figure 3.22(*b*)). Unfortunately, simultaneously the heights of ED maxima noticeably lower and their width increases, though the position of the latter remains unchanged.

In order to eliminate the Fourier series truncation effect Tsirelson *et al* (1985b) suggested a method that is free from the σ factor drawback. These authors have used the spline expansion (i.e. the approximation by cubic polynomials) of the experimental ED presented as a convolution of the 'correct' ED with a resolution function. The latter is the Fourier image of a sphere in reciprocal space, the sphere radius being determined by a maximum value of scattering vector $\boldsymbol{H}_{max}$ for reflections measured. Having calculated the resolution function moments, one can calculate new coefficients for the corrected function from coefficients of a spline, approximating the experimental ED. The authors tested the approach by the use of the analytical expression for the thermal-averaged ED of spherical carbon atoms, parameters of which were determined from the results of several diffraction

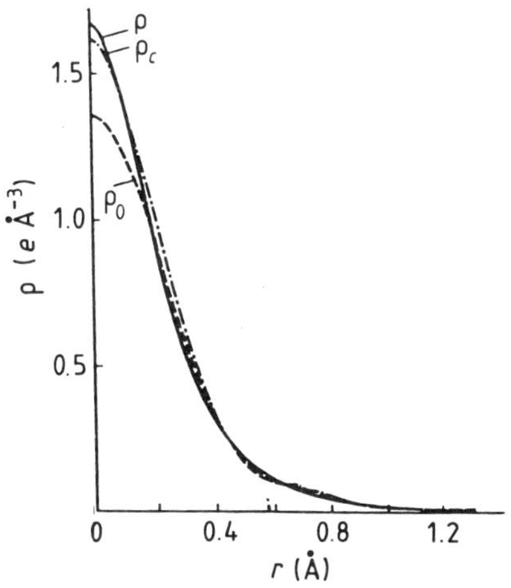

Figure 3.23 The example of the reconstruction of the carbon atom spline-corrected electron density: ρ is the electron density at indefinite resolution calculated from the analytical approximation to the experimental electron density; ρ_0 is the electron density at experimental resolution ($\sin\theta/\lambda = 1.0$ Å^{-1}); ρ_c is the spline-corrected electron density.

measurements. Then ρ_0 was approximated by a smoothing spline over a uniform grid of nodes within the $0 \leqslant r \leqslant 1.5$ Å interval and the corrected ρ_c function was calculated. Both these functions are shown in figure 3.23, where they are compared with the 'accurate' function ρ which is known in this example. The method gives a result that is very close to the correct one, especially near the atomic nucleus, where the series truncation effect is very prominent. Simultaneously, reliable ED reconstruction is provided in the other regions of an atom as well. Discrepancy factors $R = \sum_i |\rho_i - \rho_i'| / \sum_i \rho_i$ ($\rho' = \rho_0$ and ρ_c') for functions ρ_0 and ρ_c, calculated at spline nodes, turned out to be equal to 0.113 and 0.078, respectively. The spline correction method of ED is a one-step method insensitive to small errors in the initial data. It can be used at different stages of the experimental ED analysis. It is useful, in particular, in elimination of the series truncation effect in evaluating electron charges of cations in ionic crystals by numerical integration of ED. For example, in the NaF crystal the direct integration of ED gives the Na^+ ion net charge equal to $+1.8\ e$, whereas after spline correction the charge becomes equal to $+0.95\ e$. The latter result is obviously closer to reality.

There exists a powerful possibility of increasing the amount of structural information extracted from incomplete x-ray diffraction data. This possibility relates to the use of the maximum-entropy method (MEM). From a formal point of view the ED, ρ, is the same distribution function as those used in information theory. The entropy corresponding to ρ can be written as follows

(Collins 1982, 1993):

$$S = -\sum \rho'(\boldsymbol{r}) \ln\left[\frac{\rho'(\boldsymbol{r})}{\tau'(\boldsymbol{r})}\right]. \tag{3.131}$$

This expression includes the relative probability density $\rho'(\boldsymbol{r}) = \rho(\boldsymbol{r})/\sum_r \rho(\boldsymbol{r})$ and the relative prior probability density $\tau'(\boldsymbol{r}) = \tau(\boldsymbol{r})/\sum_r \tau(\boldsymbol{r})$. Sakata and Sato (1990) have introduced two sorts of limitation of χ^2 criterion type, which are imposed in the entropy maximization. One of them corresponds to the case of phase-known structure amplitudes

$$C_1 = (1/N_1) \sum_{\boldsymbol{H}} |F_{model}(\boldsymbol{H}) - F(\boldsymbol{H})|^2/\sigma^2 \tag{3.132}$$

and the other one is the case of phase-unknown structure amplitudes

$$C_2 = (1/N_2) \sum_{\boldsymbol{H}} ||F_{model}(\boldsymbol{H})| - |F(\boldsymbol{H})||^2/\sigma^2. \tag{3.133}$$

Here N_1 and N_2 are the numbers of reflections in each case and $F_{model}(\boldsymbol{H}) = V \sum_r \rho(\boldsymbol{r}) \exp(-2\pi i \boldsymbol{H}\cdot\boldsymbol{r})$. The expected values of C_1 and C_2 are unity. The problem is then reduced to the search for the extremum of the functional

$$D(\lambda_1, \lambda_2) = -\sum_{\boldsymbol{r}} \rho(\boldsymbol{r}) \ln[\rho(\boldsymbol{r})/\tau(\boldsymbol{r})] - (\lambda_1/2)C_1 - (\lambda_2/2)C_2 \tag{3.134}$$

in which λ_1 and λ_2 are Lagrange multipliers. Following Collins (1982) one has

$$\rho(\boldsymbol{r}) = \exp\Bigg[\ln \tau(\boldsymbol{r}) + (\Lambda_1/N_1) \sum_{\boldsymbol{H}} \sigma^{-2}\{F(\boldsymbol{H}) - F_{model}(\boldsymbol{H})\} \exp(-2\pi i \boldsymbol{H}\cdot\boldsymbol{r}) + (\Lambda_2/N_2) \sum_{\boldsymbol{H}} \sigma_F^{-2}\{|F(\boldsymbol{H})| \exp(i\varphi(\boldsymbol{H})) - F_{model}(\boldsymbol{H})\} \exp(-2\pi i \boldsymbol{H}\cdot\boldsymbol{r})\Bigg]. \tag{3.135}$$

Here $\Lambda_1 = \lambda_1 N_1$, $\Lambda_2 = \Lambda_2(\lambda_2)$, where N_1 and N_2 are the number of reflections with phase-known and phase-unknown structure amplitudes. Structure amplitudes are supposed to be known in the absolute scale. Expression (3.135) provides the MEM estimation of ED that is sought by an iterative procedure beginning from some initial distribution $\tau(\boldsymbol{r})$. The procrystal is the best choice for $\tau(\boldsymbol{r})$ (de Vries *et al* 1996).

Sakata and Sato (1990) implemented this approach for the silicon crystal using 30 reflections very precisely measured by Saka and Kato (1986) by the pendellösung method (see chapter 4). The ED was well reconstructed by expression (3.135) with the initial approximation in the form $\tau(\boldsymbol{r}) = \langle \rho \rangle_{unit\ cell}$. This is seen from table 3.20, where structure amplitudes F_{model} of some low-angle reflections, calculated from the MEM-ED, are compared with their experimental analogues. The discrepancy factor which was equal to 0.0005 can serve as the measure of agreement. Sakato and Sato calculated also

Table 3.20 Structure factors of some reflections of silicon calculated by the MEM compared with measured values (Sakata and Sato 1990).

h	k	l	F_{model}	F_{meas}
0	0	0	112.0000	—
1	1	1	−59.9391	−60.1312
2	2	0	−67.5070	−67.3432
1	1	3	−43.6146	−43.6336
4	0	0	−56.2169	−56.2344
3	3	1	38.2215	38.2239
4	2	2	49.0980	49.1056
3	3	3	32.8345	32.8329
5	1	1	32.9463	32.9410
4	4	0	42.8816	42.8848
5	3	1	28.8156	28.8143
6	2	0	37.5681	37.5872

Table 3.21 Values of the structure factors $F(222)$. $F(442)$, and $F(622)$ for silicon calculated by the MEM compared with measured values.

Reference	Structure factor values	
	$F(222)$	
MEM (Sakata and Sato 1990)	1.527	
Exp. (Alkire *et al* 1982)	1.456(8)	
Exp (Roberto and Batterman 1970)	1.46(4)	
Exp (Fujimoto 1974)	1.500(15)	
	$F(442)$	$F(622)$
MEM (Sakata and Sato 1990)	−0.0349	−0.0112
Exp. (Trucano and Batterman 1972)	−0.035(2)	—
Exp. (Mills and Batterman 1980)	−0.042(3)	±0.005(4)
Exp. (Tischler and Batterman 1984)	−0.0370(23)	+0.0088(11)

non-measured structure amplitudes of symmetry-allowed reflections (up to $h^2+k^2+l^2=300$) and structure amplitudes of reflections (222), (442), and (622), forbidden in the space group $Fd3m$. Comparing the latter with the data of most accurate experimental measurements (table 3.21) one can conclude that MEM allows us to extrapolate experimental quantities into the region of values for which the measurements are impossible. At the same time, the MEM did not clarify contradictions between the (622) reflection phase data found in the literature. Such weak reflections are beyond the MEM possibilities, apparently.

MEM increases the ED map resolution by a factor of two approximately. This is especially important for regions close to nuclear positions, where the Fourier maps yield highly underestimated ED values due to the limited number of high-angle reflections. So, in the Si crystal the ED peak at the nucleus position on the MEM map was found to be equal to 208 e Å^{-3}, whereas on the usual Fourier map ($\sin\theta/\lambda \leqslant 1.0$) the height of this peak was only 98 e Å^{-3}. Nevertheless, the MEM gives only about one-third of the true ED value at the Si atom position, which should be about 600 e Å^{-3}.

The correct summation of the Fourier series is complicated by the presence of random errors in structure amplitudes. As a result, the calculation of series (3.129) is an ill posed problem and the theory of regularization should be used for summation. Applying the same consideration as in the case of regularized LS (see appendix J) we can rewrite series (3.129) as follows:

$$\rho(\boldsymbol{r})=\frac{1}{V}\sum_{\boldsymbol{H}} M_{\boldsymbol{H}}F(\boldsymbol{H})\exp(-2\pi \mathrm{i}\boldsymbol{H}\cdot\boldsymbol{r}). \tag{3.136}$$

Filtering factors M_H are similar to factors (J.11). They smoothly depend on the level of error in the x-ray diffraction data and can be presented in the form (Streltsov *et al* 1985):

$$M_{\boldsymbol{H}}=\begin{cases}\dfrac{|F(\boldsymbol{H})|^2}{|F(\boldsymbol{H})|^2+\sigma^2|F(\boldsymbol{H})|} & |F(\boldsymbol{H})|\geqslant(\beta+1)\sigma(|F(\boldsymbol{H})|)\\[2ex] \dfrac{|F(\boldsymbol{H})|^2}{|F(\boldsymbol{H})|^2+\sigma^2|F(\boldsymbol{H})|}\left(\dfrac{|F(\boldsymbol{H})|}{\sigma(|F(\boldsymbol{H})|)}-\beta\right) & \text{for}\quad \beta\sigma(|F(H)|)<|F(H)|<(\beta+1)\sigma(|F(H)|)\\[2ex] 0 & |F(\boldsymbol{H})|<\beta\sigma(|F(\boldsymbol{H})|).\end{cases} \tag{3.137}$$

If the random errors in a set of structure amplitudes are distributed according to the Gauss law (that can be verified by means of the Abrahams–Keve (1971) test), then $\beta\leqslant 3$. Note that Davis *et al* (1978) obtained for the filtering factor M_H an expression similar to (3.137) by minimizing the difference $\rho_{obs}-\rho_{model}$.

Since $M_H<1$, the filtration results in the lowering of ED peak heights which is proportional to errors in the experimental data. For the same reason the ED variance $\sigma^2(\rho)$ is always decreased after filtration. The numerical estimates of this lowering are given below.

The non-completeness of Fourier series and statistical errors in experimental data must be taken into consideration both in ED reconstruction and in calculation of its integral or differential characteristics represented as Fourier series (Davis and Maslen 1978). Let us consider the calculation via Fourier series of the Laplacian of ED, $\nabla^2\rho(\mathbf{r})$, a topological characteristic of ρ, which has been discussed above in section 2.4. Note that the sign and value of $\nabla^2\rho$ do not correlate with ED maxima and minima positions, but measure the ρ curvature only. Function ρ, reconstructed from the x-ray diffraction data, is distorted by the Fourier series truncation effect and by statistical errors. Thus, the problem consists of finding from the distorted initial information a curvature of the ρ function as close to the correct one as possible. The second derivative of ρ does exist, in fact. However a formal differentiation of (3.129) with respect to x, y, z variables results in divergence of the Fourier series for $\nabla^2\rho$. To find $\nabla^2\rho$ in this case, one can use the Lanczoc (1957) method of calculating the derivatives of finite Fourier series by averaging over a 'false' wave period. Writing down the second derivative of the Fourier series in the finite-difference form, we arrive at the following expression for the Laplacian of the electron density:

$$\nabla^2\rho(\mathbf{r}) = -\frac{4\pi^2}{V}\sum_{hkl} M_{hkl}L_{hkl}F(hkl)\exp[-2\pi \mathrm{i}(hx+ky+lz)] \qquad (3.138)$$

with

$$L_{hkl} = \sum_i \left\{ \sin^2\left[\frac{\pi h_i}{H_i+1}\right] \Big/ \left[\frac{\pi a_i}{H_i+1}\right]^2 \right\}. \qquad (3.139)$$

Here $a_i = a$, b, c are crystal unit cell parameters; the other notations were introduced above. It is easily seen that the presence of L_{hkl} factors actually provides the convergence of series (3.138). The filtering factors M_H in this

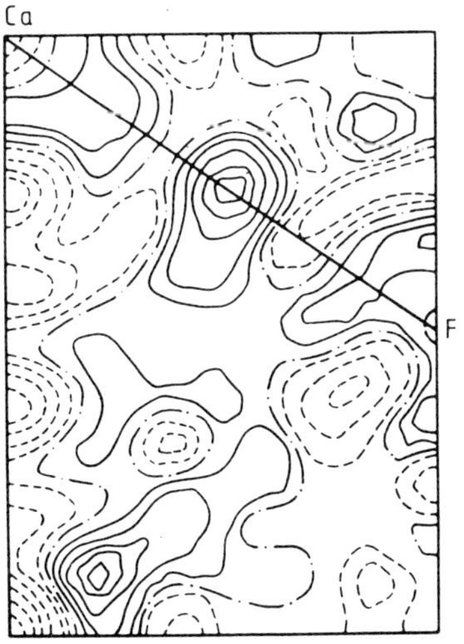

Figure 3.24 The Laplacian of the ED in CaF_2, (110) plane: the solid line connects the points corresponding to positive values of $\nabla^2\rho$. The line intervals are 10 e Å^{-5}.

series have the form of (3.137). In such a manner one can calculate the best approximation to a 'true' Laplacian of ρ from the x-ray structure amplitudes and reproduce correctly the ED curvature in the internuclear space.

The map of the Laplacian of ρ in the (110) plane of the fluorite crystal, CaF_2, calculated by formula (3.138), is shown in figure 3.24. The negative influence of the Fourier series truncation cannot be avoided completely in calculation of $\nabla^2\rho$ ($\sin\theta/\lambda \leqslant 1.164$ Å^{-1}); however, there are not typical 'false' waves around atoms in this map. The map contains all typical features of the ionic bond. So, the region of negative $\nabla^2\rho$ values is displaced along the internuclear vector towards the more electronegative F^- ion. The EDs of anions in neighbouring, differently oriented tetrahedra, are polarized with respect to each other. At the centre of the internuclear distance of Ca–F there exists a well localized region of positive $\nabla^2\rho$ values. In accordance with the general theory (section 2.4) the depletion of electrons is observed here. Thus, by using a correctly calculated Laplacian of ρ one can actually reveal the atomic interaction features which are not seen on total ED maps.

3.6.2 *Valence and deformation electron densities*

The other, widespread method of extraction of the chemical information from ED consists of calculating various difference ED functions in accordance with a general expression

$$\Delta\rho(\boldsymbol{r}) = \rho(\boldsymbol{r}) - \tilde{\rho}(\boldsymbol{r}). \tag{3.140}$$

If the term $\tilde{\rho}(\boldsymbol{r})$ corresponds to the ED of atomic cores, which are supposed to be non-disturbed by chemical interactions and calculated from tabulated atomic wave-functions (or from corresponding partial atomic scattering functions in reciprocal space), then $\Delta\rho$ describes the distribution of valence electrons: $\Delta\rho = \rho_{val}$. This function allows us to establish the regions of higher electron concentration in valence shells and their distribution details caused by a chemical bond. For example, the ρ_{val} map, reconstructed by Pietsch *et al* (1986b) from the x-ray data (figure 3.25(*a*)) shows that in a semiconductor GaAs with zincblende structure the valence electrons are displaced from the Ga atom towards a more electronegative As atom. The regions of high electron concentration on all bonds are connected by ED bridges that can be treated as the manifestation of a strong delocalization of valence electrons in the GaAs crystal. This results in the enhancement of metallic character of a bond. The regions of lowest electron concentration lie in the [001] direction from the As atom, in the region of its antibonding orbital direction. As can be seen, the valence ED map still contains some distortions caused by the Fourier series truncation, so negative regions exist on the map. Thus, the structure amplitude set used should be considered as insufficient for complete reconstruction of ρ_{val}.

The deformation ED function contains much more chemical detail than

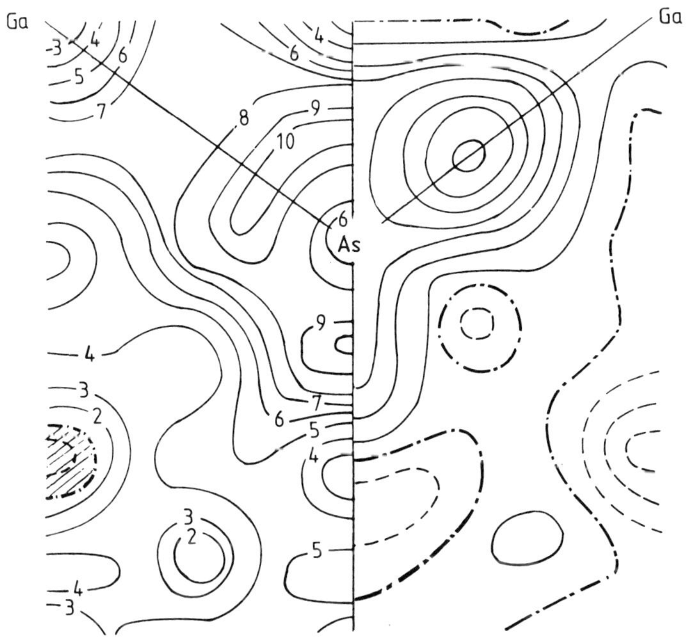

Figure 3.25 The valence (left) and deformation (right) Fourier EDs in GaAs, plane (110). The line intervals are 0.084 and 0.0234 e Å^{-3} (Pietsch *et al* 1986b). The shaded area corresponds to $\rho_{val} < 0$.

ρ_{val} (figure 3.25(*b*)) and is less affected by Fourier series cut-off. By definition, this function describes the ED redistribution that takes place during the formation of a bonded system from free (located at an infinite distance from each other) spherical atoms: $\tilde{\rho} = \sum \rho_{at}$. In terms of structure amplitudes

$$\delta\rho(\boldsymbol{r}) = \frac{1}{V} \sum_{\boldsymbol{H}} M_{\boldsymbol{H}} \{F(\boldsymbol{H}) - \tilde{F}(\boldsymbol{H})\} \exp(-2\pi i \boldsymbol{H} \cdot \boldsymbol{r}). \tag{3.141}$$

The $\tilde{F}$ values should be calculated using high-angle (or neutron) positional and thermal atomic parameters or the same parameters obtained with flexible structural models. The relativistic HF atomic scattering amplitudes for spherical atoms are, as a rule, applied to $\tilde{F}$ calculation. Strictly speaking, these functions must be calculated also taking account of electron correlation; however, this effect is essential for light atoms exclusively and influences very low-angle reflections only. Oddershede and Sabin (1982) have studied the influence of relativistic corrections and of electron correlation on the atomic amplitude of x-ray scattering for Al. Figure 3.26 shows the $\Delta f = f - f_{HF}$ differences, as a function of $\sin\theta/\lambda$, for two cases. In the first of them the atomic amplitude, f, was calculated from the wave-function of the Al atom that includes a twofold excitation $3s^2 3p^1 \rightarrow 3s^0 3p^3$ and corresponds to an energy 0.0182 au lower than that calculated by the HF method. In the second

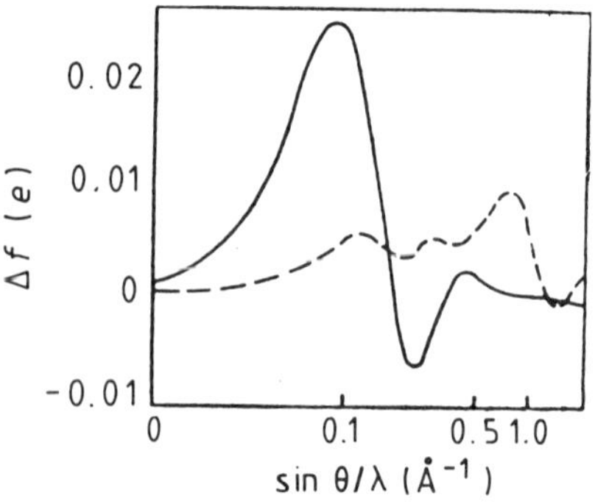

Figure 3.26 The influence of relativistic effects and electron correlation on atomic scattering amplitudes for Al. The solid line shows the $\Delta f = f_1 - f_{HF}$ function where the f_1 scattering amplitude has been calculated from the two-configurational wave-function. The dashed line shows the $\Delta f = f_2 - f_{HF}$ function where the f_2 scattering amplitude has been calculated with relativistic effects taken into account. f_{HF} is the scattering amplitude calculated from the HF wave-function. (Reprinted with permission from Oldershede and Sabine 1982.)

case f was calculated taking account of relativity. The electron correlation manifests itself only at low-angle values of scattering vector where intensities scattered are practically inaccessible for experimental measurement. On the other hand, the relativistic effects change the atomic scattering amplitude value at high scattering angles, where the experimental data contain information on distribution of inner electrons. This change is small for the Al atom but for 3d atoms it reaches 3–5% and can essentially change the $\delta\rho$ map. Thus, the relativistic effects must be taken into account in accurate studies of compounds containing atoms with $Z > 10$. As far as the electron correlation is concerned, in studying the Be single crystal Tsirelson (1985) has applied the atomic scattering amplitude calculated by Benesh and Smith (1970) from a correlated wave-function that includes 17 natural spin orbitals with s, p, d, and f symmetry and allows us to take into account 93.06% of the correlation energy. He found the discrepancies between experimental and theoretical structure amplitudes to be slightly decreased in this case. However, the changes in $\delta\rho$ maps did not exceed one estimated standard deviation (e.s.d). Hence, the electron correlation effect is difficult to be taken into account in the experimental deformation ED analysis.

The crystal ED in close vicinity of nuclei only slightly differs from the superposition of atomic EDs. Therefore, the deformation ED suffers less from the Fourier series truncation effect. This is clearly seen from figure 3.25: the $\delta\rho$ map for the GaAs crystal, which has the same resolution as the ρ_{val} map ($\sin\theta/\lambda \leqslant 0.613$ Å), does not have any clear distortion due to series cut-off.

The series (3.141) consists of a sum of members containing random errors, and the regularization theory must be used for its correct summation. In this case the sum (3.141) below may contain the structure amplitudes of all reflections available: the filtration will reject from the summation those reflections which do not contain any information on the atomic ED asphericities. In the first approximation the summation of series (3.141) can be carried out over low-angle reflections only. In this case the series truncation manifests itself in chemical bond and electron lone-pair regions in lowering $\delta\rho$ peak values. Lehmann and Coppens (1977) have found (figure 3.27) that, as the structure amplitudes corresponding to higher $\sin\theta/\lambda$ values are included in the series, the values of $\delta\rho$ peaks tend to their asymptotic values at an infinite resolution. Note that the application of σ factors for elimination of the series truncation effect in $\delta\rho$ is not correct, because the latter smoothes the ED relief, thus eliminating electron distribution details.

Filtering factors, M_H, in expression (3.141) depend more on variance $\sigma^2(|\delta F|)$, where $|\delta F|=|F-\tilde{F}|$, rather than on variance $\sigma^2(|F|)$ (Schevyrev and Simonov 1981). They retain the form of (3.137); one should only replace $|F|$ by $|\delta F|$ and $\sigma^2|F|$ by $\sigma^2|\delta F|$). One should emphasize once again that in refining the crystal structural model and in constructing Fourier series the application of the regularization theory gives rise to formally the same filtration procedure. The M_H factors in the expression for $\delta\rho$ contain quite distinct physical information: they allow us to separate the deviation of the 'experimental' ED from the ED of the set of spherical atoms which exceeds random error. The numerical experiments, carried out by Lobanov (1986) in calculating deformation ED for various crystals, show that for an x-ray intensity measurement precision of 1–2% one can recommend using the

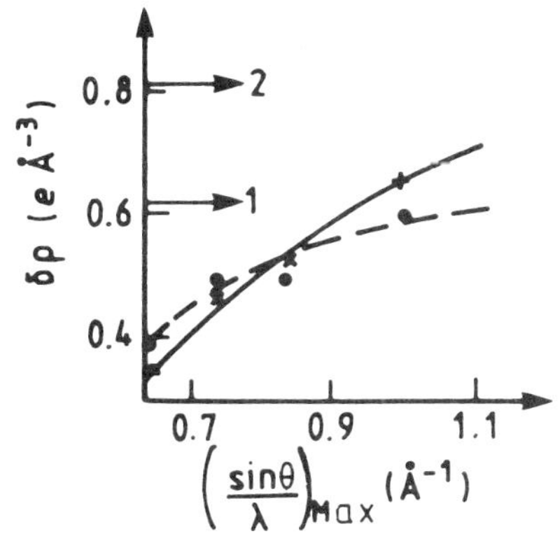

Figure 3.27 The dependence of deformation ED peak heights on the cut-off value of $\sin\theta/\lambda$ for *p*-nitropyridine-N-oxide, $C_5H_4N_2O_3$. The solid line corresponds to average peak values for C–C and C–N bonds; the dashed line corresponds to average O atom lone-pair peak values. The arrows indicate the peak values at indefinite resolution (1, on the bonds; 2, on electron lone pairs). (Reprinted with permission from Lehmann and Coppens 1977.)

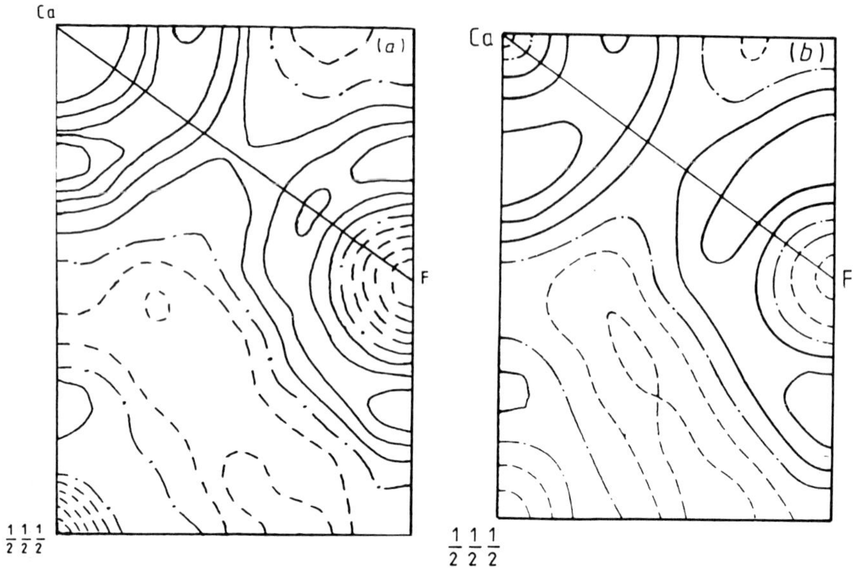

Figure 3.28 The influence of the regularization (filtration) procedure on deformation ED in CaF_2 in the plane (110): (*a*) $\beta=0$; (*b*) $\beta=1$ (β is the regularization parameter).

parameter $\beta \sim 1$ in (3.137). Even for $\beta \approx 0.67$ the $\delta\rho$ maps describe the ED details of the value of $\sigma(\delta\rho)$ with a statistical reliability of at least 50%.

Let us consider how the filtration procedure influences the deformation ED of a fluorite, CaF_2, crystal (figure 3.28). The structure of this compound was described in subsection 3.4.4. The deformation ED in CaF_2 was calculated with the same set of structure amplitudes which was used for analysis of this crystal within the $\varkappa$-model framework; however, the set of F values was later corrected for the effect of anharmonicity of thermal vibrations of F atoms. The (100) plane sections of deformation ED $\delta\rho$, calculated without filtration ($\beta=0$, 80 reflections) and with it ($\beta=1$, 25 reflections) are shown in figure 3.28. One can see that the filtration provides smoother relief and retains all main features of $\delta\rho$. The $\delta\rho$ bridge of height $\sim 0.2\ e\ \text{Å}^{-3}$, that connects Ca^{2+} and F^- atoms, becomes smaller after filtration. The anion ED polarization is revealed more prominently along the line connecting anions belonging to neighbouring, differently oriented tetrahedra. The mean distance from maxima of $\delta\rho$ peaks to an anion centre is the same on both maps and equal to 0.75 Å. The $\delta\rho$ picture agrees very well with the map of the Laplacian of ρ for CaF_2 in figure 3.24.

We have seen that the x-ray diffraction data only are sufficient, in principle, for deformation ED map calculation, if theoretical (quantum chemical) information about ED of atoms is added. Such a calculation method is called the (X–X) scheme. However, there are situations where the set of structure amplitudes measured does not contain the data corresponding to sufficiently high scattering angles. This may result in biased estimates of thermal and positional atomic parameters of a crystal structural model due to both incomplete separation of chemical bond and thermal motion effects, and due to strong correlation between various parameters during optimization of the structural model. Besides, the x-ray diffraction does not allow us to fix the H atom positions correctly due to their small scattering factor and strongly non-spherical ED. In this case the additional structural information, obtained by means of neutron diffraction (appendix K) is quite useful.

The nuclear scattering of thermal neutrons is spherically symmetric due to the small size of nuclei, and the coherent scattering amplitude, b, is defined by nucleus properties only. It does not depend on either scattering vector value or direction and does not increase monotonically with atomic number, as the x-ray atomic scattering amplitude does. For this reason the light atoms, which weakly scatter x-rays, are easily fixed by neutrons (though the opposite situations can also occur). The neutron thermal and positional parameters of a crystal structural model can be used further for calculation of the Fourier series (3.141). The x-ray reflection set can be restricted in this case by low- and middle-angle reflections. Corresponding $\delta\rho$ maps, called (X–N) maps, have been rather widely applied (Ozerov and Datt 1975, Coppens 1978).

The basic assumption of the (X–N) approach consists in the fact that within the Born–Oppenheimer approximation framework thermal vibrations of nuclei and electron shells do not differ from each other. This is the case in the first approximation, of course. However, the modern precision of diffraction experiments is such that the non-adiabaticity effect may turn out to be accessible for measurement. In particular, the difference in thermal motion of nuclei and highly delocalized valence electrons and, moreover, conduction electrons can be noticeable (Reid 1987). The joint application of different diffraction methods may turn out to be of great importance for establishing the limits of applicability of the Born–Oppenheimer approximation—one of the main simplifying steps of quantum mechanics of a crystal.

The (X–N) method is especially useful in studying chemical bonds formed by H atoms. So, in organic molecules, due to strong electron cloud displacement from an H atom to a partner atom, the differences of lengths of C–H, N–H, and O–H bonds, established by means of x-ray and neutron diffraction, reach 0.15–0.20 Å (Bacon 1975). To study these bonds one must describe position and thermal motions of the H atom as accurately as possible. The x-ray diffraction gives also the wrong positional parameters of terminal atoms which have lone-pair electrons. As seen from table 3.22, the corresponding difference between neutron and x-ray positions of these atoms essentially exceeds the statistical error of their determination. Ignorance of this fact may distort the interpretation of the ED pattern obtained.

Table 3.22 The values of atomic position shifts between x-ray and neutron diffraction results (Coppens 1974).

Tetracyanoethylene oxide	$\uparrow$ O / \\ C C	0.013(4)
Tetracyanoethylene	$-C \rightarrow N$	0.008(1)
Oxalic acid	$C\text{-}O \nearrow$	0.008(2)
Sucrose	H / C-O $\uparrow$ \\ and O H / \\ C C	0.008(2) 0.007(2)
Ammonium oxalate H_2O	$\uparrow$ O / \\ H H	0.013(3)
Hexamethylenetetramine	/ N / : \\ C C C	0.018(6)
Cyanuric acid	$C{=}O \rightarrow$	0.005(1)

The neutron diffraction data processing has much in common with procedures described in section 3.3 (Coppens 1978). There are, however, some essential differences which should be taken into account in joint application of x-ray and neutron diffraction in accurate studies.

There are many factors which influence the (X–N) method accuracy. So, in neutron experiments samples of large size are usually used due to relatively low neutron flux from a reactor as compared to the flux of a modern x-ray tube. This does not allow us, as a rule, to carry out both experiments on the same sample. This results in the different manifestations of the extinction effect. Besides, as shown in appendix G, the TDS contributions in x-ray and neutron scattering are different, in particular, due to anharmonicity. Scheringer *et al* (1978) and Blessing (1995) have also noticed that the additional discrepancy between data of the two experimental techniques results from different resolution functions and from different scattering amplitudes. The TDS contribution to neutron scattering is larger, which results in about 10% lowering of neutron thermal parameters as compared to x-ray ones. This influences significantly the scale factor (of $\sim 3\%$) and the deformation ED ($\sim 100\%$). To decrease these errors Coppens

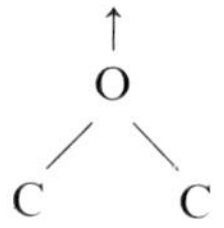

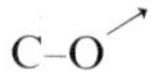

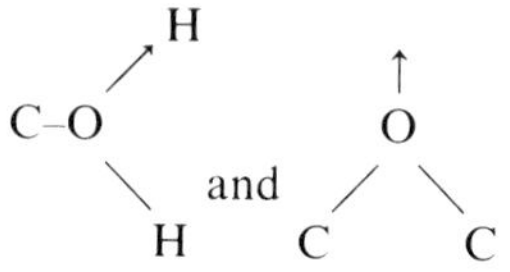

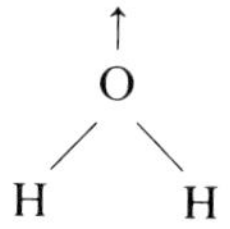

(1978) recommended multiplying neutron thermal parameters by the coefficient $(\sum_i U_{ii})_X / \sum_i (U_{ii})_N$. The latter includes components of high-angle x-ray (X) and neutron (N) tensors of thermal vibrations. This allows us, besides, to decrease the error due to background determination specificity and due to possible differences in temperatures of x-ray and neutron experiments, which are difficult to eliminate technically.

However, the differences in the x-ray and neutron thermal parameters discussed are not always explained. These differences are related, apparently, to incomplete correction of systematic errors (anisotropic ones first of all), which manifest themselves differently in the two diffraction methods. It is possible to eliminate in part the discrepancy of these different experimental methods by using x-ray and neutron structure amplitudes simultaneously in crystal structural model refinement (Coppens *et al* 1981a). Such a procedure, called the (X + N) scheme, is useful for refinement of non-centrosymmetric structures especially. Besides, it has filtering influence on the deformation ED, as was shown by the comparison of $\delta\rho$ maps for the H_2O molecule reconstructed by Coppens *et al* (1981a) within the framework of (X − X), (X − N), and (X + N) schemes. The disadvantage of this approach is the dependence of results on the precision of both data sets whose completeness is often not the same. Whereas the x-ray data must be as complete as possible (in the valence electron scattering region at least), the neutron ones must provide the proper statistical precision of determining structure parameters only.

Now we shall return to the features of calculation of function $\delta\rho$ (3.141). In centrosymmetric crystals the phases of the structure amplitudes F and $\tilde{F}$, which are equal to zero or π, are the same to very high certainty. Therefore, the high-angle refinement of parameters of the rigid-atom superpositional structural model is sufficient for finding these phases. The erroneous calculation of phases in this way due to random errors in the data is probable for only very weak reflections. However, for non-centrosymmetric crystals, where structure amplitude phases may take, in the general case, any value in the range from zero to 2π (figure 3.29), the phases calculated for $\tilde{F}$ cannot be assigned to experimental structure amplitude F (Maslen 1968, Coppens 1974). In these crystals $|F - \tilde{F}| \neq |F| - |\tilde{F}|$, and phases φ and $\tilde{\varphi}$ may essentially differ sometimes. For this reason when determining the phases of experimental structure amplitudes one should use a more flexible crystal structural model than the atomic superpositional model. In the simplest case the phases φ can be calculated from the results of refinement of parameters of a rigid atomic superpositional structural model over the whole set of structure amplitudes while phases $\tilde{\varphi}$ are obtained from the high-angle refinement or from the neutron diffraction (Thomas *et al* 1975). This allows one to decrease model errors in $\delta\rho$, which can reach 0.05–0.20 e Å^{-3}, according to Mullen and Scheringer (1978). Using this approach Ozerov *et al* (1981) have found that the phases of structure amplitudes F and $\tilde{F}$ for

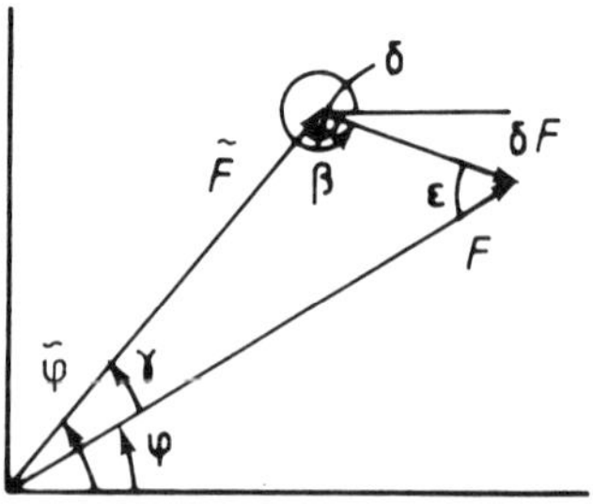

Figure 3.29 Vector diagram for a pair of the structure amplitudes in the general case (non-centrosymmetrical crystal); β is the phase of the δF vector.

lithium formate monodeuterate, $LiCOOH \cdot D_2O$, differ by 2–3° on average though for some reflections the phase difference reached 30° and more. That conclusion coincides with Coppens' (1974) opinion. In calculation of the Fourier series (3.141) for $LiCOOH \cdot D_2O$ with and without taking account of the phase difference, the changes in $\delta\rho$ were within the limit of one e.s.d., 0.06 e Å^{-3} in the internuclear space. Similar results were obtained by Voloshina *et al* (1985) in the study of ED in potassium titanyl phosphate, $KTiOPO_4$, using the bond charge model. In both compounds the values of maxima and minima of $\delta\rho$ are equal to $\pm(0.2–0.6)\,e$ Å^{-3}. These are typical for $\delta\rho$ maps. As a result, the structure amplitudes $|F|$ and $|\tilde{F}|$ do not much differ from each other on average, and the refinements of various crystal structural models result in close values of phases φ and $\tilde{\varphi}$. One should note, however, that in compounds where the ED asphericity reaches the value of $\sim 1\,e$ Å^{-3}, the discrepancies in φ and $\tilde{\varphi}$ values may be significant, and flexible structural models are needed.

3.7 Accuracy of the experimental electron density

The accuracy analysis of the ED and its parameters, recalculated from the x-ray diffraction experiment, is of primary importance both for a qualitative treatment of ED from the viewpoint of establishing the chemical bond character and for calculating various crystal properties. This is why this problem has always attracted much attention. Formally, $\rho(\boldsymbol{r})$ is a statistical function that describes the probability density of electron distribution in a system under consideration. For this reason the probability theory and mathematical statistics can be used (see appendix H), in the language of which the correctness of the experimental ED can be evaluated.

In the general case the variance of any function reconstructed from the experiment within the framework of a model is defined by the following relation:

$$\sigma_{tot}^2 = \sigma_s^2 + \sigma_m^2 + \text{cov}(\sigma_s, \sigma_m). \tag{3.142}$$

Here σ_s^2 takes into account the statistical error, estimated within the accepted model framework. It includes both a measurement random error and a randomized data processing error. σ_m^2 takes into account the systematic error of a model (estimator) itself, whereas the third term takes into account the correlation between errors s and m. The two last terms are difficult to evaluate, and often it is believed that $\sigma_{tot}^2 \approx \sigma_s^2$. This supposition is not true in the general case. This is why to decrease the uncertainty arising during the experimental data processing one tries to minimize the two last terms in (3.142). For this purpose one must use the correct models only and, while working with the latter, one should remain inside the field of their applicability. This concerns both the introduction of corrections into the intensities of diffraction reflections, which reduce measured values to the kinematic diffraction case, and to models used for finding crystal structural parameters, as well as to techniques and tactics of optimization of these models. All these problems have been considered above fully enough. Nevertheless, following this method one can only decrease a model error, not eliminate completely its influence on ED obtained. Since statistical methods are not applicable for estimating σ_m^2, we shall try to evaluate this part of a total error by comparing various experimental EDs obtained for the same compound and also by comparing experimental and theoretical EDs.

The definitions used from statistics are as follows (Schwarzenbach *et al* 1989).

Accuracy is the closeness of agreement between the value of an estimate, derived from a physical measurement, and the true value of the quantity estimated. In practice accuracy cannot be exactly evaluated. Experiments try to gain insight into physical reality through interpretation of measurements using models. This is based on the implicit assumption that a disagreement between observations and the corresponding calculated model quantities indicates inaccuracy. However, a good agreement between observations and model quantities is not evidence of accuracy. A model that does not take into account all available evidence and prior experience may give apparently precise, but inaccurate (wrong) results.

Precision is the closeness of agreement between the values of a measurement or of an estimate obtained by applying a strictly identical experimental procedure several times. It is expressed numerically by a standard deviation or variance. The precision of a diffraction intensity is often inferred from only one or maybe two measurements by invoking Poisson statistics for the count rates, and/or using the experience gained from earlier diffraction experiments. Precise estimates are not necessarily accurate.

Repeatability is the closeness of the agreement between the results of successive measurements of the same quantity carried out under the same conditions, namely the same method of measurement, the same observer, the same measuring instrument, the same location, and repetition over a short period of time.

Reproducibility is the closeness of agreement between the results of measurements of the same quantity where the individual measurements are carried out changing conditions such as method of measurement, observer, measuring instrument, location, or time.

Uncertainty characterizes the lack of knowledge of the true value of a parameter that includes the effects of systematic error (model inadequacy), as well as lack of statistical precision.

3.7.1 *Statistical error analysis*

The initial data for the statistical estimation of the experimental ED precision are the sets of absolute values of experimental structure amplitude $\{|F|\}$ (3.64) and their variances $\{\sigma^2(|F|)\}$ (3.66), (3.67). The latter, as described above, include a randomized uncertainty of models of the x-ray data processing.

There exists a number of papers dealing with the estimation of the experimental ED precision. Cruickshank (1949) expressed the experimental ED variance for a centrosymmetric crystal as a function of variances of the structure amplitude moduli, $\sigma^2(|F|)$. Maslen (1968) has generalized this method for a crystal without the centre of symmetry and has accounted for an error in a scale factor. Coppens and Hamilton (1968) have extended this approach to the deformation ED. An estimation of the errors in a promolecule ED was made by Becker *et al* (1973). Rees (1977a, b, 1978) has obtained and investigated relations for a spatial distribution of errors in a centrosymmetric crystal by using a covariance analysis of ED in a position space.

Unfortunately, even if one neglects the model error σ_m^2 in (3.142), one cannot perform a total statistical analysis of errors in ED, which would take into account mutual influence of contributions originating from different sources. The reason for this consists of the absence of a correct method for the estimation of the covariances between errors of different nature and of the mutual influence of corrections which provide the transition from intensities to kinematic structure amplitude values. Therefore, we are forced to consider the members of the $\{|F|\}$ and $\{\sigma^2(|F|)\}$ sets as statistically independent, excluding the correlation dependence due to error in scale factor. The latter dependence can be revealed explicitly, resulting from analysis of model determination of scale factor.

We shall follow Lobanov *et al* (1990) and write ED, reconstructed from x-ray diffraction data, as a Fourier series, as follows:

$$\rho(\boldsymbol{r})=\frac{1}{V}\sum_{\boldsymbol{H}=-\infty}^{\infty} k|\mathfrak{F}(\boldsymbol{H})|\cos(2\pi\boldsymbol{H}\cdot\boldsymbol{r}-\tilde{\varphi}). \tag{3.143}$$

Here $\mathfrak{F}\boldsymbol{H})$ is the structure amplitude on an arbitrary scale and k is the scale factor: $F(\boldsymbol{H})=k\mathfrak{F}(\boldsymbol{H})$. We designate the quantity $(2\pi\boldsymbol{H}\cdot\boldsymbol{r}-\varphi)$ by α and use a variance operator on ρ. As a result, omitting arguments, one obtains

$$\begin{aligned}\sigma_s^2(\rho)=\frac{1}{V^2}\Bigg\{&\sum_{\boldsymbol{H}=\boldsymbol{H}'=-\infty}^{\infty}\Bigg[\left(\frac{\sigma^2(k)}{k^2}|F|^2+\sigma^2(|F|)+2\frac{|F|}{k}\operatorname{cov}(\boldsymbol{k},|F|)\right)\cos^2\alpha\\
&+|F|^2\sigma^2(\alpha)\sin^2\alpha-2\left(\frac{|F|^2}{k}\operatorname{cov}(k,\alpha)+|F|\operatorname{cov}(|F|,\alpha)\right)\sin 2\alpha\Bigg]\\
&+2\sum_{\boldsymbol{H}}^{\infty}\sum_{\boldsymbol{H}=-\infty}^{\infty}\Bigg\{\left[\frac{\sigma^2(k)}{k^2}|F||F'|+\operatorname{cov}(|F|,|F'|)\right]\cos\alpha\cos\alpha'\\
&+|F||F'|\operatorname{cov}(\alpha,\alpha)\sin\alpha\sin\alpha'-\frac{1}{2k}[|F||F'|\operatorname{cov}(k,\alpha)\cos\alpha\sin\alpha'\\
&+|F'||F|\operatorname{cov}(k,\alpha)\cos\alpha'\sin\alpha]-\tfrac{1}{2}[|F'|\operatorname{cov}(|F|,\alpha')\cos\alpha\sin\alpha'\\
&+|F|\operatorname{cov}(|F'|,\alpha)\cos\alpha'\sin\alpha]\Bigg\}\Bigg\}.\end{aligned} \tag{3.144}$$

In the linear approximation of an error analysis, the intensities and phases of reflections are considered as statistically independent. Therefore, all covariances in (3.144), which are not related to the scale factor, should be equal to zero. For the group of symmetry $P\bar{1}$ this expression can be further simplified due to antisymmetric properties of a $\sin\alpha$ function with respect to the argument. As a result, one has

$$\begin{aligned}\sigma_s^2(\rho)=\frac{2}{V^2}\Bigg\{&\sum_{\boldsymbol{H}>0}^{\infty}[\sigma^2(|F|)\cos^2\alpha+|F|^2\sigma^2(\alpha)\sin^2\alpha]\\
&+\frac{\sigma^2(k)}{2k^2}\left[2\sum_{\boldsymbol{H}>0}|F|\cos\alpha\right]^2\Bigg\}.\end{aligned} \tag{3.145}$$

It has been taken into account here that the error in a scale factor gives rise to a correlation dependence between experimental structure amplitudes. Therefore, the last term in (3.145) includes both diagonal $|F|^2$ and off-diagonal $|F||F'|$ elements.

If the experimental measurements were not carried out over the total sphere of reflections, then, supposing summation over the hemisphere of a reciprocal space, the expression for an estimated variance of ED can be reduced to the form

$$\sigma_s^2(\rho)=\frac{2}{V^2}\left\{\eta\sum_{H>0}[\sigma^2(|F|)\cos^2\alpha+|F|^2\sigma^2(\alpha)\sin^2\alpha]\right.$$
$$\left.+\frac{2\sigma^2(k)}{k^2}\left[\sum_{H>0}|F|\cos\alpha\right]^2\right\} \tag{3.146}$$

where η is the ratio of a maximal allowed number of symmetrically equivalent reflections to the actually measured one.

In order to write down the expression for an estimate of variance of deformation ED, one should express the variance of the value $|\delta F|=|F-\tilde{F}|$ and the phase of difference structure amplitude $\boldsymbol{\delta F}=|\delta F|\,e^{i\beta}$, β, in terms of $\sigma^2(|F|)$ and $\sigma^2(|\tilde{F}|)$. Note, first of all, that the angles designated in figure 3.29 are related by the relation $\gamma+\delta+\varepsilon=\pi$. As follows from a sine theorem $\delta=\tan^{-1}[|F|\gamma/(|F|-|\tilde{F}|\cos\gamma)]$. Relationships $\beta=\tilde{\varphi}+\delta$ or $\beta=\varphi+\delta+\pi$ are valid depending on the mutual orientation of vectors $\boldsymbol{F}$ and $\tilde{\boldsymbol{F}}$. The sign of $\sin\gamma$ also depends on this orientation. Since in the following consideration the absolute value of δF is used, π can be omitted. Thus, the relation $\beta=\tilde{\varphi}+\tan^{-1}[|F|\sin\gamma/(|\tilde{F}|-|F|\cos\gamma)]$ is valid always.

Let us define $a=|F|\sin\gamma/(|\tilde{F}|-|F|\cos\gamma)$, $b=[k(|\tilde{F}|-|F|\cos\gamma)]^{-1}$ and, assuming that $\sigma^2(\tilde{\varphi})=\sigma^2(|\tilde{F}|)/|\tilde{F}|^2$ and that, as for infinite Fourier series summation, $\sigma^2(\alpha)=\sigma^2(\beta)$, we can write in the linear approximation

$$\sigma^2(\alpha)=\sigma^2(F)X^2+\sigma^2(\tilde{F})Y^2+\sigma^2(k)Z^2. \tag{3.147}$$

Here $X=[b/(1+a^2)][(a+1)\sin\gamma+(a-1)\cos\gamma]$, $Y=[kb^2/(1+a^2)][(|F|/|\tilde{F}|)\times(\cos\gamma-a\sin\gamma)-a]+1/|\tilde{F}|$, $Z=-ab|\tilde{F}|/(1+a^2)$.

Using the cosine theorem, one can express $|\delta F|$ in terms of $|F|$ and $|\tilde{F}|$ (see figure 3.29) and, introducing notations $A=(1/k|\delta F|)[|F|-|\tilde{F}|\times(\cos\gamma+\sin\gamma)]$, $B=(1/|\delta F|)[|\tilde{F}|-|F|(\cos\gamma-\sin\gamma)]$, $C=(|F|/k|\delta F|)(|\tilde{F}|\cos\gamma-|F|)$ one obtains

$$\sigma^2(|\delta F|)=A^2\sigma^2(|F|)+B^2\sigma^2(|\tilde{F}|)+C^2\sigma^2(k). \tag{3.148}$$

As was noted in a previous section, in non-centrosymmetric crystals the angle γ for the majority of reflections is equal to 2–3°. Therefore, one can suppose that $a\approx 0$ and this leads subsequently to $Z\approx 0$. Then the estimate of variance of deformation ED will be described by the expression

$$\sigma_s^2(\delta\rho)\cong\frac{2}{V^2}\left\{\eta\sum_{H>0}M_H[\sigma^2(|\delta F|)\cos^2\alpha+\delta F^2\sigma^2(\alpha)\sin^2\alpha]\right.$$
$$\left.+2\sigma^2(k)\left[\sum_{H>0}M_H C\cos\alpha\right]^2\right\}. \tag{3.149}$$

Here, as before, the filtering weight factors M_H are introduced; these take into account the statistical significance of each structure amplitude.

Thus, the value of a statistical estimate of ED variance depends on the precision of measurement of reflection intensities, on the adequacy of models used in recalculating them into $|F|$ values, on the error in scale factor k, and on possible cross-dependences arising in the determination of k by the LS, as well as on the term $\sigma^2(\tilde{F})$ caused by random errors in structural parameters. In the linear approximation of the theory of error analysis one has

$$\sigma^2(\tilde{F})=\sum_{\mu}\left[\left(\frac{\partial \tilde{F}}{\partial f_{\mu}}\right)^2 \sigma^2(f_{\mu})+\sum_{j}^{9}\left(\frac{\partial \tilde{F}}{\partial U_j^{\mu}}\right)^2 \sigma^2(u_j^{\mu})+\sum_{k}^{3}\left(\frac{\partial \tilde{F}}{\partial X_k^{\mu}}\right)^2 \sigma^2(X_k^{\mu})\right] \tag{3.150}$$

where f_{μ}, U_j^{μ}, and X_j^{μ} are atomic scattering functions, thermal parameters, and coordinates of the atoms, respectively. Their variances can be obtained by converting the matrix of the LS normal equation. Note that the weight matrix in the LS is usually diagonal, and the mentioned variances provide, generally speaking, biased estimates of precision of the results. As shown by Rees (1978), the correlation between refined parameters of the structural model leads to the fact that the random error in a deformation ED either does not change practically at all (on light atoms and in the interatomic space) or decreases (on heavy atoms).

Let us consider a spatial distribution of various contributions into the statistical error in the deformation ED of the centrosymmetric bcc vanadium crystal (space group $Im3m$). The profile of filtered deformation ED of this crystal along the [111] direction (figure 3.30) shows that during the crystal formation from atoms some of the electrons are displaced into the internuclear

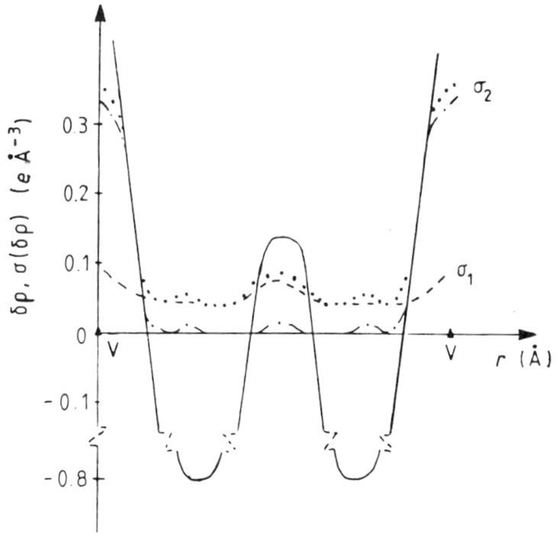

Figure 3.30 The profile of deformation ED in the V crystal along the (111) direction (solid line), the e.s.d. of $\delta\rho$, $\sigma_s(\delta\rho)$ (dotted line), and contributions from statistical errors in structure amplitudes, including random errors in structural and thermal atomic parameters, σ_1, and from uncertainty in scale factor, σ_2; $\sigma_s=\sigma_1+\sigma_2$.

space. Simultaneously, a rather high contraction of ED near nuclei is observed. Lobanov *et al* (1990) have calculated the contributions to $\sigma_s(\delta\rho)$ for V caused by errors in experimental F and model (promolecule) $\tilde{F}$ structure amplitudes (the first term in (3.149)) and uncertainty in a scale factor refined (the last term in (3.149)), as well as their sum. The first component determines the general level of error over a unit cell and is characterized by small maxima on closed symmetry elements. The error of a scale factor determination manifests itself, mainly, in positions of atoms, where it reaches a relatively high value. The filtration procedure at the filter parameter $\beta=0.75$ has decreased $\sigma(\delta\rho)$ at V atom positions from the value of 0.9 e Å^{-3} obtained for $\beta=0$ down to 0.35 e Å^{-3}. The error in the internuclear space also decreased from 0.08 e Å^{-3} down to 0.06 e Å^{-3}. Comparing these estimates with the deformation ED of the V crystal one can conclude that in the internuclear space, i.e. in the region where $\delta\rho$ is usually interpreted from the chemical bond viewpoint, the details of $\delta\rho$ distribution exceed the $\sigma(\delta\rho)$ value by a factor of 1.5 to 2. In other words, in this crystal the $\delta\rho$ details are determined with a reliability of about 90%.

If the set of structure amplitudes used for calculation of $\delta\rho$ is incomplete, then various fragments of $\delta\rho$ maps may have various uncertainties. Maslen (1988) recommended supplementing the $\sigma_s(\delta\rho)$ map by a set of χ^2 indices, calculated for various $\{F\}$ subsets, and comparing these indices on the basis of statistical criteria.

The estimated standard deviation of ED in non-centrosymmetrical crystals includes the contribution from error in structure amplitude phase determination. Strictly speaking, this component of total error (3.142) has systematical character. However, in the case of using a rather flexible crystal structural model, such as the multipole model, one may consider this error to be randomized and calculate the corresponding variance by means of relation (3.147). As experience shows, non-centrosymmetricity increases the $\sigma_s(\delta\rho)$ value by a factor of 1.5 to 2 approximately.

The additional contribution to $\sigma_s^2(\delta\rho)$ caused by uncertainty in values of coherent scattering of neutrons b_μ is significant for (X–N) deformation ED near atomic centres. Tsirelson and Nozik (1982) have found that this contribution can be estimated by means of expression

$$\sigma^2_{\mathrm{X-N}}(\delta\rho) \cong \sum_\mu [\sigma^2(b_\mu)/b_\mu^2]\rho_\mu^2(\mathbf{r}) \tag{3.151}$$

where ρ_μ is the ED of atom μ, which rapidly decreases with the distance from the nucleus position. If $\sigma(b)/\sigma \sim 1\%$, then the error value from this source may turn out to be comparable with the error caused by uncertainty in scale factor determination by refinement.

When the crystal ED is reconstructed within the framework of some crystal structural model rather than by means of the Fourier series, then the estimate of a standard deviation of ρ or of $\delta\rho$ can be obtained from variances of refined model parameters. This estimate is then more realistic when the pseudoatom ED asphericity and thermal motion effects are quite separated in a refinement of the structural model and correlations between model parameters are minimized. In any case, such an estimate is more valid for dynamic, rather than for static, ED. It is clear that the estimates of model parameters obtained by the LS, as well as the characteristics of properties calculated from these parameters (the electric field characteristics inside a crystal, for example) should be considered with some scepticism: one cannot eliminate the mutual influence of model and random errors described by the $\mathrm{cov}(\sigma_s, \sigma_m)$ term in (3.142).

3.7.2 *Model errors*

The most adequate models which are used for x-ray diffraction data treatment randomize the systematic errors. In this case the total error is estimated by variance σ_s^2. Unfortunately, the models available do not always reflect the real experimental situation fully enough and, thus, introduce some distortions into the results. Even if no systematic errors are revealed (and we suppose that the methods described above would actually provide such a situation), then the real ED error value can noticeably differ from a statistical estimate.

Let us first consider the possible sources of errors related to the diffraction theory approximations used. The thermal diffuse scattering, if it was not taken into account properly, results in overestimating the ED value near nuclear positions due to distortion of values of atomic thermal vibration parameters (see subsection 3.3.2). This gives rise to underestimation of $\delta\rho$ minimum depths in these regions (Helmholdt and Vos 1977). The extinction may influence the deformation ED both near nuclei and in the chemical bond region. The first effect has been illustrated in figure 3.9, the second one was presented in figure 3.10. A special difficulty when properly considering extinction consists in the fact that this effect is strongly determined by the defect structure of a crystal sample which may noticeably vary from sample to sample. Van der Wal *et al* (1987) have encountered just the same situation in studying three samples of forsterite, Mg_2SiO_4. The extinction was revealed by these authors from the results of careful experiments using x-ray radiation of various wavelengths. However, the refinement of extinction correction with the Stewart multipole model has shown that in one of the samples the extinction is isotropic and described by the Lorentzian mosaic spread function, whereas in two other samples it is anisotropic and described by the Gaussian mosaic spread function. Accordingly, the multipole deformation EDs also differ from each other: qualitative agreement was observed; however, the quantitative discrepancies reached the value of 0.15–0.20 e Å^{-3}. Thus, the error due to extinction needs much attention.

The Fourier series truncation influences the deformation ED much less than the total ED. However, it still remains as a distorting factor, especially near nuclear positions, where the ED curvature is rather high. Breitenstein *et al* (1982) studied this problem in detail. They calculated at a near-HF level the ED of organic molecules, taking into account the smearing of ED by thermal motion. Then the effect of Fourier series truncation on deformation ED was simulated by a double Fourier transform. It was found that at $(\sin\theta/\lambda)_{max}=(0.8\text{–}1.0)$ Å^{-1} the value and shape of peaks in the internuclear space on a dynamic deformation ED maps are indistinguishable, in fact, from those obtained at an infinite resolution. Near the nuclear positions, however, the depths of $\delta\rho$ minima for the resolution mentioned are underestimated by the value of (0.1–0.2) e Å^{-3} for N and O atoms. For heavier atoms a larger underestimation of $\delta\rho$ can be expected. The influence of the Fourier series truncation on static deformation ED is essentially stronger near nuclei.

The ED distribution near nuclei strongly depends also on the correct determination of a scale factor, k, in (3.49). As has already been pointed out in section 3.3 this factor is usually refined simultaneously with the crystal structural model parameters. The structural models used essentially influence the precision of obtained ED and its parameters, scale factor in particular. Rees (1978), while comparing the results obtained within the framework of various models of the $Cr(CO)_6$ crystal, came to the conclusion that the corresponding model error is $[\sigma(k)/k]_{model}=0.002$. This estimate is applicable if flexible structural models are used. At the same time, there is no common opinion so far as to how the scale factor should be determined, if a rigid-atom-type superpositional model is used. In this case the scale factor is often fixed at the value obtained from the high-angle refinement of a model. Figure 3.31 shows how the deformation ED value of the boron carbide crystal, $B_{13}C_2$, depends on the method of scale factor determination. This dependence is weak in the internuclear region, whereas near nuclear positions the changes in $\delta\rho$ may be drastic. If the scale factor value, obtained by means of a multipole model, is assumed to be the best estimate, then the k value from the high-order refinement is indeed closest to this value. Will *et al* (1979) have found that the artificial change of k by 1% increases the estimated standard deviation near nuclei by the value of 0.10–0.15 e Å^{-3}; however, $\sigma(\delta\rho)$ remains practically unchanged in the internuclear region.

If the crystal contains atoms with essentially different numbers of electrons then, after completion of high-angle refinement, the scale factor is sometimes calculated by scaling $|F|$ over the total set of structure amplitudes $|F_{model}|$ (Savariault and Lehmann 1980).

In order to analyse the role of crystal models let us compare the filtered $\delta\rho$ maps for fluorite, CaF_2, calculated with the same initial set of structure amplitudes, but scaled and phased after the use of the harmonic rigid-atom superpositional model (figure 3.32), and of the anharmonic $\varkappa$-model (figure 3.28(b)). The latter map is corrected for the anharmonicity of the third order of thermal vibrations of atom F. The corresponding correction is small and there is no essential anharmonicity influence on $\delta\rho$. This fact indicates that

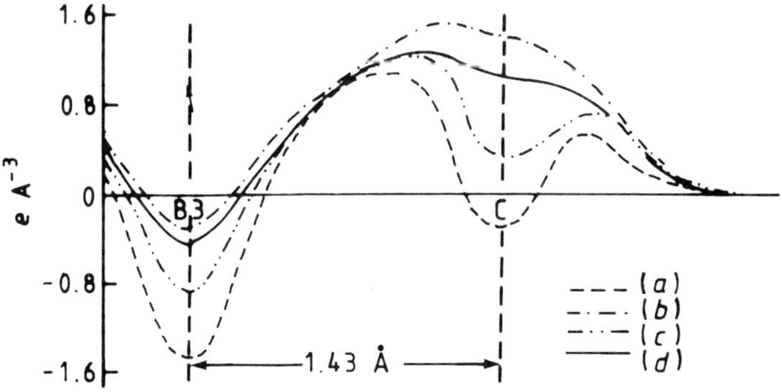

Figure 3.31 Fourier dynamic deformation density in boron carbide, $B_{13}C_2$, calculated from refinement in the high-angle approximation with the following parameters and scale factors: (*a*) all data refinement (0.479); (*b*) high-angle refinement (0.450); (*c*) high-angle refinement adjusted to low-angle data (0.461); and (*d*) multipole model refinement (0.453) (Kirfel *et al* 1979).

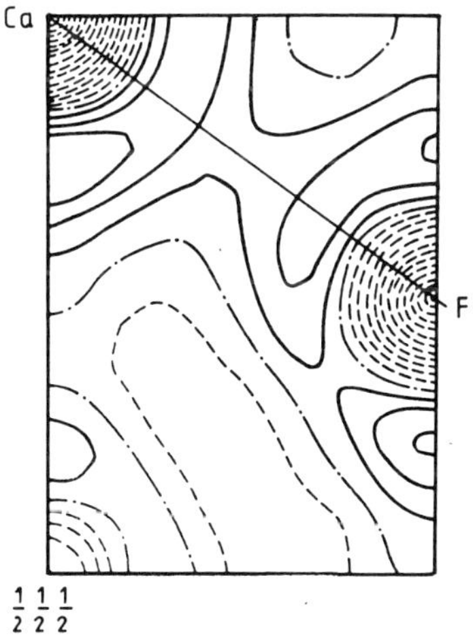

Figure 3.32 Deformation ED in CaF_2 in the (110) plane calculated with the filtration procedure ($\beta = 1$) with parameters from high-angle harmonic refinement of the rigid superpositional structure model. The line intervals are 0.1 e Å^{-3}.

in crystals with low anharmonicity of atomic thermal vibrations the results of high-angle refinement of a rigid atomic superpositional model with harmonic description of thermal motion can be used in construction of $\delta\rho$ maps. At the same time, less deep negative $\delta\rho$ regions near the positions of nuclei in figure 3.28 resulted from a more accurate scale factor determination from a more flexible anharmonic $\varkappa$-model as compared to the harmonic high-angle model. One can also see a smoother character of electron distribution anisotropy around the atomic positions on $\delta\rho$ maps constructed with the $\varkappa$-technique data. This can be explained by the fact that within the framework of a model which takes into account spherical deformation only, parameters $\varkappa_\mu$ absorb real anisotropic ED details. This results in smoothing of some $\delta\rho$ features. In general, both types of map are in satisfactory quantitative agreement in the internuclear space.

The application of two different flexible formalisms, Hirshfeld (1971) and Hansen–Coppens (1978) multipole models, for processing the same x-ray diffraction data resulted in EDs which are in excellent agreement, according to Baert *et al* (1982) and Delaplane *et al* (1985). This can be seen by analysing figure 3.18 which shows the $\delta\rho$ maps obtained with both these models for the $NaHC_2O_4 \cdot H_2O$ crystal at 120 K. The deviation of these maps does not exceed 0.1 e Å^{-3}.

Most difficulties are met in estimating the uncertainty in experimental data that is related to the diffraction quality of a crystal sample under study and to the measurement procedure reliability. This problem was studied in detail in 1982–1984 by the Commission on Charge, Spin and Momentum Densities of the International Union of Crystallography. The crystal of α-oxalic acid dihydrate, $C_2H_2O_4 \cdot 2H_2O$, was chosen for analysis in different laboratories all over the world; different crystals and different experimental data processing techniques were used. Coppens (1984), who was the head of this project, published a report in which the results obtained were generalized. It was found that in the crystal studied the positional parameters of nuclei obtained with the rigid-atom superpositional model in the high-angle approximation are reproduced to a deviation of 0.0005 Å. However, the deviation in harmonic parameters of thermal vibrations of atoms reached 25–30% for some investigations. These discrepancies could be attributed to inadequate account of extinction, which strongly depends on sample quality for this crystal. As a result, the deviations of the deformation ED maps reached 0.15 e Å^{-3}, though these maps were in full qualitative agreement with each other.

By comparing the results of studies of the same fluorite, CaF_2, crystal carried out with two different diffractometers, Streltsov *et al* (1988b) came to the conclusion that the reproducibility of harmonic atomic thermal parameters in this crystal is 10% approximately. The discrepancies in the $\delta\rho$ maps, constructed in a unique manner, did not exceed 0.1 e Å^{-3}.

Li *et al* (1988) have demonstrated that the model uncertainty in deformation ED can be decreased by averaging the maps over chemically equivalent fragments, as suggested by Rees (1976). They have studied the ED in two specimens of bis(pyridine)(mesotetraphenylporphinato)iron (II), $FeN_6C_{54}H_{38}$, (space group $P\bar{1}$) with the same x-ray autodiffractometer at 110–120 K. The completeness of experimental data for the two specimens was different: I, $\sin\theta/\lambda < 0.91$ Å^{-1}, 5672 symmetrically independent reflections with

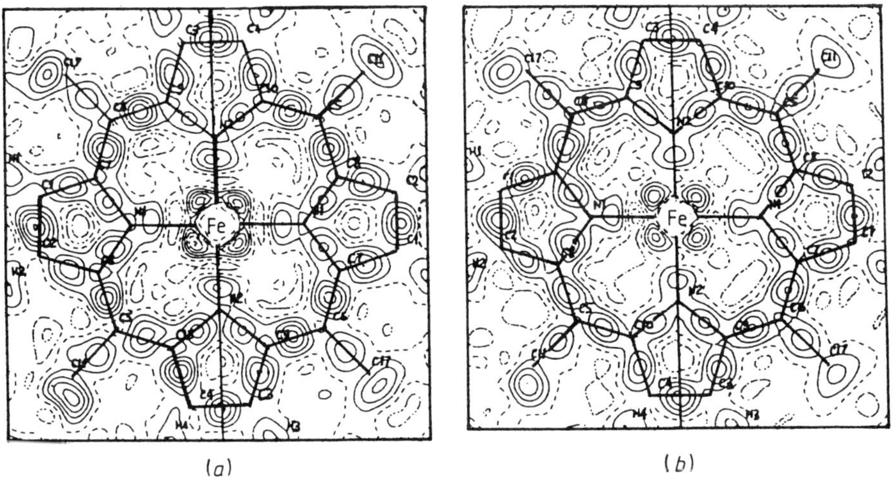

Figure 3.33 A comparison of deformation EDs in the porphyrin plane of two $FeN_6C_{54}H_{38}$ crystals before (links from solid vertical lines) and after averaging over chemically equivalent fragments: (*a*) crystal I; (*b*) crystal II (Li *et al* 1988).

$|F|>6\sigma(|F|)$; II, $\sin\theta/\lambda<1.15$ Å^{-1}, 8497 independent reflections with $|F|>6\sigma(|F|)$. The $\delta\rho$ maps for both specimens before and after averaging over chemically equivalent fragments in the porphyrin plane are given in figure 3.33. Before averaging the discrepancies in values of $\delta\rho$ maxima and minima were about 0.1 e Å^{-3} in accordance with the above estimations. The topography of $\delta\rho$ distribution was the same in both cases. After averaging the discrepancies became lower than the step of isolines (0.1 e Å^{-3}) on $\delta\rho$ maps. After averaging the estimated standard deviation of $\delta\rho$ decreases, i.e. the reliability of determination of $\delta\rho$ details increases.

The electron populations of 3d orbitals, recalculated from the multipole expansion parameters for both specimens of Fe-porphyrins, are in satisfactory agreement as well (table 3.23).

Thus, some results can be summarized now. The random error in the deformation ED, reconstructed from the x-ray diffraction data, is uniformly distributed over the crystal unit cell on average. When the filtration procedure with filter parameter $\beta\sim 1$ is used, this error is equal to (0.04–0.06) e Å^{-3} for centrosymmetric and (0.06–0.09) e Å^{-3} for non-centrosymmetric structures. The peaks of error due to scale factor uncertainty are imposed near nuclear positions on this general background. These peaks have the value of (0.1–0.2) e Å^{-3} for light atoms ($Z\leqslant 17$) and the value of (0.2–0.4) e Å^{-3} for 3d elements. When the filtration is not applied, the error of $\delta\rho$ near nuclei is increased by a factor of two to three. Moreover, the $\sigma(\delta\rho)$ peaks are always observed on closed symmetry elements in a unit cell. The model error in the deformation ED $\sigma_m(\delta\rho)$ can be estimated at a value of $\sim$0.1 e Å^{-3}. This estimate can be taken as an accuracy of deformation ED map determination.

Table 3.23 d electron orbital populations of Fe atom in $FeN_6C_{54}H_{38}$ (percentages in parentheses).

	Populations	
Orbital	Crystal 1	Crystal 2
$d_{x^2-y^2}$	0.57(8.4)	0.35(4.8)
d_{z^2}	0.83(12.2)	1.05(14.4)
d_{xz}	1.78(26.1)	1.93(26.5)
d_{yz}	1.78(26.1)	1.93(26.5)
d_{xy}	1.86(27.3)	2.02(27.7)
Total	6.82	7.29

The error can be lowered, of course, in very careful experiments and on applying correct procedures of experimental data processing.

3.7.3 *Experimental versus theoretical electron densities*

The comparison of x-ray experimental and theoretical electron densities can serve as an objective indicator of ED determination reliability. However, the direct comparison of the two EDs is not as trivial a task as it may seem at first sight. The theoretical ED is static, calculated for a system at the equilibrium nuclear configuration (usually), and the experimental ED is dynamic, i.e. averaged over all possible nuclear configurations. Besides, the experimental ED maps are subjected to various distortions caused both by incomplete elimination of deviations from the kinematic diffraction theory considered above and by limitations inherent in the experiment. Theoretical ED maps also suffer from model errors (see section 2.3).

Strictly speaking, for a correct comparison of the two electron distributions one must solve the Schrödinger electron equation in the Born–Oppenheimer approximation for each of the possible nuclear configurations, then obtain wave-functions of electrons, to calculate the ED. After that ED averaging over all these nuclear configurations should be performed in accordance with the contribution of each nuclear configuration to the total vibrational energy of a system. Corresponding weights are determined by Boltzmann factors W_a. With the notations introduced in section 3.1, the dynamic analogue of the theoretical ED can be written as

$$\rho(\boldsymbol{r}) = \sum_a W_a \int \chi_a^*(\boldsymbol{R}_a)\rho(\boldsymbol{r}, \boldsymbol{R}_a)\chi_a(\boldsymbol{R}_a)\, \mathrm{d}\boldsymbol{R}_a. \tag{3.152}$$

The direct calculations were carried out according to this method for simple molecules only, such as diatomics H_2, CH, BeH, CO, N_2 (Rozendaal and Ros 1982) and triatomics CO_2 and H_2O (Rozendaal and Baerends 1986). In practice, however, two other approaches are used for comparing the ED for the two types of molecule.

The first one consists of obtaining static (quasistatic) multipole or other model EDs from the experimental ED (Hirshfeld 1976). These EDs are obtained by calculation with refined electron parameters of some crystal structural models, when atomic thermal vibration parameters are believed to be zero. In doing so the experimental random errors are partially filtered and the ED map resolution is improved (Abramov *et al* 1995, Moss *et al* 1995, Souchassou *et al* 1995). At the same time, however, the maps become more strongly dependent on the ED model used.

The second approach consists in thermal smearing of theoretical ED in the convolution approximation (3.24) (Scheringer and Reitz 1976, Stevens *et al* 1977, Azavant *et al* 1994). The distribution function for atomic positions $\tilde{p}_\mu(\boldsymbol{u}_\mu)$, which appears in (3.24), must take into account all types of atomic motion. The thermal motion of molecules in a crystal presents the most general case. It may be presented as a combination of two motions: internal vibrations of atoms, which change the instantaneous geometry of a molecule, and translational and librational motions of a molecule as a whole (external vibrations). The frequencies of these vibrations noticeably differ: whereas in the first case they exceed 300 cm^{-1}, in the second case they are lower than 100 cm^{-1}; the amplitudes of external vibrations are essentially higher than those of internal vibrations. The diffraction experiments give the total characteristics of thermal vibrations of atoms. In order to separate the two types of vibration, one must use additional information, the spectroscopic data for example (Scheringer and Fadini 1979). Besides, the motion of atoms in molecules is correlated and, as a result, the internal vibrations manifest themselves at low temperatures only. This fact allowed Hirshfeld (1976) to postulate the rigidity of lengths of bonds in a molecular crystal as a criterion of physical significance of the atomic thermal parameters obtained from the structural model refinement. According to Hirshfeld, the differences in the values of components of tensors of atoms' thermal vibrations along their bond lines which exceed 0.001 Å^2 indicate an unsatisfactory description of a thermal motion. One should also keep in mind that when the atoms are described as an ensemble of independent oscillators the molecular fragment rotation is not included in the description (Dunitz *et al* 1988). This fact may involve considerable errors in description of dynamic ED in crystals with non-rigid molecules. Nevertheless, Rozendaal and Baerends (1986), analysing CO_2 and H_2O molecules, have found that ignorance of internal vibrations within the limits of the HFS method accuracy results only in underestimating the peak values very close to the nuclei. Hence, in molecular crystals one may take into account, in the first approximation, translational and librational vibration effects only.

Further, one should impose characteristic experimental restriction on the dynamic ED, for example, take into account the Fourier series truncation effect. This can be done most simply by using the double Fourier transform of a theoretical ED (Stevens *et al* 1977). In this case the thermal motion can be taken into account in a simple manner by introducing corresponding temperature factors in reciprocal space. Besides, when the theoretical calculations are made for separate molecules then the influence of crystal environment should be simulated. The latter decreases mainly the heights of peaks on $\delta\rho$ maps corresponding to lone-pair electrons, and changes their shape (Breitenstein *et al* 1983).

The thermal smearing deconvolution of experimental ED is not always achieved. Epstein *et al* (1982) have found that the static multipole $\delta\rho$ maps of imidazole, obtained from x-ray diffraction data measured at different temperatures, have essential discrepancies. The same conclusion has been drawn by Hermansson and Olousson (1984) in studying the $LiNO_3 \cdot 3H_2O$ crystal. The main cause of these discrepancies consists, apparently, in inadequate description of thermal motion. Besides, in reconstructing a multipole static electron density it always remains unknown how well the function ρ obtained corresponds to an infinite resolution due to ignorance of higher multipoles. This is why one cannot always establish the reason for discrepancies in compared static EDs. This allowed Stevens *et al* (1977) to suppose that the comparison of EDs is more valid in the case when the experimental limitations are introduced into the theoretical ED.

There are a few works in which both methods of comparison of experimental and theoretical maps of deformation ED were utilized (Breitenstein *et al* 1982, Stevens *et al* 1977, Coppens and Stevens 1977, Stevens 1980, Kutoglu *et al* 1982). Let us consider the results of one of these works which dealt with studying the α-oxalic acid dihydrate (Stevens 1980). Theoretical calculations were carried out in an extended basis set, which included polarization functions (basis set A), and in the two-exponential basis set 4-31G (B). Both basis sets provide reproducibility of main features of electron distribution in the compound, but the values and shape of deformation ED peaks differ from each other. As table 3.24 shows, basis set B leads to greater overestimation of peaks of lone-pair electrons and to greater underestimation of peaks on the bonds; this effect was discussed above in section 2.3. The comparison shows that the dynamic experimental and theoretical $\delta\rho$ maps are in a better mutual agreement than the static ones. This statement especially concerns the peaks of lone-pair electrons where the thermal smearing smooths the effect of insufficient flexibility of the basis set. In the internuclear region the shape of $\delta\rho$ peaks on experimental maps differs from the shape of similar peaks on theoretical maps. This can be attributed to the form of analytical functions used in multipole models, though this fact may also reflect the ignorance of internal thermal vibrations of atoms. The static experimental $\delta\rho$ demonstrate a 'sharper' picture as compared to dynamic ones: on these maps the $\delta\rho$ peaks of the lone-pair electrons are higher than on dynamic maps. The authors of other works, cited above, have come to similar conclusions.

Table 3.24 A comparison of static and dynamic multipole deformation ED peaks in the α-oxalic acid molecule (e Å^{-3}) in the $(COOH)_2 \cdot 2H_2O$ crystal (Stevens 1980).

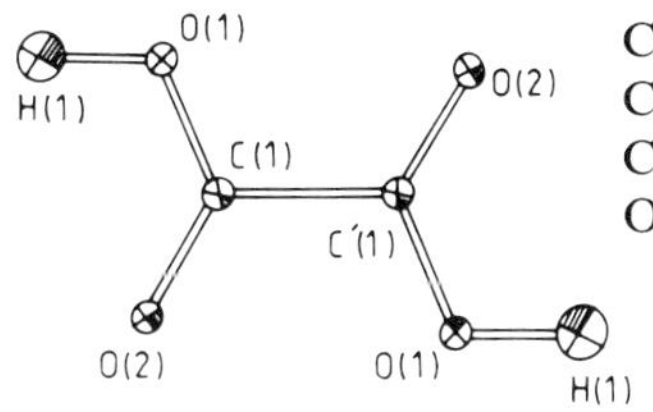

Situation of atom in molecule	Static deformation ED			Dynamic deformation ED		
		Theory			Theory	
	Exp.	A[a]	B[b]	Exp.	A	B
Bonds						
C(1)–C′(1)	0.64	0.67	0.40	0.65(3)	0.58	0.38
C(1)–O(1)	0.60	0.54	0.25	0.38(2)	0.43	0.28
C(1)–O(2)	0.85	0.70	0.54	0.49(2)	0.62	0.46
O(1)–H(1)	0.65	0.56	0.34	0.27(4)	0.40	0.23
Electron pairs						
O(1)	0.66	1.10	1.18	0.42(6)	0.60	0.73
O(2) a	0.42	1.22	1.49	0.50(8)	0.53	0.61
O(2) b	0.41	1.32	1.39	0.38(8)	0.53	0.64

[a] Extended basis.
[b] Two-exponential basis 4-31G.

If there are hydrogen bonds or strong short electrostatic or specific interactions in a crystal, then the effect of crystal environment on ED of molecules is usually not small. Dovesi *et al* (1990) have performed the *ab initio* HF calculations of ED for crystalline urea, CH_4N_2O, and for a single molecule of this compound. It has been found that in the crystal, where molecules are connected by a system of hydrogen bonds, ED appreciably deviates from the superposition of the ED of molecules. The maximal ED discrepancy achieved is about 0.1 e Å^{-3} in the region of lone-pair electrons of O atoms. The relative discrepancies between crystalline (bulk) structure amplitudes for low-angle reflections and that for the procrystal, composed of separate molecules, are between 0.20 and 3.94%. Therefore, establishing the mutual influence of structural fragments in urea lies within the extremes of the possibilities of the HF method and of the sensitivity of precise x-ray diffraction studies. This quite important conclusion will be supported by a number of facts below.

Krijn and Feil (1988a) and Krijn *et al* (1988) have shown that thermal and crystal environment effects can be properly taken into account in cluster calculations. They have calculated in the framework of DFT with a local approximation for exchange and correlation the ED in the cluster depicted

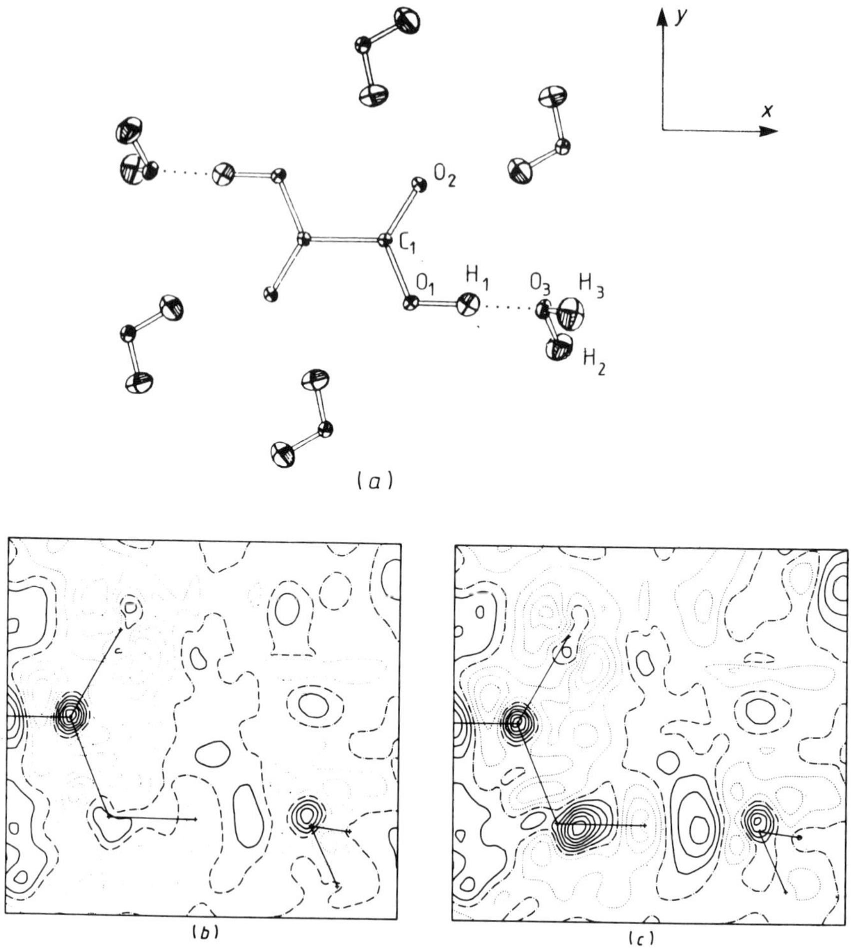

Figure 3.34 The α-oxalic acid dihydrate: (*a*) the geometry of the acid molecule and its nearest neighbours; (*b*) the experimental minus vibrationally averaged theoretical ED in the *xy* plane; the cluster depicted in (*a*) was included in the calculation; (*c*) the experimental minus vibrationally averaged superposition of free oxalic acid and free water ED (all crystal field effects are neglected) in the *xy* plane. The contour intervals are 0.0075 $e\,(\mathrm{au})^{-3}$ (Krijn and Feil 1988a).

in figure 3.34(*a*). The latter is a fragment of the crystal of α-oxalic acid dihydrate, $C_2H_2O_4 \cdot 2H_2O$. The thermal motion was accounted for in the harmonic convolution approximation. The thermal parameters obtained at 100 K by Dam *et al* (1983) were used. As shown in figure 3.34(*b*), the agreement between experimental and vibrationally averaged cluster theoretical ED is remarkable (excepting the nuclear positions). Comparing with figure 3.34(*c*), taking account of the effects of hydrogen bonding and crystal environment improves the agreement significantly. The maximal discrepancies remain in oxalic acid and water lone-pair regions. The same conclusion was reached by Breitenstein *et al* (1983) in a study of the same problem for other compounds.

As concerns crystals of inorganic compounds, the comparison of experimental and theoretical ED for silicon has shown that the first-principles calculations with DFT theory systematically underestimate the $\delta\rho$ peak value at the centre of internuclear distance (see figure 2.11). Tsirelson *et al* (1988b) have also compared the experimental deformation ED of the V crystal with that constructed from the APW calculation of Wakoh and Kubo (1980). The X_α potential was chosen in the calculation in such a manner that the Fermi surface would best describe the measurement results obtained in the de Haas–van Alphen effect experiment. However, even the thus-'corrected' calculation does not provide agreement between theory and experiment (figure 3.35). The deformation ED maps surely agree at a qualitative level: both of them similarly demonstrate the anisotropic character of deformation of electron shells of V atoms, involved in a chemical bond, and the ED contraction near nuclei. The theory, however, does not reproduce this contraction in full measure. Since the increase of α value in the X_α potential corresponds to a deeper exchange–correlation role, this must result

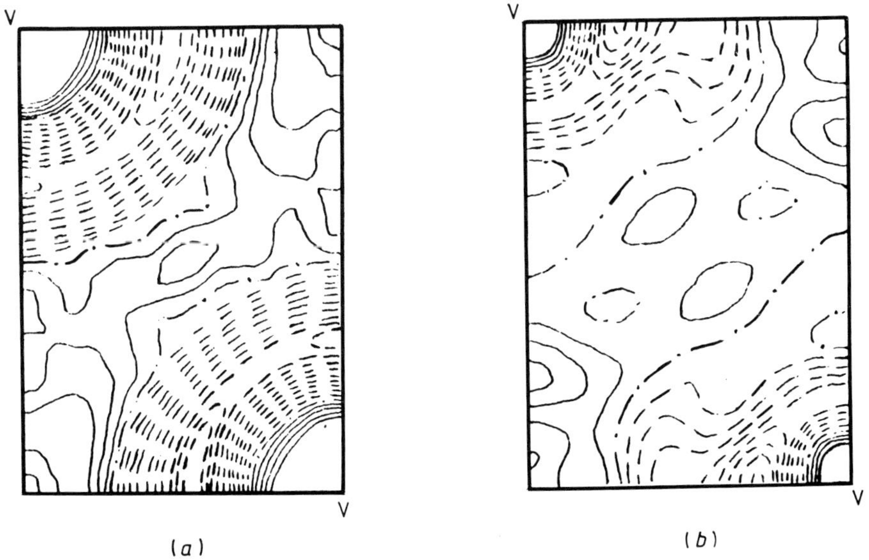

Figure 3.35 Deformation ED in the (110) plane of the V crystal at 295 K; (*a*) experimental map (the line intervals are 0.05 e Å^{-3} for positive and 0.1 e Å^{-3} for negative density), (*b*) theoretical dynamic map (the line intervals are 0.05 e Å^{-3}). Near atomic centres the lines are not depicted (Tsirelson *et al* 1988b).

in the contraction of 3d wave-functions. Probably, the underestimation of the value of α is one of the reasons for the discrepancies observed. The differences in $\delta\rho$ along the V–V line are also more than the experimental error (see figure 3.30).

Thus, one has not always been able to achieve a good accord between the theory and experiment so far. Nevertheless, the usefulness of comparison of theoretical and experimental $\delta\rho$ maps is beyond doubt. This is convincingly exemplified by Kutoglu *et al* (1982). Taking into account the analysis of discrepancies between experimental and theoretical deformation ED for thiourea they found the error which had been made by Mullen and Hellner (1978) when they took into account the background in x-ray diffraction data processing. Another example is the study of Pietsch and Hansen (1996) who used the HF calculation of GaAs in order to discriminate the results of different experimental investigations of ED distorted by model error and phase uncertainty, important in this non-centrosymmetrical crystal.

4

NEW AND COMPLEMENTARY METHODS IN ELECTRON DENSITY INVESTIGATIONS

As noted above, only diffraction methods make it possible to reconstruct the ED and related quantities in the whole unit cell volume. Along with diffractometers with the usual x-ray tubes, synchrotron radiation and γ-ray devices are also used in such diffraction experiments. The radiation from these sources is an electromagnetic radiation in the x-ray wavelength range. However, both the radiation itself and, especially, the character of sources strongly influence the equipment and technique of experiments. The question arises of in what relation are the results of experiments carried out on the same object with different sources? What are the advantages and disadvantages of differential techniques?

High-energy electron diffraction is close to the techniques mentioned. Since the electrons are scattered due to the interaction of charges, an electron diffraction experiment can give information on the EP distribution, provided the electron–electron exchange is negligible. Because the charge and potential distributions are interrelated, it is of interest to discuss the relationship between the results of x-ray and electron diffraction studies of the same quantities.

Finally, there have appeared new methods providing some complementary opportunities in the study of ED.

The problems mentioned will be discussed in this chapter.

4.1 Synchrotron radiation in x-ray diffractometry

The electromagnetic radiation arising during accelerated motion of charged particles, electrons and/or protons, in synchrotrons is called synchrotron radiation. Accelerating charged particles up to energies of 1–10 GeV, synchrotrons generate radiation which covers a wide range of wavelengths from infrared to short x-rays. The usual synchrotron radiation spectrum decreases slowly towards the long-wavelength region and decreases rapidly from the short-wavelength side. The position of that cut-off depends on the total energy of the accelerated particles. In third generation synchrotrons constructed especially for radiation production, the problem of shifting of the cut-off position to shorter wavelengths is solved by the introduction in the path of charged particles of special devices which undulate their trajectory (bending magnets, wigglers, undulators): the undulation changes the trajectory curvature and positively influences the spectrum (Catlow 1990,

Harding 1995, Margaritondo 1995). In this way moderate-energy synchrotrons can generate radiation for use in solid state investigations.

Synchrotron radiation possesses some specific features which must be taken into account when using it in solid state physics and chemistry.

First, synchrotron radiation is six to eight orders of magnitude higher than the radiation of modern x-ray tubes. Therefore, the experiment requires significant protection to provide personal safety for the operators. On the other hand, such high intensity permits one to use crystal samples of a very small size ($r \approx 5 \times 10^{-5}$ m).

Second, the photons are emitted within a narrow angular range along the tangent to a trajectory. The beams are led out of a synchrotron along tangential channels, along which the experimental installations are set up.

Third, the radiation spectrum is 'white'. This opens the possibility of cutting out the monochromatic radiation within the wavelength range from 10^{-1} to 10 Å. At the same time, the use of a crystal monochromator (germanium or silicon single crystal, as rule) results in the appearance of higher-order reflections ($n\lambda = 2d \sin \theta$, $n = 2, 3, \ldots$) making a parasitic contribution to the intensity of diffraction reflections. This problem is well known in neutron diffraction analysis (Bacon 1975, Nozik *et al* 1979).

Fourth, the wave front of a collimated beam is nearly planar, the radiation is nearly non-divergent, and the beam intensity decreases with path length much more slowly than in the case of point radiation sources. Further, synchrotron radiation is highly polarized and is not continuous: it consists of individual bursts separated by time intervals of 0.85 μs (on the CHESS synchrotron, according to Nielsen *et al* 1986). Finally, in modern accelerators some drift of the radiation intensity both in time and space exists.

All these features should be taken into account in dealing with synchrotron radiation diffraction.

The development of synchrotron crystallography has greatly stimulated the achievements in single-crystal (Becker 1988, Krick 1988, Coppens 1992, Hasnain *et al* 1994, Harding 1995) and powder diffractometry (Cox *et al* 1986, 1988, Cox 1991, Finger 1989, Catlow 1990) in biological crystallography (e.g. Sweet and Woodhead 1989, Helliwell 1992, Sakabe *et al* 1992) as well as in inelastic and/or incoherent scattering experiments (Dorner and Peisl 1983, Dorner *et al* 1987, Schülke 1988, Batterman 1990). Let us consider here the application of synchrotron radiation to an accurate single-crystal structure analysis.

Single-crystal diffractometers are installed now on the CHESS synchrotron at Cornell University, Ithaca, NY, USA, at Brookhaven National Laboratory, Upton, NY, USA, at the Photon Factory in Tsukuba, Japan, at HASYLAB synchrotron, Hamburg, Germany, at Synchrotron Radiation Source, Daresbury, England, etc. These devices differ only slightly from each other, and are supplied with cooling equipment (Larsen 1995). The principal scheme of the HASYLAB diffractometer, which is used in accurate diffraction studies,

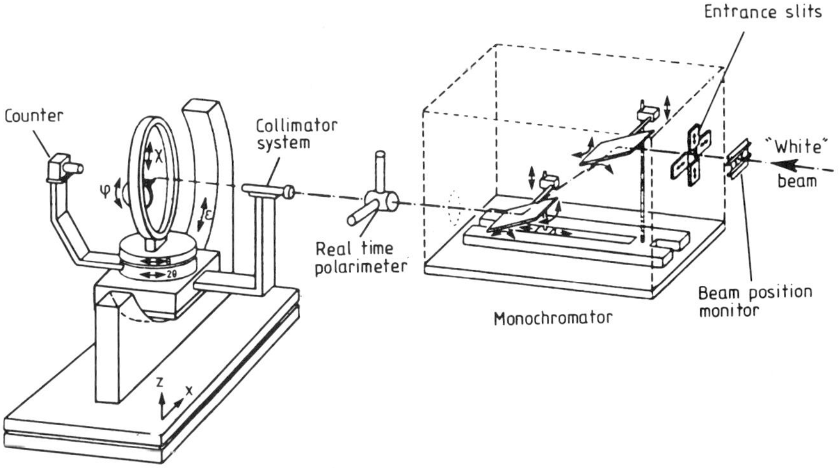

Figure 4.1 The layout of the five-circle diffractometer for synchrotron radiation in the wavelength range from 0.3 to 2.2 Å installed at DESY-HASYLAB in Hamburg, Germany (Kupcik *et al* 1986).

is shown in figure 4.1 (Kupcik *et al* 1986). The white radiation beam emerging from the synchrotron through the tangential channel falls on the first crystal monochromator. This crystal, whose position does not change during the measurement, is cooled with water to remove the heat released during absorption of white radiation of high intensity. The reflected monochromatic radiation is directed to a second crystal monochromator which reflects the radiation with the same wavelength into the primary white beam direction. Such a monochromator unit design allows one to change the incident radiation beam wavelength without considerable displacement of the diffractometer. The double reflection slightly decreases the beam intensity but the loss is fairly small due to low beam divergency. The detector of low efficiency is used for monitoring the photon beam drift. Fine adjustment of both the second crystal and the diffractometer as a whole is provided.

The synchrotron diffractometers have one more fitting parameter along with the four usual variable angles. In the HASYLAB it is the fifth angle accounting for the rotation of the whole diffractometer around the monochromatic beam; the diffractometer becomes five circle. In other cases this additional fitting parameter can be related to planar diffractometer displacement in the horizontal plane, which follows the monitor indications of the beam drift. The installation is completely operated by an on-line computer.

The first publication in the field discussed was presented by Nielsen *et al* (1986). The hexacyanochromate hexaammoniochromate, $[Cr(NH_3)_6][Cr(CN)_6]$, designated as Cr–Cr hereafter, was chosen for the investigation with a spherical single-crystal size of about 0.1 mm at the CHESS synchrotron. The isomorphic crystals Co–Co (Iwata and Saito 1973) and Co–Cr (Iwata 1977) have been studied earlier with a conventional four-circle x-ray diffractometer. The synchrotron wavelength of 0.302 Å was significantly lower than those used in x-ray studies. The beam monochromation was accomplished by two silicon single crystals. Due to the low divergence the shape of reflections differs from the conventional one: a very narrow (half width of about 0.010°) main reflection had a slowly diminishing 'tail'. To avoid the second-order contributions, an amplitude analyser was introduced into the detector circuit. These were, in general, 0.2–0.3%; however, the corrections to the four weak reflections, reached in F^2 of approximately 10%, were applied. Due to the small size of crystal studied and short wavelengths used the corrections for extinction and absorption have been neglected.

The total number of reflections whose intensity was able to be measured with synchrotron radiation does not exceed a similar number in a conventional experiment (see e.g. Ohgaki *et al* 1992). However, the number of weak high-angle reflections is essentially higher here.

The positional and thermal structural parameters and, consequently, bond lengths and angles in synchrotron radiation crystallography were found to be very close to traditional x-ray results. The same conclusion was reached for deformation density maps. Figure 4.2 depicts the deformation ED map reconstructed from synchrotron radiation structure amplitudes in the

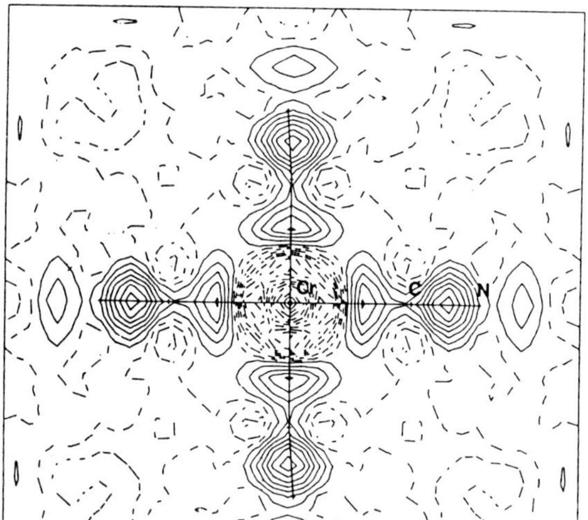

Figure 4.2 X–X deformation map in the $Cr(CN)_4$ plane based on parameters from the high-order synchrotron refinement ($\sin\theta/\lambda > 0.7$ Å^{-1}), including all reflections with $\sin\theta/\lambda < 0.8$ Å^{-1} (contours are at 0.05 e Å^{-3}; negative contours are broken) after averaging over chemically equivalent regions (Nielsen *et al* 1986).

Table 4.1 A comparison of some characteristics of experiments studying corundum (ruby) with synchrotron and conventional x-ray radiation.

	Al_2O_3 (<0.05% Cr)	Al_2O_3 (0.46% Cr)	Al_2O_3	
Reference	Lewis *et al* (1982)	Tsirelson *et al* (1985a)	Kirfel and Eichhorn (1990)	
Radiation source	tube	tube	synchrotron	
Radiation	Ag Kα + filter (Pd)	Mo Kα monochrom.	0.5599 Å	
Temperature (K)	~300	~300	~300	
Crystal size and shape (mm)	sphere, 0.13	sphere, 0.152	sphere, 0.200	
$(\sin\theta/\lambda)_{max}$ ($Å^{-1}$)	1.495	1.25	1.024	
Total number of reflections measured	8923	5336	1758	
Number of non-zero independent reflections ($T > 3\sigma(T)$)	618	408	259	
R(intern) (F)	—	0.019	0.014	
Refinement			SAMSY[a]	MUSY[b]
R	0.042	0.0069	0.0107	0.0078
R_w	0.044	0.0099	0.141	0.0102
S	1.14	1.055	1.39	1.00
Extinction	isotropic	isotropic	—	
Taking into account absorption	yes	yes	yes	
Taking into account TDS	no	yes	yes	
$T/\bar{T}$	$\sim\frac{1}{15}$	$\sim\frac{1}{13}$	~1/6.5	

[a] Superpositional atom model.
[b] Multipole model.

$Cr(CN)_4$ plane of the Cr–Cr compound. This map averaged over all chemically equivalent regions shows qualitative agreement with those from x-rays. The peaks on all chemical bonds as well as lone-pair electrons of nitrogen atoms are clearly seen. The noise level is small and flat.

A few years later, Darovsky *et al* (1996), using advanced imaging plate technology (Bolotovsky *et al* 1995), performed a new synchrotron study of the same Cr–Cr compound at 50 K. They reproduced all the main features of the map depicted in figure 4.2. They have also calculated additionally the 3d electron populations of Cr atoms in agreement with theory.

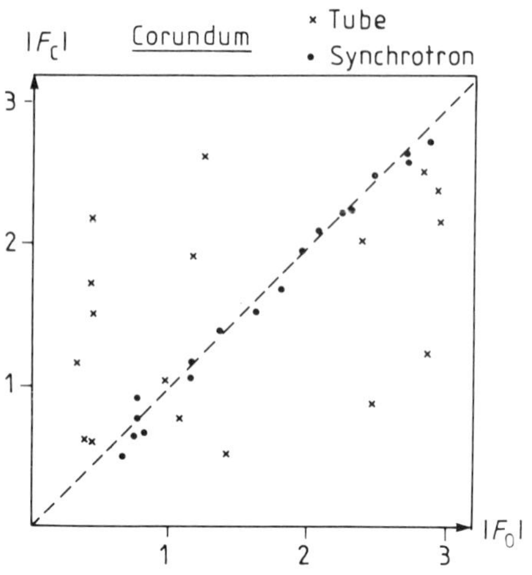

Figure 4.3 A comparison between calculated and observed structure factors of α-Al_2O_3 deduced from synchrotron radiation (Kirfel and Eichhorn 1990) (black circles) and sealed x-ray tube data (Lewis *et al* 1982) (crosses).

In order to improve the experimental conditions for reflection measurements a toroidal computer-controlled glass mirror coated with a thin layer of platinum has been introduced in the monochromated beam at the BL14A beamline at Photon Factory in Tsukuba (Stow and Iitaka 1989). This has made the beam more intense and, at the same time, convergent. The reflection profile became more regular. In addition, the use of the mirror almost completely eliminated second-order wavelength contamination. Unfortunately, the mirror is ineffective with wavelengths shorter than 0.5 Å.

A detailed comparison of the results obtained with conventional x-ray diffraction and with synchrotron radiation has been performed for the corundum crystal by Kirfel and Eichhorn (1990). Their results are given in table 4.1. Since the synchrotron radiation wavelength chosen, $\lambda \approx 0.56$ Å, is close to that generally used, all standard corrections were introduced in the data. The intensities of the symmetry-allowed weak reflections were measured as statistically significant: there were no reflections with $T < 3\sigma(T)$ in the synchrotron data whereas in the x-ray case the number of insignificant reflections is large.

The authors compared the intensities of weak reflections measured with the calculated ones. Figure 4.3 presents such a comparison for two radiations: conventional x-ray tube (data of Lewis *et al* 1982) and synchrotron radiation. The correspondence is much better in the latter case. Kirfel and Eichhorn (1990) presented also the results of studying the Cu_2O crystal which had earlier been studied with x-rays by Restori and Schwarzenbach (1986). The resolution achieved in synchrotron work was larger and 23 weak forbidden reflections were additionally measured. The latter are especially important, since along with the atomic ED deviation from sphericity, the thermal vibration anharmonicity of copper atoms gives a great contribution to the forbidden reflections in this crystal.

The results presented allow one to draw some conclusions on the possibility of using synchrotron radiation in accurate structure analysis. The use of synchrotron radiation does not meet any principal methodological difficulties though some new experimental features should be taken into account. At the same time, it becomes possible to make more reliable the correction for extinction. Indeed, the use of small crystals (10–100 μm) can reduce the severity of extinction as a factor limiting the precision of the experiment. Small crystal size limits the statistical precision if a conventional x-ray tube source is used but not in the case of synchrotron radiation. Measurements at several wavelengths provide additional information on extinction (Maslen and Streltsov 1992, Hester *et al* 1993) while simultaneously providing a check on the reproducibility of the measured ED (Maslen and Streltsova 1992).

The conditions of measuring weak and forbidden reflections are greatly improved. The latter are, apparently, the most important in applying synchrotron radiation: these reflections contain important structural information and greatly influence the noise level on ED maps. They are also responsible for small but important contributions to crystal properties, for instance, for details of the electric field gradient as calculated from the experimental ED distribution.

4.2 Gamma diffractometry

After being irradiated by neutrons in a nuclear reactor, some materials become radioactive with a lifetime of the order of hours. They can be used as a γ-radiation source and then used in diffractometry. Two isotopes at least should be mentioned in this respect. The ^{198}Au isotope emits γ quanta with an energy of 412 keV and wavelength $\lambda = 0.0301$ Å and the ^{153}Sm isotope emits γ quanta with an energy of 103 keV and $\lambda = 0.12$ Å.

Two γ diffractometers were described by Schneider (1974a, b, 1975a, b, 1976, 1977). One of them was constructed at Laue–Langevin Institute (Grenoble, France) specially for studying the mosaicity of large single crystals used as monochromators in neutron experiments. This diffractometer was used for precisely measuring the structure amplitudes of some reflections from a copper single crystal (Schneider *et al* 1978). The other γ diffractometer was designed at Hahn–Meitner Institute (HMI, Berlin, Germany) specially for accurate single-crystal diffractometry. There are other γ diffractometers designed for different applications too, for example, the γ diffractometer of Missouri University, USA (Alkire and Yelon 1981, Alkire *et al* 1982), which

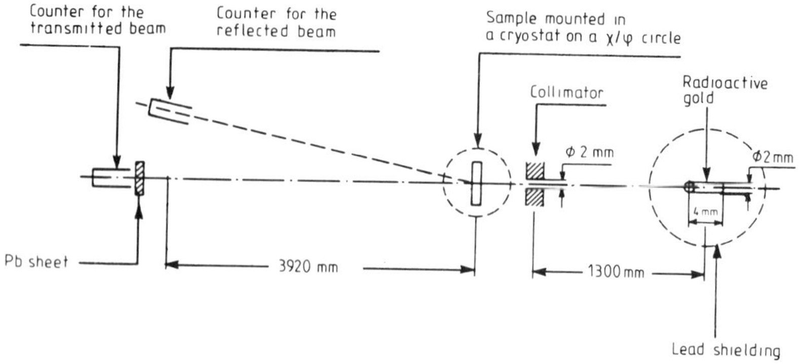

Figure 4.4 The schematic set-up of the HMI gamma diffractometer (Schneider *et al* 1978).

utilizes the ^{153}Sm isotope. Due to the strong wavelength dependence of crystal reflectivity, very-short-wavelength radiation can be used in studying only heavy-element compounds, whereas the Sm source provides the possibility of working with light atoms in a crystal.

The scheme of the HMI diffractometer is shown in figure 4.4. A gold wire, 4 mm long and 2 mm in diameter, irradiated in a reactor with thermal neutron flux of $\sim 9\times10^{13}$ neutrons cm^{-2} s^{-1} over 4 d, is used as the source. After irradiation the wire activity is 100 Ci. Some difficulties take place during the transportation of the irradiated material from the reactor (Schneider *et al* 1979). The half-decay period of ^{198}Au is 2.7 d and, hence, this source is rather intense, stable, and relatively long-living. On a freshly irradiated gold specimen the Ge detector registers about 6000 counts s^{-1} with a background of 0.03 counts s^{-1}. The amplitude discriminator makes the radiation free from contamination, with spectrum line width $(\Delta E/E)\sim 10^{-6}$. The goniometer itself is standard.

Since the wavelength is small, the Ewald sphere radius in the γ diffractometry is large, so a greater number of reciprocal lattice points are inside the sphere with large density. In order to reach the resolution required, careful collimation is needed. However, even in this case one cannot avoid the overlap of neighbouring reflections. As a result, accurate γ diffractometry can be applied mainly to crystals with small unit cell dimensions. However, even in this case not all the points in the Ewald sphere can be measured. This is because the crystal reflectivity for crystals of various degrees of perfection depends on the wavelength as λ^2 or λ^3, and therefore for short wavelength it is much lower as compared to the usual x-ray range. This allows one to measure only a limited number of Ewald sphere points.

Some other circumstances should also be mentioned when applying such short wavelengths. The analysis has shown, in particular, that the Born approximation is satisfied here to a rather high accuracy (Schneider *et al*

1978). Anomalous dispersion is absent since the photon energy is much higher than the binding energy of electrons in the majority of atoms. The γ beam is ideally unpolarized, so that the polarization correction can easily be taken into account. Multiple scattering should be high, but a small reflectivity makes second and higher-multiplicity processes weak; small rotation of a specimen around the scattering vector allows one to exclude this effect completely (Schneider 1975b, Alkire and Yelon 1981). The TDS contribution is much lower due to a very small width of diffraction reflections. Even for high-order reflections, the Bragg scattering angles are small, and the geometry of the experiment can be taken into account rather easily. The latter circumstance together with the Laue geometry allows one to obtain a structure amplitude in the absolute scale.

Let us consider the results of γ diffractometric study of a copper single crystal (Schneider *et al* 1981, Schneider 1988b). For $\lambda = 0.03$ Å the extinction length noticeably exceeds the linear size of an average coherent volume in copper crystals with mosaicity of about 10′ (this fact was established specially), and the primary extinction is absent. At such a short wavelength the extinction parameter y is very close to unity, thus the correction for the secondary extinction is beyond the limits of experimental error. Multiple reflections and the influence of TDS were estimated and found to be insignificant.

The 19 structure amplitudes (or 19 points on the atomic amplitude curve) were measured up to values of $(\sin\theta/\lambda) = 1.6$ Å^{-1}. A precision of $\sim 0.5\%$ was achieved at low angles. This data set for the Cu crystal which is, probably, the most complete and accurate, has been used for verifying the theoretical

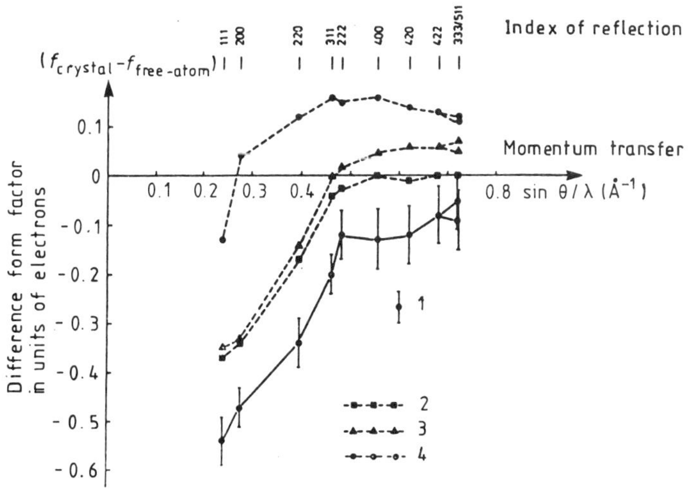

Figure 4.5 The deviation of experimental and theoretical atomic scattering amplitudes of copper from free-atom values (Schneider 1988b). 1, x-ray experimental data. Theoretical SCF calculation within the LDA of the DFT; 2, LCAO LDA (Bagayoko *et al* 1980); 3, LAPW (MacDonald *et al* 1982); 4, cellular method (Eckardt *et al* 1984).

models used for description of electronic properties of this crystal. The comparison of theoretical and experimental structure amplitudes is presented in figure 4.5. All theoretical values are higher than the experimental ones by about 0.8% but, nevertheless, the agreement especially in the high-angle region is good. The discrepancy is larger in the low-angle region where the deformation of the electron shell due to chemical bond formation is pronounced.

Another metal whose crystals were studied by γ diffractometry is beryllium. The atoms of this metal contain only four electrons, two of which are valence ones. That is why a large number of both theoretical and experimental studies have been devoted to beryllium metal (for a list of references see Schneider 1988a). The beryllium atom has low scattering power and a longer wavelength is, thus, preferable for ED study. This is why Hansen *et al* (1987) had to pass from radiation of ^{198}Au ($\lambda = 0.03$ Å) to radiation of the ^{153}Sm ($\lambda = 0.12$ Å) isotope. This made it possible, over the period of 36 d, to measure the intensity of the 37 independent structure amplitudes (from the total number of 500 reflections) up to values of $(\sin\theta/\lambda) \leqslant 0.85$ Å and, additionally, 14 structure amplitudes with higher values of scattering angles. The use of radiation with two wavelengths allowed then the reliable estimation of the contribution of a secondary extinction ($\sim$3.4% to the strongest reflection) and elimination of its influence on the atomic ED asphericity. This study allowed the precise scaling of the x-ray structure amplitudes measured by Larsen and Hansen (1984) and provided a further study of the ED properties of Be crystal using the maximum entropy method (Iversen *et al* 1995).

Investigations of the ED in crystals with more complicated structure than Cu and Be (rutile- and perovskite-like) with γ-diffractometry in the range of temperatures have been described recently by Palmer (1993) and Palmer and Jauch (1993).

The combination of x-ray and neutron diffraction seems to be fruitful for studying many fine structural effects (see subsection 3.6.2). The results, however often depend not only on the chemical composition of some particular specimen but also on its size and real structure. This explains, in particular, the fact that the initial wide application of the X–N technique was substituted later by the X–X technique. On the other hand, the large specimen size in γ diffractometry (as compared to conventional x-ray structure analysis) and the lowering of this size in the neutron time-of-flight technique (see appendix K) made it possible to carry out the experiment by two methods on the same crystal sample. Such a study of the MnF_2 compound by two methods at two temperatures using the same specimen—a spherical single crystal about 3.27 mm in diameter—was performed by Jauch *et al* (1989). It was found that the two methods provide practically coincident results at room temperature. After lowering the temperature below the Néel point ($T_N = 67$ K) it was found by γ diffractometry that fluorine atoms are shifted in the x

parameter by $\delta x = -(52 \pm 9) \times 10^{-5}$. This fact apparently explains the jump in the optical linear birefringence when the antiferromagnetic ordering occurs which had been observed by Borovik-Ramonov *et al* (1973). A noticeable difference in position parameters of fluorine atoms obtained by the two methods at low temperature, against the very good agreement at room temperature, led the authors to suppose the displacement of electron shells took place with respect to nuclei during the transition into the magnetically ordered state. These small, but measurable effects led to a new look at the microscopic magnetic ordering model in MnF_2.

4.3 Precise structure amplitude determination by the pendellösung effect

An overwhelming majority of diffractometric studies of ED distribution have been carried out on imperfect crystals. The description of diffraction in such crystals is based on the kinematic scattering theory, in which the interaction between primary and secondary waves is taken into consideration with the help of the introduction of the extinction correction (see subsection 3.3.3). A crystal is usually considered as imperfect if its integrated reflecting power R, measured at the given Bragg angle with radiation at the given wavelength, is close to the Q_{0kin} value calculated within the kinematic theory and much larger than the Q_{0dyn} value calculated within the dynamical theory for a 'thick' crystal: $Q_{0kin} \geqslant Q_0 \gg Q_{0dyn}$. The dynamical theory of diffraction deals just with the interaction of primary and secondary waves.

The dynamical theory of x-ray scattering began with the works of Darwin, Ewald, and Laue and is outlined in books by James (1948) and Cowley (1975). This theory has been further developed by Kato (1976, 1979, 1980, 1988). As applied to neutron diffraction the dynamical effects have been described by Rauch and Petraschek (1978).

The interference of primary and scattered waves is greater the more prominent their coherency is, i.e. the more perfect the crystal is. This is why the dynamical theory relates mainly to perfect crystals. Since the latter are scarce both in nature and in modern technology, the dynamical theory is not so widely applied to the accurate structural analysis. At the same time,

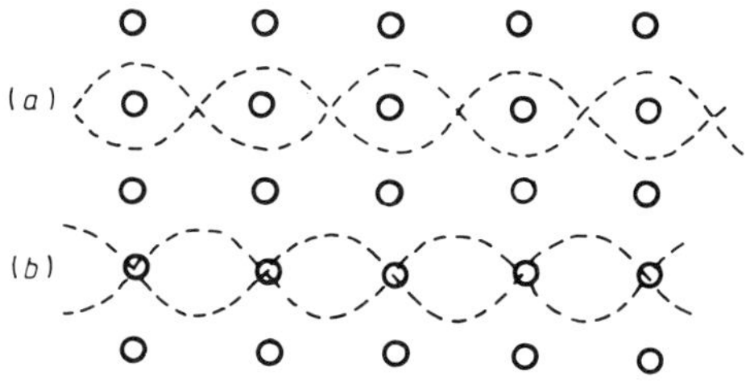

Figure 4.6 The scheme of Bloch standing waves in an ideal crystal.

the dynamical effects allow one to measure structure amplitudes with a high degree of accuracy owing to the so-called pendellösung effect. Its essence is as follows. As a result of interaction between primary waves and those reflected from crystallographic planes, a certain wave field is created in a crystal. If only the single reflected wave is intense enough and considered in the treatment, the approximation is called two wave. In the symmetric Laue diffraction position, the wave field is produced by a superposition of two Bloch waves whose periodicity is equal to the interplanar distance. The nodes of one standing wave coincide with crystallographic planes; the nodes of the other one lie between these planes (figure 4.6). Their amplitudes are the same, but wave-vectors and refraction indices slightly differ from each other. Along the crystal thickness t_0 this difference in phase may reach some value upon which the intensity of reflected and transmitted beams will be mutually dependent. Since the initial phase difference between two beams is equal to π, the maximum intensity for a reflected wave corresponds to the transmitted wave minimum, and vice versa. Thus, energy transfer from one beam into another takes place, in a fashion similar to oscillations of two linked pendula. That is why this phenomenon was called the pendellösung effect (Renninger 1973a, b). The period of intensity variation of reflected (and transmitted) waves over a crystal thickness is equal to the extinction length $\Lambda = \pi V \cos\theta_B/(|F(\boldsymbol{H})|\lambda$, where V is the unit cell volume. In the simplest case (the scattering of non-polarized thermal neutrons with their low absorption) the integral reflection factor R_i is a function of effective path length u inside a crystal expressed in units Λ, i.e. $u=(2\pi t/\Lambda)=2t|F(\boldsymbol{H})|\lambda/(V\cos\theta_B)$ (t is the wave path length). In x-ray scattering a large number of other factors must be taken into account, complicating the description of the effect and not touching its essence. The measurement of periodicity of R_i as a function of u allows one to determine the structure amplitude moduli $|F(\boldsymbol{H})|$.

Most frequently the experiment consists of measuring the intensity of the x-ray beam coherently reflected in the Laue geometry from some particular system of planes depending on crystal thickness or on wavelength. However, a more effective method was proposed by Utemisov *et al* (1980, 1981). These

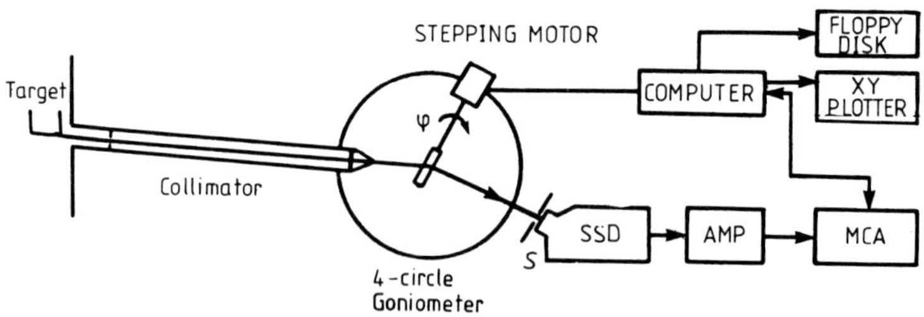

Figure 4.7 The experimental set-up of apparatus for pendellösung effect measurements. SSD, solid state detector; AMP, amplifier; MCA, multichannel analyser (Saka and Kato 1986).

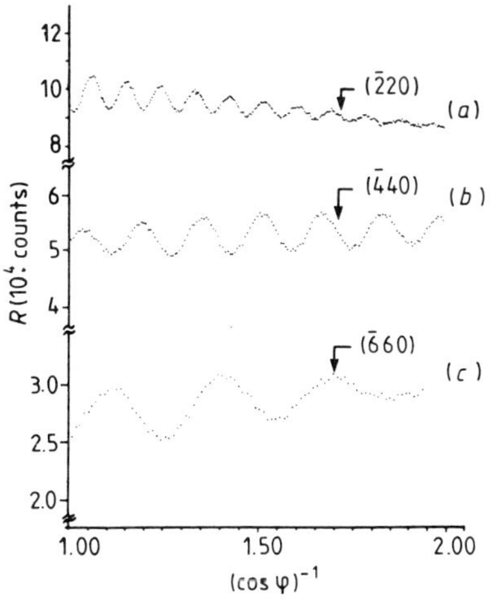

Figure 4.8 Typical examples of $R(\varphi)$ for lower- (*a*) middle- (*b*) and higher-order (*c*) reflections (Saka and Kato 1986).

authors suggested changing the path length t of the radiation in a crystal by turning it through angle φ around the scattering vector; then $t = t_0(\cos\varphi)^{-1}$. Substituting t into the expression for u and measuring the periodic R_i against u dependence one can determine $|F(\boldsymbol{H})|$ from the oscillation's period.

As an example of such a measurement we can present the work of Saka and Kato (1986). Figure 4.7 shows the scheme of their experiment. A white x-ray spectrum from the usual x-ray tube was used in order to carry out measurements at various wavelengths. The goniometer allowed them to turn the sample through angle φ to vary the value of t. The solid state energy-dispersive Ge detector cut the wavelength (mostly $\lambda = 0.4$ Å) from a continuous spectrum. Figure 4.8 depicts the R_i against $(\cos\varphi)^{-1}$ dependence for the silicon single crystal. The values were determined from the oscillation period as described above.

Due to the fact that the atomic amplitude and temperature factor are present in $F(\boldsymbol{H})$ as a product, each multiplier can be determined provided the other one is already known. For homoatomic crystals the knowledge of the Debye temperature, measured by any other independent method, allows one to determine, in the harmonic approximation, the atomic scattering curve versus $\sin\theta/\lambda$ dependence. This allows one, in turn, to verify most reliably the adequacy of existing solid state theories as applied to the crystals which can be treated in such a manner.

Of most importance is the fact that the structure amplitudes determined by the pendellösung method do not depend on extinction, TDS, multiple scattering, etc. These factors drastically decrease the accuracy of structure amplitudes determined by other methods.

At the same time, there exist some other factors complicating the method discussed. First of all, the influence of sample microimperfection should be mentioned. Schneider (1988a) discussed the influence of oxygen content in Si samples on the pendellösung effect. Cummings and Hart (1988) have studied the influence of internal stress gradients in an Si crystal on this effect. Other fine effects also become noticeable, including the real part of the anomalous dispersion, the anharmonicity of atom vibrations, the nuclear scattering, etc.

Silicon is an ideal substance for the growth of nearly perfect single crystals. The pendellösung effect in silicon has been studied in a great number of works. Hattori *et al* (1965), Aldred and Hart (1973a, b), Takemura and Kato (1972), Treworte and Bonse (1984), and Graf and Schneider (1986) have measured a limited number of reflections to an accuracy of up to 0.1–0.5% in $|F(\boldsymbol{H})|$. In the most recent and already cited work of Saka and Kato (1986) the number of amplitudes increases and reaches 30. The set is quite valuable not only for theoretical verifications but also for new data treatment method developments. Subsection 3.6.1 describes the ED reconstruction based on the MEM. The accurate structure amplitudes of Saka and Kato (1986) for the Si crystal were used to work out and test this method.

The accuracy of the determination of $|F(\boldsymbol{H})|$ and $f(\boldsymbol{H})$ reaches a value at which very fine effects become noticeable. The latter include the real part of the anomalous dispersion, the anharmonicity of atomic vibration, the nuclear scattering, etc. At the same time, the use of pendellösung data has not revealed to date any new qualitative changes in the ED maps in comparison with ordinary x-ray experiments, with the exception of regions in the close vicinity of nuclei. For example, Takama *et al*'s (1990) pendellösung measurements in diamond did not change the shape of the $\delta\rho(r)$ curve but led to a decrease of the covalent peak in the interatomic space in comparison

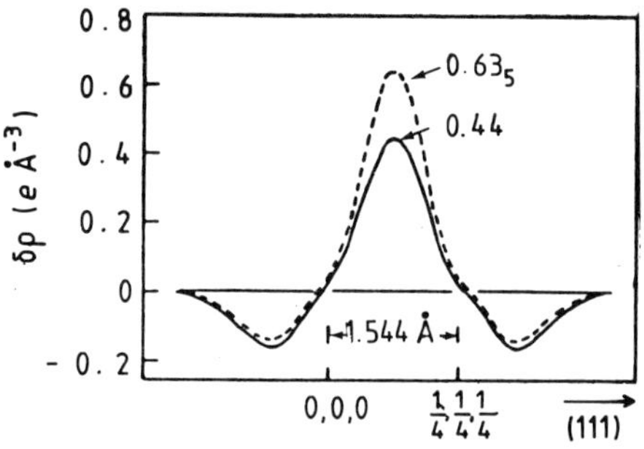

Figure 4.9 The distribution of deformation ED $\delta\rho$ in diamond plotted along the [111] direction through two nearest-neighbour atoms located at (000) and $(\frac{1}{4}, \frac{1}{4}, \frac{1}{4})$. The solid line is obtained from the work of Takama *et al* (1990) and the broken line from the publication of Dawson (1967b) in which Göttlicher and Wolfel (1959) data had been used.

with Dawson's (1967b) result (figure 4.9). Note that the treatment of the same structure amplitude set in the manner described above has led to agreement with pendellösung $\delta\rho$ result (Varnek *et al* 1982).

Similar investigations have been performed on the Ge single crystal. The seven reflections have been measured by the pendellösung effect by Takama and Sato (1981) and combined with the structure amplitudes, previously measured by several authors in the usual way, by Brown and Spackman (1990).

The pendellösung effect is pronounced with neutrons as well. Low absorption and small scattering amplitude enhance the effect with neutrons. Shull (1968) managed to measure the coherent scattering amplitude of silicon for thermal neutrons to an accuracy of 0.05%, whereas in the other methods the accuracy was about 1% only. The dynamic oscillations have also been observed in magnetic scattering of neutrons by magnetically ordered crystals α-Fe_2O_3 and $FeBO_3$ (Kvardakov *et al* 1988, Zalepuchin *et al* 1989).

4.4 High-energy electron diffraction

The high-energy (20 keV–1 MeV) electron diffraction allows us to investigate the thermally smeared charge and EP distributions in crystals. Owing to the small exchange effect, high-energy electrons are scattered by the crystal EP which is related to charge density by the Poisson equation (2.95). The interaction constant in electron diffraction is 100–1000 times more than in x-ray diffraction hence very small samples can be studied with this method. The other feature of electron diffraction consists of the small wavelengths used: these are 10–50 times smaller than in x-ray crystallography. Therefore, the radius of the Ewald sphere is very large, diffraction beams are directed 'forward' and scattering angles are less than 0.1 rad.

The Mott formula, an analogue of the Poisson equation in reciprocal space, relates the atomic x-ray and electron scattering amplitudes, f_x and f_e respectively: $f_e=(me^2)(Z-f_x)/(2h^2 \sin\theta/\lambda)$; Z and e are nuclear and electron charges, h is the Planck constant (Cowley 1991). The peculiarities of the atomic potential distribution (see appendix D) result in a fast decrease of the power of electron scattering with scattering angle. At the same time, the valence electrons contribute to the scattering amplitude f_e significantly. This means that nearly all electron reflections contain information on the EP distribution in internuclear space.

The Mott formula allows one to estimate experimentally the electron transfer by measuring $Z-f_x$, deriving f_e from an experiment, and calculating f_x in the framework of a crystal structural model. Fourier reconstruction and modelling of the EP, as it takes place in x-ray diffraction, can be used in practice as well. However, electron diffraction had not been widely used in crystal structure investigations until recently. First of all, large dynamical effects are manifest in electron scattering: the interaction of electrons with a crystal is very strong as compared with that of x-rays; therefore, multiwave processes influence the scattering strongly. These are considered in the frame of scattering perturbation theory (see subsection 3.1.1) and the experimental data are reduced to the case of kinematical diffraction theory as usually takes place in x-ray data processing. Corresponding methods were developed by Bethe (1928) and Blackman (1939) and generalized by Pinsker (1953), Vainstein (1964), and Cowley (1975).

Accurate electron beam intensity measurement was for many years another difficult problem in electron diffraction. Only very recently has the precise electrometric method of electron beam intensity measurement been introduced in classical transmission electron diffraction studies of polycrystalline compounds by Avilov (1976) and Orechov and Avilov (1989). The approach of Dvorjankina and Pinsker (1958) was also used by Avilov *et al* (1984, 1989) in order to sum up all waves scattered by small crystallites, the latter being spread over all orientations and positions in the film sample studied. Let us consider the results which are obtained with this approach.

One of the first investigations with the transmission electron diffraction technique was performed by Vainstein and Dvorjankin (1956) on the Li_2O crystal. Later, the MgO, CsBr, MnO, and CuI crystals were investigated by Storozhenko *et al* (1984), Avilov *et al* (1989), Avilov and Purmon (1990), Avilov *et al* (1994), and Abramov *et al* (1995). In the typical 75 keV experiment on an MgO thin film (Abramov *et al* 1995), 55 reflections up to $\sin\theta/\lambda = 1.279$ Å^{-1} were measured with a statistical precision of 1%. The two-wave scattering processes were taken into account according to Blackman. The high-angle refinement of the ionic superpositional structural model resulted in $R = 0.047$. Some characteristics of MgO obtained in this study are presented in table 4.2. The good agreement of electron diffraction thermal atomic parameters with accurate x-ray results should be noted first of all. The Debye characteristic temperature calculated from these parameters

Table 4.2 The results of MgO study by high-energy electron diffraction: thermal parameters, B, electrostatic potentials near nuclei, φ, and diamagnetic susceptibility, χ_d.

	B_{Mg} (Å^2)	B_O (Å^2)	φ_{Mg} (V)	φ_O (V)	$-\chi_d$ (10^{10} m^3 mol^{-1})
Electron diffraction	0.33(2)	0.35(2)	1179(2)	767(2)	2.1(1)
X-ray diffraction	0.323(6)	0.336(7)	1175 (1115[a])	832 (787[a])	2.2(1) (2.31[b])

[a]Non-empirical HF values.
[b]Experimental value.

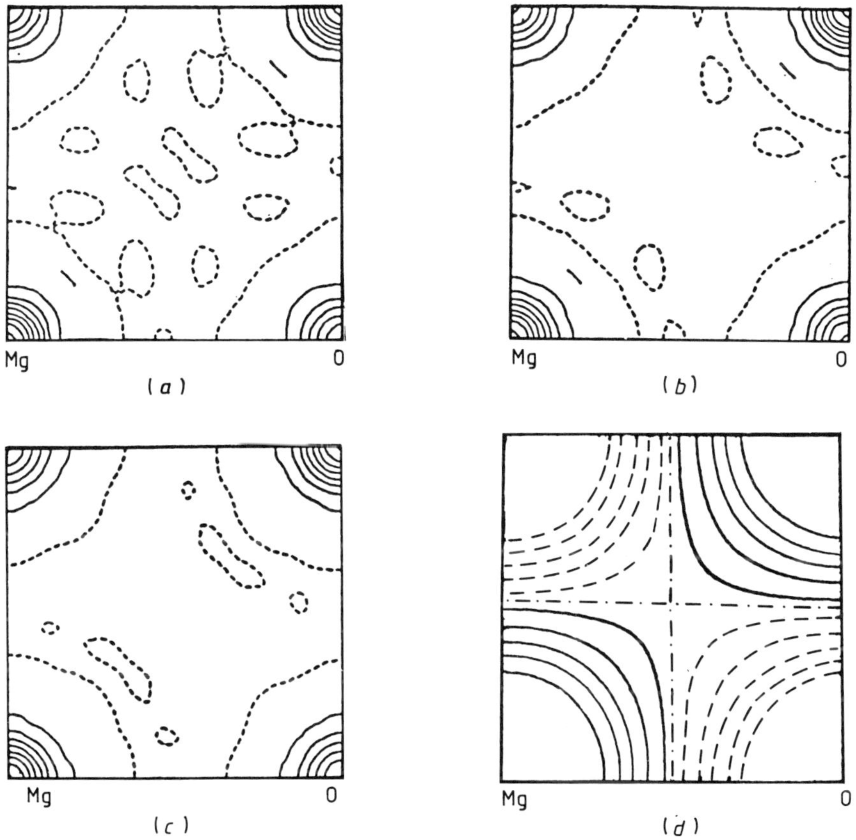

Figure 4.10 EP distribution $\varphi(xy0)$ in MgO calculated from high-energy electron diffraction data (Avilov *et al* 1989).

was equal to 710 K; this is close to the experimental value of 697 K. The values of EP at nuclear positions, reduced to zero temperature, agree very well too. Note that these values are averaged over a sphere with a radius of about 0.1 Å because the true quantum mechanical value of the EP on nuclei is negative.

In figure 4.10(*a*) the potential distribution map $\phi(xy0)$ in MgO calculated as a Fourier transform of kinematic electron diffraction structure amplitudes is depicted. Its comparison with x-ray and non-empirical HF EP distributions (figure 4.10(*b*) and (*c*)) demonstrates the quantitative agreement between EP as reconstructed from different sources. This is an important result. It means that experimental EP can be in practice used for more realistic modelling of the crystal potential in Schrödinger and Kohn–Sham equations.

At the same time, it is interesting to note that the experimental picture of EP is very different from the one which results from a formal $\pm 2\,e$ distribution of atomic charges (figure 4.10(*d*)). Meanwhile, these charges are usually used in crystal chemistry modelling of crystal energy.

The electron diffraction data allow also the satisfactory estimation of the diamagnetic susceptibility (table 4.2). This topic will be discussed in chapter 7 in detail.

In recent years other methods for accurate structure factor and reflection phase measurements by electron scattering have been develolped. These include the critical voltage technique (Nagata and Fukuhara 1967, Lally *et al* 1972, Sellar *et al* 1980, Thomas *et al* 1974, Fox 1984, Fox and Fisher 1986, 1987, 1988, Rocher and Jouffrey 1972, Arii *et al* 1973), the intersecting Kikuchi line method (Gjonnes and Hoier 1971, Mitsuhata *et al* 1984), and convergent-beam electron diffraction (Goodman and Lehmful 1967, Zuo *et al* 1988). All these methods are used in studies of bonding in crystals. Smart and Humphreys (1980) have suggested combining the accurate low-angle electron structure amplitudes with accurate high-angle values from other sources. This has subsequently been pursued by Zuo *et al* (1988), Fox *et al* (1990, 1991), and Tabbernor and Fox (1990), who have generated, by using the Mott formula, the deformation ED distributions for Be, Al, Si, Ge, Fe, Ni, Cu, Zn, GaAs, and β-NiAl. The structure amplitudes determined by the critical voltage technique and high-order theoretical structure amplitudes were used together to cover the absent diffraction information.

The obtained deformation EDs for the Al and Cu metals and β-NiAl alloy are presented in figures 4.11–4.13, respectively. The zero contours in the (110) plane of the Al map are ticked toward regions of electron depletion. This

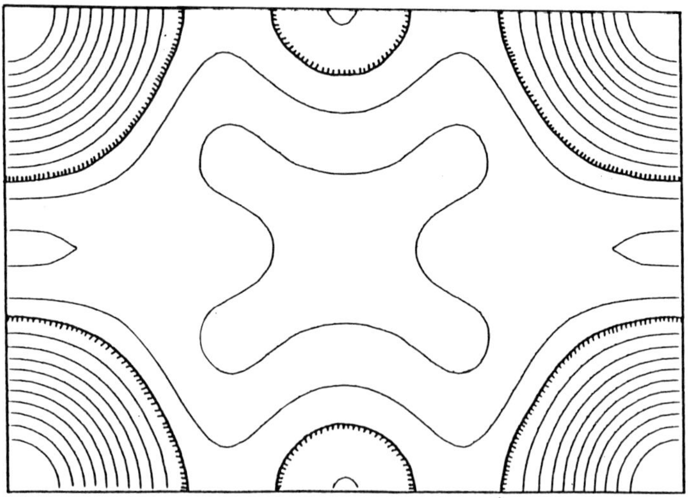

Figure 4.11 The (110) deformation ED distribution for Al. The contour spacing is 0.005 $e\,\text{Å}^{-3}$. The zero contours are ticked toward regions of electron depletion and the atomic sites can be clearly seen (Fox *et al* 1990).

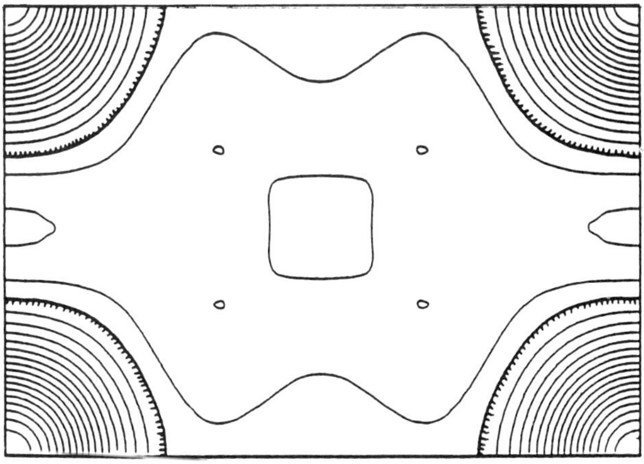

Figure 4.12 The (110) deformation ED distribution for Cu with contour spacing 0.04 e Å^{-3}. The zero contours are ticked as in figure 4.11 (Tabbernor *et al* 1990).

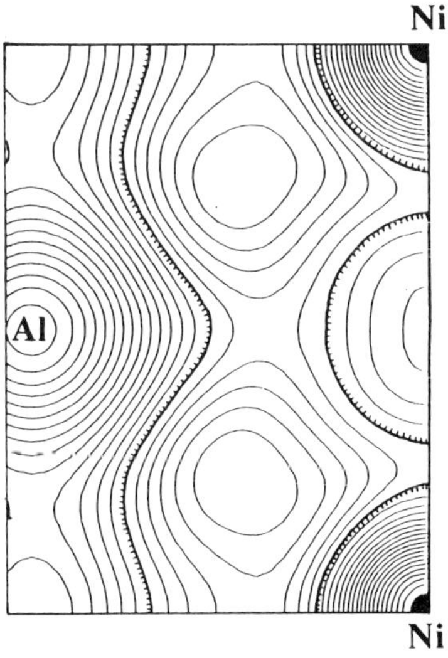

Figure 4.13 The (110) deformation ED distribution for β-NiAl with contour spacing 0.02 e Å^{-3}. The zero contours are again ticked toward regions of electron depletion (Fox and Tabbernor 1991).

indicates the close-to-spherical depletion of electrons from atom sites with build-up of ED between atoms such that there are maxima at the midpoints between nearest-neighbour atoms along ⟨110⟩. The density between second-nearest-neighbour atoms (along the ⟨100⟩ direction) is much lower.

The bonding in the Cu crystal (figure 4.12) is very similar to that in Al in spite of the fact that more electrons take part in the bonding (note that the contour spacing is much larger here). This is not surprising since several of the 3d and $4s^1$ electrons contribute to bonding in Cu, whereas for Al only $3s^2$ and $3p^1$ are valence electrons. Probably, this bonding scheme is characteristic of the fcc structures where nearest-neighbour interaction dominates.

The map in the (110) plane for β-NiAl (figure 4.13) suggests that bonding in this alloy arises as a result of close-to-spherical depletion of electrons from atomic sites with depletion between second-nearest-neighbour atoms in the $\langle 100 \rangle$ direction together with a large build-up of electrons around the midpoint between atoms along the $\langle 111 \rangle$ direction. The bonding in this compound is predominantly covalent with some metallic features.

We can conclude that electron diffraction allows us to generate important bonding information about different materials. It should be added that recently Zuo *et al* (1989a, b, 1990) have succeeded in accurately measuring structure factor phases. Therefore, the opportunities of electron diffraction are now approaching those of x-ray diffraction in the field of experimental determination of accurate charge densities and EP in crystals.

The electron scattering on molecules in the gaseous phase is beyond the scope of this book. The monograph edited by Hargittai and Hargittai (1988) and especially the review by Shibata and Hirota (1988) in that monograph cover this gap.

4.5 Schwinger scattering of neutrons

Along with the usual well known nuclear and magnetic interactions of neutrons and atoms (appendix K), some other much weaker interactions take place. The spin–orbit interaction between a moving neutron and the atomic electric field gives rise to a weak polarization-dependent scattering term, in quadrant with the nuclear scattering—the Schwinger scattering (Schwinger 1948). The Schwinger scattering is proportional to the difference of the nuclear charge, Z, and the conventional atomic x-ray scattering amplitude, $f(s)$ (Blume 1964). In centrosymmetric structures the Schwinger effect only gives rise to a polarization-dependent cross-section by interference with the imaginary part, b', of the nuclear scattering amplitude. The contribution to the observed flipping ratio in a polarized beam experiment is then of the order of $(b'/b)^2(m/m_N)$, where b is the real neutron nuclear scattering amplitude and m and m_N are the masses of the electron and neutron, respectively. Because of its small size, the Schwinger scattering has been of significance up to now only as a correction to small magnetic scattering amplitudes measured by polarized neutron techniques in cases where either b' is large or b is small. If, however, the structure is non-centrosymmetric, the Schwinger scattering gives a polarization-dependent cross-section by interference with the imaginary part of the nuclear structure factor and is only reduced over the normal nuclear scattering by a factor of the order of m/m_N (1/1833). A high sensitivity of the polarized neutron technique is then quite sufficient to allow reasonably precise measurements of the Schwinger structure factors to be obtained.

The Schwinger effect properties have been used in the study of charge distribution in gallium arsenide (Brown and Forsyth 1988). Accurate x-ray diffraction study of the ED in the heavier III–V compounds is difficult because of small interatomic electron transfer and high x-ray absorption. For neutrons, on the other hand, the absorption is generally small, and in the low-angle reflections, important for probing the valence state, the Schwinger form factor $Z - f(s)$ is primarily due to just the valence electrons. In the zincblende structure the nuclear structure factors are of two types, those with even $(h+k+l)$ being real and the remainder complex. The Schwinger structure factors are all complex, since the acentric components of atomic electron distributions give imaginary components to the electronic structure factor for the $(h+k+l)$ even reflections, so long as none of (hkl) are zero.

The neutron diffraction experiment has been carried out using a polarized neutron beam in the high-flux reactor at the Laue–Langevin Institute in Grenoble with a single-crystal sample at ambient temperature in a magnetic field of 0.56 T. Measurements of the flipping ratio of selected Bragg reflections were made at a neutron wavelength of 0.84 Å. The deviations of the flipping ratios from unity $(R-1)$ are given in the ideal case by the relation $R-1=4\,\mathrm{Re}(N^*X_S)/N^2$, where N^* is the complex conjugate of the nuclear structure factor, N is its modulus and, X_S is the Schwinger structure factor.

The set of real $(N^*X_S)/N^2 \times 10^6$ values, after corrections for extinction and calibration of R (with respect to the main reflection (200)), is given in table 4.3.

Inspection of the table shows that for all reflections, except (222), the free-atom calculation gives a larger value of absolute magnitudes than was observed. This corresponds to an increased Schwinger scattering factor for Ga and a reduced one for As relative to neutral atoms; in other words, this corresponds to charge transfer from Ga to As in accordance with some degree of ionicity. The degree of ionicity is difficult to quantify since the isolated ionic form factors do not give a realistic description of the charge

Table 4.3 Schwinger structure amplitudes for GaAs.

hkl	F (10^{-6})	F_{calc} (free atom form factor)
111	−156(22)	−253
113	+296(18)	+345
331	−342(57)	−460
222	−776(67)	0

density in a solid. The authors estimate that 0.1 electrons transferred from gallium to arsenic would be sufficient to account for the differences between the observations and calculation based on the free-atom form factors. The accuracy of observations corresponds to some 0.02 electrons in the difference between the Ga and As form factors at the (111) reflection.

Pietsch (1981) has presented for GaAs a bond charge model refined with x-ray diffraction data (see subsection 3.5.3). He introduced a bonding charge with an appropriate form factor in the 4e positions of the space group $F43m$ with a single positional parameter x. The corresponding charge density in the simplest form of his model is given by the Gaussian distribution $\rho(r) = A \exp(-a^2 r^2)$. The best fit was achieved with $x = 0.161$, $A = 0.36\ e\ \text{Å}^{-3}$, and $a = 1.24$ Å. The use of this model with these parameters made it possible to calculate the Schwinger term for the (222) reflection, giving -6.11×10^{-4} compared with the observed value of -7.76×10^{-4}. This bond charge also contributes to the (hkl) odd reflections and for the (111) reflection the use of Pietsch's parameters reduces the absolute magnitude of the calculated Scrawnier term down to 0.59×10^{-4}, which is much smaller than the observation.

The above comparisons demonstrate the extreme sensitivity of the Schwinger term to the bonding density details in III–V compounds. Finally, Brown and Forsyth (1988) have emphasized that this method, being sensitive to $Z - f(s)$ rather than to $f(s)$ and not being limited by a photoelectric absorption, would enable the charge density data to be obtained even for rather heavy compounds with the zinc-blende structure, such as InSb or even HgTe, which are hardly accessible by the x-ray technique.

5

MAGNETIZATION AND SPIN DENSITY

The magnetization and spin density (MD and SD, correspondingly) are another manifestation of an electronic system. The total MD can be presented as

$$\boldsymbol{m}(\boldsymbol{r}) = \int \Psi(X_1 \ldots)\hat{\boldsymbol{m}}(\boldsymbol{r})\Psi(X_1 \ldots)\, \mathrm{d}X_1. \tag{5.1}$$

The unit for $\boldsymbol{m}(\boldsymbol{r})$ is μ_B Å^{-3}, μ_B being the Bohr magneton. The MD operator, $\hat{\boldsymbol{m}}(\boldsymbol{r})$, consists of two parts: one is associated with electron spins and the other describes the orbital states of electrons. Accordingly,

$$\hat{\boldsymbol{m}}(\boldsymbol{r}) = \hat{\boldsymbol{m}}_S(\boldsymbol{r}) + \hat{\boldsymbol{m}}_L(\boldsymbol{r}). \tag{5.2}$$

The first term here is a local operator,

$$\hat{\boldsymbol{m}}_S(\boldsymbol{r}) = -2\sum_i \hat{\boldsymbol{s}}_i \delta(\boldsymbol{r} - \boldsymbol{r}_i) \tag{5.3}$$

where $\hat{\boldsymbol{s}}_i$ is the spin operator for the ith electron whose position is $\boldsymbol{r}_i$. The second term gives the orbital MD. It is a non-local operator of the form

$$\hat{\boldsymbol{m}}_L(\boldsymbol{r}) = (cr)^{-1} \int_r^{\infty} \xi(\boldsymbol{r}/r) \times \boldsymbol{J}(\xi \boldsymbol{r})\, \mathrm{d}\xi \tag{5.4}$$

where ξ is the variable of integration and $\boldsymbol{J}(\boldsymbol{r})$ is the orbital current density with the form

$$\boldsymbol{J}(\boldsymbol{r}) = -(2\pi c/h)\sum_i [\boldsymbol{p}_i \delta(\boldsymbol{r} - \boldsymbol{r}_i) + \delta(\boldsymbol{r} - \boldsymbol{r}_i)\boldsymbol{p}_i] \tag{5.5}$$

where $\boldsymbol{p}_i$ is the momentum of the ith electron. The orbital term can be pictured as arising from the current density due to the circulation of all the electrons about the point being considered.

The MD reduces to the SD when the orbital magnetic moment of electrons is zero. This is almost the case for the transition ions where the orbital moment of the d electrons is most often 'quenched' by the crystal field, but not at all for the f electrons of many rare earths and actinides, where the orbital magnetization is as important as the spin magnetization.

The SD represents directly the ED of unpaired electrons. If the electron charge density is a sum

$$\rho(\boldsymbol{r}) = \sum_i [\rho(\boldsymbol{r}_i s^{\alpha}) + \rho(\boldsymbol{r}_i s^{\beta})] \tag{5.6}$$

then the z component of SD is a difference

$$\rho_{s,z}(\boldsymbol{r})=\sum_i [\rho(\boldsymbol{r}_i s^{\alpha})-\rho(\boldsymbol{r}_i s^{\beta})] \tag{5.7}$$

the x component being equal to

$$\rho_{s,x}(\boldsymbol{r})=2\sum_i \rho(\boldsymbol{r}_i s^{\alpha}, \boldsymbol{r}_i s^{\beta}). \tag{5.8}$$

Here indices α and β indicate the different directions of spin vectors along the z axis.

Being a periodical function in a crystal lattice, magnetization density $\boldsymbol{m}(\boldsymbol{r})$ can be expanded into a Fourier series as

$$\boldsymbol{m}(\boldsymbol{r})=\frac{1}{V}\sum_{\boldsymbol{H}} \boldsymbol{F}_m(\boldsymbol{H})\exp(\mathrm{i}2\pi\boldsymbol{H}\cdot\boldsymbol{r}) \tag{5.9}$$

where $\boldsymbol{F}_m$ are the magnetic neutron structure amplitudes. The sum is over all vectors $\boldsymbol{H}$, including $\boldsymbol{H}(000)$ which represents the total magnetization of the unit cell.

The direct methods of measuring the MD and SD distributions are the magnetic neutron scattering techniques and, especially, the polarized neutron magnetic diffraction (PND). There are well developed theories and practices for neutron magnetic scattering, advantages and drawbacks of which are described in detail by Izyumov and Ozerov (1969), Marshall and Lovesey (1971), Bacon (1975), Forsyth (1979), Brown (1980), Schweizer (1980), Izyumov *et al* (1991), and others. Appendix K presents the essence of the modern PND technique and measurement procedures.

Metals and metal oxides were the first to be studied with these techniques. A wide range of magnetic order has been found. Analysing the atomic neutron scattering amplitudes (form factors) a large variety of electronic properties of metal atoms have been established: for example, the nearly ideal spherical symmetry of atomic 3d electron distribution in Co; the predominant orbital e_g electron occupancy in Fe and t_{2g} orbital occupation in Ni; the 'quenched' and nearly 'quenched' orbital contribution in d metals and significant orbital contribution in f elements.

Special computational methods were employed to study the MD distribution in the internuclear space in iron and some intermetallic compounds. Magnetization of oxygen ions by the influence of magnetic ordering was observed in $Y_3Fe_5O_{12}$. The participation of magnetic electrons in covalent bonding was also investigated. All of these findings provide a significant contribution to solid state physics and chemistry in the 1960–1970s.

We will restrict ourselves to the relatively new aspects of SD investigation in 3d metal molecular complexes.

The 3d electrons are of special interest in this respect. They take part in so-called exchange interactions (direct and indirect) which provide the magnetic ordering in crystals. At the same time, 3d electrons take part in chemical bonding. These two phenomena are closely connected (see e.g. Forsyth 1979). The PND method is very sensitive to the unpaired 3d electrons, whereas x-rays 'feel' all atomic electrons, core included. This makes PND a sensitive and precise tool for the investigation of 3d metal compounds. The range of 0.02–0.06 e Å^{-3} can be accepted as the accuracy of SD determination.

The experimental procedure of MD investigations consists of aligning the paramagnetic moments in the crystal sample at a very low temperature using a strong external magnetic field. The magnetic structure amplitudes, the coefficients of the series (5.9), are then measured by means of the PND technique. Quantitative estimates of the moment contribution of specific electron orbitals (in the framework of a defined electron model) to the MD and/or SD are the final results of such experiments.

The data treatment in the framework of the multipole expansion model is mostly used (see e.g. Chandler *et al* 1982a). The expansion coefficients are determined by LS fitting to the experimental magnetic structure amplitudes, derived from the PND measurements after correction for a number of effects which influence the magnetic scattering process, similar to those discussed in section 3.3. Having indirect physical and chemical meaning, the expansion coefficients can be transformed into orbital population parameters according to the procedures suggested, e.g. by Varghese and Mason (1980) and Holladay *et al* (1983). On the other hand, the experimentally determined expansion coefficients can be used to present the results in the form of SD and/or MD maps. These maps can be compared with theoretical ones. The thermal smearing may be taken into account in such a comparison.

Two cases can be considered: first, the fully 'quenched' orbital moment, and, second, the orbital contribution to the total atomic magnetic moment exists. In the latter case, the coupling between the spin and orbit magnetic moments (appendix K) leads to the total magnetization density.

The orbital contribution is usually small in the first-row transition metals. In this case there is only the z component of $\boldsymbol{F}_M$ measured by the flipping ratio (formula (K.10)). For considerable orbital contribution, the flipping ratio depends also on $F_{M,x}$ and/or $F_{M,y}$ in a complicated manner (Barnes *et al* 1989).

If $F_{M,x}$ and/or $F_{M,y}$ are small (this is valid for an orbitally non-degenerate ground state) the expression (K.10) produces a value of $\boldsymbol{F}_M$ which is very close to the magnitude of $F_{M,z}$. In other words, if the effect of the orbital magnetization is to reorient the total magnetization vector away from the external magnetic field direction to a minor extent, this misorientation being field and temperature dependent, the conventional PND data treatment produces a magnetic structure factor component in the direction of the magnetic field only. In this (dipole) approximation, the influence of the orbital contribution can be accounted for quantitatively by introducing the multipler $g/2$ which is close to unity in the majority of cases. In other cases it may be necessary to take the orbital contribution into account in a more sophisticated manner.

If there is no orbital angular momentum component associated with the ground term then the SD is, in the absence of magnetic order, a vector aligned along the magnetic field direction. Its value, but not its direction, varies from point to point in space. The orbital magnetization is, in general, not oriented in the direction of the magnetic field. The coupling between the spin and orbital parts of the MDs leads to a spin component which is also not aligned with the magnetic field. Even if the total magnetic moments are aligned with the field, the densities may not be: thus the moments have both collinear and non-collinear components. In the presence of orbital magnetization, the total magnetization varies in magnitude and direction in space. This fact has important implications for interpretation of the PND experiment as the usual PND analysis assumes only collinear MDs and scalar structure factors.

The number of transition-element complexes studied so far by the PND technique is not very large. There are compounds which contain the $[CoCl_4]^{2+}$ ion (Figgis *et al* 1987b, 1980a, c, 1989, Figgis and Reynolds 1990, Chandler *et al* 1982a, b, Chandler and Philips 1986, Barnes *et al* 1989, Mason 1986), Tutton salts of 3d metals with $M(H_2O)_6$ groups (Fender *et al* 1986a, b, Figgis *et al* 1990, Deeth *et al* 1989), and others.

A detailed treatment of the SD of some ions was performed by Figgis and Reynolds (1986) (table 5.1). The increasingly sophisticated levels of the description were considered: (*a*) crystal-field theory; (*b*) simple empirical ligand-field or Wolfsberg–Helmholtz (1952) LCAO model.

In the crystal-field model, where orbital overlap between the central atom and ligands is ignored, the effective charge sets up an electric field at the central metal ion position. This polarizes it so as to place ED away from the metal–ligand axes. According to crystal-field theory, the Cr^{3+} ion in the octahedral crystal field should have $t_{2g}^3 e_g^0$ electron configuration. For $Cr(CN)_6^{3-}$ (Figgis *et al* 1985, 1987a), the electronic configuration observed by x-ray, $t_{2g}^{2.5(2)} e_g^{1.2(2)}$, is still somewhat similar to $t_2^3 e_g^0$ (table 5.1). Charge and spin transfers are less than 0.5 *e* per ligand. However large diffuse charge and spin populations were found by PND which are labelled as '4s'. In the strongly bonded case of $Ni(NH_3)_4(NO_2)_2$ (Figgis *et al* 1981, 1983a, Delfs and Figgis 1990) the 3d radius of the Ni atom is contracted, while in the weaker bond case of $Cr(CN)_6^{3-}$ (Figgis *et al* 1985, 1987a), it is expanded.

The crystal-field model only gives a rough explanation of the data. The experimental accuracy of PND is sufficient for a more realistic theoretical account of the experimental data. This can be done with the MO theory. In empirical MO theory, a metal–ligand σ bonding wave-function in octahedral symmetry (which is commonly found) is written as a linear combination of the metal 3d orbitals and a set of ligand orbitals χ:

$$\varphi = (1/N)\{|3d_\sigma|\rangle + \lambda_\sigma |\chi\rangle\}. \tag{5.10}$$

There is a corresponding orthogonal antibonding, σ^*, orbital and accordingly

Table 5.1 Some experimental charge and spin populations and transfers for a range of simple transition metal molecules and ions compared with crystal-field theory predictions (Figgis and Reynolds 1986).

		X-ray results				PND results				
		$Cr(CN)_6^{3-}$	$Ni(NH_3)_4(NO_2)_2$	$FeCl_4^-$	$Ni(tu)_4Cl_2$	$Cr(CN)_6^{3-}$	$Ni(NH_3)_4(NO_2)_2$	$CoCl_4^{2-}$	$FeCl_4^-$	CrF_6^{3-}
Metal 3d configuration	expt	2.5(2)	5.0(1)	3.9(1)	5.0(2)	2.30(3)	−0.03(8)	2.86(3)	2.5(*j*)	2.66(5)
	t_{2g} theory	3	6	3	6	3	0	3	3	3
	exp	1.1(2)	2.6(1)	1.7(1)	2.5(2)	0.12(4)	1.71(8)	−0.22(3)	1.7(*j*)	−0.06(5)
	e_g theory	0	2	2	2	0	2	0	2	0
Metal 3d radius		0.93(3)	1.01(1)	1.062(4)	1.01(1)	1.082(4)	0.92(1)	0.961(5)	1.01(1)	1.001(7)
Ligand/charge spin transfer (per ligand, to metal)	axial or entire (*k*)	0.27(7)	0.38(8)	0.37	0.5(2)	0.04(8)	−0.11(1)	−0.078(6)	−0.22	0.00(1)
	equatorial	—	0.11(6)	—	0.2(2)		−0.08(1)	—	—	—
Metal '4s'		1.4(4)	1.5(2)	1.6(3)	1.0(7)	0.83(7)	0.1(1)	0.06(5)	0.1(4)	0.4(1)

bonding and antibonding π molecular orbitals. A total of four parameters arises, two covalence parameters, λ_σ and λ_π, and two overlap integrals, $S_\sigma = \langle 3d_\sigma | \chi \rangle$ and similarly defined S_π. These parameters are treated as empirical and measured to fit the experimental data. For the Cr compounds listed in table 5.1 the covalence order Cr–CN > Cr–F was found in accordance with expectations (Figgis and Reynolds 1986).

Although this empirical parametrization procedure produces more satisfactory explanations of results than does the crystal-field model, it still has severe limitations. The covalence parameters evaluated for the molecule $Ni(II)(NH_3)_4(NO_2)_2$ from PND on the one hand and from charge densities on the other do not agree well; the 'spin' covalence is significantly less than the 'charge' covalence (Figgis *et al* 1983a). The spin densities also do not show negative regions. The empirical MO models put spin in a single or possibly degenerate molecular spin orbital (φ_i), and thus the SD,

$$\rho_s = \langle \varphi_i | \varphi_i \rangle \tag{5.11}$$

is positive everywhere in the model. In addition, spin appears in places where it is expected to be zero (e.g. the 3d(e_g) orbitals of the $[Cr(III)(CN)_6^{3-}]$ ion). For these reasons one must regard such simple treatments as unsatisfactory, even for the parametrization obtained.

As far as quantum chemistry is concerned, many all-electron calculations on relevant systems of the restricted Hartree–Fock (RHF), unrestricted Hartree–Fock (UHF) (in which some decoupling of spins on the same orbital is allowed), and X_α (both 'muffin-tin' and discrete-variation (DV) types) have been made. Calculations including configurational interaction from RHF initial wave-functions are available for some small molecules but not yet for transition-metal systems. In table 5.2 RHF and DV X_α results for the $[Cr(CN)_6^{3-}]$ ion (Figgis *et al* 1985, 1987a) are listed in comparison with PND experiments.

It is clearly seen from the table that the RHF calculations give a system that is much too ionic. In addition, RHF calculations necessarily give an SD which is positive everywhere, in contradiction to the experimental observations. These latter seem to show that electron correlation effects, as demonstrated by the presence of negative spin, are as important as covalent spin transfer. This confirms the conclusion, already developing from fits to spectroscopic data, that RHF calculations are a rather poor description of the bonding for transition-metal systems. The UHF calculations are an improvement and give spin populations in qualitative agreement with observations (Chandler *et al* 1982b), but bonds are still too ionic, particularly in the more covalent cases such as $Ni(NH_3)_4(NO_2)_2$. The DV X_α calculations appear to give a more covalent description, although this conclusion is based solely on the case of the $[Me(CN)_6^{3-}]$ ions (Me = Cr, Fe) (Figgis *et al* 1985, 1987a, Daul *et al* 1988).

It is obvious that in the future good-quality calculations must allow for

Table 5.2 *Ab initio* theoretical and experimental spin and charge distribution results for $Cr(CN)_6^{3-}$ (Figgis and Reynolds 1986).

	Spin			Charge		
	Expt	DV X_α	RHF	Expt	DV X_α	RHF
Cr $3d(t_{2g})$	2.30(3)	2.62	2.94	2.5(2)	2.88	2.97
$3d(e_g)$	0.12(4)	0.30	0.0	1.1(2)	1.28	0.81
'4s'	0.83(7)	0.12	0.0	1.4(4)	0.62	0.63
C σ	−0.044(8)	−0.07	0.0	−0.26(6)	−0.17	−0.45
C π	−0.044(8)	−0.03	0.0			
N σ	−0.034(7)	0.00	0.0	−0.47(5)	−0.53	−0.34
N π	−0.087(4)	0.09	0.01			

spin polarization in order to reproduce the negative-spin regions in transition-metal complexes.

Detailed examination of the $[CoCl_4]^{2-}$ ion has been made in a series of works cited above. The influence of basis sets and the method used in the theoretical SD calculations was investigated by Chandler and Phillips (1986), The charge and spin densities were calculated as well as bond lengths and structure factors. In all cases the expectation value of $\langle \hat{S}^2 \rangle$ differed from the ideal value of 3.75 by at most 0.2%, indicating that the spin contamination introduced by the use of the UHF method is small.

Figure 5.1 illustrates the influence of the computational peculiarities on the SD. Figure 5.1(*a*) represents the scheme of the $CoCl_4^{2-}$ ion and the planes examined. The map of figure 5.1(*b*) is an SD section by the σ_v plane calculated by the RHF method with an extended basis set. Figure 5.1(*c*) presents the result with the same basis set but in the UHF approximation. Figure 5.1(*d*) presents the same UHF calculation but in more flexible basis set. The differences can clearly be seen in comparison of these maps. Finally, figure 5.1(*e*) presents the σ_d section of the SD with the same calculation scheme as in figure 5.1(*d*).

On the other hand, figure 5.1(*f*) depicts the experimental σ_v plane SD map for Cs_3CoCl_5 crystal. This map has been produced from a multipole expansion treatment of experimental data and the determination of expansion coefficients by LS fitting. The calculated magnetic structure amplitudes have then been used to construct the Fourier transform to give the SD maps. The overall agreement of experiment with UHF theory can be seen by comparing the experimental map with the theoretical calculations of different levels. However, there are significant discrepancies. They consist mostly of some positive SD regions in the experimental maps which are not confirmed in the theoretical ones. The positive SD region is larger behind the Cl atom than in front of it, whereas the opposite is true in the theoretical pictures. A difference is also seen in the Co–Cl separation region.

Chandler and Phillips (1986) stated that there is agreement between the theoretical and experimental SD maps at the qualitative level only. It was found also that the SD is very stable to basis set change. At the quantitative level considerable disagreement was found which is attributed to inadequacies in the theoretical treatment. The major deficiencies in order of the probable importance for the $CoCl_4^{2-}$ ion are (1) neglect of the effect of neighbouring ions in the crystal, (2) neglect of correlation effects, and (3) an inadequate compensation for spin–orbital coupling. These circumstances may and should be taken into account more properly.

A further step in MD investigation consists of the PND measurement of orbital participation in the overall magnetization. The flipping ratio (K.10) in this case depends on $F_{M,x}$ and $F_{M,y}$ components. Figure 5.2 presents the MD maps for the $CoCl_4^{2+}$ ion in Cs_3CoCl_5 (Barnes *et al* 1989). The z component of spin magnetization in the plane σ_d (see figure 5.1(*a*) and figure 5.2(*a*)) is reconstructed in dipole approximation ignoring the $F_{M,x}$ and $F_{M,y}$ contribution to the flipping ratio, i.e. it predominantly presents the SD distribution. Whereas this picture is very similar in shape to the usual d_{xy} orbital density, there are no true nodes, only minina, in the x and y directions. Figure 5.2(*b*) presents the difference map obtained by subtracting a theoretical 'spin-only' contribution from the map of figure 5.2(*a*). This means that this picture shows the changes in the density brought about by spin–orbit coupling. The changes are quite small, corresponding to rearrangement of about 0.3 e, and so are not significant within the error limits of the experimental data. The corresponding orbital MD can be seen in figure 5.2(*c*).

Another illustration of PND studies of chemical bond peculiarities is provided by the results obtained for the series of Tutton salts referenced above. The crystals contain the octahedral $[Mn(II)(H_2O)_6]^{2-}$ units. In some cases deuterated crystals were used to reduce the background from hydrogen incoherent scattering of thermal neutrons. The M(II) ions are located at centrosymmetric (000) and $(\frac{1}{2}\frac{1}{2}0)$ positions (space group $P2_1/a$) so reflections with $h+k$ odd contain no contribution from moments centred on these points. The fact that small but measurable flipping ratios are observed provides direct evidence that the magnetic moment distribution of the free ion is perturbed in the solid and that some spin delocalization occurs.

All the data for these compounds have been collected in a magnetic field of 4.6 T at temperatures of 2–4.2 K. The nine-parameter model was used to describe the MD. The parameters related to 3d, 4s, and '4d' orbitals of the 3d ion, sp^2 hybrid orbitals of oxygen atoms, 1s H atom orbital, the M–O orbital overlap, and the radial expansion/contraction. The $(ND_4)_2Ni(SO_4)\cdot 6D_2O$ investigation results are listed in table 5.3, where the population numbers are presented against each orbit (Fender *et al* 1986a).

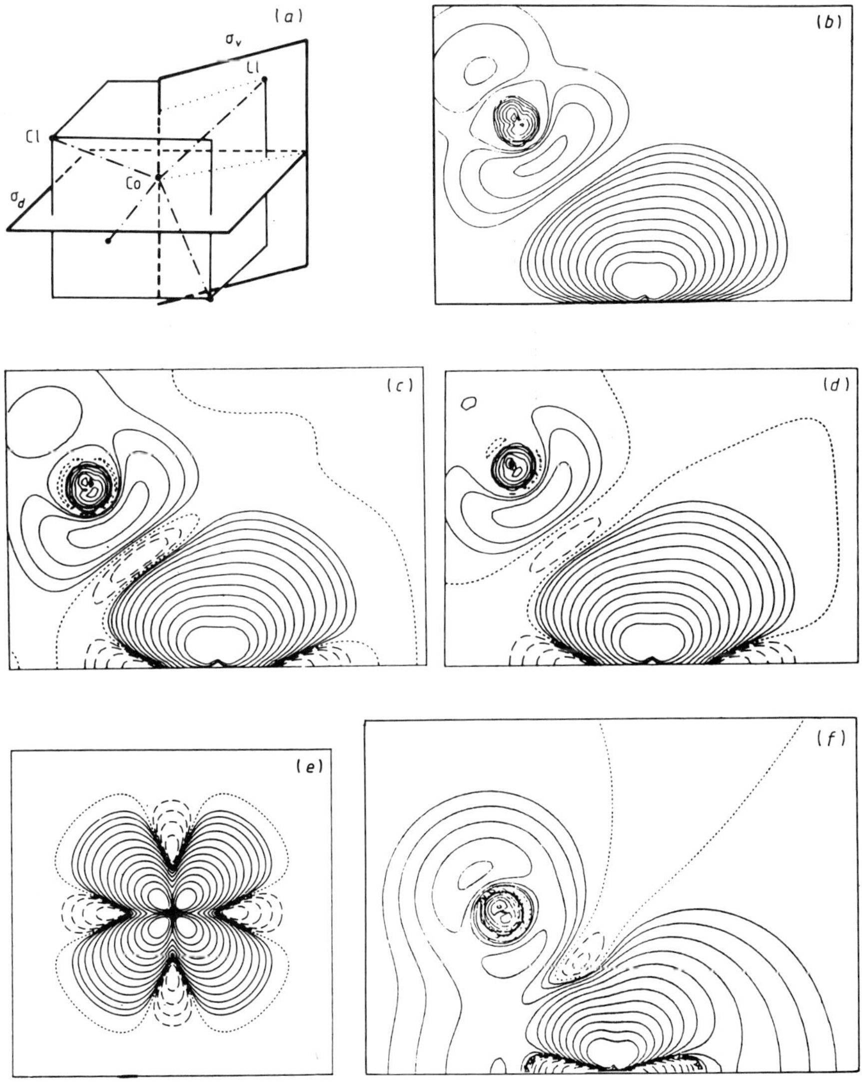

Figure 5.1 The SD distribution in the $CoCl_4^{2-}$ ion in Cs_3CoCl_5. (*a*) Planes in the $CoCl_4^{2-}$ ion for which density plots are given. (*b*) Calculated spin density in plane σ_v; full lines represent α spin density, long-dashed lines β spin density, and the short-dashed line is the zero contour; RHF method; the basis set is for Co^{2+}, 14s 9p 5d/8s 4p 2d and for Cl, 12s 5p/6s 4p. (*c*) The same as (*b*), UHF method. (*d*) The same as (*c*) with a better basis set: for Co^{2+}, 14s 10p 6d/8s 5p 3d; for Cl, as in (*b*). (*e*) The same as (*d*), section by plane σ_d (Chandler and Phillips 1986). (*f*) The experimental SD in the dipole approximation; the contours are logarithmic (12th $\approx$ 3400 spins nm^{-3}, first $\approx$ 1.64 spins nm^{-3}, a factor of two between each) (Chandler *et al* 1982b).

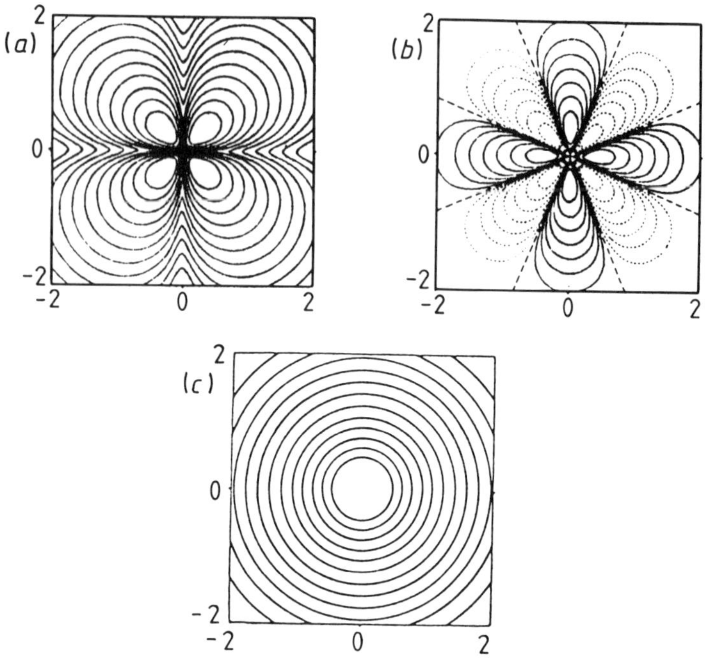

Figure 5.2 The experimental MD of the $CoCl_4^{2+}$ unit in Cs_3CoCl_5. Lengths are in au (appendix A). The contour levels increase by a factor of two with the 10th level 0.125 μ_B $(au)^{-3}$. The zero level is not shown. The magnetic field is directed along the *z* axis. The section corresponds to the σ_d plane presented in figure 5.1(*a*). (*a*) The spin MD; (*b*) the difference map: the theoretical 'spin-only' component has been subtracted from figure 5.2(*a*); (*c*) the orbital MD (Barnes *et al* 1989).

A summary of the bonding features in some of the 3d metal Tutton salts has been given by Figgis *et al* (1990) and by Deeth *et al* (1989) as follows. The compounds are paramagnetic, even at low temperature, due to the smallness of the interatomic magnetic exchange. V(II) and Mn(II) atoms have a $3d(t_{2g})$ predominant population and substantial in-plane π spin delocalization leading to spin transfer to diffuse metal and O-N(D) orbitals. Ni(II) and Mn(II) have a $3d(e_g)$ population, but σ spin transfer in antibonding orbitals; the occupation of diffuse metal and oxygen lone pairs is less. The out-of-plane π interaction is negligible for all ions.

As far as the iron (II) compounds are concerned (Figgis *et al* 1990), a substantial orbital contribution to the total magnetic moment has been found and the canting of these moments away from the external magnetic field direction has been established. The canting angles are 37 and 45° with respect to *b* and *c** axes. The compound is also paramagnetic, even at very low temperature. The orbital moment is approximately fixed with respect

Table 5.3 Parameters for the $Ni(D_2O)_6^{2+}$ ion in $(ND_4)_2Ni(SO_4)_2 \cdot 6D_2O$ based upon the $F_M(hkl)$ data set. Estimated standard deviations are given in parentheses. (Reprinted with permission from Fender *et al* 1986.)

Atom	Orbital	Population/spin
Ni	t_{2g}	−0.06(4)
	e_g	1.64(3)
	4s	0.38(15)
	κ_{3d}	0.99(1)
O	$(sp^2)_1$	0.037(2)
	$(sp^2)_{2,3}$	0.009(2)
	p_y	0.001(2)
D	1s	0.003(7)
'overlap'	Gaussian	−0.047(10)
Other quantities		
$R=0.113$, $R_w=0.061$, $\chi=1.98$		

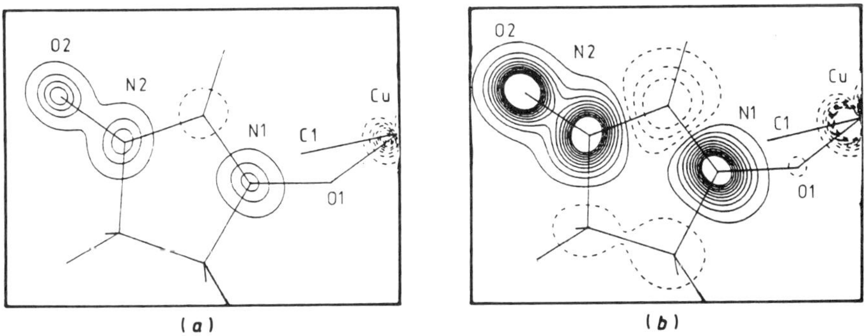

Figure 5.3 The SD distribution in $CuCl_2 \cdot C_{26}H_{36}N_4O_4$.

to the oxygen framework with strong spin–orbit coupling. Substantial delocalization of the MD away from Fe atoms has been observed, 6.5% onto the D_2O ligand and 4.5% into the metal–oxygen overlap region. Both σ and π bonding exist, with σ dominant, between oxygen and deuterium atoms. A 19% 3d electron contraction from the free state has also been found. Many of the observations mentioned are consistent with the data of other methods.

Recently, some paramagnetic free-radical compounds have been studied by the PND method. The investigation of the SD distribution in two dialkylnitroxides and in two nitronylnitroxide copper complexes should be mentioned in this respect (Ressouche 1991). The investigations were performed at 2–13 K in a field of 4.6 T. It was found that the main SD was produced by the N–O groups in all compounds studied with several hundred $m\mu_B\,Å^{-2}$ in maxima in two-dimensional maps. Figure 5.3 presents a two-dimensional projection of the SD distribution in copper (II) dichloro(2-phenyl-4,4,5,5-tetramethylimidazoline-1-oxide), $CuCl_2{\cdot}C_{26}H_{36}N_4O_4$. The essential difference was found in oxygen atom spins. Attempts have been made to analyse the difference from the point of view of the magnetic interactions.

A number of free radical organic compounds have been investigated further by PND (Bordeaux *et al* 1993, Boucherle *et al* 1992, 1993, Ressouche *et al* 1993a, b). magnetically ordered purely organic substances have been discovered among them (Zheludev *et al* 1994a, b).

6

ELECTRON DENSITY AND THE CHEMICAL BOND

6.1 Concepts of the chemical bond

When substances are supposed to consist of small particles (atoms, molecules) it is necessary to explain the existence of forces which can hold a set of these particles together. In ancient times, Lucretius Titus Carus (see Carus, translation of 1951, and e.g. Mackay 1990) had referred to Epicure in his attempt to describe an interparticle bonding in matter with the help of 'hooks' as holders. Boscovich (1758) also considered matter to consist of small particles with interactions between them. He suggested, besides, that these interactions are described by a law which has an asymptocity $\sim R^{-2}$ at large R, by analogy with Coulomb's law. The notion of the chemical bond appears when the interatomic interactions in stable many-atomic systems are described. This notion was first introduced about two centuries ago, after the appearance of Dalton's atomistic hypothesis, and it has been revised periodically as new experimental facts became available. Berzelius first guessed that the main cause of interatomic bonding is the mutual neutralization of oppositely charged particles. With the development of structural organic chemistry in the mid-nineteenth century (Kekule, Van't Hoff, Butlerov) this idea was completed when atomic directed forces were proposed, which resulted in the concept of the localized chemical bond between particular pairs of atoms. At the same time the capacity of an atom to be bonded to a finite number of other atoms had been established, yielding the notion of valence. At this stage the chemical bond was considered to be a result of the mutual saturation of a single valence from each interacting atom which is localized between the interacting atoms. It was proposed that multiple bonds were caused by the saturation of several valences from the two interacting atoms. The bond was depicted diagramatically as a single line for each valence between the chemical element symbols; correspondingly, molecules were depicted by structural formulae in which a number of atoms were linked by single or multiple bonds. Later, the valence of an atom was related to its position in the periodic table. Werner in 1893 showed that the valence model failed to explain the bonding in inorganic compounds. He developed the concept of the coordination complex, in which the central atom can form more bonds with ligands than would be expected from its formal valence. The coordination number notion was introduced in this model.

The discovery of the electron by Thomson in 1897 and the development of the planetary atomic model by Rutherford and Bohr in 1911–1913 forced the return to the early Berzelius hypothesis. Kossel (1916) in his first electrostatic valence theory postulated that a chemical bond is created by the displacement of electric charge between atoms. He proposed that the electronic structure of interacting atoms should have inert atom configurations. As a result, stable substances should consist of positively and negatively charged particles, resulting in an ionic model. This model led to the idea that ions have well defined radii, and to the concept of close packing of ions in crystals. Taking ions as undeformed, charged spheres, Born developed a theory which permits one to calculate the crystal energy and some energy-dependent crystal properties (see e.g. Born and Huang 1954). Further, the theory was modified by Fajans in 1924 in order to take into account an ionic polarization in the crystal field.

In the ionic model the atoms from the left-hand side of the periodic table (groups Ia–IIIa) and transition elements are assumed to lose their electrons, which are accepted by the atoms in the right-hand side of this table. As a result, the atoms are assumed in the ionic model to be charged particles: ions Na^+, Mg^{2+}, Al^{3+}, Ni^{2+}, F^-, O^{2-}, etc. The formal charges are assumed to have only integer values. The ions of transition elements and lanthanoids can have variable valence (e.g. Fe^{2+} and Fe^{3+}) while some elements, such as sulphur, can form both positive and negative ions in the ionic model.

The simple ionic model can describe properly general features of only binary compounds, whereas the properties of multiply charged ions in coordination compounds cannot be described in this framework (Levin *et al* 1969). The charges of ions in complexes are non-integer with values less than $+2\,e$ for cations and $-1\,e$ for anions. At the same time, a significant ED is concentrated between nuclei in coordination compounds. This effect has quantum mechanical (not electrostatic) origin and is not predicted by the ionic model. As was mentioned earlier, electrostatics only are not able to explain the stable state of a system of charges (Ernshow theorem) or, moreover, to describe chemical bond formation (Teller 1962).

Lewis (1916) and, later, Langmuir (1919) and Sidgwick (1927) completed the development of the electronic theory of valence. Lewis suggested that atoms can be represented in the form of the stable core and variable valence parts. Each bond is formed by pairs of valence electrons which belong concurrently to both bonded atoms. Each electron pair can be shared between two atoms in equal portions (covalent bond) or displaced entirely (ionic bond) or partly (polar covalent bond) to one of them. Each single valence corresponds to a single atomic valence electron. The valence of an atom is equal to the number of its electrons which take part in bonding in the compound considered; this means that the valence depends not only on the kind of atom, but also on its environment in a given molecule or crystal. Some rules, such as the octet rule, the 18-electron rule, etc, were formulated in order to describe the behaviour of atoms in different compounds. Thus, two limiting cases of the chemical bond can be defined: the covalent bond and the ionic bond. There are, however, many intermediate cases.

Thus, it was found in the 1920s that the chemical bond can be described in terms of certain values. These are the electrostatic atomic charges, which may be regarded as effective ones, the parameters which describe the deviation of the atomic shape from sphericity, the degree of polarity of the bond, and the sizes (radii) of ions, the relationship between which determines the crystal and molecular structure of a compound. Pauling introduced later one more notion—the electronegativity, which is a measure of the ability of an atom to attract electrons to itself (see e.g. Pauling 1960). The physical nature of the bonding, however, was still vague.

Heitler and London (1927) were the first to illuminate this problem. They used the quantum mechanical perturbation theory and solved the problem of describing the change in the energy of two hydrogen atoms when they approach each other, reaching the distance typical for the H_2 molecule. That was the starting point of quantum chemistry. The method used by Heitler and London was later called the valence bond method (subsection 2.1.7). The wave-function of the H_2 molecule was approximated by a linear combination of isolated atomic, one-electron wave-function products (VB structures); the atomic interaction potential was used as the perturbation. It was shown that the energy minimum is reached when both electrons belong to both nuclei, their spins being antiparallel. Due to the atomic wave-function overlap, the probability density for finding electrons between two H atoms increases in comparison with superposition of the spherical ED of free atoms.

In heteropolar molecules, the many-electron wave-function should contain the VB structures which describe the localization of electrons near one of the nuclei. The ionicity of a bond is taken into account in this way. Of course, the ionicity is only a rough measure of electron displacement to one of the bonded atoms. Besides, the atomic idea is absent from the final description of a bonded system, therefore the charges of covalently bonded atoms are undefined without additional suppositions. In order to preserve the atomic idea, it is necessary to have a method which permits one to subdivide a molecule or a crystal into parts which can be identified as atoms. That is why the effective charges of such atoms are strongly dependent on the partitioning method.

Pauling (1931) found that the many-electron wave-function of simple molecules can be approximated by a single VB structure if the AO is directed toward the bonding partners. Such AOs may be constructed as a linear combination of orbitals with the same principal quantum number n and different orbital quantum numbers l. They are called hybrid AOs and are often used in structural chemistry for the explanation of a molecule's form.

Mulliken (1928), Hund (1929), and Lennard-Jones (1929) developed the MO method. This has permitted the start of a systematical analysis of the electronic structure of molecules and crystals. A many-electron system is presented in this method as a set of multicentred, one-electron functions, MOs, which, in turn, are constructed as a linear combination of the one-electron atomic wave-functions. The MOs describe both the ground and excited states of the electron system, the degenerate states included. They are delocalized over the whole system and, provided the set of AOs is complete enough, are only slightly dependent on the form of the AOs. The MO language is nowadays the language of modern chemistry. The MOs provide the interpretation of many of the physical and chemical properties of substances such as electronic spectra, magnetic and electric characteristics, etc, and are successfully used in the interpretation of the data of spectroscopic methods (Flygare 1978).

The MO characteristics can be related to notions accepted in classical structural chemistry and in electronic valence theory. So, from canonical MOs, the MOs localized on specific bonds can be constructed (Chalvet *et al* 1975). Further, the ED in a closed-shell system can be expressed via MOs as

$$\rho(\boldsymbol{r})=2\sum_{\mu\nu} P_{\mu\nu}\Phi_{\mu}(\boldsymbol{r})\Phi_{\nu}^{*}(\boldsymbol{r})=2\sum_{ab}\sum_{kl} P_{akbl}\Phi_{ak}(\boldsymbol{r})\Phi_{bl}^{*}(\boldsymbol{r}) \tag{6.1}$$

which follows from the one-electron DM decomposition (2.26). The a, b indices refer to atoms, k, l to the AOs centred on them. The charge–bond order matrix $P_{\mu\nu}\equiv P_{akbl}$ is quadratic ($m\times m$), its size determined by the basis set dimension. McWeeny and Sutcliffe (1976), following Mulliken (1955), showed that the most compact description of the ED of a bonded system can be achieved with this matrix. The diagonal terms of the $P_{\mu\nu}$ matrix are the electronic population of the AO centred on an atom (the charges). The off-diagonal terms characterize the ED redistribution over AOs both inside an atom ($a=b$) and among them ($a\neq b$). Integrating the terms with $a\neq b$ over the whole space in (6.1), the orbital overlap populations can be obtained. They are closely related to bond charges. In order to evaluate the electronic charge of any atom in a molecule or crystal, all electron populations of orbitals on each centre should be summed and added to the half sum of overlap populations for all cases of different-centre AO overlap. The net atomic charge can be calculated as the difference between the nucleus charge and the total number of electrons which are assigned to the atom in question.

This procedure of $P_{\mu\nu}$ matrix element treatment is called population analysis. The main drawback of this analysis is the non-uniqueness of electronic populations and their dependence on the basis set functions. In particular, if the AOs on different centres are mutually orthogonal (Löwdin 1970) the total electronic charge appears to be divided formally between the atoms only. If AOs are non-orthogonal, the total orbital populations can appear to be more than two or their values can be negative. Therefore, cases where the population analysis leads to unreasonable pictures of chemical bonding can be met.

Fortunately, both the MO and VB methods allow the direct interpretation of the results in terms of ED. For example, in the MO theory, a chemical bond is described by bonding, antibonding, and non-bonding MOs. They are formed, correspondingly, as a result of the in-phase overlap of AOs, out-of phase overlap of AOs, and by orbitals which have predominantly atomic character. The overlap can be considered as interference of AOs. The interference is constructive in the case of the in-phase orbital overlap; the interference is destructive in the case of the out-of-phase overlap. As a result, the picture of ED distribution will be determined by the nature of the occupied MOs.

The covalent chemical bonds are usually classified on the basis of the symmetry properties of corresponding MOs (Murrell *et al* 1978). The MOs which are symmetric with respect to mirror reflection in the molecular plane are called σ orbitals. The MOs which are antisymmetric with regard to such a reflection are called π orbitals; they have a nodal plane where the wave-function is zero. There are MOs which have two (δ MOs) and more nodal planes. This classification is strictly correct only for diatomic molecules, but is used for multi-atomic molecules as well. In diatomics, owing to the restrictions of the Pauli exclusion principle, the formation of a single σ and two π bonding MOs is possible, for example, for p-type AOs. The antibonding σ^* and two π^* p MOs can also be formed. In multi-atomic systems the same MOs can have constructive and destructive character in the different space regions, depending on the nature of AO overlap. The ED will concentrate in bonding regions of occupied MOs and deplete in empty antibonding regions and near nodal surfaces of MOs. In non-bonding, well localized MOs (for example, in electron lone-pair MOs) and in some regions of fully occupied antibonding MOs the ED can also be increased. When different types of atom are interacting, the displacement of the ED toward a more electronegative atom takes place: this is the case for the polar covalent bond. Thus, the features of ED distribution can be associated with certain types of chemical bond.

Figure 6.1 illustrates the possible variants of s-, p-, and d-type AO overlap resulting in σ, π, and δ MO formation. The spatial distributions of valence MOs, describing the chemical bond in the CO molecule, is taken as an example of MO formation in a real system (figure 6.2). Note that the polar character of the bond manifests itself in the increase of the AO oxygen atom weights in the corresponding bonding MO.

A quite different non-orbital approach to ED distribution analysis is given by the Bader (1990) topological theory (subsection 2.4.2). In this theory the non-evident chemical ED properties and their connection with energy characteristics is established by the analysis of scalar functions $\rho(\boldsymbol{r})$ and $\nabla^2\rho(\boldsymbol{r})$ and vector field $\nabla\rho(\boldsymbol{r})$. The notion of a bonded atom, which occupies some volume and has a certain charge, the notion of bond paths and critical points, which are characteristics of the chemical bond, and others naturally appear in this theory.

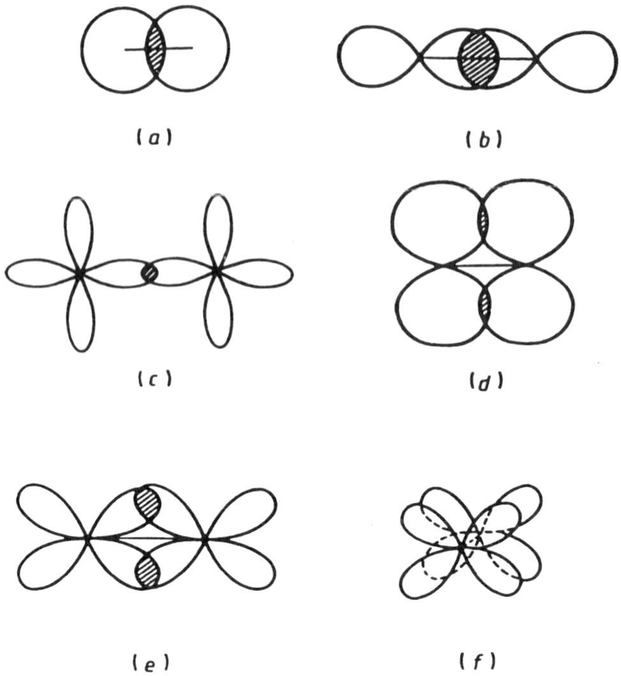

Figure 6.1 Different combinations of AO overlap, describing chemical bond formation: (*a*) σ(s–s); (*b*) σ(p–p); (*c*) σ(d–d); (*d*) π(p–p); (*e*) π(d–d); (*f*) δ(d–d).

Thus, one can see that there are different levels of and different approaches to chemical bond considerations. They use different formal languages but they are unified together by the following circumstance. All these approaches are nothing other than different ways of characterization of the features of ED distribution in various compounds and the energetical factors related to formation of a many-electron system from a set of atoms or atomic fragments. Having the detailed ED picture in one's hands, as reconstructed either from the x-ray diffraction experiment or from the quantum chemical calculation, it is natural to try to obtain the characteristics of the chemical bond directly from ED, which is a quantum chemically valid function. We will present possible ways for solving this problem below.

6.2 Atomic charges and pseudoatomic moments

The net atomic charge is one of the most widely used notions in crystal chemistry, solid state physics, etc. It plays an important role in the ionic crystal model and connects with many characteristics of the electron structure

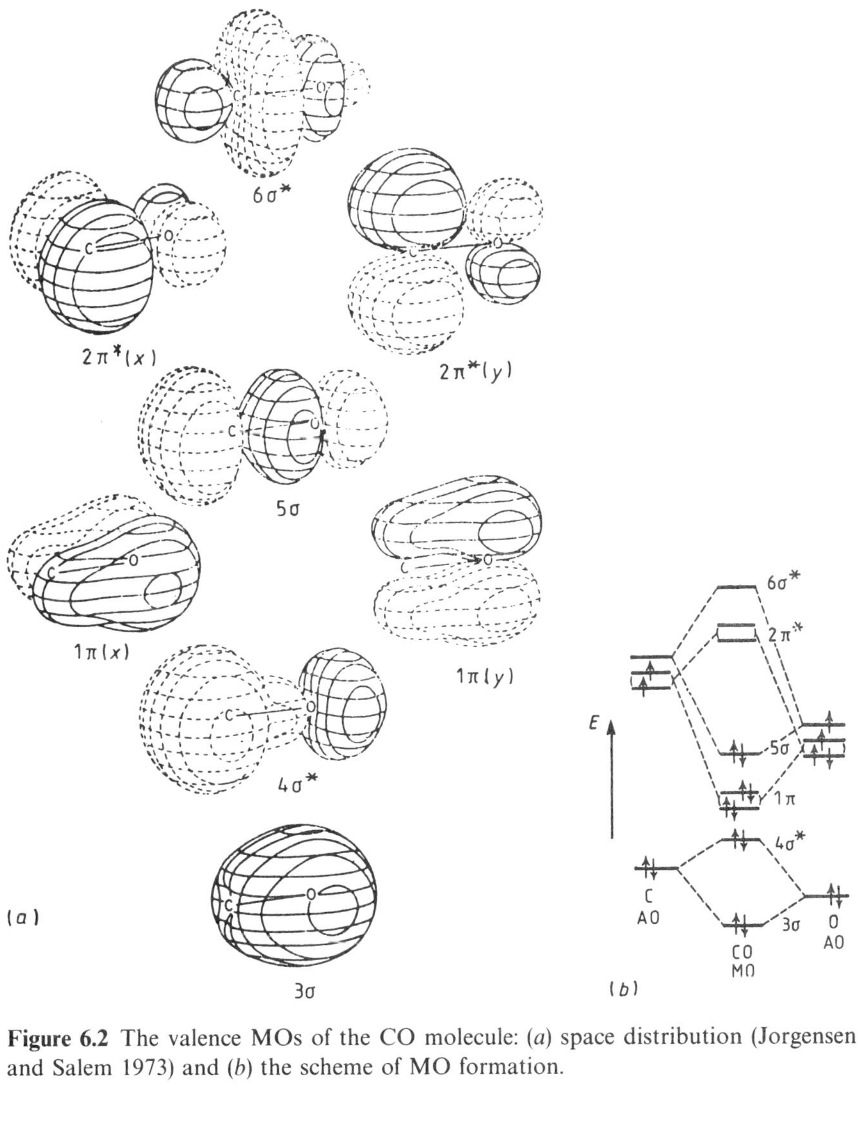

Figure 6.2 The valence MOs of the CO molecule: (*a*) space distribution (Jorgensen and Salem 1973) and (*b*) the scheme of MO formation.

by correlational equations, for example, with the exponential parameter in the AO radial part (see figure 3.16). However, there are serious methodological reasons to criticize this conceptual value. The atomic charge concept results from the simplest method of ED approximation. As has already been noted, it is impossible to determine unambiguously in a bonded system the atomic boundary inside which the atomic charge is located. The charge values appear to be dependent on the form and the size of integration volume (or on type of AO used); they are indirectly measured values in contrast to ED and only approximate the charge distribution. Nevertheless, an approach which permits us to understand the crystal properties in terms of atoms and their interactions is very attractive. The traditional quantum mechanics which operates with wave-functions is unable to do this because it can explain the properties of the whole system but not its parts. The atomic crystal chemical models, nevertheless, are widely used: the atomic parameters are used in the pseudopotential method, the atomic and ionic wave-functions are successfully applied as initial guesses in quantum chemical calculations, etc. Fortunately, quantum chemical description of a bonded atom appears to be possible via ED but not the wave-function. This has been performed by Bader (1990) in the framework of quantum topological theory of atoms in molecules (see subsection 2.4.2). In this theory both the free and the bonded atom is represented as the unity of a nucleus and its electron basin. The whole system is presented as a sum of non-overlapping atomic fragments. Identification of a bonded atom in some region of real space is realized by topological analysis of the ED. According to Bader, the bonded atom (pseudoatom) is the spatial region which contains a nucleus with its electrons and which is comprised by a unique surface of zero flux of $\nabla\rho$.

Thus, if the form of the regions which are assigned to atoms and values of the corresponding electronic charges are close to those in quantum topological theory, the atom notion is useful for investigation of physical and chemical properties of molecules and crystals. Consider now how one can estimate the net atomic charges and atomic volumes in crystals from the experimental ED. There are two methods of approximate solution of this problem. The first one consists of sharing, in the framework of some scheme, the continuous ED in pieces with volume Ω_i which are attributed to an atom. The integration over this volume gives the atomic electronic charge value. The second approach consists of refinement of atomic electron subshell populations by LS; corresponding methods have already been discussed in chapter 3. These are the refinement of the flexible atomic-type superposition $\varkappa$-model, refinement of the electron populations of monopoles in the multipole model, and recalculation of atomic charges from parameters of quantum chemical models. Therefore, we will discuss here only the methods which use spatial integration.

The question arises first of all of how to draw most simply the atomic boundary in a bonded system because the direct application of the Bader condition (2.66) is rather tedious. It is clear that the boundary location is dependent on the chemical bond nature. It is also well known that the radial dependence of ED in a crystal is different in bonding and non-bonding directions; the atomic interaction character differs as well (see e.g. O'Keefe and Hyde 1981). Even in the simplest compounds, such as crystals of the NaCl type, which are considered usually as ionic, part of the binding energy is related to covalent interaction. At the same time, different methods show that formation of alkali halide crystals corresponds to nearly full ionization of atoms. Analysing the results of many x-ray ED investigations, Tsirelson *et al* (1984c) came to the conclusion that the ED values in the internuclear space in alkali halides are approximately 0.1 e Å^{-1}. This means that ions are well separated in position space and the crystal is stable mainly because of electrostatic forces. The EDs of ions are nearly spherical, with a small hexadecapole-type distortion on the periphery of anions. The multipole analysis of the NaCl crystal (Tsirelson 1985) has shown, that the ED of the Na^+ ion is somewhat expanded whereas that of the Cl^- ion is contracted along the internuclear Na–Cl line (figure 6.3). Asphericity of the ED of ions has also been found in other halides (Henderson and Maslen 1987). Nevertheless, this effect is small and the ions can be treated approximately as spheres in some cases. It is valid especially for positively charged ions, cations, in ionic crystals. Thus, the radial density functions, $D(r)$, can be used in order to define atoms (ions) in these crystals. It is expressed via structure amplitudes as

$$D(r) = \frac{4\pi r^2}{V} \sum_H F(H) \frac{\sin 2\pi Hr}{2\pi Hr} \tag{6.2}$$

where r is the distance of point r from a nucleus. This function is strongly dependent on the Fourier series termination. This effect is usually eliminated by introducing into (6.2), in addition to the experimental structure amplitudes, calculated ones related to large H values. Such radial functions for the cations in the crystals MgO, CaO, SrO, and BaO with NaCl-type structure were calculated by Vidal-Valat *et al* (1978). They are presented in figure 6.4 along with their theoretical analogues calculated using ionic scattering functions tabulated by Ibers and Hamilton (1974). It can be seen that each cation is separated from its surroundings by a distinct minimum in the radial ED curve, followed by a steep rise due to the nearest neighbours. The structure of crystals considered can, thus, be described as a close packing of touching anions. As compared to the ionic superpositional model, represented by the

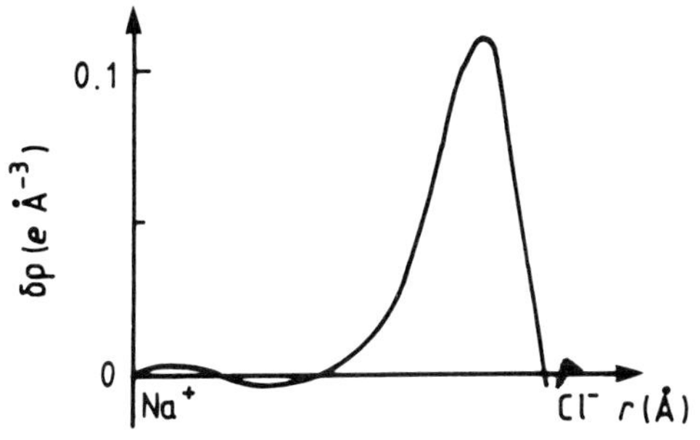

Figure 6.3 The $\rho_{multip} - \rho_{ion}$ profile for the NaCl crystal.

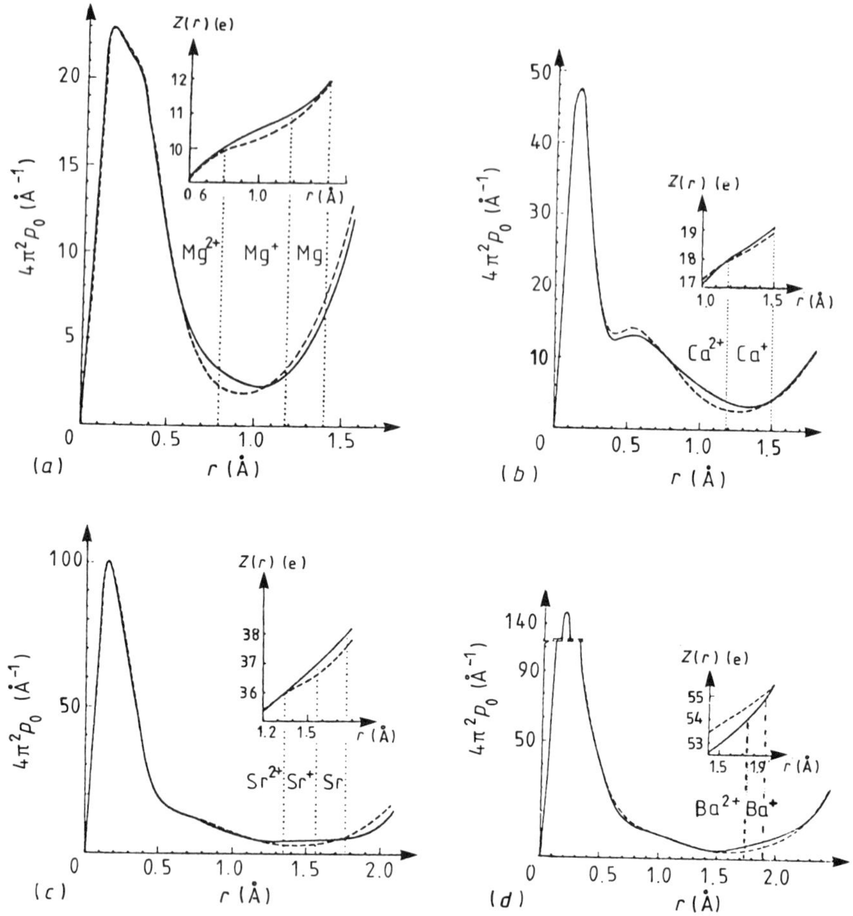

Figure 6.4 Radial ED and electron counts around cations in alkali-earth oxides (Vidal-Valat *et al* 1978): (*a*) Mg^{2+} in MgO; (*b*) Ca^{2+} in CaO; (*c*) Sr^{2+} in SrO; (*d*) Ba^{2+} in BaO. Solid lines correspond to experimental data, dashed lines to the ionic superpositional model.

theoretical curves, the Mg^{2+} and Ca^{2+} ions show a slight extension whereas the Sr^{2+} and Ba^{2+} ions remain nearly the same in the crystal.

The existence of the minimum of ED around each ion in ionic crystals leads to the notion of the separation radius, R_s, which differs from the usual atomic and ionic radii (Sasaki *et al* 1979). R_s is the distance from the nucleus to the point with minimal radial ED on the internuclear line. The well pronounced space separation of ions permits us to calculate the ionic electron charge by ED integration over the sphere with radius R_s. In this sense one can speak of ionic net charges. In compounds with predominantly ionic character of the chemical bonds the net charges and R_s are useful characteristics of electron structure. The R_s values for some ionic crystals derived from a series of x-ray diffraction studies are listed in table 6.1. The experimental values of R_s exceed, as a rule, the classical ionic radii in the case of cations (the Ag compounds are an exception). This is true not only for halides but for alkali-earth metal oxides and oxides of transition elements with NaCl structure type.

The $D(r)$ function behaviour in compounds with different types of chemical bond has been investigated by Sasaki *et al* (1980). It has been found that the $D(r)$ function has in the internuclear space a single minimum (Mg in MgO, Li in $LiAlSi_2O_6$), a horizontal plateau (Mn in MnO, Ca in $CaMgSi_2O_6$), or a small maximum (Si in $LiAlSi_2O_6$) (figure 6.5). In the first case the position of the $D(r)$ minimum was accepted as R_s, in the second case R_s was calculated as the average over positions of the nodes of function $\partial D/\partial r$, and in the third

Table 6.1 Geometrical characteristics of ions in some crystals with NaCl-type structure. (ls indicates the low-spin state, hs the high-spin state.)

Compound	Ion	Ionic radius (Å)[a]	R_s (Å)	Atomic radius (Å)
NaCl	Na^+	1.02	1.21	1.80
	Cl^-	1.81	1.61	1.00
KCl	K^+	1.38	1.45	2.20
	Cl^-	1.81	1.70	
AgCl	Ag^+	1.15	1.03	1.60
	Cl^-	1.81	1.74	
AgBr	Ag^+	1.15	1.13	1.60
	Br^-	1.96	1.76	1.15
MgO	Mg^{2+}	0.72	0.92	1.50
	O^{2-}	1.40	1.28	0.60
CaO	Ca^{2+}	1.00	1.32	1.80
	O^{2-}		1.19	
SrO	Sr^{2+}	1.18	1.27	2.00
	O^{2-}		1.44	
BaO	Ba^{2+}	1.35	1.49	2.15
	O^{2-}		1.24	
MnO	Mn^{2+}	0.67 (ls) 0.83 (hs)	1.15	1.40
CoO	Co^{2+}	0.65 (ls) 0.75 (hs)	1.09	1.35
NiO	Ni^{2+}	0.69	1.08	1.35

[a] Wells (1984).

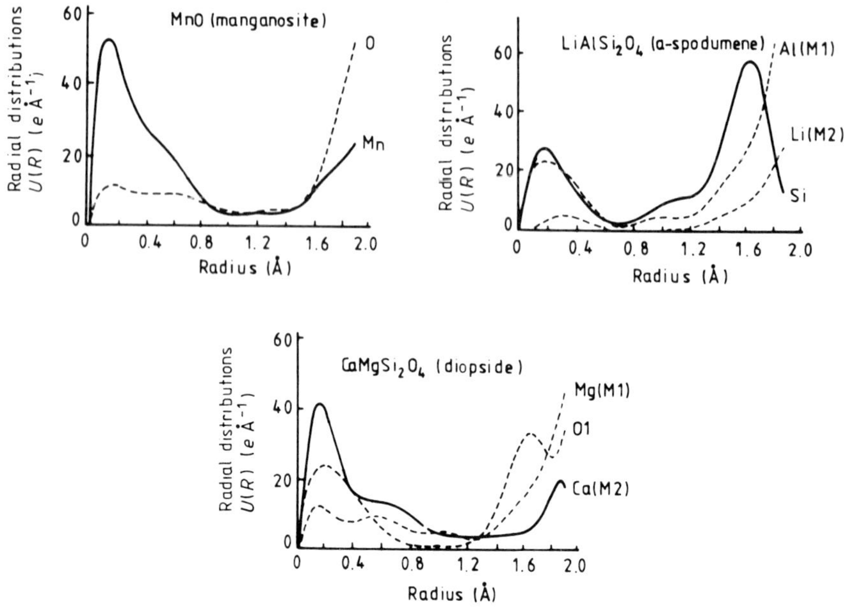

Figure 6.5 Curves of $D(R)$ for (*a*) Mn and O in MnO, (*b*) Li, Al, and Si in $LiAlSi_2O_6$, and (*c*) Ca, Mg, and O (O_1 site) in $CaMgSi_3O_6$ (Sasaki *et al* 1980).

case, which is characteristic of covalently bonded Si and O atoms, the R_s value has been undefined. Thus, the analysis of the radial density distribution does not give a uniform approach to the separation of atoms in the molecule and crystal position space. Moreover, with increasing bond covalency as well as in the cases of multiple, multicentre, and metallic bonds part of the ED is out of the atomic spheres. This results in artificial violation of the electroneutrality in the ionic model. Nevertheless, Sasaki and Takeuchi (1982) came to the conclusion that the same cation in different compounds has nearly the same separation radius (table 6.2). These radii systematically exceed the values of both classical and Shannon (1976) crystal radii. The atomic net charge values derived with these radii for some silicates are presented in table 6.3.

Coppens *et al* (1981b) have formulated some general principles for atomic ED partitioning: the total unit cell charge should be conserved and pseudoatoms should additively fill the unit cell space. Two kinds of partitioning method, which satisfy these conditions, are widely used. The first one is the generalized Wigner–Seitz method (Coppens 1975): an atom (group of atoms) is surrounded by a polyhedron, the boundaries of which

Table 6.2 The best electron density separation, R_s, ionic, R_i, and atomic, R_a, radii for Li, Mg, Al, Si, Ca, Mn, Fe, Co, and Ni atoms in some oxides and silicates (Sasaki and Takeuchi 1982).

Atom	Compound	Atom	R_i [a]	R_s [b]	R_a
Li			0.76		1.45
	$LiAlSi_2O_6$	M(2)		0.83	
Mg			0.72	⟨0.95⟩	1.50
	MgO			0.92	
	Mg_2SiO_4	M(1)		0.93	
		M(2)		0.94	
	$CaMgSi_2O_6$	M(1)		0.91	
	$Mg_2Si_2O_6$	M(1)		0.89	
		M(2)		0.96	
Al			0.535		1.25
	$LiAlSi_2O_6$	M(1)		0.91	
Si			0.40 (0.26, CN = 4)	⟨0.95⟩	1.10
	Mg_2SiO_4			0.96	
	$LiAlSi_2O_6$			0.96	
	$Mg_2Si_2O_6$			0.91	
	$Mg_2Si_2O_6$			0.95	
Ca			1.00		1.80
	$CaMgSi_2O_6$	M(2)		1.27	
Mn			0.67 ls 0.83 hs	⟨1.15⟩	1.40
	MnO			1.15	
	Mn_2SiO_4	M(1)		1.15	
		M(2)		1.16	
Fe			0.78	⟨1.12⟩	1.40
	Fe_2SiO_4	M(1)		1.10	
		M(2)		1.12	
	$Fe_2Si_2O_6$	M(1)		1.11	
		M(2)		1.16	
Co			0.65 ls 0.75 hs	⟨1.09⟩	1.35
	CoO			1.09	
	Co_2SiO_4	M(1)		1.08	
		M(2)		1.07	
	$Co_2Si_2O_6$	M(1)		1.09	
		M(2)		1.14	
Ni			0.69		1.35
	NiO			1.08	

[a] ls is the low-spin state, hs is the high-spin state, CN is the coordination number.
[b] ⟨ ⟩ means average values.

Table 6.3 Net atomic charges for silicates determined for cations by integration within spheres with $r = R_s$, and for anions by the $\varkappa$-technique. The values in parentheses are e.s.d.s. (Tsirelson *et al* 1990.)

Compound	Net atomic charge (in electrons)					
Olivine-like	Si	M(1)	M(2)	O(1)	O(2)	O(3)
Mg_2SiO_4	2.11(3)	1.76(3)	1.74(3)	−1.52(7)	−1.29(7)	−1.40(7)
Fe_2SiO_4	2.43(6)	0.85(8)	1.45(7)	−1.13(11)	−1.21(11)	−1.24(1)
Ni_2SiO_4	2.4(1)	1.8(1)	1.7(1)	−1.5(1)	−1.5(1)	−1.5(1)
Co_2SiO_4	2.2(1)	1.6(1)	1.5(1)	−1.2(1)	−1.2(1)	−1.4(1)
Mn_2SiO_4	2.28(5)	1.21(6)	1.49(6)	−1.27(12)	−1.13(12)	−1.29(12)
Orthopyroxenes[a]	Si	M(1)	M(2)	O(1)	O(2)	O(3)
$Mg_2Si_2O_6$	⟨2.28(4)⟩	1.84(4)	1.79(3)	⟨−1.44⟩	⟨−1.40⟩	⟨−1.26⟩
$Fe_2Si_2O_6$	⟨2.19(11)⟩	1.14(12)	1.10(13)	⟨−1.13⟩	⟨−1.15⟩	⟨−1.05⟩
$Co_2Si_2O_6$	⟨2.28(7)⟩	1.29(11)	0.61(12)	⟨−1.12⟩	⟨−1.14⟩	⟨−1.47⟩
Clinopyroxenes						
$LiAlSi_2O_6$	2.4(1)	2.4(1)	0.7(1)	−1.3	−1.4	−1.3
$CaMgSi_2O_6$	2.56(1)	1.44(1)	1.39(2)	−1.33	−1.28	−1.35

[a] For all Si atoms and O atoms in orthopyroxenes, averaged value of atomic charges in independent positions are given.

are determined by an equation

$$\frac{(\boldsymbol{r}-\boldsymbol{r}_A)\boldsymbol{R}_{AB}}{R_A}=\frac{(\boldsymbol{r}-\boldsymbol{r}_B)\boldsymbol{R}_{BA}}{R_B} \tag{6.3}$$

(discrete boundary partitioning). Here $\boldsymbol{r}_A$ and $\boldsymbol{r}_B$ are vectors (see figure 6.6) drawn from the origin to A and B atoms with $\boldsymbol{R}_{AB}$ being an internuclear vector drawn from the A atom to B. The second method is the Hirshfeld (1977) 'stockholder' scheme, which uses the weight function

$$w_\mu(\boldsymbol{r})=\rho_\mu(\boldsymbol{r})/\rho_p(\boldsymbol{r}). \tag{6.4}$$

Here $\rho_\mu(\boldsymbol{r})$ and $\rho_p(\boldsymbol{r})$ are the ED of the spherical μth atom and the promolecule, correspondingly. Being multiplied by $\rho(\boldsymbol{r})$, the function w_μ separates in a unit cell the μth pseudoatom, proportional to the free-atom analogue contribution to the promolecule:

$$\rho_\mu^{pseudo}(\boldsymbol{r})=w_\mu(\boldsymbol{r})\rho(\boldsymbol{r}). \tag{6.5}$$

In order to calculate the pseudoatom electronic charge the integral of ρ_μ^{pseudo} could be taken over the whole space. In practice, however, the integral is done over the volume which is surrounded by a surface with small values of w. Such pseudoatoms have fuzzy boundaries and overlap; however, they are less diffuse than pseudoatoms in the multipole model. Both methods described can be applied to the dynamic and to the static ED, independently of the ED presentation either in the form of multipole expansion or as Fourier series. In Hirshfeld's method the net atomic charges and atomic moments are calculated by integration of deformation ED.

The drawback of the Coppens method is in the fact that the procedure contains the *a priori* values of atomic radii R_A and R_B. In order to reduce

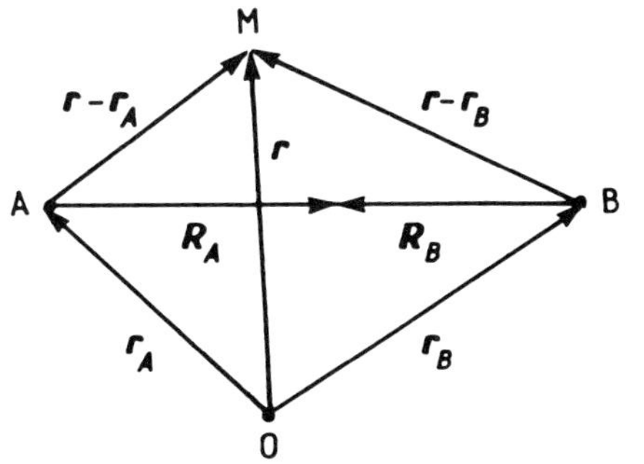

Figure 6.6 The vector disposition in ED partitioning into pseudoatoms with discrete boundaries. O, the origin; A and B, positions of neighbouring atoms. The ED at a point is assigned to the A pseudoatom if $[(\boldsymbol{r}-\boldsymbol{r}_A)/R_A]<[(\boldsymbol{r}-\boldsymbol{r}_B)/R_B]$.

the arbitrariness arising Ivanov-Smolenskii *et al* (1984) suggested that one replace the radii R_A and R_B in (6.3) by the values $\langle r_A \rangle / Q_A$ and $\langle r_B \rangle / Q_B$, $\langle r_A \rangle$ and Q_A being, correspondingly, the average distance over electrons from the nucleus A and the atomic electronic charge of an atom. These values have been calculated by iterations to self-consistency and they have led to good accordance of the crystal diamagnetic and electric characteristics with experiment. In order to reduce the boundary diffuseness effect, the weight function

$$w_\mu(\boldsymbol{r}) = \rho_\mu^N(\boldsymbol{r}) / \rho_p(r) \tag{6.6}$$

has also been analysed. It proved to be that, in halides with NaCl-type structures, the values of Q_A, $\langle r \rangle_A$, and diamagnetic susceptibilities are practically independent of N for $N > 6$.

Both methods can also be applied to the separation of molecules in a crystal. Examples can be found in the works of Moss and Coppens (1980) and Moss (1982).

The significant drawback of any total and valence ED partitioning method applied to net atomic charge determination has been noted by Maslen and Spackman (1985). Even in a promolecule composed of neutral atoms, the overlapping of atomic ED results in the fact that calculation of the atomic charges by the integration of ρ_p within Bader's fragments gives non-zero values of net atomic charges (Δq_s^p values in table 6.4). Therefore, net atomic charges as calculated by volume integration of molecular ED (Δq^{mol} in table 6.4) not only characterize interatomic charge transfer but also contain atomic ED contributions. For example, the oxygen negative net atomic charge in diatomic molecules has the value $-(1.0\text{–}1.5)\, e$ (table 6.4) and cannot be accepted as realistic. In order to avoid this drawback it is better to calculate the net atomic charges by integration of the deformation ED instead of total or valence ED. This is done easily in Hirshfeld's stockholder scheme (Δq^H in table 6.4). Maslen and Spackman (1985) modified also the net charge calculation procedure in the framework of the Bader partitioning scheme: they determined the atomic net charges, Δq^B, as the difference between Δq^{mol} and Δq^p. Such net charges correspond to these, calculated by integration of specific difference ED; their meaning is close to that of Δq^H values. As can be seen from table 6.4, the Δq^B and Δq^H values are indeed closer to each other. The operation with $\delta\rho$ instead of ρ decreases the series termination effect as well.

At the same time, Maslen and Spackman (1985) have found that the Bader scheme of separating pseudoatoms meets some practical obstacles. Only theoretical EDs calculated with a very extended basis set provide an appropriate wave-function behaviour at the molecule periphery. The calculations of LiF and CO molecules, using several near-HF-limit-level molecular wave-functions of the same quality with the energy deviations from the measured value ~ 0.085 eV, have shown noticeable differences in net atomic charges (see table 6.4). These differences can be attributed to the

Table 6.4 Atomic charges obtained from the partitioning schemes of Bader and Hirshfeld (Maslen and Spackman 1985). All Δq values represent electron transfer from A to B in the molecule AB.

Molecule	Bader Δq^P	Bader Δq^{mol}	Bader Δq^B	Hirshfeld Δq^H
LiH	0.644	0.911	0.267	0.414
BeH	0.750	0.868	0.118	0.193
BH	0.720	0.753	0.033	0.075
CH	0.019	0.032	0.013	−0.016
NH	−0.236	−0.322	−0.086	−0.091
OH	−0.428	−0.584	−0.156	−0.164
HF	0.585	0.759	0.174	0.228
NaH	0.356	0.810	0.454	0.413
MgH	0.508	0.796	0.288	0.282
AlH	0.613	0.825	0.212	0.228
SiH	0.627	0.795	0.168	0.125
PH	0.488	0.580	0.092	0.034
SH	0.123	0.094	−0.029	−0.050
HCl	0.108	0.240	0.132	0.124
LiF	0.650	0.937	0.287	0.624
LiF	0.607	0.940	0.333	0.619
BeF	0.777	0.945	0.168	0.328
BF	0.962	0.940	−0.022	0.118
CF	0.748	0.781	0.033	0.080
NF	0.324	0.439	0.115	0.112
NaF	0.425	0.941	0.516	0.677
AlF	0.711	0.974	0.263	0.357
LiO	0.651	0.932	0.281	0.580
BeO	1.151	1.692	0.541	0.647
BO	1.216	1.552	0.336	0.376
CO	1.220	1.346	0.126	0.139
CO	1.241	1.363	0.122	0.138
NO	0.380	0.495	0.115	0.086
MgO	0.680	1.413	0.733	0.678
SiO	1.184	1.633	0.449	0.461
LiN	0.666	0.916	0.250	0.542
BeN	0.731	1.236	0.505	0.415
BN	0.979	0.836	−0.143	−0.001
CN	0.810	1.123	0.313	0.198
PN	1.310	1.741	0.431	0.289
LiC	0.599	0.883	0.284	0.444
BeC	0.574	0.853	0.279	0.233
LiB	0.515	0.761	0.246	0.257
BeB	0.334	0.438	0.104	0.086
LiCl	0.624	0.926	0.302	0.553
NaCl	0.428	0.915	0.487	0.620

improper choice of the pseudoatomic boundaries far away from nuclei (~9 au). In the stockholder partitioning scheme discrete boundaries are absent and the corresponding changes are very small. In the analysis of the experimental ED the drawback of the Bader partitioning procedure mentioned will take place as well: it will influence the results mostly in intermolecular space.

The refinement of pseudoatom monopole electron populations in the framework of the ϰ-model or multipole model has some drawbacks also. First, there is the net atomic charge dependence on the crystal model chosen (table 6.5) and on the basis functions used in the radial part of the pseudoatoms. Second, the correlation between model parameters at the stage of refinement can also lead to biased parameter estimates. This approach has, however, some advantages: there are, for example, no questions regarding the pseudoatom volume choice. This is the decisive factor in the net charge determination in crystals with partly covalent bonds. At the same time, in such compounds uncertainty arises in the intercomparison of the results derived by the ϰ-technique and by valence ED integration. Stewart and Spackman (1981) compared the net atomic charges in corundum, α-Al_2O_3, and in α-quartz, SiO_2, both by multipole population analysis and by ED integration within the different radii spheres. According to LS refinement, net charges are in α-SiO_2, $q_{Si} = 1.027(94)\,e$ and $q_O = -0.514(48)\,e$, and in α-Al_2O_3, $q_{Al} = 0.41(23)\,e$ and $q_O = -0.274(65)\,e$. It could be concluded that the chemical bond in quartz has a more ionic character. However, a larger shift of ED toward the O atoms was revealed on the valence ED map in α-Al_2O_3 than in α-SiO_2. In order to derive the same net charge values for

Table 6.5 Net atomic charges for pyridinium-1-dicyanomethylide, $C_8D_5N_3$, at 118 K according to charge density refinement with multipole models (Baert *et al* 1982). The total charge is constrained to neutrality.

Atom numbering	Atom	Hirshfeld (1971) multipole model	Hansen and Coppens (1978) multipole model
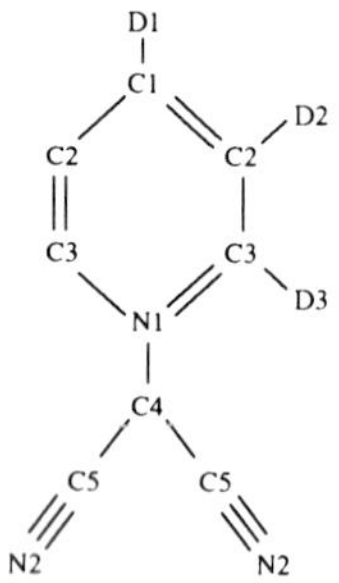	C1	−0.03	−0.44
	C2	+0.81	−0.45
	C3	+1.29	−0.35
	C4	−2.01	−0.08
	C5	−0.20	−0.00
	N1	−1.00	+0.06
	N2	+1.00[a]	−0.22
	D1	−0.55	+0.54
	D2	−0.55	+0.52
	D3	−0.55	+0.49

[a] Hydrogen atoms constrained to have identical charges.

Al and Si atoms, the integration volumes should be taken as spheres with radius 4.0 Å. On the other hand, the formal valence charge value +3 could be derived for the Al^{3+} ion by ED integration within the 0.8 Å radius sphere. For the Si^{4+} ion the analogous sphere radius proved to be 0.58 Å. It is clear that electrons of the Al and Si atoms outside the spheres indicated will screen the net charges and will lead to their effective value reduction.

Brown and Spackman (1991) have investigated how the simplest flexible $\varkappa$-model (two parameters for an atom accounting for the spherical deformation of a bonded atom only) describes the real asphericity of an atom in a crystal. They used the ED of 28 diatomics, whose wave-functions are known at the HF limit level. The deformation ED of these molecules has been partitioned according to the Hirshfeld approach using the weight factor w_μ (6.4)

$$\delta\rho_\mu(\boldsymbol{r}) = w_\mu(\boldsymbol{r})\delta\rho(\boldsymbol{r}). \tag{6.7}$$

The $\delta\rho_\mu$ fragments were then spherically averaged and used for pseudoatom construction according to the equation $\rho_\mu^{pseudo}(\boldsymbol{r}) = \rho_\mu^{at}(\boldsymbol{r}) + \langle\delta\rho_\mu(\boldsymbol{r})\rangle_{spher}$. These pseudoatoms were consequently approximated by $\varkappa$-pseudoatoms whose parameters were determined by LS. The results are presented in figure 6.7. As can be concluded from these data, quite satisfactory general descriptions of monopole atomic deformations in molecules are achieved with the $\varkappa$-technique. At the same time, for the second-row elements the real ED deformation is more complicated. The molecules AlH, SiH, PH, SH, ClH, LiH, and NaCl are examples of this. The best fit is achieved for atoms in molecules with s-type chemical bonding (Li in LiH, LiO, LiCl; Na in NaH, NaF, NaCl; Be in BeH, BeO; Mg in MgH, MgO).

Hydrogen, which has no core electrons, exhibits a monopole deformation which is due unequivocally to the effects of bonding on the valence ED; its deformation density is relatively simple. It is noteworthy that hydrogen is well modelled in each of a wide variety of chemical environments, which the first- and second-row hybrides represent. In particular, the $\langle\delta\rho_\mu\rangle_{spher}$ functions reflect, in a very obvious way, the electronegativity differences between each atom in the diatomic and also between the different first- and second-row elements.

The atomic net $\varkappa$-charges, $\Delta q^\varkappa$, disagree sometimes with the charges derived by the direct integration of deformation ED within pseudoatomic fragments. This can be seen from table 6.6 where charge transfer values are presented. This especially concerns the pseudoatoms with non-monotonic $\langle\delta\rho_\mu\rangle_{spher}$ function behaviour. Thus, the atomic ED deformation in molecules and crystals can be adequately described by the $\varkappa$-technique only in the first approximation, though it gives the correct estimates of ED transfer.

As far as $\varkappa$ values are concerned, their quantities for atoms with negative net charges (positive charge transfer) correspond to expansion of the valence

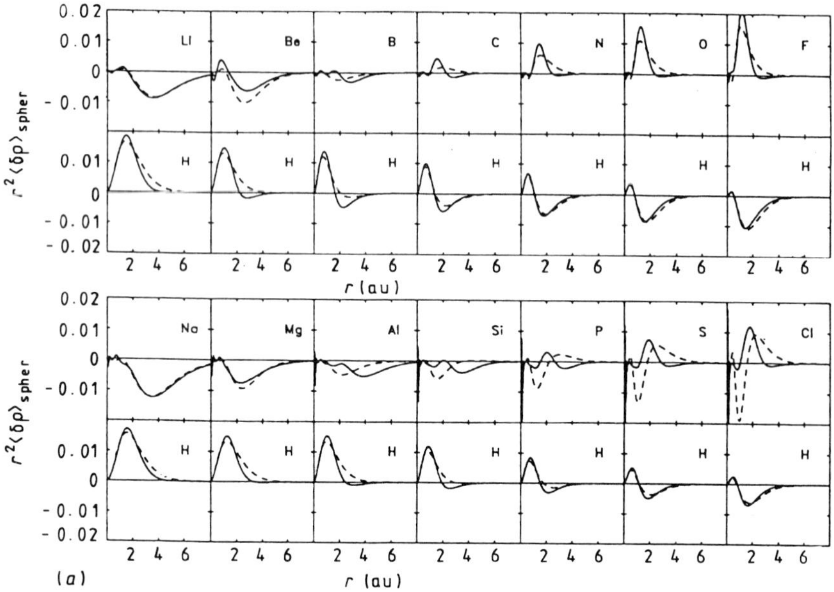

Figure 6.7 The radial distribution of spherically averaged deformation ED function $r^2\langle\delta\rho_\mu\rangle_{spher}$ and its $\varkappa$-analogues for the first- and second-row elements in diatomic molecules: (*a*) hydrides; (*b*) alkali halides, oxides, and oxide analogues; (*c*) miscellaneous diatomics. The solid line is the $r^2\langle\delta\rho_\mu\rangle_{spher}$ function; the dashed line is the $\varkappa$-approximation of that function (Brown and Spackman 1991).

shells, $\varkappa < 1$, as compared with free atoms. Atoms with positive net charge have their valence shells contracted, with $\varkappa > 1$. Exceptions to this are observed when the valence shells of bonded atoms are marginally more diffuse than in the isolated atoms (N_2, NO, MgH, and AlF molecules), or are slightly contracted (BeH and BH molecules). The greatest effect is observed for the Li atom and in the LiO molecule, where there is a 24.9% contraction of the valence electron density on Li relative to the isolated atom.

Brown and Spackman (1991) compared the $\varkappa$-value work with energy-optimized exponential factors for atoms in polyatomic hydrides, obtained with Gaussian expansion of STOs (Hehre *et al* 1969, 1970) (see appendic C). They found that for H in CH_4, NH_3, H_2O, and HF the energy-optimized exponents are 1.16, 1.23, 1.26, and 1.23, respectively, generally larger than corresponding $\varkappa$-values of 1.144, 1.171, 1.183, and 1.182 for the diatomic hydrides. For H bonded to elements from Na to Cl in second-row compounds the energy-optimized exponent ranges from 0.77 (NaH) to 1.20 (HCl); the refined $\varkappa$-values display similar behaviour, but span a narrower range, 0.892 (NaH)–1.122 (HCl). It is also possible to deduce energy-optimized $\varkappa$-values

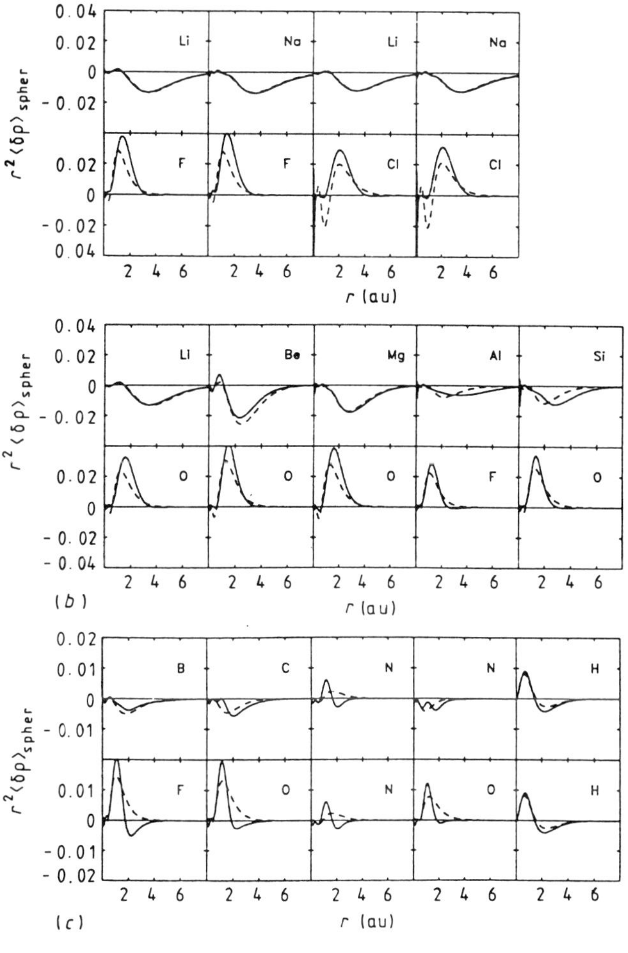

Figure 6.7 *Continued.*

for heavy atoms from the atomic and molecular energy-optimized exponents reported by Hehre and coworkers. These energy-opyimized values display the same trends as, but often more exaggerated than, the $\varkappa$-refined (density-fitted) values. For example, for the heavy atoms Na to Cl, Hehre *et al* obtained scale factors of 1.40 (NaCl), 1.43 (MgH_2), 1.26 (AlH_3), 1.18 (SiH_4), 1.04 (PH_3), 1.02 (H_2S), and 1.00 (HCl), compared with $\varkappa$-refined values of 1.061 (NaH), 0.984 (MgH), 0.999 (AlH), 0.990 (SiH), 0.984 (PH_3), 0.979 (H_2S), and 0.976 (HCl). The two different estimates of valence expansion and contraction upon bonding suggest that the effect is small in covalent molecules and large in ionic systems, and typically less than 20%. However, energy optimization weights the region near the nucleus heavily, whereas ED fitting

Table 6.6 Kappas and charge transfers Δq^{κ} (in electrons), from κ-refinement of $\delta\rho_{\mu}(r)$ spherically averaged about the nuclear position of atom μ (Brown and Spackman 1991). The charge transfer from one atom to another is determined by the population parameter, P, in (3.118)

$$\Delta q^{\kappa} = n^{val}(P - 1.0)$$

where Δq^{κ} is the electron transfer from the atom in question and n^{val} is the number of valence electrons in the isolated atom. Note that Δq^{κ} refers to the transfer of negative charge (i.e. electrons), so that a positive Δq^{κ} indicates that an atom has a net negative electric charge. Charge transfer values, Δq^{H}, are from integration of the partitioned deformation density (Maslen and Spackman 1985), and are given only for the second atom in each diatomic.

Diatomic	κ	Δq^{κ}	Δq^{H}	Diatomic	κ	Δq^{κ}	Δq^{H}
Li	1.129	−0.407		Li	1.249	−0.607	
H	0.907	+0.491	+0.414	O	0.972	+0.420	+0.581
Be	1.048	−0.348		Be	1.136	−0.854	
H	1.020	+0.275	+0.193	O	0.962	+0.534	+0.646
B	1.005	−0.069		Mg	1.025	−0.644	
H	1.100	+0.134	+0.075	O	0.963	+0.480	+0.678
C	0.994	+0.036		Al	0.998	−0.243	
H	1.144	−0.002	−0.016	F	0.981	+0.357	+0.357
N	0.989	+0.115		Si	1.001	−0.320	
H	1.171	−0.105	−0.091	O	0.971	+0.468	+0.461
O	0.987	+0.192		Li	1.214	−0.629	
H	1.183	−0.183	−0.164	F	0.974	+0.448	+0.620
F	0.986	+0.252		Na	1.057	−0.652	
H	1.182	−0.266	−0.228	F	0.974	+0.441	+0.677
Na	1.061	−0.400		Li	1.138	−0.566	
H	0.892	+0.467	+0.413	Cl	0.966	+0.301	+0.551
Mg	0.984	−0.304		Na	1.025	−0.598	
H	0.957	+0.356	+0.282	Cl	0.965	+0.322	+0.617
Al	0.999	−0.150		B	1.008	−0.141	
H	1.003	+0.305	+0.228	F	0.988	+0.231	+0.118
Si	0.990	−0.096		C	1.006	−0.112	
H	1.049	+0.174	+0.125	O	0.987	+0.234	+0.139
P	0.984	−0.040		N	0.994	+0.040	
H	1.080	+0.056	+0.034	N	0.994	+0.040	0.000
S	0.979	+0.012		N	0.998	−0.065	
H	1.104	−0.048	−0.050	O	0.991	+0.138	−0.086
Cl	0.976	+0.049		H	1.110	+0.031	
H	1.122	−0.139	−0.124	H	1.110	+0.031	0.000

procedures such as $\varkappa$ refinement depend more strongly on regions further from the nucleus. Thus, energy-optimized exponents are not necessarily appropriate for use in LS fitting of the ED.

The net charges should correspond to the polarity of the corresponding bond. Maslen and Spackman (1985) have found that the net atomic charges obtained by the Hirshfeld stockholder partitioning scheme for 39 diatomics correlate with the valence of point charges which formally form the dipole (figure 6.8). The point charge value was found to be μ/R, where R is the internuclear distance. The dipole moment, μ, was calculated from the molecule wave-function. The correlation coefficient is 0.939, and the slope of the LS line of best fit in figure 6.8 is 0.644, with an intercept of 0.019. This means that the Hirshfeld-type atomic charges compose approximately 65% of the values which give the true dipole moment in the point charge approximation.

There are examples which demonstrate the usefulness of the net atomic charge concept in the study of bonding in molecules and crystals. One of them is the dicyanodiamide, $C_6H_4N_4$, investigated at 83 K by Hirshfeld and Hope (1980). This simple molecule (figure 6.9(a)) consists of atomic groups

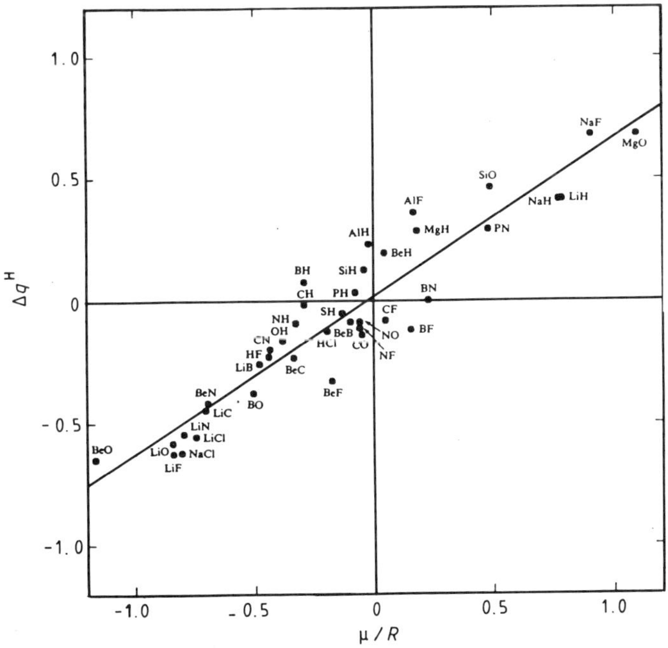

Figure 6.8 Plot of the ED transfer Δq^H against μ/R, the charge predicted by the molecular dipole moment. The line through the data points is the best fit to all 39 points. The Δq^H values represent electron transfer from the heavy to the light atom (Maslen and Spackman 1985).

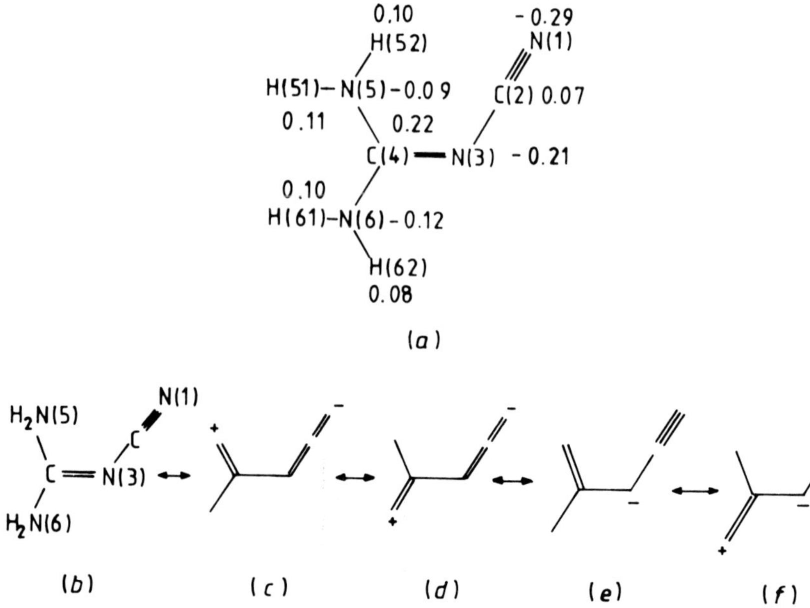

Figure 6.9 Dicyanodiamide, $C_2H_4N_4$: (*a*) structural formula and net atomic charges obtained from multipole model refinement; (*b*)–(*f*) most probable resonance structures (Hirshfeld and Hope 1980).

with distinctive electronic properties. The actual bonding in the molecule is not clear *a priori* (Rannev *et al* 1966) since three C–N bond lengths around the C4 atom are nearly equal: they are 1.339, 1.341, and 1.333 Å. Such a structure of the molecule can be explained by resonance between the three main structures (*b*), (*c*), and (*d*) (figure 6.9). Therefore, a negative charge on the N1 atom and positive charges on the amino groups may be expected. Indeed, the Hirshfeld multipole model refinement resulted in such a distribution of atomic charges (see figure 4.9(*a*)). At the same time, a large negative net charge on the N3 atom was found. This suggests that the two additional resonance structures (*e*) and (*g*) may be as important as the others.

Craven and Weber (1983) in their zwitterion γ-aminobutyric acid, $C_4H_9NO_2$, study have found that the three tetrahedral C atoms are in fact neutral, whereas the trigonal carboxylate C4 atom has a significant negative net charge, $-0.17\,e$ (table 6.7). This charge appears to be an inductive effect due to carboxylate O atoms. All H atoms in the methylene chain are electrically neutral, except atom H6 which carries significantly positive charge, 0.11 *e*. Analysis of the atomic structure of the γ-aminobutyric acid shows that atom H6 is within short intramolecular distances of the two electronegative N (2.64 Å) and O1 (2.54 Å) atoms; the corresponding sums of the van der

Table 6.7 Atomic charges in γ-aminobutyric acid from the multipole model.

Atom numbering	Atoms	Atomic charge
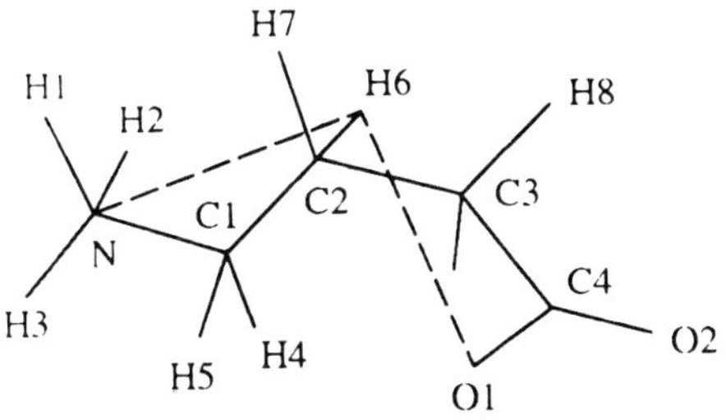	N	−0.23(3)
	C1	0.05(4)
	C2	−0.04(4)
	C3	0.03(4)
	C4	−0.17(3)
	O1	−0.15(2)
	O2	−0.27(2)
	H1	0.18(2)
	H2	0.19(2)
	H3	0.20(2)
	H4	0.01(2)
	H5	0.03(2)
	H6	0.11(2)
	H7	−0.03(2)
	H8	0.04(2)
	H9	0.04(2)

Waals radii are 2.7 and 2.6 Å. These short distances come from the folding of the molecular backbone into a twisted conformation with torsion angle $\pm 72.6^\circ$ about the C2–C3 bond. It may be imagined that the positive charge at the H6 atom is induced as the molecule folds, and that the resulting H bridge $N^+ \cdots G^+ \cdots O^-$ helps to stabilize the observed molecule in crystal conformation.

In alkali-earth metal oxides with NaCl-type structure, where EDs of ions are well separated, the ED integration within spheres with radius R_s has shown that the net oxygen charge values are $-1.0 \pm 0.5\ e$. The cation charges are increased from MgO to BaO (Vidal-Valat *et al* 1978). The cation charges in NaCl-type 3d transition-metal oxides MnO, CoO, and NiO are 1.51(1), 1.40(1), and 0.91(1) e correspondingly (Sasaki *et al* 1979). It should be noted that in deformation ED maps, the increase of the distance between the cation nucleus and the nearest excessive $\delta\rho$ peak along the [100] direction correlates in the Ni–Co–Mn series with R_s values.

Thus, in spite of the overall pessimism concerning the vagueness of the notion of the atomic charges and their physical sense, which we share, these values roughly reflect the interatomic charge transfers and interactions in simple systems. Nevertheless, even in such systems a conclusion about the character of the chemical bond should be made with caution, if atomic charges are used as the initial point. For example, the lead chalcogenides PbX (X = S, Se, Te) with the same NaCl-type structure are narrow-gap semiconductors. This suggests that the chemical bond in these compounds should be predominantly covalent. However, Noda *et al* (1983) have not found typical covalent peaks on Pb–X lines in $\delta\rho$ maps: the $\delta\rho$ distribution is characterized by a smeared positive $\delta\rho$ distribution which is usually observed in a metallic-type bond. Such a character of the chemical bond has led to $\varkappa$-charge values for the Pb atoms of -1.2, -0.9, and $-2.7\ e$, in PbS, PbSe, and PbTe correspondingly. These do not correlate with electronegativities of atoms and do not describe the ED picture in the compounds in question.

The more rigorous and flexible way to describe the ED uses the 'outer' pseudoatomic moments in addition to net atomic charges (Hirshfeld 1977). These values allow one to describe, quantitatively, the main features of ED of a bonded atom (i.e. to describe its valence state) with a limited number of parameters. The general expression for pseudoatomic moments via ED is

$$\mu_{\alpha\beta\ldots\eta} = \int_\Omega \hat{\mu}\rho(\boldsymbol{r})\,\mathrm{d}\boldsymbol{r} \qquad \hat{\mu} = r_\alpha^n r_\beta^m \cdots r_\eta^l \qquad \alpha = x, y, z \tag{6.8}$$

the integral being taken over the pseudoatomic volume, Ω, and radius vector $\boldsymbol{r}_\alpha$ originating at the nucleus of the pseudoatom in question. The orders $n, m, \ldots, l$ for 'outer' moments are positive integer numbers, or zeros (for 'inner' moments, such as electrostatic potential, electric field, or its gradient, they are integer negative numbers). For $\hat{\gamma} = 1$ expression (6.8) gives the number of electrons of the pseudoatom, for $\hat{\gamma} = r_\alpha$ it gives the α component of the pseudoatomic dipole moment, for $\hat{\gamma} = r_\alpha r_\beta$ it gives the $\alpha\beta$ component of the pseudoatomic second moment, etc. If the negative sign of ED is taken into account, then a negative value of the dipole moment component means that the pseudoatomic ED is shifted along the corresponding positive direction of the local coordinate system. A positive sign of the second moment shows the pseudoatom to be contracted relative to the spherical atom.

The outer moments are mainly determined by valence ED and only slightly depend on the completeness of x-ray diffraction data in the high-angle region. However, they strongly depend on the integration volume, Ω. The pseudoatomic moments can be calculated both by direct numerical integration of ED (Moss and Coppens 1980, 1981, Moss 1982) or deformation ED (Hirshfeld 1977), regardless of whether experimental or theoretical, static or dynamic densities are used. The use of low-temperature x-ray diffraction data or static experimental model ED is preferable since it decreases the negative thermal smearing effect. In the case of ED presentation by a Fourier series, equation (6.8) is transformed for first outer moments as

$$\begin{aligned}\mu_{\alpha\beta\ldots\eta} &= \frac{1}{V}\sum_{\boldsymbol{H}} F(\boldsymbol{H}) \int_\Omega w(\boldsymbol{r}) r_\alpha r_\beta \cdots r_\eta \exp(-2\pi\mathrm{i}\boldsymbol{H}\cdot\boldsymbol{r})\,\mathrm{d}V \\ &= \frac{1}{V}\sum_{\boldsymbol{H}} F(\boldsymbol{H}) B(\Omega, r_\alpha r_\beta \cdots r_\eta, \boldsymbol{H}).\end{aligned} \tag{6.9}$$

The volume, Ω, of a pseudoatom with discrete or fuzzy boundaries may be chosen by any of the methods described above ($w(\boldsymbol{r})$ is the weight function (6.4)). The uncertainty of volume choice remains, of course. As this volume is, generally, irregularly shaped, the subdivision of it into large numbers of parallelepipeds simplifies the numerical integration. In the Cartesian coordinate system the integration over parallelepipeds with equal volumes and with the same orientation, provided each of them has its origin in $\boldsymbol{r}_k$ and $w(\boldsymbol{r})=1$, gives

$$B=\sum_k B_k=\sum_k \exp(-2\pi i\boldsymbol{H}\cdot\boldsymbol{r}_k)\int_{-\delta_x}^{\delta_x}\cdot\int_{-\delta_y}^{\delta_y}\cdot\int_{-\delta_z}^{\delta_z} r_\alpha r_\beta\cdots r_\eta \exp(-i2\pi\boldsymbol{H}\cdot\boldsymbol{r}_k)\,\mathrm{d}V. \tag{6.10}$$

The δ_α ($\alpha=x, y, z$) values are the parallelepiped half widths along the corresponding directions; usually $\delta<0.1$ Å. The expressions arising from calculation of various outer moments are presented in table 6.8.

The dipole as well as other odd higher-order moments of spherical atoms are zeros. This is not the case for even moments and, if the pseudoatomic moments are derived using standard deformation ED, then the contribution from the promolecule should be taken into account. Corresponding values for the second moments, for example, as was noted by Spackman (1992), can be easily calculated as $\mu_{ii}=\mu_{ii}(\delta\rho)+\frac{1}{3}\langle r^l\rangle_{spher.atom}$ by using atomic values $\langle r^2\rangle_{spher.atom}$ tabulated by Boyd (1977b) and McLean and McLean (1981). The use of $\delta\rho$ decreases the Fourier series termination effect and the negative influence of the atomic thermal vibrations.

When ED in a unit cell is presented in the form of a multipole expansion, the $\rho(\boldsymbol{r})$ in expression (6.8) can be replaced by $\rho_M(\boldsymbol{r})$ and the integration can be performed over the whole space (Stewart 1972). This allows one to obtain the analytical expressions for outer and inner point moments of pseudoatoms. The outer point multipole moments also describe the deformation of the pseudoatom ED due to chemical bonding. If Cartesian spherical harmonics and a multipole radial function of (3.89) type are used, the general expression for such moments is (Spackman *et al* 1988, Coppens and Becker 1992)

$$\begin{aligned}\mu_l&=-C_{lm\pm}\int_0^\infty r^{l+2}R_n(r)\,\mathrm{d}r\int_0^\pi\int_0^{2\pi}[y_{lm\pm}(\theta,\varphi)]^2\sin\theta\,\mathrm{d}\theta\,\mathrm{d}\varphi\\&=-C_{lm}N(n,l,m)\alpha^{-l}.\end{aligned} \tag{6.11}$$

The minus sign arises due to the negative value of electron charge; α and C_{lm} are parameters of a multipole model. The explicit form of constant $N(n, l, m)$ depends on the form of normalization factor in the radial multipole function and on the specific form of its angle part. For example, for the Stewart (1976) multipole model $N(n, l, m)=[(n+l+2)!(1+\delta_{m,0})(l+m)!]/[(n+2)!(4l+2)(l-m)!]$.

Let us see how the pseudoatomic moments can be used in ED analysis, taking formamide, CH_3NO, as an example (Moss 1982). Two molecules in

Table 6.8 Formulae for pseudoatomic moment calculations from electron density (Moss and Coppens 1981). $\mathbf{\Omega}_k$ is the volume of the parallelepiped, j_1 and j_0 are spherical Bessel functions; $\alpha, \beta, \gamma = x, y, z$.

$\hat{\gamma}$	Moment	Expression for integral in (6.10)
1	monopole moment (charge)	$\mathbf{\Omega}_k j_0(2\pi h\delta_x) j_0(2\pi k\delta_y) j_0(2\pi l\delta_z)$
r_α	dipole moment	$-\mathrm{i}\mathbf{\Omega}_k \delta_\alpha j_1(2\pi h_\alpha \delta_\alpha) j_0(2\pi h_\beta \delta_\beta) j_0(2\pi h_\gamma \delta_\gamma)$
$r_\alpha r_\beta$ $\alpha \neq \beta$	second moment (non-diagonal element)	$-\mathbf{\Omega}_k \delta_\alpha \delta_\beta j_1(2\pi h_\alpha \delta_\alpha) j_1(2\pi h_\beta \delta_\beta) j_0(2\pi h_\gamma \delta_\gamma)$
$r_\alpha r_\beta$ $\alpha = \beta$	second moment (diagonal element)	$-\mathbf{\Omega}_k \delta_k^2 \left\{ \frac{j_1(2\pi h_\alpha \delta_\alpha)}{\pi h_\alpha \delta_\alpha} - j_0(2\pi h_\alpha \delta_\alpha) \right\} j_0(2\pi h_\beta \delta_\beta) j_0(2\pi h_\gamma \delta_\gamma)$

the unit cell of this crystal are bonded by two hydrogen bonds forming the centrosymmetrical dimer. The third hydrogen bond connects the dimers in the chains. The outer pseudoatomic moments calculated for fuzzy boundary pseudoatoms from the experimental ED are presented in table 6.9. As follows from net atomic charges and values of pseudoatom dipole moments, the ED in C–H and N–H bonds of formamide is shifted from the H atoms towards their more electronegative partners. The second moments show that most of the atoms are isotropically contracted in comparison with free atoms, except the O atom, which is expanded in the direction perpendicular to the NCOH fragment plane. This pattern coincides with the evidence from $\delta\rho$ maps calculated by Stevens (1978), where a shift of the O atom electron lone pairs from this plane is also observed.

Schwarz (1991) has noted that for singly, doubly, and triply bonded C atoms in different compounds there is a linear correlation between the quadrupole pseudoatomic parameter and the electronegativity of atom partners. Thus the shape of bonded atoms in molecules and crystals, together with the bond lengths and bond angles, characterizes the bond.

The molecular dipole, μ_i, and second, μ_{ij}, moments, which are measured quantities for free molecules, can be derived from outer pseudoatomic moments. According to Hirshfeld (1977) these moments are given, correspondingly, in the same coordinate system by

$$\mu_i = \sum_\alpha (r_{\alpha i} q_\alpha + \mu_{\alpha i}) \tag{6.12}$$

$$\mu_{ij} = \sum_\alpha [r_{\alpha i} r_{\alpha j} q_\alpha + r_{\alpha i}\mu_{\alpha j} + r_{\alpha j}\mu_{\alpha i} + \mu_{\alpha ij}] \tag{6.13}$$

$i, j = x, y, z$, and α indicates an atom in the system. The nuclear charges are assumed to be added to the monopole terms. The traceless quadrupole moment tensor, $\theta_{ij} = \frac{1}{2}\int \rho(\mathbf{r})[3r_i r_j - r^2\delta_{ij}]\,\mathrm{d}V$, which is a sensitive measure of the anisotropy of molecular charge distribution, is also often used (Buckingham 1959). Their components are derived from molecular second moments as follows:

$$\theta_{ii} = \mu_{ii} - \tfrac{1}{2}(\mu_{jj} + \mu_{kk}) \qquad \theta_{ij} = \tfrac{3}{2}\mu_i. \tag{6.14}$$

Dipole and quadrupole moments of the formamide molecule in the crystal, determined by different methods, are presented in table 6.10. They satisfactorily agree with each other and with the experimental values. Note that the non-zero, z component of the dipole moment, perpendicular to the molecule plane, can be related to the corresponding hydrogen bond in the crystal. Thus, the pseudoatom moments give important information on a formamide molecule in a crystal.

The complete list of data concerning the dipole and quadrupole moments, derived from the x-ray diffraction data, has been given by Spackman (1992).

Table 6.9 Pseudoatomic fuzzy boundary deformation moments determined by direct integration for formamide, CH_3NO, referred to local coordinate frames (Moss 1982).

Local coordinate frames

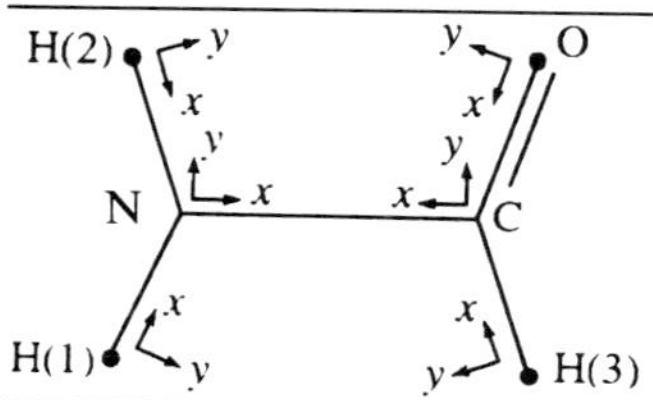

Atom	$\mu_0 = q$ (e)	μ_x (e Å)	μ_y (e Å)	μ_z (e Å)	μ_{xx} (e Å^2)	μ_{yy} (e Å^2)	μ_{zz} (e Å^2)
O	−0.32	−0.047	0.013	0.012	0.068	0.072	−0.120
N	−0.19	−0.088	0.027	0.004	0.123	0.150	0.020
H(2)	0.21	−0.079	−0.028	0.003	0.038	0.153	0.085
C	0.01	−0.035	−0.051	−0.075	0.114	0.137	0.183
H(1)	0.16	−0.072	−0.015	0.020	0.033	0.135	0.080
H(3)	0.13	−0.099	−0.004	0.028	0.083	0.101	0.098

Table 6.10 Molecular dipole (in Debye) and quadrupole (in e Å^2) moments for formamide, CH_3NO (Moss 1982, Spackman 1992). The x axis of the molecular coordinate system which has been used for dipole moment determination is directed from the N to the C atom. The atom O has positive y coordinate in the xy plane. The α is an angle between the dipole vector direction and the negative x axis direction. The quadrupole moment components are presented in the principal inertial axes. A, discrete boundary direct integration of ED; B, fuzzy boundary direct integration of ED; C, $\varkappa$-technique; D, multipole model refinement; E, experimental measurements; F, calculation with extended basis set.

	μ_x	μ_y	μ_z	μ	α	θ_{xx}	θ_{yy}	θ_{zz}
A	−4.02	−2.84	0.48	4.94	35.7	—	—	—
B	−3.25	−2.31	0.44	4.01	35.4	—	—	—
C	−2.88	−3.31	0.15	4.39	39.0	−0.04(2)	0.57(11)	−0.53(9)
D	−3.56	−3.26	0.11	4.83	42.5	—	—	—
E	−2.85	−2.38	0.0	3.71	39.6	−0.06(4)	0.70(8)	−0.64(16)
F	−3.06	−2.70	−0.49	4.07[a]	41.9	−0.27	0.78	−0.51

[a] The calculation was done for a geometry in which H atoms bonded to N atoms lie behind the molecular plane.

He also discussed the accuracy of determination of these quantities and their use for the study of physical properties of molecular crystals.

Thus, the language of pseudoatomic moments, in which net atomic charges should be included, is useful and relatively simple in chemical bond description.

6.3 Deformation electron density

In the simplest and most widespread meaning the deformation ED, $\delta\rho$ (1.15), is the difference between the ED of a real system and the ED of some hypothetical reference state: a set of spherical atoms, placed in real atomic positions. This 'standard' deformation ED describes the total redistribution of the electrons while the bonded system is formed from the set of spherical atoms. The intermediate stages of the atomic ED changes while atoms are approaching each other are not considered here at all.

There are other possibilities of reference state choice. For example, the promolecule from atoms in certain valence states can be used. The reference state in which the valence σ orbitals of bonded atoms, directed along the internuclear line, are occupied by a single electron, and the rest of the electrons are distributed among π, δ, etc AOs uniformly, is also used (Bader 1981). For a better understanding of the information which is contained in the deformation ED, let us consider more closely the electron structure of atoms being subtracted from the total ED.

The atomic electron structure has now been studied in much detail (see e.g. Fraga and Malli 1968). The atoms in a promolecule should be, in principle, in the lowest-energy ground states. The wave-function of a free atom may, however, be spatially non-degenerate as, for example, in the H(^{2}S), Li(^{2}S), and N(^{4}S) atoms or threefold spatially degenerate as in B(^{2}P), C(^{3}P), O(^{3}P), and F(^{3}P) atoms. In the presence of degeneracy, the spherical symmetry of the ground state of atoms with partly filled valence electron p, d, f,... subshells is violated: atomic ED gain the quadrupole and higher multipole components. There are, however, no experimental methods to determine the distribution of electrons over degenerate orbitals in such atoms. Therefore, their ED should be considered as a statistical average of EDs of different pure (ground) states, the electron configurations of which are dependent on the temperature and orbital populations can be non-integer. The statistical operator (2.17), which presents the atomic state as a mixture of pure states with the same energies, but with different valence electron orbital configurations, should be applied for description in this case. Precisely speaking, the spin–orbital coupling eliminates the spatial degeneracy. The corresponding energy level splitting for light atoms ($Z \leqslant 17$) is too small in comparison with the binding energy to be taken into account; it may be neglected. For transition elements or, moreover, lanthanoids, the level splitting is significant and should be taken into account (Schwarz *et al* 1989).

Thus, though the molecules and crystals are formed from atoms, the promolecule (and procrystal) is a fictive system only. The promolecule cannot be described by any antisymmetrical wave-function and the virial and Hellmann–Feynman theorems as well as Pauli principle are not valid here. Nevertheless, the promolecule composed of spherical atoms is used regularly nowadays. Due to the sphericity of the atomic ED there is no necessity to consider the atomic orientations in this model, which is not a trivial problem in the case of complicated heterodesmic compounds. Besides, Hirshfeld and Rzotkiewicz (1974) found that the classical electrostatic energy of the spherical atom interaction is always negative, excepting H atoms. Hence the forces acting on the nuclei in a promolecule from the side of ED are binding and the promolecule is always stable relative to the set of spherical atoms removed to infinity. Finally, the accurate x-ray scattering functions, tabulated by Ibers and Hamilton (1974) and Su and Coppens (1994), have been calculated for spherical atoms.

In order to look at the ED redistribution when the two atoms approach each other the Rayleigh–Schrödinger perturbation theory can be used (see Kaplan 1985). At large distances the interaction energy can be decomposed into (2.90)-type contributions. At zeroth order in perturbation the long-range atomic interaction leads to the degeneracy removal: the energy of the atomic system is nearly unchanged, but each atom orients in the field of the neighbouring atoms in such a manner that the force acting on a nucleus appears to be directed at the partner which produces the electric field. At the same time, a similar effect results from the 'rigid'-atom electrostatic interactions (the first order in perturbation theory) and from polarization interactions (the second order in perturbation theory). The two latter effects lead to AO hybridization and to promotion of atoms into a valence state. After approaching nearer than $\leqslant 7$ Å, the quantum effects become significant: the wave-function's interference, the ED delocalization (charge transfer included), and the exchange. In this way, atoms are 'prepared' for chemical bond formation.

Because of such ED redistribution in the prepared atoms with less than half-filled p, d, f,... valence electron subshells (such as boron and carbon atoms), the ED along the internuclear vector is higher, but in the perpendicular direction it is lower than that in a spherically averaged atom. For prepared atoms with more than half-filled valence electron subshells (such as oxygen and fluorine atoms) the opposite is true (Ransil and Sinai 1972). Therefore, the appearance of positive and negative features in the standard $\delta\rho$ maps in σ, π, and δ bond regions is dependent on the ratio of atomic spherical cloud overlap and on quantum mechanical effects. In particular, the Pauli repulsion in a bond formed by atoms with more than half-filled shells can lead to ED decrease in the internuclear space when the number of electrons in the atomic valence subshells increases.

The quantum effect of constructive interference, at the same time, leads to only small changes in ED in the same regions. The resulting ED redistribution along the internuclear vector is, thus, less than the sum of the spherical atom EDs. Breitenstein *et al* (1982) found that the value of the positive deformation ED in the internuclear space decreases in the series N_2–O_2–F_2, even being negative for the last member (table 6.11). At the same time, molecule F_2 is stable. This finding is confirmed by numerous theoretical and experimental investigations of more complex molecules. The negative deformation ED has been observed on N–N bonds in diformylhydrazine (Hope and Otterson 1979, Tanaka 1979), tetraformylhydrazine (Otterson *et al* 1982), and carbohydrazine (Otterson and Hope 1979), on the long O–O bond in oxygen peroxide (Savariault and Lehmann 1980), on the C–F bond in tetrafluoroterephtalodinitrile (Dunitz *et al* 1983), on O–O, N–N, C–O, and C–N bonds in 1,2,7,8-tetraaza-4,5,10,11-tetraoxatricyclo[4.4.1.1]tetradecane (Dunitz and Seiler 1983, Kunze and Hall 1987), and also in M–M bonds (M = Cr, Mn, Mo) where the $\delta\rho$ peaks are small or are completely absent (Benard *et al* 1980, Hino *et al* 1981, Martin *et al* 1982).

Hirshfeld and Rzotkiewicz (1974) underlined the influence of the molecular wave-function nodes on the ED distribution. When the number of nodes in σ orbitals increases, the requirements of the orthogonality of the overlapping wave-functions together with the Pauli exclusion principle decrease the ED on the σ bond. At the same time, the relative contribution of π and δ electrons in the stabilization field at nuclei positions increases. Apparently, because of this effect the covalent bonds with participation of C, N, and O atoms are stronger than those with Si, P, and S atoms. For the same reason, a more complicated picture of $\delta\rho$ can be expected in transition-metal polynuclear complexes away from the metal–metal line (Benard 1982).

Low and Hall (1990) have analysed the influence of various factors on the chemical bond picture. They put the effects of an atom's orientation, atomic polarization, hybridization, and promotion into one group generally named hybridization effects. The charge transfer and delocalization were put into another (delocalization) effect group. The constructive interference, therefore, was separated from the other effects responsible for deformation of the atomic electron shells. In order to distinguish these effects the generalized valence bond (GVB) method was used. The GVB pair orbitals represent the atoms

Table 6.11 Theoretical EDs in the centre of the internuclear distance in diatomic molecules (Breitenstein *et al* 1982).

Molecule	Bond	ρ_{mol} (e Å^{-3})	$\delta\rho$ (e Å^{-3})
N_2	N≡N	4.85	1.34
O_2	O=O	3.70	0.32
F_2	F–F	0.50	−0.05

after hybridization and delocalization, but before the electrons are spin coupled to form the covalent bond. Therefore, subtraction of the density of the singly occupied GVB pair orbitals from the total GVB molecular ED (or from the same-quality HF molecular ED) will show the changes in ED due to constructive interference only. The truncation of the GVB pair orbitals removes the effect of delocalization and results in GVB hybrid orbitals. Therefore, the GVB hybrid orbitals contain only the effects of hybridization and polarization. The various difference and deformation density maps which describe the different effects mentioned are listed in table 6.12 and are illustrated in figure 6.10.

In the GVB extended basis set calculations each A–H bond was described by a GVB pair orbital, ϕ_a, obtained from the two GVB natural orbitals of each bond pair as

$$\phi_{a,b}=(c_1/(c_1-c_2))^{1/2}\phi_1\pm(-c_2/(c_1-c_2))^{1/2}\phi_2. \tag{6.15}$$

c_1 and c_2 are constructive interference coefficients of the GVB natural orbitals ($c_1>0$, $c_2<0$); ϕ_1 and ϕ_2 are the strongly and weakly occupied GVB natural orbitals. Atomic and hybrid valence state orbitals, as well as the total molecular ED, should be calculated in the same basis set.

In the simplest case of the H_2 molecule, the spinless part of the GVB wave-function in terms of the GVB pair orbitals is

$$\Psi_{GVB}=N[\boldsymbol{a}(1)\cdot\boldsymbol{b}(2)+\boldsymbol{b}(1)\cdot\boldsymbol{a}(2)] \tag{6.16}$$

where N is the normalization constant and $\boldsymbol{a}$ and $\boldsymbol{b}$ are GVB pair orbitals originating on H1 and H2 atoms, respectively. At the equilibrium bond

Table 6.12 Different density maps in the Low–Hall (1990) partitioning scheme. (Reprinted with permission from Low and Hall (1990). © 1990 American Chemical Society.)

$\Delta\rho$[a]	ρ	$\tilde{\rho}$
Standard deformation ED	Total molecular ED (HF or GVB)	Spherical atom promolecule
GVB hybrid deformation ED (delocalization + constructive interference)	Total (HF or GVB) molecular ED	GVB hybridized atom promolecule
Constructive interference deformation ED	Total (HF or GVB) molecular ED	GVB pair promolecule
Delocalization difference ED	GVB pair promolecule	GVB hybridized atom promolecule
GVB hybrid difference ED (hybridization)	GVB hybridized atom promolecule	Spherical atom promolecule

[a] $\Delta\rho=\rho-\tilde{\rho}$.

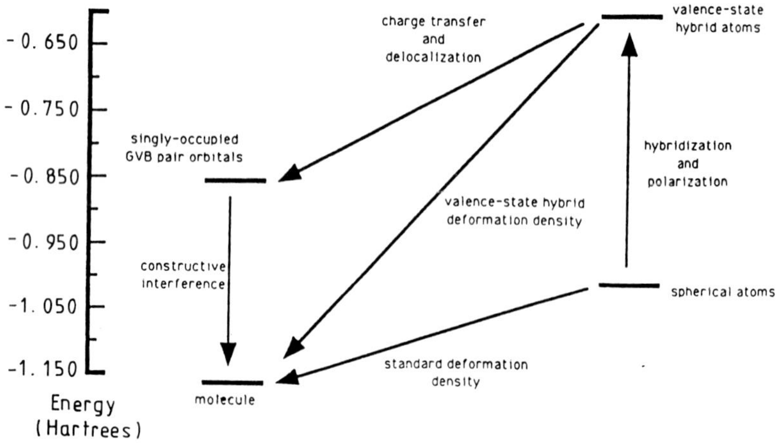

Figure 6.10 A chart commenting on the Low and Hall (1990) partitioning scheme. The energy scale shown on the left is based on the respective energies calculated for the hydrogen molecule and its different promolecules. The spherical atom energy is twice the energy of a spherical H atom. The energy of the valence state hybrid atoms was twice the energy of the first SCF cycle of an atomic calculation of H starting from the valence state hybrid orbital. The energy of the singly occupied GVB pair orbitals was twice the energy of the first SCF cycle of a molecular calculation on H_2^+ starting from a singly occupied GVB pair orbital.

distance, the ***a*** orbital will also contain a significant amount of character of the H2 atom and the ***b*** orbital will contain a significant amount of character of the H1 atom. The GVB valence state hybrid orbitals are calculated from these orbitals by truncating the functions on the other atoms. The truncated orbitals, ***a***′ and ***b***′, are then renormalized to give the hybrid orbitals:

$$\phi_{hyb}^{a} = N'(\boldsymbol{a}') \qquad \phi_{hyd}^{b} = N'(\boldsymbol{b}'). \tag{6.17}$$

The GVB pair orbitals may be expressed as consisting of the truncated orbital on the main atom plus a fraction of an orbital on the other atom. They are, with N'' as a normalization constant,

$$\boldsymbol{a} = N''(\boldsymbol{a}' + \lambda \boldsymbol{b}'') \qquad \boldsymbol{b} = N''(\boldsymbol{b}' + \lambda \boldsymbol{a}''). \tag{6.18}$$

The ED of the H_2 molecule from the GVB wave-function is

$$\rho_{GVB} = 2\int \Psi_{GVB}^2 \, dV_2. \tag{6.19}$$

Putting (6.16) into (6.19) and integrating, one arrives at the following expression:

$$\rho_{GVB} = 2N^2[\boldsymbol{a}^2(1) + \boldsymbol{b}^2(1) + S_{ab}\{\boldsymbol{a}(1)\cdot\boldsymbol{b}(1) + \boldsymbol{b}(1)\cdot\boldsymbol{a}(1)\}] \tag{6.20}$$

where S_{ab} is an overlap integral. Then, by accounting for (6.18), one obtains

$$\rho_{GVB} = 2N^2N''^2[\boldsymbol{a}'^2 + \boldsymbol{b}'^2 + (\lambda\boldsymbol{a}'')^2 + (\lambda\boldsymbol{b}'')^2 + \lambda(\boldsymbol{a}'\cdot\boldsymbol{b}'' + \boldsymbol{b}''\cdot\boldsymbol{a}' + \boldsymbol{b}'\cdot\boldsymbol{a}'' + \boldsymbol{a}''\cdot\boldsymbol{b}') + S_{ab}\{(\boldsymbol{a}'\cdot\boldsymbol{b}'' + \boldsymbol{b}''\cdot\boldsymbol{a}' + \boldsymbol{b}'\cdot\boldsymbol{a}'' + \boldsymbol{a}''\cdot\boldsymbol{b}') + \lambda^2(\boldsymbol{a}''\cdot\boldsymbol{b}'' + \boldsymbol{b}''\cdot\boldsymbol{a}'')\}]. \tag{6.21}$$

The ED of the valence state hybrid and the GVB pair promolecules can be expressed as

$$\rho_{hyb} = N'^2(\boldsymbol{a}'^2 + \boldsymbol{b}'^2) \tag{6.22}$$

and

$$\rho_{pairs} = \boldsymbol{a}^2 + \boldsymbol{b}^2 = (N'')^2[\boldsymbol{a}'^2 + \boldsymbol{b}'^2 + (\lambda\boldsymbol{a}'')^2 + (\lambda\boldsymbol{b}'')^2 + \lambda(\boldsymbol{a}'\cdot\boldsymbol{b}'' + \boldsymbol{b}''\cdot\boldsymbol{a}' + \boldsymbol{b}'\cdot\boldsymbol{a}'' + \boldsymbol{a}''\cdot\boldsymbol{b}')]. \tag{6.23}$$

As results from the expressions (6.21)–(6.23), the ED of molecules can be split into four parts. The first two terms in (6.21) correspond, with a change in normalization constants, to the ED of the GVB hybrid orbitals. The pair of terms $(\lambda\boldsymbol{a}'')^2$ and $(\lambda\boldsymbol{b}'')^2$ correspond to the ED change due to charge transfer. The next term, $\lambda(\boldsymbol{a}'\cdot\boldsymbol{b}'' + \boldsymbol{b}''\cdot\boldsymbol{a}' + \boldsymbol{b}'\cdot\boldsymbol{a}'' + \boldsymbol{a}''\cdot\boldsymbol{b}')$, corresponds to the ED change due to charge delocalization. This term can be viewed as part of the overall constructive interference of the molecule. However, with a very large basis set, it would be possible to achieve this degree of delocalization without using functions from the other atoms. The sum of the first five terms in (6.21) corresponds, again with a change in normalization constants, to the density of the GVB pair orbitals. The last term in (6.21) corresponds to the ED change due to constructive interference of the GVB pair orbitals.

The difference ED maps for the classical example, the hydrogen molecule, which make up the partitioning scheme, are shown in figure 6.11. The standard deformation density map (figure 6.11(*a*)) shows the accumulation of ED in the nuclear region and its depletion in the outer non-bonding regions. The bond accumulation completely surrounds the hydrogen nuclei. The GVB hybrid difference ED map (figure 6.11(*b*)) shows that the GVB valence state hybrid orbitals are contracted relative to the spherical atom. The ED accumulation regions about the nuclei are extended toward the centre of the molecule due to some H p_z mixing into the GVB hybrid. The GVB hybrid deformation ED map (figure 6.11(*c*)) reveals the ED accumulation region in the centre of the molecule and ED depletion behind the H nuclei.

The deformation ED of the hydrogen molecule results mainly from constructive interference and delocalization effects. The delocalization term in (6.21) represents the major contribution to the standard deformation ED and is responsible for about half the ED accumulation at the bond midpoint (figure 6.11(*d*)). At the same time, a constructive interference accumulation region located between the two H nuclei (figure 6.11(*e*)) is the smallest of all the ED accumulation regions in all the maps. The ED regions are adjacent

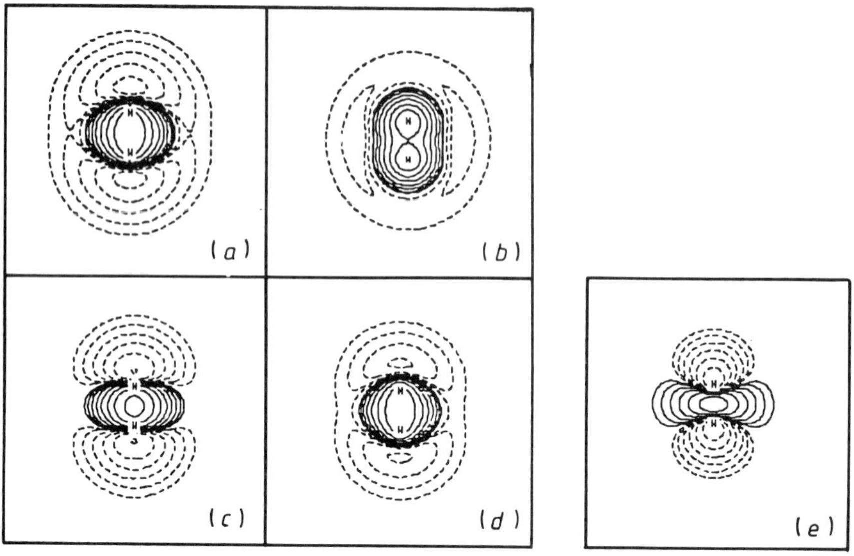

Figure 6.11 Difference density maps for H_2: (*a*) standard deformation density; (*b*) GVB hybrid difference density; (*c*) hybrid deformation density; (*d*) delocalization difference density; (*e*) constructive interference density. Density maps are contoured geometrically with each contour differing by a factor of two. The smallest positive and negative contours have a value of $\pm 2^{-12}$ electrons $(au)^{-3}$ (2.4414×10^{-4}). Negative contours are dashed (Low and Hall 1990).

to the H nuclei. The remaining accumulations consist of two nearly equal contributions from hybridization and polarization, mainly contraction of the hybrid orbitals.

The partitioning scheme used clearly shows the different effects contributing to the standard deformation ED map for the hydrogen molecule. This molecule, however, is unique in many respects: there are no nodes of its electron-filled orbital in the internuclear space, the electrostatic energy of the two H spherical atoms is positive, etc. Therefore, the H_2 results cannot be applied in full to other, more complex, systems. In order to work out an objective view of the role of various factors on chemical bond formation, let us consider an application of the same partitioning scheme to the first-row element hydrides (Low and Hall 1990). The type of bond between atoms ranges in this series from ionic (LiH) to covalent (CH_4) and polar covalent (HF). The standard deformation ED maps of these molecules are shown in figure 6.12. The largest ED accumulation region in the LiH is nearly centred on the H nucleus with a polarization toward the Li atom. The features around the Li atom differentiate this map from others: there is a dipolar deformation of ED, with an accumulation region on the side of the Li atom

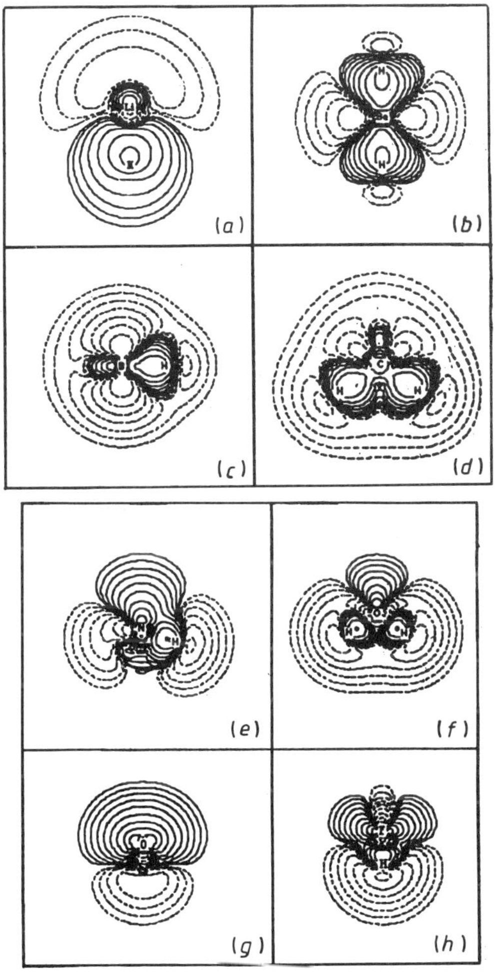

Figure 6.12 Standard deformation density maps: (*a*) LiH, a plane containing the molecule; (*b*) BeH_2, a plane containing the molecule; (*c*) BH_3, a plane perpendicular to the molecular plane containing the boron and one of the hydrogen atoms; (*d*) CH_4, a plane containing the carbon atom and two of the hydrogen atoms; (*e*) NH_3, a plane containing the nitrogen, one of the hydrogen atoms, and the nitrogen lone pair; (*f*) H_2O, a plane containing all of the atoms; (*g*) H_2O, the plane perpendicular to the molecular plane containing oxygen lone pairs; (*h*) HF, a plane containing the molecule. Density plots are contoured as in figure 6.11 (Low and Hall 1990).

pointing away from the H atom and an ED depletion region on the side of the Li atom closer to the H atom. Surrounding this region is another dipolar pattern of ED with loss on the side opposite the hydrogen and gain on the other side which combines with the accumulation region around the H atom.

These patterns can be attributed to polarization of the Li is core away from the H atom in response to the negative charge around the H atom, and transfer of the Li 2s ED closer to the H atom, respectively. The transfer of ED from one nucleus to another and the polarization of the ED on both the anionic nucleus and on the cationic nucleus has been defined by Bader (1981) as characteristic of ionic bonding.

A progressive series of changes occurs from the BeH_2 to the HF molecule: the peak of the internuclear ED accumulation, shifted to the H atom, moves from the H to the A atom in agreement with the increase of electronegativity. In BeH_2, BH_3, and CH_4, the largest ED accumulation occurs along the A–H bond axis. In NH_3, H_2O, and HF, however, the largest accumulations in the $\delta\rho$ maps are associated with the lone pair(s) and not the A–H bond. Small regions of ED depletion exist in the three last molecules along the bond direction adjacent to the A nucleus. This can be attributed to the growth of ED of the p orbitals of spherical N, O, and F atoms in the promolecule and the polarization of the 2s electron pair away from the A–H bonds. The large p_σ AO population in HF produces a region of ED depletion on the back side of the F nucleus. The large ED accumulation in the lone-pair regions in NH_3, H_2O, and HF and the large depletion regions perpendicular to the bond axis in BH_3 and BeH_2 can also be attributed to the choice of the spherical atom promolecule.

A small region of ED deficit on the back side of the H atom, distant from the H nucleus, increases in area and value and moves towards the H nucleus from BH_3 to HF. Starting from NH_3, the depletion region begins to bend around the back of the H nucleus. In HF this depletion region is found adjacent to the H atom. Analogous deficit regions, again observed by Bader (1981) in diatomics, can be attributed to movement of the ED from the back side of the H atom to form the A–H bond.

The GVB hybrid difference ED maps (figure 6.13) allow one to analyse the effects of hybridization. So, the 2s–$2p_z$ mixing upon formation of the GVB hybrid of the Li atom in LiH results in a rising region of ED depletion on the non-bonding side of the Li atom with a corresponding ED accumulation on the opposite side which merges with the accumulation region about the H atom. The region closest to the Li atom is one of ED deficit with a small accumulation region on the back side of the Li nucleus.

The ED accumulation around the H atom is polarized toward the Li atom in LiH. This polarization diminishes from BeH_2 to CH_4 and is polarized away from the A atom in H_2O and HF. This is caused by the fact that hybrid difference ED maps correspond to the valence state hybridization of atom A in the promolecule. The hybrid orbital, pointing along the A–H bond axis, contains one electron; therefore, the GVB hybrid promolecule will contain more ED than the spherical atom promolecule in the internuclear A–H region for BeH_2 and BH_3. The opposite picture will be observed in H_2O and HF. Thus, the internuclear region can be a region of deficit in the difference ED

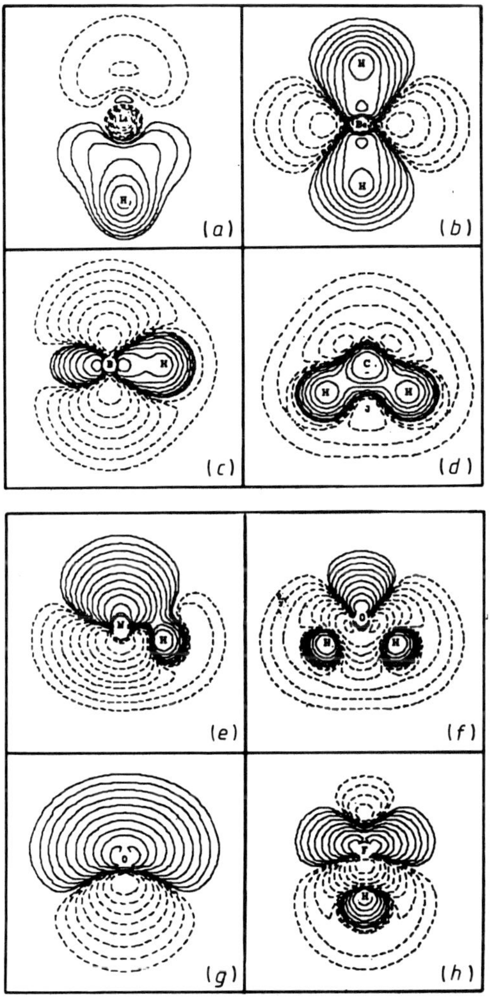

Figure 6.13 GVB hybrid difference density maps: (*a*) LiH; (*b*) BeH_2; (*c*) BH_3; (*d*) CH_4; (*e*) NH_3; (*f*) H_2O; (*g*) H_2O lone pairs; and (*h*) HF; in the same planes as described in figure 6.12. The contouring is the same as in figure 6.11 (Low and Hall 1990).

maps discussed. Due to properties of the hybrid promolecule the large ED depletion regions perpendicular to the bond axis also appear in difference maps of BeH_2 and BH_3 and large lone-pair ED accumulations appear in the maps of NH_3, H_2O, and HF.

It can be concluded that, due to the low population of the p orbitals in the spherical Be and B atoms, the bond accumulation observed in the standard deformation ED maps of BeH_2 and BH_3 is enhanced because the effects of hybridization add density to this region. Conversely, the internuclear accumulation in these maps of NH_3, H_2O, and HF is depleted because the effects of hybridization now subtract ED from this region. Besides, the large ED deficit regions due to vacant p orbitals and the large accumulation regions due to the lone pairs should be accounted for in the valence state hybrid promolecule. Therefore, these features should not dominate the GVB valence state hybrid deformation ED maps as they do in the standard deformation ED maps.

As was the case for H_2, the GVB hybrid orbital on the H atom in all molecules has contracted relative to the spherical H atom, as indicated by an accumulation region around the H atom. The degree of contraction increases from LiH to HF.

The GVB hybrid deformation ED maps of the first-row hydrides are shown in figure 6.14. The map of LiH shows a pattern around the Li atom similar to that in the standard deformation ED map although ED accumulation here is more polarized toward the Li atom. This is the result of the contraction of the GVB hybrid orbital of the H atom. When compared to the standard deformation ED maps, the ED accumulation in the bond region is smaller for BeH_2 and BH_3, roughly equal for CH_4 and NH_3, and larger for HF, as expected from the hybrid difference ED maps. The ED accumulation region between the A and H atoms in the maps of BH_2, BH_3, and CH does not extend toward the central A atom. The area close to the A atom, just outside the A atom core, is now one of ED deficit due to the mixing of 2s and 2p orbitals upon formation of the GVB valence state hybrid atoms. The latter effect can increase the ED near the nucleus.

As in the standard deformation ED maps, the $\delta\rho$ peak maximum shifts toward the A atom as A changes from Be to F; in HF it is much closer to the F atom than to the H atom. This effect can be attributed to the ED transfer.

The large deficit ED region on the non-bonding side of the H nucleus in the GVB hybrid deformation ED map is adjacent to the H atom in the maps from BeH_2 to HF. This is in contrast to the standard deformation ED for BeH_2 and BH_3 where this region is located a significant distance away from the H nucleus. The region of ED accumulation adjacent to the H atom facing the A atom is smaller in magnitude as compared to the standard deformation ED maps. All of these features are due to the contraction of the H atom valence state hybrid ED relative to the spherical ED.

In NH_3 (figure 6.14(*e*)) the lone-pair ED accumulation in the standard deformation ED map has been replaced by a region of ED depletion near the N atom and a diffuse ED accumulation in the outer regions. The origin of this change can be found by comparing the lone-pair regions of the maps in figures 6.12(*e*) and 6.13(*e*). Because the GVB lone-pair ED contracts when it is renormalized after orbital truncation, the lone-pair region in the GVB hybrid has a larger accumulation close to the N nucleus. At the same time, it has a smaller overall size than corresponding accumulation in the standard

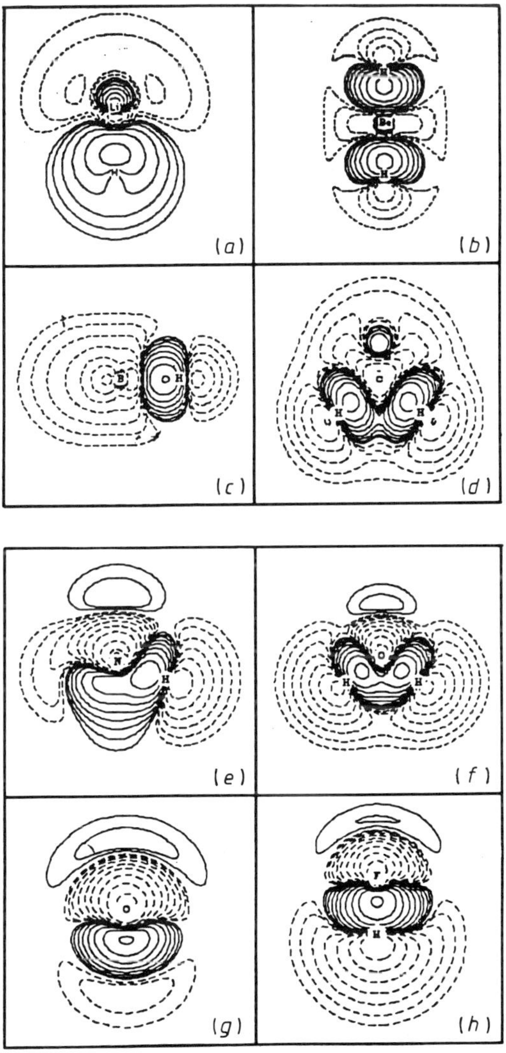

Figure 6.14 GVB hybrid deformation density maps: the same molecules as in figure 6.13, in the same planes (Low and Hall 1990).

deformation ED map. Another explanation would be N atom lone-pair orbital expansion when the bonds to the H atoms are formed. Similar patterns are observed in the lone-pair regions of H_2O and HF.

The partitioning of hybridization and polarization from the standard deformation ED maps produces maps with substantial ED accumulations between atoms. The size of the accumulation region grows from BeH_2 to HF, while it stays fairly constant throughout the series of standard deformation ED maps.

The delocalization difference ED maps are shown for the first-row hybrids in figure 6.15. A large ED accumulation region in LiH is nearly centred on the H atom. There is the dipolar-type ED accumulation on the back side of the Li nucleus and a deficit region facing the H atom surrounded by an opposite ED flow in the outer regions. The ED maximum position migrates further from the H atom in the maps of BeH_2–HF. Starting with BeH_2, areas

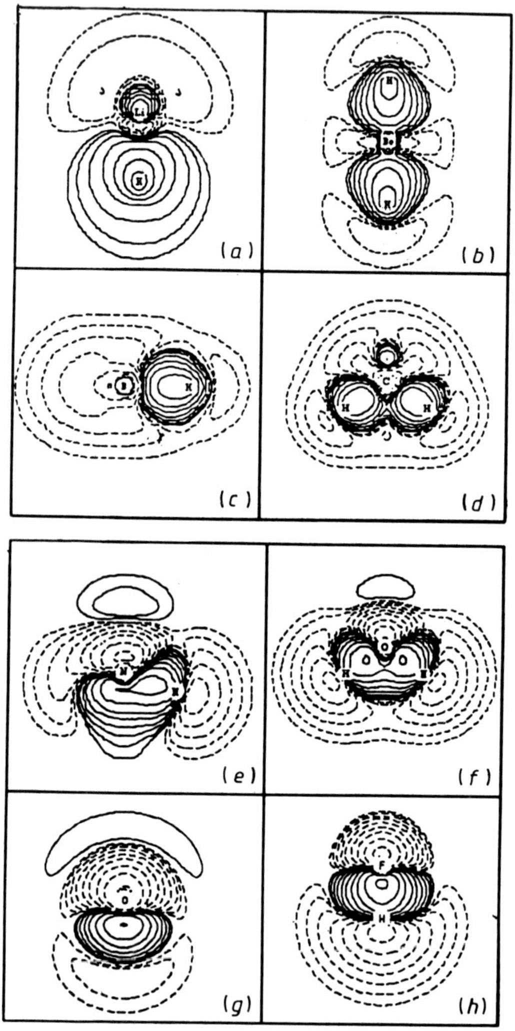

Figure 6.15 Delocalization difference density maps: the same molecules as in figure 6.13, in the same planes (Low and Hall 1990).

of ED deficit at the back of the H atom grow as atom A becomes increasingly electronegative. The lone-pair regions in the maps of NH_3, H_2O, and HF are similar in appearance to the GVB hybrid deformation ED maps.

The ED differences due to constructive interference of the GVB pair orbitals are shown in figure 6.16. The main feature in the map of LiH is a nearly centred, a lop-sided, ringlike ED accumulation region surrounding the deficit region near the H nucleus. This accumulation is larger on the side closer to

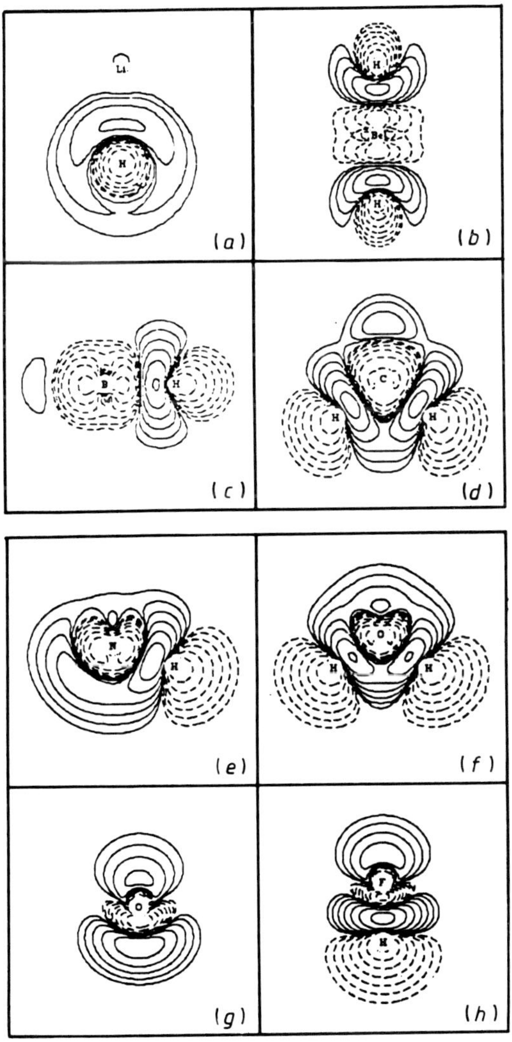

Figure 6.16 Constructive interference deformation density maps: the same molecules as in figure 6.13, in the same planes (Low and Hall 1990).

the Li atom due to Li character in both GVB pair orbitals. The bend of the ED accumulation region around the H atom is much less pronounced in BeH_2 and BH_3 and it is almost gone in CH_4. Starting with NH_3, the accumulation region begins to bend back around the central A atom; the effect is more pronounced in NH_3 and HF. The Ed deficit region near the H nucleus moves out toward the back side of the H atom in the series from BeH_2 to HF.

In the BeH_2, BH_3, and CH_4 molecules the only significant ED accumulation regions are associated with the A–H bond. The area around both the A and H atoms is one of ED deficit. There is a very small ED accumulation in the lone-pair region in NH_3. Analogous accumulation is larger in H_2O and even larger still in HF.

The pattern of ED accumulation described may be rationalized through examination of the GVB pair orbitals associated with the A–H bond. These orbitals for LiH, BH_3, NH_3, and HF are shown in figure 6.17. For LiH, the GVB pair orbital on H consists mainly of an H 1s orbital with a much smaller part on Li. The GVB pair orbital on Li, in contrast, contains a relatively large amount of hydrogen atom character. Most of the constructive interference of these two orbitals would be expected to be near the hydrogen atom. Thus, the constructive interference is bent around the H atom because it represents the interaction of two mainly hydrogen s-like orbitals. This will be characteristic of the constructive interference of an ionic bond.

A steady charge is observed in the GVB pair orbitals for BH_3 and NH_3. In the orbital on the H atom, the relative amount of A character steadily increases. The orbital on A contains an increasing amount of p character of the A atom relative to the A s and H s character as one progresses from BH_3 to NH_3. These observations can explain the differences in the bend of the constructive interference in the maps. For BH_3, the orbital on the B atom still contains a significant amount of H s character while the orbital on the H atom almost entirely consists of H s character. Therefore, when these two orbitals interact, the constructive interference should be bent back around the H atom since it still involves the interaction of two orbitals with a large amount of H s character. However, for NH_3 the GVB pair orbital on H contains a significant amount of N p character while the orbital on N consists mainly of N p character. When these two orbitals interact, the constructive interference will be bent back around the N atom since it involves the interaction of two orbitals with a significant amount of N p character.

The GVB pair orbitals of HF both contain large amounts of F atom p_z character. The orbital on H even looks quite like an F p_z orbital with some H s character mixed in. When these two orbitals interact, the internuclear constructive interference will be bent back around the F atom. However, that is not the only place these orbitals will constructively interfere. On the back side of the F nucleus, there is a large orbital lobe for both GVB pair orbitals. These orbital lobes will also produce the constructive interference on the back side of the F nucleus. Thus, the accumulation region on the non-bonding side of the F atom is due to the interaction of two GVB pair orbitals which consist of mostly F p character. A similar argument can be applied to rationalize the accumulation in the lone-pair region of the constructive interference deformation density map of H_2O.

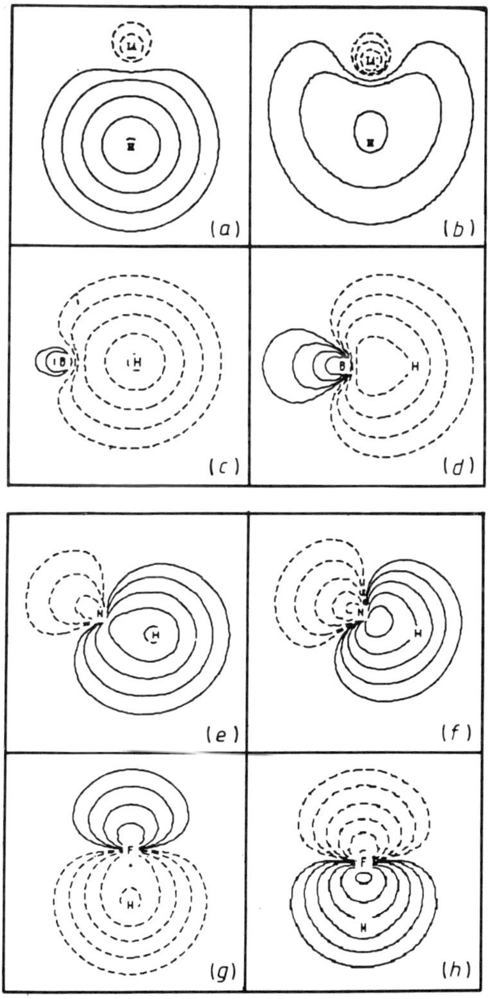

Figure 6.17 Orbital plots of the GVB pair orbitals: for LiH, (*a*) GVB pair orbital originating on H and (*b*) GVB pair orbital originating on Li; for BH_3, (*c*) GVB pair orbital originating on H and (*d*) GVB pair orbital originating on B; for NH_3, (*e*) GVB pair orbital originating on H and (*f*) GVB pair orbital originating on N; and for HF, (*g*) GVB pair orbital originating on H and (*h*) GVB pair orbital originating on F. Plots are contoured geometrically with each contour differing by a factor of two. The smallest contours have a value of $\pm 2^{-5}$ $[e\,(\text{au})^{-3}]^{1/2}$ (Low and Hall 1990).

The results of Low and Hall (1990) have been plotted here in every detail because they permit one to analyse the deformation ED maps more reliably, at least for non-transition-element compounds. It can be concluded that standard deformation ED can be divided into three parts: hybridization (and polarization), delocalization (and charge transfer), and constructive interference. Hybridization requires a certain amount of energy in order to promote the electrons and to mix the s and p orbitals. The amount of hybridization will change due to different bonding requirements and differences in the energy of the orbitals mixed. The energy required for hybridization will be compensated by the better overlap of the hybrid orbitals as compared to the spherical atoms. The hybrid deformation ED maps represent the energy difference between the valence state hybrid atoms and the molecule or crystal. Moreover, as Low and Hall have noted, the values of largest contour in bond ED accumulation on these maps seem to be qualitatively correlated to the bond energies.

The effects of delocalization are always attractive. This is due to the Coulombic attraction between the ED and both nuclei since delocalization contributes to the ED accumulation between the two nuclei, by definition.

For all of the first-row hydrides discussed the constructive interference results in the smallest internuclear accumulation. Thus, there are cases when major contributions to the standard deformation ED are related to terms other than constructive interference.

Let us consider now the experimental deformation ED maps for chemical bonds of different kinds. The silicon crystal (space group $Fd3m$) can serve as a classical example of a covalent interaction. The Si atoms are positioned in the highly symmetrical site $\bar{4}3m$ and form four equal covalent bonds with their neighbours. Such bonds are usually described by overlap of sp^3 hybrid AOs. The standard deformation ED map (figure 6.18(*a*)) shows that the bond formation is followed by the electron accumulation in the internuclear space. The $\delta\rho$ peak here is 0.22 e Å^{-3} and its shape has an axial symmetry relative to the internuclear line. There is also a small excessive $\delta\rho$ region in the nuclear site. It is twice as high as the estimated standard deviation of $\delta\rho$ and it is observed with x-rays at both Mo Kα and Ag Kα wavelengths, which may convince one that it is not due to improper account of anomalous scattering. Behind the nuclei on the continuation of the Si–Si line, where the antibonding σ orbitals are located, the small $\delta\rho$ minima of -0.09 e Å^{-3} are seen. The negative $\delta\rho$ is also observed in the crystallographic holes. Approximately 50% of covalent peak $\delta\rho$ is related to the first reflections (111), (220), and (311) and, what is very important, 30% is related to forbidden reflections, of which (222) is a predominant one (figure 6.18(*b*)). The forbidden experimental reflections used in calculation of the map in figure 6.18 were corrected for anharmonicity of thermal motion and resulted in an ED picture in excellent agreement with theoretical evidence. Thus, they really reflect the electron delocalization and constructive interference of the AOs in the Si crystal.

Figure 6.18 The ED distribution in the (110) plane of the Si crystal: experimental deformation ED (*a*), line intervals are 0.05 e Å^{-3}, and ED restored from measured forbidden reflections only (*b*) and correspondingly theoretically calculated (*c*), line intervals are 0.03 e Å^{-3}. (Reprinted with permission from Pietsch *et al* 1986a, Lu and Zunger 1992.)

The same $\delta\rho$ features are observed in crystals isomorphous to Si: in diamond (see figure 4.9), germanium (Brown and Spackman 1990), and α-Sn (Bilderback and Colella 1975). The internuclear peak values are lowered, however, with increasing atomic number; apparently this is a manifestation of the growth of electron delocalization with increasing metallicity of the chemical bond.

The essential electron delocalization on the $\delta\rho$ maps in the metal chemical bond has already been demonstrated in bcc vanadium (see figure 3.35). The maps for the crystals of other metals, Cr (Ohba *et al* 1982) and α-Fe (Ohba and Saito 1982) with the same structure, revealed an analogous pattern. $\delta\rho$ surplus peaks on the lines directed to the first, second, and third neighbours were observed. The heights of these peaks, ~0.15 e Å^{-3}, are lower than those in the covalent crystals. Significant ED accumulation in the vicinity of nuclei and ED decrease in localized pseudoatomic fragments are observed. The same picture is observed in theoretical $\delta\rho$ maps calculated by high-level approximations; however, the quantitative agreement between theory and experiment in metals is absent (see e.g. Tsirelson *et al* 1988b).

The standard $\delta\rho$ map for GaAs (see figure 3.27) exhibits the ED distribution in a polar covalent (partially ionic) chemical bond formed by atoms of different kinds. There is a rather high $\delta\rho$ peak on the Ga–As line, shifted toward the more electronegative Ga atom. As in the Si crystal, the accumulation of the ED between nuclei is accompanied by the ED depletion in antibonding σ orbital regions and in crystallographic holes.

In fluorite, CaF_2 (see figure 3.34), a high degree of ED separation between ions and significant ED accumulation on the F^- ion are observed, in accordance with the large electronegativity difference of the ions. Both anion and cation are polarized by the internal crystal field. Long-range forces polarize the $\delta\rho$ excess peaks not along the anion–cation line, but along the parallel cubic unit cell edges. Comparing the picture described above with ED maps in figure 6.16, which characterize the constructive interference in molecules, with predominantly ionic bonding (e.g. HF), one can conclude that this effect greatly influences the features of ED in CaF_2, especially in the neighbourhood of ions.

Let us consider multiple chemical bonds. Van Nes (1978) has investigated at low temperatures a series of crystalline compounds with increasing C–C bond order: ethane, C_2H_6, ethylene, C_2H_4, and acetylene, C_2H_2. Figure 6.19 depicts the $\delta\rho$ maps in sections normal to bond lines at the middle of the C–C bonds. In the single bond (ethane) the $\delta\rho$ peak form has essentially axial symmetry; there are small distortions of ED on the periphery of the peak due to intermolecular interaction. In the double C=C bond in ethylene the $\delta\rho$ peak is elongated normal to the bond line and to the molecular plane due to the presence of one π bond apart from the σ bond. One σ bond and two π bonds are in acetylene and the axial symmetry of the $\delta\rho$ peak is restored. The $\delta\rho$ values increase with increasing bond order. Schwarz (1991) has calculated the theoretical hybrid deformation ED of the molecules considered and concluded that this increase is due to the enhancement of oriented C atom orbital overlap, if each such atom has a single electron in the active valence σ orbital. As quantum effects are concerned, the ED at the midpoint of the internuclear vector decreases with bond order, due to electron delocalization and constructive interference. This can be attributed to the manifestation of the Pauli exclusion principle: the electrons are pulled out from the area of localization in doubly occupied MOs.

Two electrons can occupy the same MO, usually well localized, which is composed of valence AOs but is, nevertheless, a non-bonding MO. One describes these as lone-pair electrons. The spatial localization of such an MO depends on the character of the chemical bond. Figure 6.20 presents some typical examples of lone-pair $\delta\rho$ distributions. In dicyanodiamide, $C_2H_4N_4$, the structure of which is depicted in figure 6.9, the lone pair of the terminal atom N1 manifests itself on the standard deformation ED map in

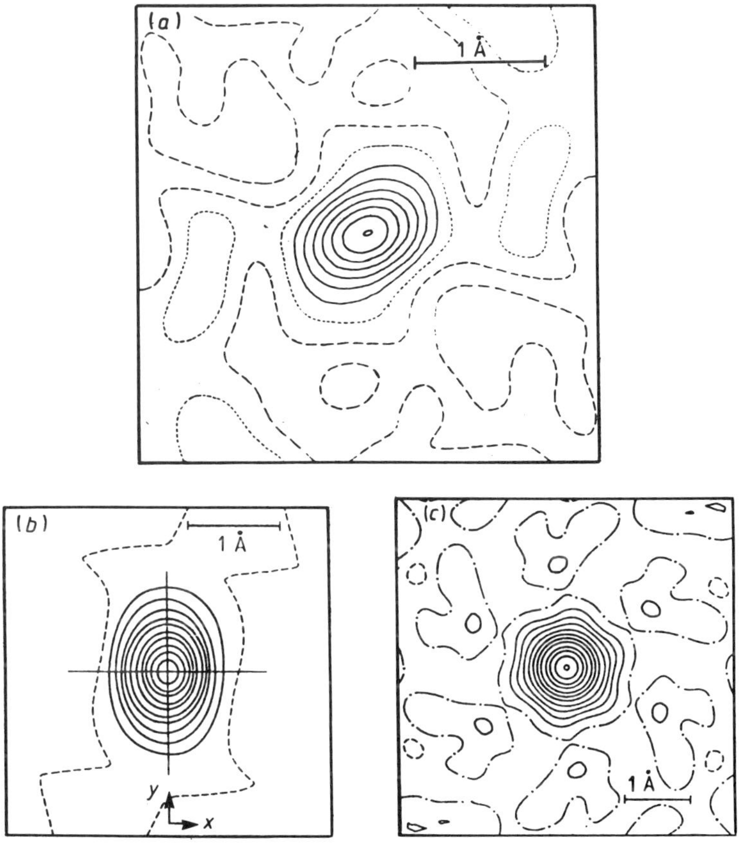

Figure 6.19 Deformation ED of multiple bonds: (*a*) ethane, C_2H_6; (*b*) ethylene, C_2H_4; (*c*) acetylene, C_2H_2. The sections are in the planes which intersect the carbon–carbon bond at the midpoint. The line intervals are 0.05 $e\,Å^{-3}$ (van Nes 1978).

a well-developed localized peak behind the N atom (figure 6.20(*a*)). It can be described by a σ^* MO or by an sp hybrid orbital. Figure 6.16(*h*) shows that this peak results also from constructive interference of orbitals in the molecule. The lone pairs of the terminal atom O1 in acetamide, C_2H_5NO (figure 6.20(*b*)) reflect themselves as $\delta\rho$ peaks which form a 120° angle with the O1–C2 line: this picture is typical for sp^2 hybridization of the O1 atom. Finally, an oxygen atom in a water molecule, which can be described as an atom in the sp^3 hybrid state, has two lone pairs smearing from the external side of atoms higher and lower than the molecular plane (figure 6.20(*c*) and (*d*)). The comparison with figures 6.14–6.16 leads to the conclusion that such a form of ED accumulation is due to hybridization of atoms and electron delocalization.

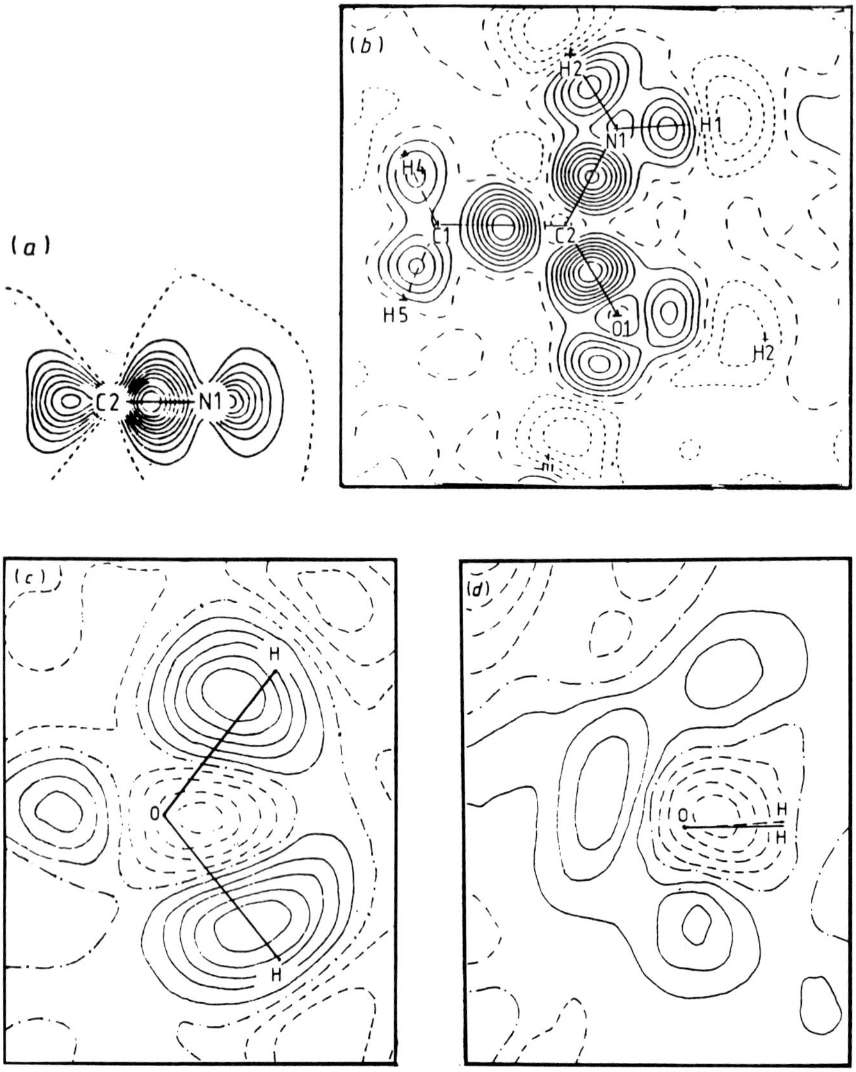

Figure 6.20 The manifestation of electron lone pairs on deformation ED maps: (*a*) dicyanodiamide, $C_6N_4H_4$, at 83 K (Hirshfeld and Hope 1980); the lone pair of the N atom is peaked behind the N nuclei; (*b*) acetamide, C_2H_5NO, at 23 K (Zobel *et al* 1991); the lone pairs of the O atom are peaked on lines which form a 120° angle with the O1–C2 line; (*c*) and (*d*) water, H_2O, in sodium nitroprusside at 153 K (Antipin *et al* 1987); the lone pairs of the O atom are seen on the sections of $\delta\rho$ in the plane of the molecule (*c*) and normal to that plane (*d*). The interval between lines is 0.1 $e\,Å^{-3}$ (*a*) and 0.05 $e\,Å^{-3}$ (*b*–*d*).

Examining the $\delta\rho$ minima and maxima in the crystal space and their position with respect to the nuclei and internuclear lines, one can establish the character of the chemical bond. The positive $\delta\rho$ areas, which are related to the covalent σ bonds, peak between bonded nuclei and behind them. The peaks of covalent π bonds are elongated normal to the bond lines. The electron lone pairs peak near the atomic position, depending on the atomic valence state. The $\delta\rho$ peak maximum in an ionic bond is displaced toward the more electronegative atom, in accordance with the conventional point of view. Negative-$\delta\rho$ areas can be attributed to out-of-phase atomic orbital overlap regions.

However, it is only a simple, near-intuitive, chemical bond treatment which can be extracted from $\delta\rho$ maps. Really, the details of the deformation ED maps characterize not only the diatomic interactions, but reflect the nature of the multi-atomic interactions. Besides, the constructive interference of the AOs and electron delocalization may result in weaker ED accumulation between nuclei, than, for example, superposition of the 'rigid' atomic clouds. The latter case appears when interacting atoms have more than half-filled valence electronic subshells.

In order to comment on the non-triviality of the treatment of the real standard deformation ED picture, let us consider heterobonds involving carbon atoms. The polarity of such bonds is not high and depends both on the kind of partner atom and on the electron effects in the whole system. If the electronegativities of the atoms bonded are significantly different, then there should be a remarkable displacement of the $\delta\rho$ maximum toward the more electronegative atom. In organic compounds, however, mutual bond interactions, conjugation, and other effects can make this displacement difficult to observe in deformation ED maps, though such a tendency is indeed often found.

The example of boron nitrilotriacetate, $C_6H_6BNO_6$ (Moeckli *et al* 1988), gives a good illustration of the standard deformation ED in such compounds. There are six different types of interatomic bond in this cage molecule (see figure 6.21(*a*)), in particular two C–O bonds: a C2=O2 double bond of length 1.206 Å and a C2–O1 ester bond (1.334 Å). The experimental (100 K) and theoretical deformation ED maps are presented in figure 6.21(*b*) and (*c*). Only on C1–N and B–N bonds are $\delta\rho$ peaks noticeably displaced toward the more electronegative N atom. The $\delta\rho$ peak is located at the centre of the C2–O1 bond and displaced even towards the C atom in the C2–O2 bond. The bonds in $C_6H_6BNO_6$ can be placed in order with respect to the lowering of the $\delta\rho$ peak asymmetry, as follows: B–N > B–O ≈ C=O > C–N > C–C ≈ C–O. Thus, judging by the standard $\delta\rho$ map, the C–O bond should be less polar than the C–N bond, and the B–O less polar than the B–N bond. However, this is in contradiction to the usual point of view on the ionicity of such bonds. Moreover, the Bader topological theory reveals the asymmetry in ED distribution in C–N, C–O, and C–F bonds: the Laplacian of ED in various

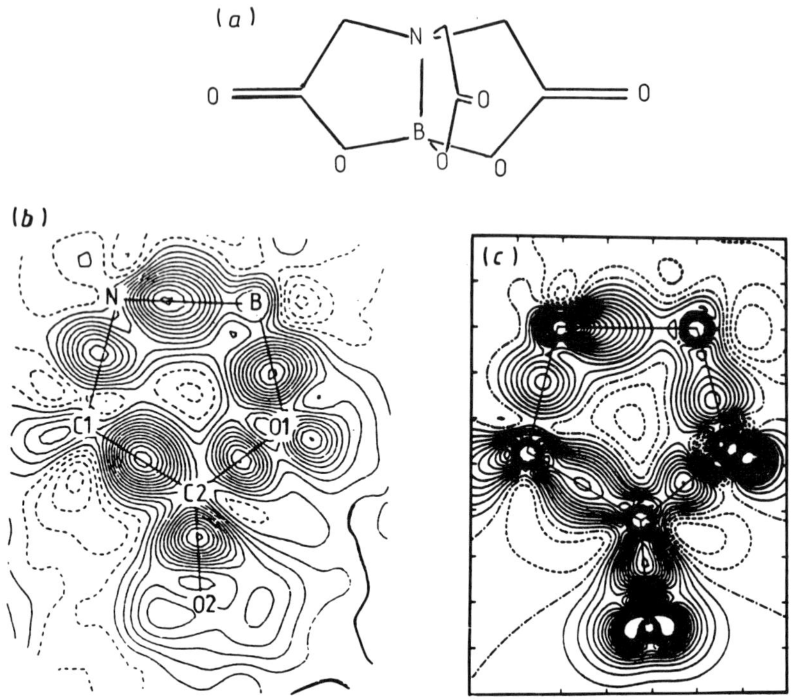

Figure 6.21 Boron nitrilotriacetate, $C_6H_6BNO_6$ (Moeckli *et al* 1988): (*a*) a view of the molecule; (*b*) multipole static deformation density; (*c*) theoretical (static) deformation density (self-consistent DFT calculation in LDA, extended basis set). Intervals are 0.05 $e\,\text{Å}^{-3}$.

molecules containing such bonds shows that charge accumulation region displacements are steadily observed (Bader *et al* 1984, MacDougall and Bader 1986).

The simple orbital scheme, which was considered by Moeckli *et al* (1988), can help us to understand the $\delta\rho$ features mentioned. Let us consider A and B as the first-row atoms with atom A less electronegative and p orbitals φ_A and φ_B oriented along A–B. The ED of the A–B bond is then

$$\rho = 2|c_A\varphi_A + c_B\varphi_B|^2/(c_A^2 + c_B^2 + 2c_Ac_BS_{AB}) \tag{6.24}$$

and the ED of the promolecule is

$$\rho_p = n_A\varphi_A^2 + n_B\varphi_B^2. \tag{6.25}$$

Here S_{AB} is the overlap integral and n is taken as one-quarter of the number of valence electrons of the atom. Using STOs with exponents from Clementi and Raimondi (1963), the following characteristics of the standard deformation ED were obtained. In the series C–C, N–N, O–O, with bond distances 1.52, 1.44, 1.47 Å, $\delta\rho_{max}$ were 0.20, 0.00, $-0.12\, e\,Å^{-3}$, respectively. The calculated trend matches the observed one for the molecule $C_6H_{12}N_4O_4$ (Dunitz and Seiler 1983), hence the model used is qualitatively satisfied. Then it was found that, if $c_A \approx c_B$, the bonding density is shifted toward the less electronegative atom A. For the B–N and B–O bonds, $\delta\rho_{max}$ is 0.19 and $0.20\, e\,Å^{-3}$, and the corresponding asymmetries are -7 and -10%, respectively. As can be seen, the reverse polarization is more pronounced for the more electron-rich and more electronegative atom B.

For $c_A < c_B$ the maximum of $\delta\rho$ is shifted toward B as expected. This shift is less pronounced if B is more electron rich. For B–N and B–O with $c_A/c_B \approx 0.47$, $\delta\rho_{max}$ values are 0.25 and $0.15\, e\,Å^{-3}$ and asymmetries are 16 and 5%, respectively. Thus, the $\delta\rho$ peak position in standard deformation ED maps depends on the sorts of atom forming the bond.

An unexpected, at first glance, feature of the standard deformation ED is the trend to depress the $\delta\rho$ peaks in the C–A bond region with increasing A atom electronegativity. Relatively low $\delta\rho$ peaks (in comparison with C–C bonds) were found in the C–O bond in trans-2,5-dimethyl-3-hexane-2,5-diol hemihydrate, $C_8H_6O_2 \cdot \frac{1}{2}H_2O$ (van der Wal and Vos 1979), in 18-crown-6, $C_{12}H_{24}O_6$ (Maverick *et al* 1980), and the lowest peak in the C–F bond in tetrafluoroterephthalodinitrile, $C_8N_2F_4$ (Dunitz *et al* 1983). Moreover, very low or even slightly negative $\delta\rho$ density in N–N bonds has been found experimentally in diformylhydrazine and its derivatives (Hope and Otterson 1979, Otterson and Hope 1979, Otterson *et al* 1982), in 4-nitropyridine N-oxide, $C_5H_4N_2O_3$ (Wang *et al* 1976), and *cis,cis*-trialkyltriaziridine, $C_{12}H_{13}O_6N_3$ (Irngartinger *et al* 1989). A negative $\delta\rho$ value has been observed in O–O bonds in hydrogen peroxide, H_2O_2 (Savariault and Lehmann 1980) and 1,2,7,8-tetraaza-4,5,10,11-tetraoxatricyclo-$[6.4.1.1^{2.7}]$tetradecane, $C_6H_{12}N_4O_4$ (Dunitz and Seiler 1983). It has been noted that deformation ED values decrease in bonds of the molecule $C_6H_6BNO_6$, in the order C=O ≈ C–C > B–N ≈ B–O > C–N ≈ C–O, and in ordinary bonds of the molecule $C_6H_{12}N_4O_4$ (figure 6.22(*a*)) in the order C–N > C–O > N–N > O–O. The same fragments of experimental and theoretical standard deformation ED for O–O bonds of the latter molecule are shown in figure 6.22(*b*) and (*c*); they coincide in all details. This agreement is evidence that the lack of ED accumulation in standard deformation ED maps is not caused by experimental uncertainties. The hybrid difference ED and hybrid deformation ED maps (figure 6.23), calculated by Kunze and Hall (1987), allow one to analyse the mechanism of the chemical bond formation in $C_6H_{12}N_4O_4$. The ‘preparation’ of spherical atoms for chemical bond formation, when the ED accumulates in the lone-pair regions of O atoms and near the C atom nucleus, can be seen in figure 6.23(*a*). The electron delocalization and constructive interference at the covalent O–O and C–O bond formation lead to the additional accumulation ED in the internuclear space (figure 6.23(*b*)). The

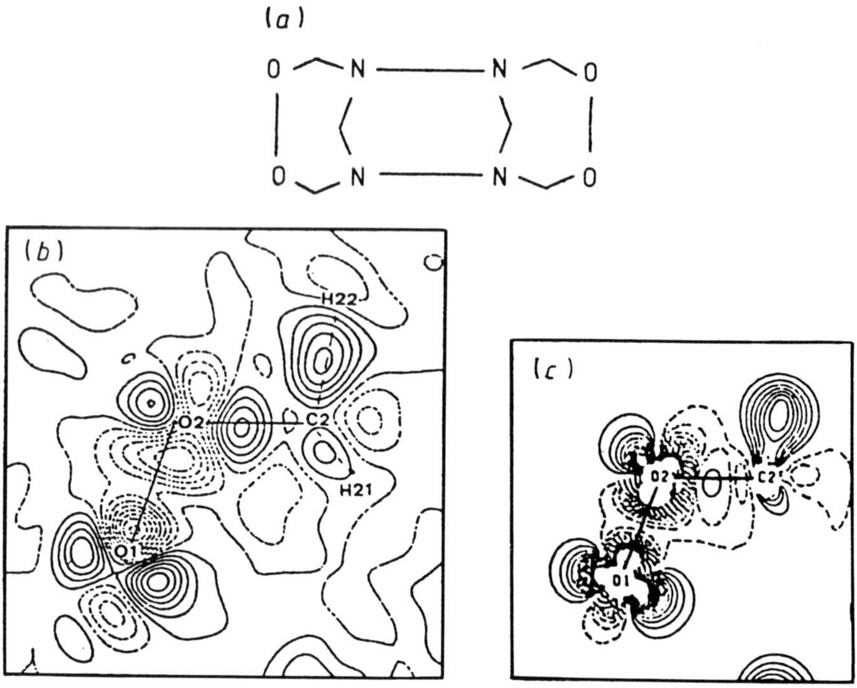

Figure 6.22 The deformation ED of tetraazatetraoxatricyclotetradecane, $C_6H_{12}N_4O_4$, in the O1–O2–C2 plane: (*a*) a view of the molecule; (*b*) the experimental $\delta\rho$ map; (*c*) the theoretical $\delta\rho$ map. The line intervals are 0.075 $e\,Å^{-3}$. (Figure 6.22(*b*) reprinted with permission from Dunitz and Seiler (1983). Figure 6.22(*c*) reprinted with permission from Kunze and Hall (1987). © 1983, 1987 American Chemical Society.)

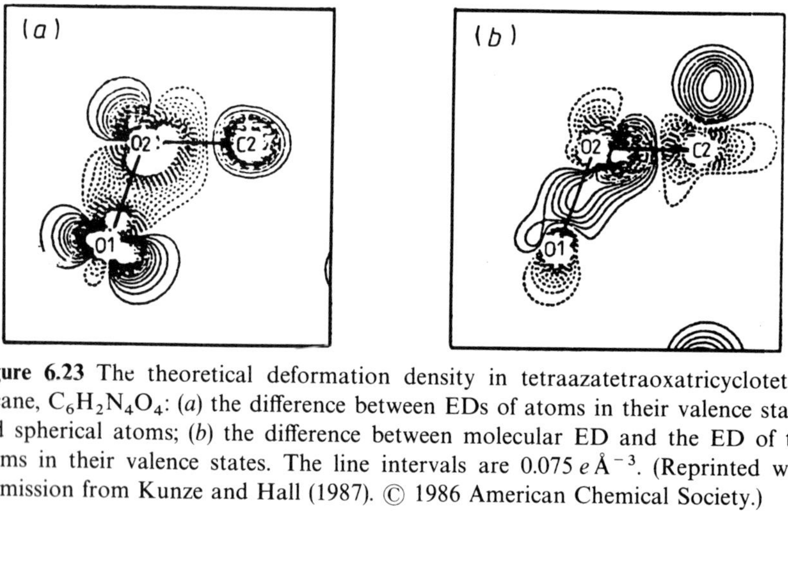

Figure 6.23 The theoretical deformation density in tetraazatetraoxatricyclotetradecane, $C_6H_2N_4O_4$: (*a*) the difference between EDs of atoms in their valence states and spherical atoms; (*b*) the difference between molecular ED and the ED of the atoms in their valence states. The line intervals are 0.075 $e\,Å^{-3}$. (Reprinted with permission from Kunze and Hall (1987). © 1986 American Chemical Society.)

same is true for the C–H bond. Besides, the minima of hybrid deformation ED, located behind the atomic nuclei, should be mentioned. Such minima, in the standard deformation ED maps, are typical for a chemical bond with predominantly p character. Hence, electron delocalization and constructive interference in $C_6H_{12}N_4O_4$ manifest themselves in ED accumulation in the internuclei space, as compared to 'prepared' atoms, and in some depletion of ED behind the atomic nuclei. The first effect can be non-evident in standard $\delta\rho$ maps, but the second is usually seen there.

The hybrid promolecule was used in x-ray diffraction investigations of electron lone pairs in $LiNO_2 \cdot H_2O$ (Okuda *et al* 1990), electron-pure C–Cl chemical bonds in C_6H_6NCl, $C_7H_5O_2Cl$, and $C_6H_3N_2O_4Cl$ (Takazawa *et al* 1989), and C–F bonds in $F_4S{=}C(CH_3)CF_3$ (Buschmann *et al* 1991). In the latter compound, the structure of which is shown in figure 6.24(*a*), the standard deformation ED has a minimum near the fluorine atoms bonded

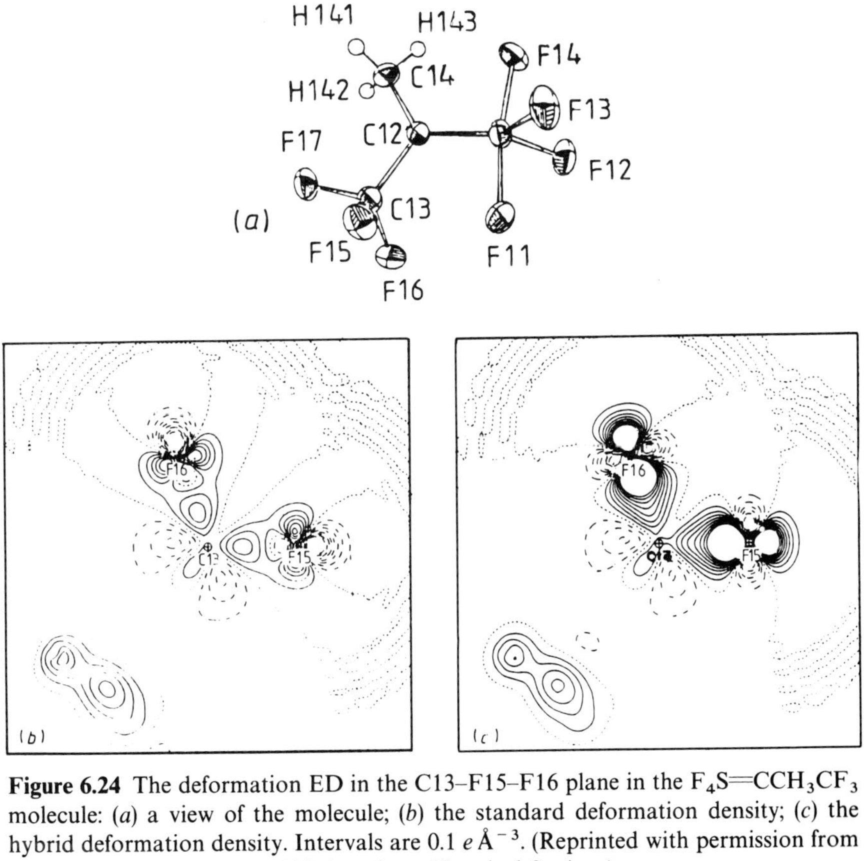

Figure 6.24 The deformation ED in the C13–F15–F16 plane in the $F_4S{=}CCH_3CF_3$ molecule: (*a*) a view of the molecule; (*b*) the standard deformation density; (*c*) the hybrid deformation density. Intervals are 0.1 $e\,Å^{-3}$. (Reprinted with permission from Buschman *et al* 1991. © 1991 American Chemical Society.)

to the carbon atoms: the ED accumulation appears only near C atoms in those bonds (figure 6.24(*b*)). When the configuration of fluorine atoms in the promolecule was chosen as $2p_\sigma^1\,2p_\pi^4$, the hybrid deformation ED showed (figure 6.24(*c*)) that electron delocalization and constructive interference result in high accumulation of the ED on the C–F bond with a shift of the maxima to F atoms. Both effects also lead to overall increase of ED behind the nuclei of these atoms.

It is interesting that in the compound discussed accumulation of ED has not been found in the S–F bond even in hybrid deformation ED maps. The trigonal bypyramidal environment of the S atom was described in the promolecule by sp^2, d_{xz}, p_z, and d_{y^2} orbitals for equatorial σ and π as well as the axial σ bonds, respectively. After subtracting the corresponding aspherical reference ED, the charge-depleted region seems to be unaffected. One reason for this is the large ED depletion in the region around of S atom that exceeds the corresponding decrease of ED in the valence state S atom in the hybrid promolecule. Thus, there are cases where a more sophisticated reference state is needed for analysis of the chemical bond.

The hybrid atom promolecule is difficult to apply for systems with atoms in higher valence states. In this case no simple oriented atom reference ED can be established. Schwarz *et al* (1989), and Ruedenberg and Schwarz (1990) proposed to construct a reference state from non-spherical ground state atoms optimally oriented in accordance with their local surroundings in the crystal. The orientational parameters of non-spherical atoms are included in the crystal structural model and are optimized along with the other structural and electronic parameters. The difference between the experimental (or calculated) ED and the ED of the promolecule from atoms oriented in such a way was called the chemical deformation ED by Schwarz. It reflects small effects of interatomic electron redistribution when a molecule or a crystal is formed and only in a minimal degree depends on intra-atomic interactions. Some results of such calculations for theoretical and experimental EDs of simple molecules were presented by Schwarz *et al* (1988, 1992) and Menschung *et al* (1989) and Niu (1994).

Tsirelson *et al* (1986) and Downs and Gibbs (1987) carried out the experimental investigation of ED in the phenakite, Be_2SiO_4, crystal. The structure of phenakite (figure 6.25(*a*)) is composed of corner sharing chains, each consisting of one SiO_4 tetrahedron and two BeO_4 tetrahedra. These chains extend along the *c* axis and are combined in threefold bonds linked together by common oxygen corners, thereby forming hexagonal channels extending along the *c* axis. The chemical bonds formed by the elements Be and Si in phenakite can be considered as coordination ones, with coordination number four. The standard deformation ED in the SiO_4 tetrahedron (figure 6.25(*b*) and (*c*)) demonstrates the typical picture of a partially covalent multicentre chemical bond with a delocalization of the electron cloud along the O–Si–O fragments. The peaks of the $\delta\rho$ are shifted toward O atoms and

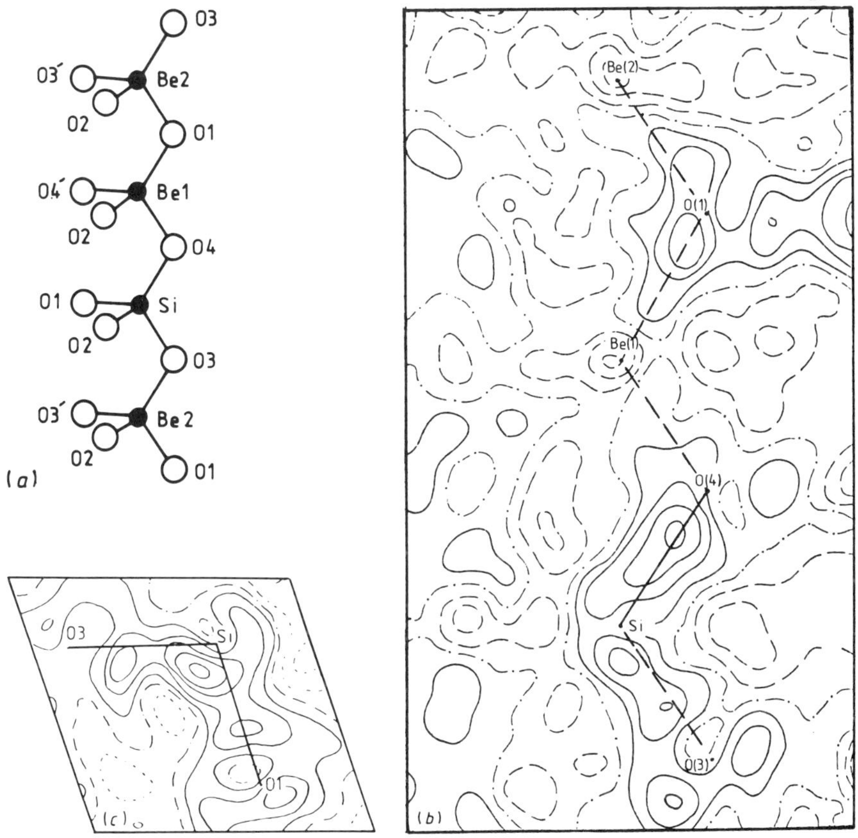

Figure 6.25 The deformation ED of phenakite, Be_2SiO_4: (*a*) a view of the structure; (*b*) the plane passing through O3–Si–O4–Be1 atoms; (*c*) the plane passing through a fragment of the SiO_4 tetrahedron. The line intervals are 0.1 $e\,Å^{-3}$ (Tsirelson *et al* 1986).

displaced from the Si–O line. Their heights vary in the range 0.18–0.45 $e\,Å^{-3}$, due to asymmetry in the second and higher coordination spheres of the Si atom. The highest peak, of 0.57 $e\,Å^{-3}$, is observed in the Si tetrahedron near the Si atom; it is shifted toward the tetrahedron face which abuts onto the hexagonal channel.

In both Be tetrahedra the pattern of ED is qualitatively the same. The $\delta\rho$ peaks, 0.1–0.4 $e\,Å^{-3}$ in height, are strongly shifted toward the O atoms. Such ED distribution reflects the large ionic component of the bond in the Be tetrahedra.

There are negative deformation ED bands in phenakite which are parallel to the (0001) basal plane. They may be attributed to the layered arrangement of the O atoms along this plane in the structure. These bands reflect the anisotropy of the cation–environment interaction as well as the largest dimension of the O–O edges, which are parallel to the *z* axis. Such polarization of cations should be taken into account when modelling the properties of Be_2SiO_4.

It should be noted that Collins *et al* (1983), and Seiler and Dunitz (1986), and Seiler (1991), who studied the phenakite-like crystal of lithium tetrafluoroberyllate(II), Li_2BeF_4, have come to the same conclusion concerning ionicity of the corresponding bonds.

A quite different case is found for the chemical bond in coordination compounds with a transition element as the central atom. Various approximations are used for chemical bond description in these compounds (Bersuker 1976). The simplest description is given by crystal-field theory. It explains the electron structure of a complex in terms of perturbation of transition ion AOs by the electrostatic field of surrounding ligands. The perturbation is assumed to be small and it leads to the splitting only of d and f energy levels of the elements. The character of this splitting depends on the crystal-field symmetry (figure 6.26). The distribution of valence electrons over split levels depends on the atom's nature and on the values of repulsion of these electrons from ligands, assumed to be negative point charges, as well as on intra-atomic electron interaction. If the spin–orbital

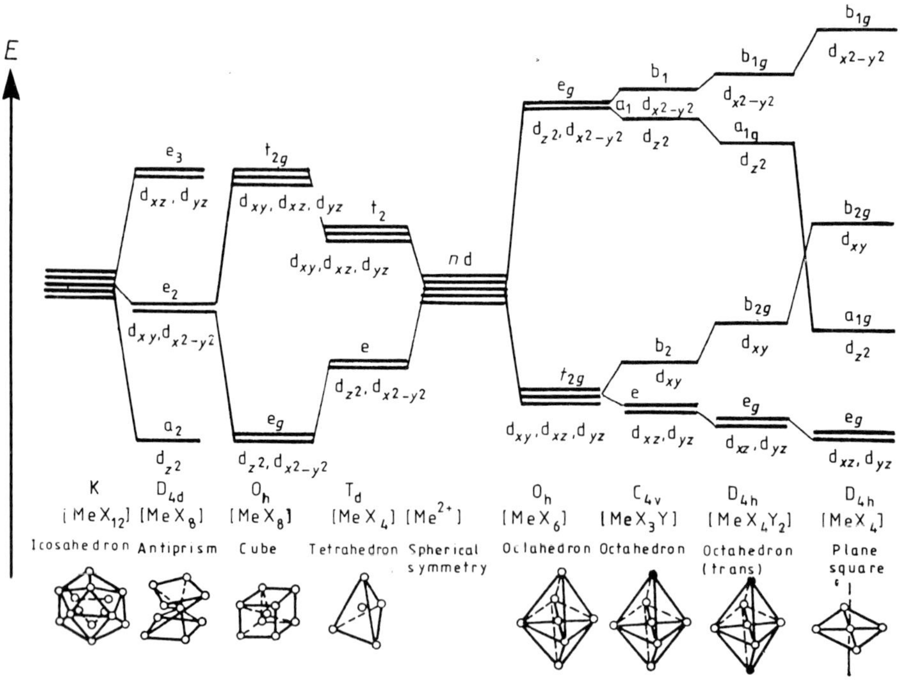

Figure 6.26 The splitting of the energy levels in crystal fields with different symmetry.

interaction in the central atom is great (which is true for 4d, 5d, 4f, and 5f elements), the crystal field just weakly perturbs the energy level structure of the ion. In this case, one may consider the valence electrons, for example, for the central d element as occupying d orbitals which are predominantly of atomic type. If one directs a local coordinate system axis along the fourth-order symmetry axis in an ideal octahedral complex, then the angular parts of the two d AOs corresponding to the same energy levels ($d_{x^2-y^2}$ and d_{z^2}) will be oriented to ligands in octahedral corners and another three AOs, describing the same, but lower, energy levels (d_{xy}, d_{xz}, d_{yz}) to octahedron faces. The latter levels will be the first occupied by electrons if these approximations are accepted as valid. Corresponding maxima on deformation ED maps will be directed towards the faces of the octahedron. In the tetrahedral complex the level occupation order will become reversed. In any case the electron distribution over the energy levels follows Hund's rule, the number of unpaired d electrons is maximal and these complexes are called high spin.

If the electron–electron repulsion in the central atom is large and the ligand influence on its d electrons overcomes the spin–orbital interaction (the strong-field case), then crystal-field theory is not valid for chemical bond description. Now the central atom and ligand wave-function overlap should be taken into account, that is why one must use MO theory. Group theory makes it possible to select without calculation orbitals of the central atom and ligands, the symmetry of which permits them to overlap, forming the system of σ and π bonds. The selection rules for these orbitals are well known (see e.g. Bersuker 1976). The orbitals are labelled in accordance with the irreducible representations of point group symmetry of the complex: one-dimensional representations are labelled as A and B, two-dimensional ones as E, and three-dimensional as T. The corresponding orbitals are labelled as a, b, e, and t. Linear combinations of ligand orbitals, transforming according to irreducible representations of any point group, give the group orbitals. Six σ orbitals of a metal ion and ligand group orbitals in an octahedral complex allowed by symmetry are shown in figure 6.27. Figure 6.28 illustrates π orbital formation in this complex. The full set of symmetry-allowed orbitals forming the σ and π bond system in the octahedral complex is shown in table 6.13. The analogous tables for complexes with other symmetries can be found in the textbook by Bersuker (1976).

There exists one more approach to the description of chemical bonding in complexes. It assumes that the bonds result from electron pair transitions from ligands to the vacant orbitals of the central ion and that the VBM can be used for bond description. The central ion is assumed to be in some state of hybridization: d^2sp^3 in octahedral and sp^3 in tetrahedral complexes, for example. This hybridization state usually corresponds to directed bond formation and determines the geometry of the complex. The valence d orbitals and vacant p orbitals play the main role in d complexes. For example, the Cr^{3+} ion in an octahedron has six orbitals ($3d_{z^2}$, $3d_{x^2-y^2}$, 4s, $4p_x$, $4p_y$, $4p_z$),

Table 6.13 Ligand group orbitals for an octahedral (O_h) complex.

Irreproducible representations of O_h group	AO of central atom	σ type	π type
A_{1g}	s	$(1/\sqrt{6})(\sigma_1+\sigma_2+\sigma_3+\sigma_4+\sigma_5+\sigma_6)$	
T_{1u}	p_x	$(1/\sqrt{2})(\sigma_2-\sigma_5)$	$\frac{1}{2}(\pi_1-\pi_4+\pi_3-\pi_6)$
	p_y	$(1/\sqrt{2})(\sigma_3-\sigma_6)$	$\frac{1}{2}(\pi_1-\pi_4+\pi_2-\pi_5)$
	p_z	$(1/\sqrt{2})(\sigma_1-\sigma_4)$	$\frac{1}{2}(\pi_2-\pi_5+\pi_3-\pi_6)$
E_g	$d_{x^2-y^2}$	$\frac{1}{2}(\sigma_2+\sigma_5-\sigma_3-\sigma_6)$	—
	d_{z^2}	$(1/\sqrt{12})(2\sigma_1-2\sigma_4-\sigma_2-\sigma_5-\sigma_6-\sigma_3)$	—
T_{2g}	d_{xy}	—	$\frac{1}{2}(\pi_2+\pi_5+\pi_3+\pi_6)$
	d_{xz}	—	$\frac{1}{2}(\pi_1+\pi_4+\pi_2+\pi_5)$
	d_{yz}	—	$\frac{1}{2}(\pi_1+\pi_4+\pi_3+\pi_6)$
T_{1g}	—	—	$\frac{1}{2}(\pi_1+\pi_4-\pi_3-\pi_6)$
	—	—	$\frac{1}{2}(\pi_2+\pi_5-\pi_1-\pi_4)$
	—	—	$\frac{1}{2}(\pi_3+\pi_6-\pi_2-\pi_5)$
T_{2u}	—	—	$\frac{1}{2}(\pi_1-\pi_4-\pi_2+\pi_5)$
	—	—	$\frac{1}{2}(\pi_3-\pi_6-\pi_1+\pi_4)$
	—	—	$\frac{1}{2}(\pi_2-\pi_5-\pi_3+\pi_6)$

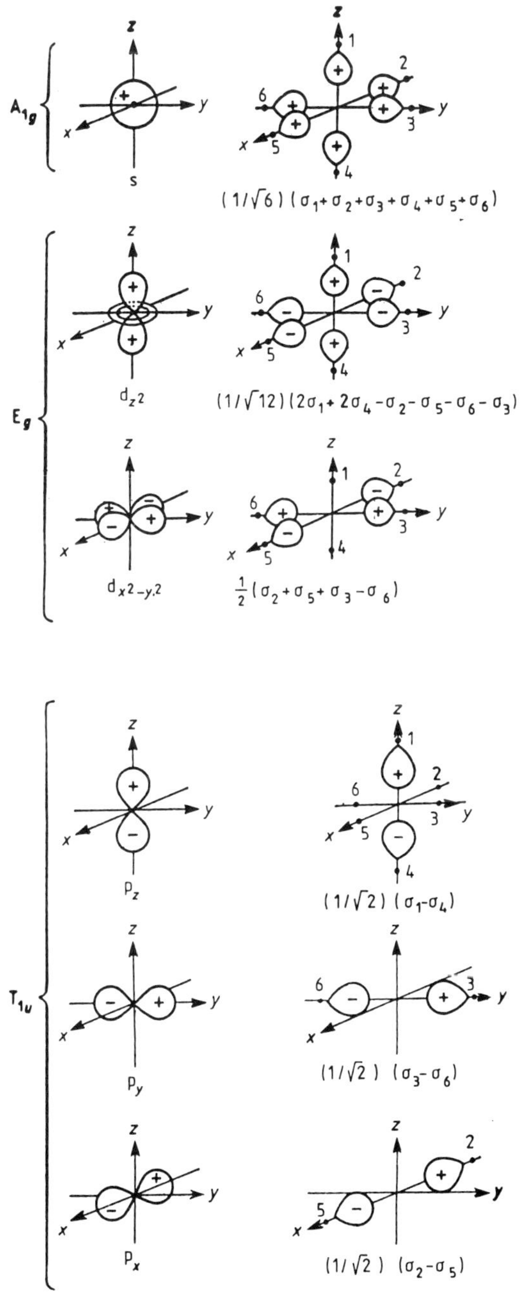

Figure 6.27 The central atom orbitals and σ ligand group orbitals for an octahedral complex.

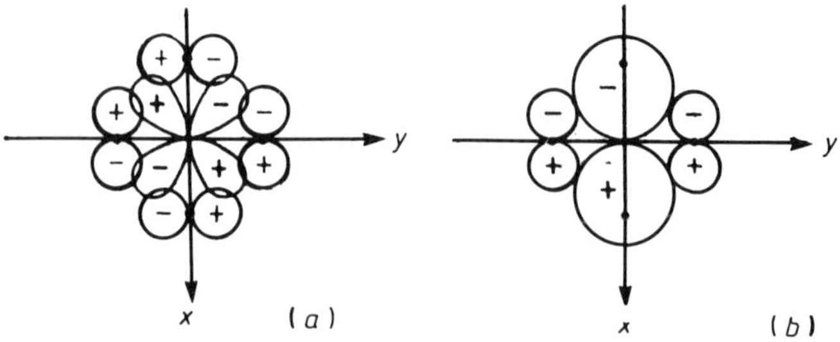

Figure 6.28 The space overlap of ligand π orbitals with t_2 (*a*) and t_1 (*b*) orbitals of a transition ion.

linear combinations of which give six vacant equivalent d^2sp^3 hybrid AOs directed along the local coordinate axes. In the complex cation $[CrL_6]^{3+}$, where L denotes ligands, double-filled ligand orbitals interact with these vacant orbitals resulting in six Cr–L σ bond formations. The angular parts of $3d_{xy}$, $3d_{xz}$, $3d_{yz}$ orbitals bisect the coordinate axes, therefore these orbitals may overlap with ligand π orbitals (see figure 6.28) which are p- or d-type AOs in this case. The π bond formation in a complex is more typical for the transition elements having double-filled d_π AOs and being donors for vacant π orbitals of the ligand (back-donation). The opposite type of interaction is also possible. If there are no π bonds in a complex the corresponding orbitals of a central atom preserve their atomic nature.

This short review gives the key to deformation ED map interpretation. Indeed, having chosen the local coordinate system on the central atom and comparing spatial characteristics of the atomic valence orbitals with positions of maxima and minima on a $\delta\rho$ map, one may obtain a qualitative description of these maps in terms of orbital representations. This kind of information has, of course, only approximate character because molecular (or crystal) orbitals are not the exact superposition of a small number of valence atomic orbitals.

Let us discuss a rather general case of the chemical bond in complexes, taking an organometallic compound—chromium hexacarbonyl, $Cr(CO_6)$—as an example. The Cr atom is at the centre of an almost ideal octahedron of O atoms in this compound and forms both σ and π bonds with ligands. There are 10 valence electrons in the CO molecule and the ground state electron configuration of this molecule is $(\sigma_{2s})^2(\sigma^*_{2s})^2(\pi_{2p_{x,y}})^4(\sigma_{2p_z})^2$ (see figure 6.2). The main contribution to the filled non-bonding σ^*_{2s} MO is from the C atom AO. It manifests itself in the accumulation of ED behind this atom in the $\delta\rho$ profile (figure 2.9). The chemical bond formation in the $Cr(CO)_6$ complex usually accounts for σ interaction of field σ^*_{2s} MO ligand orbitals with corresponding vacant d orbitals of the central atom. The latter are e_g orbitals (d_{z^2} and $d_{x^2-y^2}$ if the local coordinate system axis coincides with interaction lines). Simultaneously the overlapping of filled t_{2g} orbitals (d_{xy}, d_{xz}, d_{yz}) of the Cr atom with the vacant π^* MO of the ligands takes place. This pattern is illustrated in figure 6.29. As a result, electron displacement toward the metal ion via the σ bond (σ donation or donor–acceptor bond) and partial reversed electron displacement toward the ligand via the π bond (back-donation) take place in $Cr(CO)_6$. The picture described is in good agreement with x-ray photoelectron spectroscopy data (Jolly *et al* 1977).

Let us analyse how this picture looks in deformation ED maps. Rees and Mitschler (1976) carried out an x-ray diffraction investigation of the $Cr(CO)_6$ crystal at 78 K. Their standard $\delta\rho$ map averaged over chemically equivalent sections is depicted in figure 6.30. The σ bond region along the Cr–Co line has a minimum of $\delta\rho$ of $-0.30\,e\,\text{Å}^{-3}$ at a distance of 0.65 Å from the Cr atom; there is also the maximum $\delta\rho$ at about $0.4\,e\,\text{Å}^{-3}$ near the C atom. This maximum is elongated normally to the Cr–CO line and is connected with the $\delta\rho$ maximum $0.10(3)\,e\,\text{Å}^{-3}$ lying on the octahedron threefold symmetry axis 0.75 Å away from the Cr atom in the π bond region. The

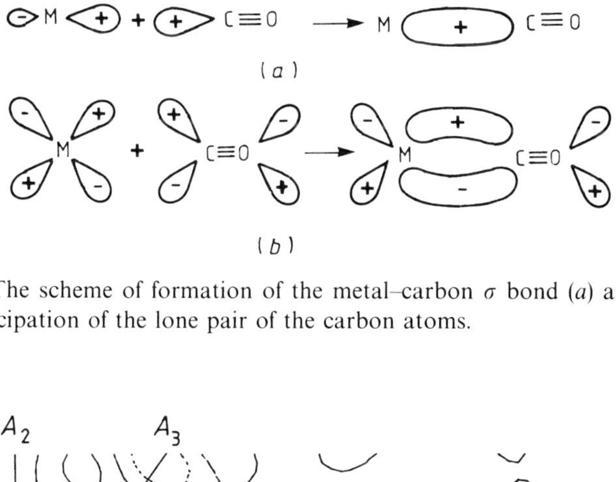

Figure 6.29 The scheme of formation of the metal–carbon σ bond (*a*) and π bond (*b*) with participation of the lone pair of the carbon atoms.

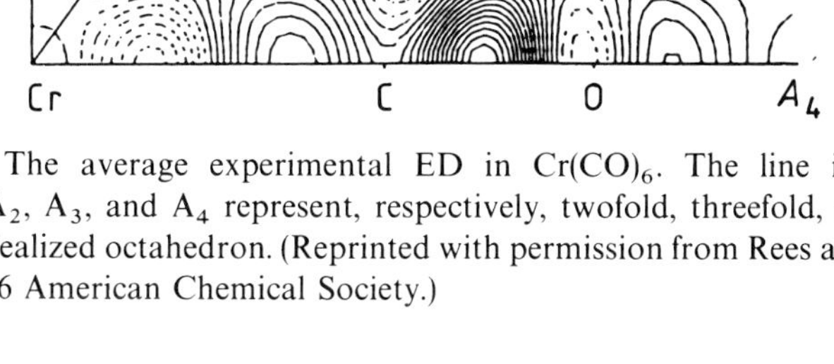

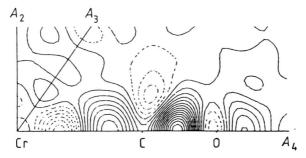

Figure 6.30 The average experimental ED in $Cr(CO)_6$. The line intervals are $0.05\,e\,\text{Å}^{-3}$. A_2, A_3, and A_4 represent, respectively, twofold, threefold, and fourfold axes in the idealized octahedron. (Reprinted with permission from Rees and Mitschler 1976. © 1976 American Chemical Society.)

experimental $\delta\rho$ map is in good agreement with the chemical bond picture described above. The electron populations of Cr 3d orbitals, recalculated from parameters of the multipole model by Holladay *et al* (1983) (table 3.14), also showed that the t_{2g} AO of this ion had a larger electron population than the e_g AO. However, it should be admitted that donor–acceptor bond specificity does not manifest itself in the standard $\delta\rho$ map. For example, one may ask the question of what part of the ED accumulation revealed between Cr and C atoms is due to bond formation and what is due to the lone pair of the C atom. Searching for the answer, Sherwood and Hall (1983) have calculated both the standard deformation ED map and the fragment deformation ED map. The latter was calculated as a difference between the ED of $Cr(CO)_6$ and $Cr(CO)_5$ and CO fragments of the same geometry and must reflect just the donor–acceptor interaction. Both maps are presented in figure 6.31. The ED accumulation of interest on the fragment deformation map is essentially smaller and its maximum is slightly shifted toward the middle of the Cr–C line. It is also elongated normal to the Cr–CO line, i.e. it shows the presence of a π component in the interaction of the central atom with ligands.

Hall (1986) has suggested an orbital explanation of the difference between the fragment and the standard deformation ED maps of the donor–acceptor bond. The doubly occupied M–L bonding σ MO can be described as

$$\varphi = \Phi_L + \lambda\Phi_M \tag{6.26}$$

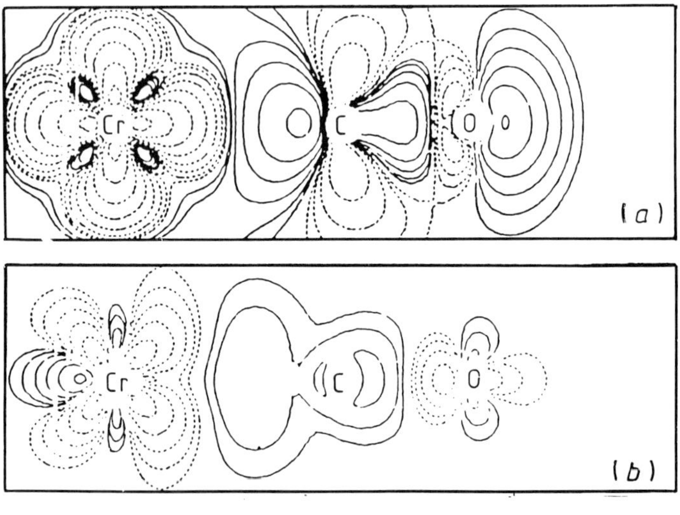

Figure 6.31 The theoretical ED of chromium hexacarbonyl, $Cr(CO)_6$: (*a*) the standard deformation density (molecular density minus a promolecule composed of spherical atoms); (*b*) the fragment deformation density (molecular density minus $Cr(CO)_5$ and CO fragments) (Sherwood and Hall 1983).

where Φ_L is the ligand (L) lone-pair orbital and Φ_M is the metal (M) acceptor orbital. The ED of that MO is

$$\rho_{LM} = 2(1 - 2\lambda S_{LM})(\Phi_L + \lambda\Phi_M)^2 \tag{6.27}$$

where S_{LM} is the metal–ligand overlap integral. The coefficient λ is small for the donor–acceptor bond. Expanding (6.27) in a series over λ, one can describe the fragment deformation ED of such a bond in a simple approximate form

$$\delta\rho_{FD} \simeq 4\lambda\Phi_L(\Phi_M - S_{LM}\Phi_L) \tag{6.28}$$

(it is assumed that the promolecule consists of a doubly occupied Φ_L AO and an empty Φ_M AO). If the overlap is also small then

$$\delta\rho_{FD} \simeq 4\lambda\Phi_L\Phi_M. \tag{6.29}$$

The $\delta\rho$ for the standard deformation ED is

$$\delta\rho \cong \Phi_L^2 - \Phi_M^2 + 4\lambda\Phi_L\Phi. \tag{6.30}$$

Thus, the difference between $\delta\rho_{FD}$ and $\delta\rho$ along the M–L line in the donor–acceptor bond is dependent on the relationship between the characteristics and interference of orbitals Φ_L and Φ_M. As was found by Figgis *et al* (1983a, b) for $Ni(NH_3)_4(NO_2)_2$ and Kok and Hall (1985) and Wang *et al* (1987) for $Cr(C_6H_6)(CO)_3$, which are typical donor–acceptor complexes, the $\delta\rho_{FD}$ value is, as a rule, smaller than $\delta\rho$ in that region. Therefore, the σ interactions in these compounds are almost entirely interactions of the lone pairs of ligands with the vacant d orbitals of transition metals.

If the complex geometry differs from the ideal, it becomes hard to give an orbital interpretation of the deformation ED. In these cases McIntyre *et al* (1990) recommended one use the multipole model and take into consideration the ED of the central pseudoatom in addition to the complete ED picture in a crystal, which is calculated by summing the contributions of a great number of pseudoatoms. If the covalency of the central atom–ligand bond is not large it gives one the opportunity to remove the superpositional contribution due to the pseudoatomic surroundings.

From this point of view the example of tetragonal nickel sulphate hexadeuterate, $NiSO_4 \cdot 6D_2O$, (McIntyre *et al* 1990) is rather instructive. In this compound the six O atoms of water molecules form a slightly deformed octahedron in which O3 and O3′ are situated at opposite corners whereas O1 and O1′ and similarly O2 and O2′ atoms are adjacent to one another. The octahedron is elongated in the O3–Ni–O3′ direction; the nickel atoms are located at the 4(a) site with symmetry 2. If this distortion is neglected, the expected electron configuration of the Ni^{2+} ion is $d_t^6 d_e^2$, where d_t orbitals are diagonally directed (the d_{z^2} orbital points along the O3–Ni–O3′ line). In standard deformation ED maps (figure 6.32(*a*)) the minima corresponding to d_e orbitals are clearly seen around the Ni^{2+} ion, but the manifestation of the $\delta\rho$ maxima of d_t orbitals is not obvious due to octahedral distortion:

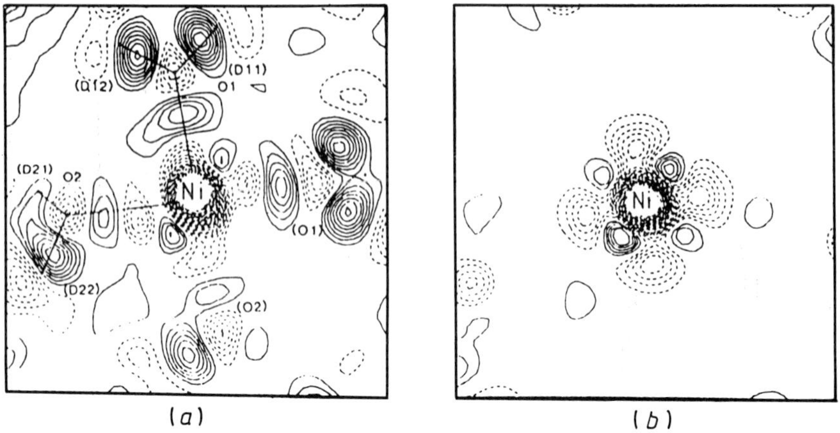

Figure 6.32 The static multipole deformation density of the $Ni(D_2O)_6$ crystal: (*a*) the standard deformation density map; (*b*) the Ni atom only deformation density. The line intervals are 0.05 $e\,Å^{-3}$ (McIntyre *et al* 1990).

these orbitals are really not equivalent. The pattern for the single Ni pseudoatom (figure 6.32(*b*)) is clearer. The $\delta\rho$ minima along the three independent Ni–O directions now appear more or less equivalent, likewise the $\delta\rho$ peaks in the diagonal directions. These observations agree with the Ni^{2+} high-spin configuration with both d_e orbitals half full as opposed to the spin-paired alternative with the d_{z^2} orbital full and the $d_{x^2-y^2}$ orbital empty. The deformation ED, thus, has nearly ideal octahedral symmetry, even though only crystallographic site symmetry 2 was imposed on the Ni deformation functions in the refinement. Later study of $NiSO_4{\cdot}6H_2O$ at 25 K confirmed the results described (Ptasiewicz-Bak *et al* 1993).

One more case of chemical binding is the hydrogen bond. It plays a fundamental role in governing the structural and chemical behaviour of many organic and inorganic compounds, of most biological substances, etc. Compared to the typical energy of a covalent bond, the hydrogen bond energy is at least an order of magnitude less ($\sim$10–40 kJ mol^{-1}) but is still stronger than the van der Waals type of interaction. There are a few non-conflicting definitions of a hydrogen bond. In general, a hydrogen bond R–X––H ··· Y–R′, where R are rest groups, is formed if X and Y are atoms with electronegativity greater than that of hydrogen, especially when Y has an unshared pair of electrons. The Y atom is usually referred to as the acceptor, whereas the X–H group is called the donor of the hydrogen bond. The generalization of the geometry features of the hydrogen bond demonstrates that there is a large variety of hydrogen bond lengths ranging from 2.4 to 3.4 Å. They are often divided into groups: short strong bonds (2.4–2.5 Å), intermediate bonds (2.5–2.7 Å), and long weak bonds ($>$2.7 Å).

The quantum chemical treatment of the H bond deals with the electrostatic, polarization, charge transfer, exchange, and dispersion (or correlation) interaction components (see section 2.5). Analysis of the large number of experimental and theoretical data shows (see e.g. Olovsson 1980) that, to the first approximation, the energy contributions of the last four components approximately cancel for weak and intermediate H bonds. Furthermore, the variation in electrostatic energy follows the same trend as the total H bond energy. As each of the other four separate contributions is less sensitive to the relative orientation of the interacting molecules, this rougly explains the empirical rule that most geometrical features of H bonding can be explained simply by an electrostatic model. The latter corresponds to the classical Coulomb interaction between undisturbed constituent molecules as they are brought together into the actual position in the H-bonded complex, without any deformation of the original molecular ED distribution and without charge transfer. This pattern is applicable not only to the total ED but to deformation ED as well. As a result the $\delta\rho$ distribution in weak and intermediate H bonds can be considered as a simple superposition of the $\delta\rho$ of the constituent molecules. This has been illustrated by McIntyre *et al* (1990) who compared the multipole static $\delta\rho$ distribution through the O_w–D...O_w group with the O–O distance 2.797 Å in nickel sulphate hexadeuterate, $NiSO_4{\cdot}6D_2O$, (figure 6.33(*a*)) and the composition of the two

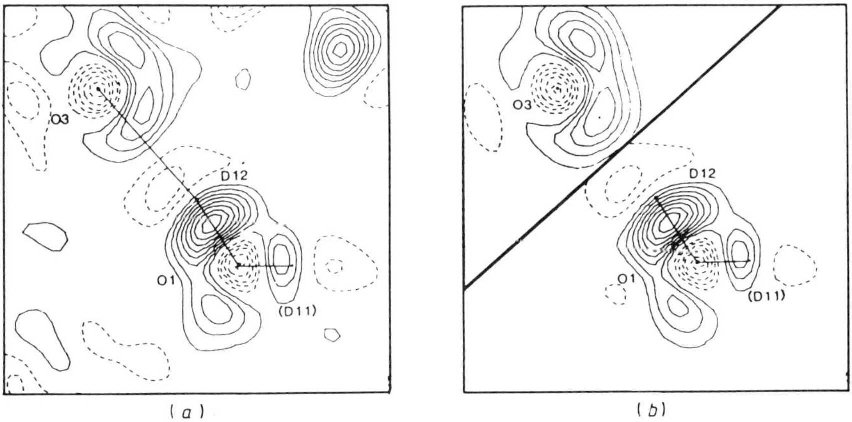

Figure 6.33 Static deformation ED maps through O1–D12···O3 atoms (bond length 2.797 Å) in the nickel sulphate hexadeuterate, $NiSO_4{\cdot}6D_2O$ (McIntyre *et al* 1990). (*a*) The experimental map calculated by multipole modelling of all atoms; (*b*) a map which is a composite of the two maps with only the deformation functions of the atoms in D_2O included in the calculation of the region near D_2O and only the deformation functions of the O3 atom included in the calculation of the region near O3. Reflections up to $\sin\theta/\lambda = 1.03$ Å were used. Contours are drawn at intervals of 0.05 $e\,Å^{-3}$. The heavy line in (*b*) indicates the border between constituent atomic groups.

maps corresponding to the isolated water molecule and the oxygen atom (figure 6.33(*b*)). The similarity of the pictures is obvious. Some small difference in the oxygen lone-pair region is due to a small expansion and deformation of the isolated $\delta\rho$ picture in comparison with the real one (the lack of the mirror symmetry plane bisecting the lone-pair lobes is clearly seen too). This can be attributed to the lone-pair ED polarization by action of the environment; full cancellation of the four contributions mentioned is not the case here.

More detailed examination shows that some particular contribution becomes significant in many cases of H bonding. A study of the polarization, exchange, correlation, and charge transfer components as well as the influence of the crystal environment has been performed by Krijn and Feil (1987a, b, 1988a–c) on the examples of the water molecule, the water dimer, and α-oxalic acid dihydrate in the LDA of the DFT. Taking the effects mentioned above into account, the authors found a theoretical result closer to the experimental one. The ED redistribution suggests the acceptor molecule becomes a better donor, whereas the donor becomes a better acceptor of a second hydrogen bond. Charge transfer between donor and acceptor amounts to 0.09 *e*, increasing to 0.14 *e* when taking the crystal environment into account.

An example of the ED rearrangement along the hydrogen bond as a function of its length (energy) is given by Hermansson (1987) (figure 6.34). The consideration begins with the short (but not symmetric) bond in $KH(HCOO)_2$ and ends with the relatively long bond in $NaHC_2O_4 \cdot H_2O$. The trends in $\delta\rho$ distribution are clearly seen in the picture.

The typical deformation ED distribution for a weak and intermediate H bond (figure 6.34) consists of an electron excess in the X–H donor bond, a slight electron deficiency in the $H \cdots Y$ bond close to the hydrogen atom and an electron excess closer to the acceptor atom Y. There are a lot of compounds with the same $\delta\rho$ features.

As one proceeds from the long H bond, where the ED is mainly the superposition of the EDs of constituent atomic groups, to a much shorter bond a definite ED redistribution due to monomer interactions can take place. The electron shift should increase as the interaction gets stronger.

In a short (strong) H bond the electron excess in both the donor and acceptor region is less pronounced. The electrostatic approximation is no longer so applicable for these bonds and it becomes conceptually more difficult to separate the ED between the constituent systems. The partitioning of the bond energy into the components mentioned above also becomes invalid. Some new assumptions should be introduced in order to classify the ED changes. In very short H bonds of type $O \cdots H \cdots O$ with O–O distance around 2.4–2.5 Å the deformation ED maps have a rather different appearance: there is a much less pronounced charge build-up in the $O \ldots H$ as well as in $H \ldots O$ regions and the ED is, in general, somewhat more

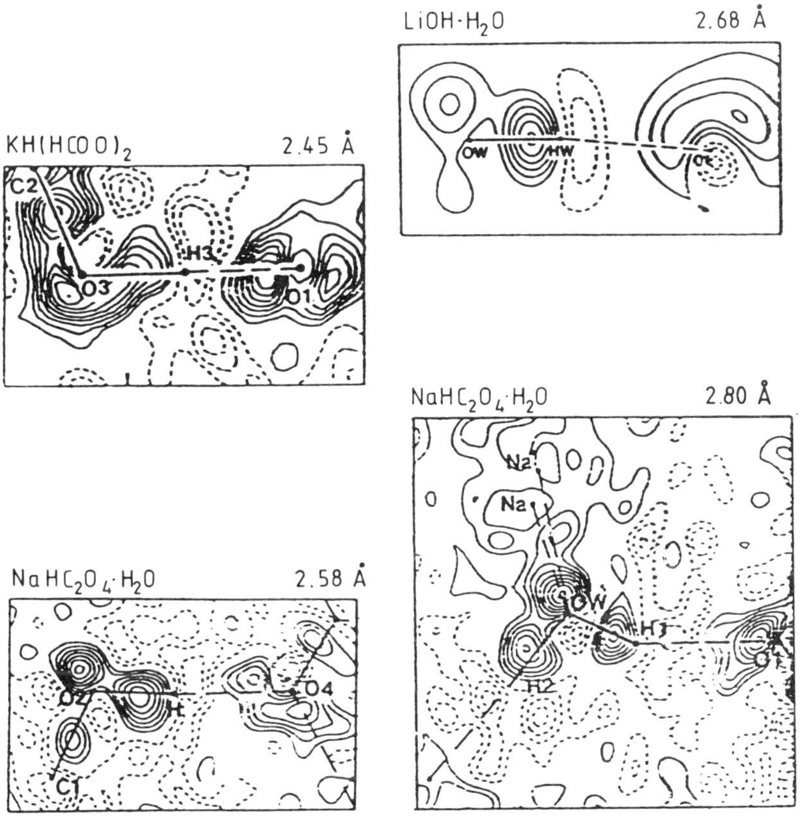

Figure 6.34 Experimental dynamic deformation ED maps for O–H···O hydrogen bonds of different lengths. Contour level intervals are 0.05 $e\,Å^{-3}$. (Reprinted with permission from Hermansson 1987.)

symmetrically distributed around the midpoint. An example of such an H bond is depicted in figure 6.34 where the dynamic deformation ED distribution in the short symmetric H bond in $KH(HCOO)_2$ is presented. $\delta\rho$ peaks appear near the O atoms with a trough centred at a symmetry inversion point.

The water molecule plays an essential role in H bond formation because of its two hydrogen atoms and lone electron pairs of the oxygen atom. It can thus act both as donor and acceptor. This especially concerns the crystalline hydrates where water molecules essentially stabilize their crystal structure. There are numerous examples of different coordinations of water molecules in which different properties are exhibited. The questions about the participation of the water molecule in H bonding, its coordination in the crystalline hydrates and in solutions, the influence of the neighbourhood on the ED, the partioning of the water molecule bonding energy, the role of different contributions to the total energy, etc have been carefully considered (see e.g. Nozik *et al* 1979, Olovsson and Jonsson 1976, Olovsson 1980, Hermansson and Olovsson 1984, Hermansson 1985, 1987, Krijn and Feil 1987a, b, 1988a–c).

6.4 Quantum topological theory and the chemical bond

As has already been shown in section 2.4, Bader's (1990) quantum topological theory gives a unique way of partitioning molecules or crystals into pseudoatomic fragments. Each of them contains a single attractor, e.g. the atomic nucleus, surrounded by an electronic basin. The boundary of fragments is the zero-ED-gradient flux surface which surrounds the nucleus. The pseudoatoms determined in such a way do not overlap and fill all of space. The pseudoatomic interactions determine the chemical bond in a system.

Important characteristics of pseudoatomic interactions in many-electron systems are described by the number and type of ED critical points, points where $\nabla\rho(\boldsymbol{r})=0$. In particular, there are certainly the saddle critical points of (3, −1) type on the boundary surfaces between neighbouring basins. In the cyclic and cage fragments of molecules and crystals there appear the ring critical points of (3, +1) type and the minimum critical point of (3, +3) type, respectively. Therefore, in order to describe the chemical bond it is necessary to take the topological characteristics of ED at these critical points into account. Besides, the nuclei in neighbouring basins are linked by gradient path lines passing through the (3, −1) critical point. The ED has diminishing values under any sideways shifts from these lines, the bond paths. The existence of the bond path in a stable system provides the necessary and sufficient condition for bond formation (Bader 1985, Tsirelson *et al* 1995).

The ED curvature is characterized by the Hessian matrix (2.69). This matrix can be diagonalized and three of its eigenvalues, λ_i, characterizing the ED curvature, as well as three corresponding mutually orthogonal eigenvectors, $\boldsymbol{u}_i$, directed along the principal curvatures, can be obtained. In the bond critical point (3, −1) two of these eigenvalues, λ_1 and λ_2 ($|\lambda_1|<|\lambda_2|$), are always negative. They correspond to the eigenvectors normal to the bond path. The third eigenvalue, λ_3, is always positive and the corresponding eigenvector is directed along the bond path (figure 6.35). This means that the ED curvature is always positive along the bond path and is always negative normal to the bond path. At the ring critical point of (3, +1) type, two of the eigenvalues

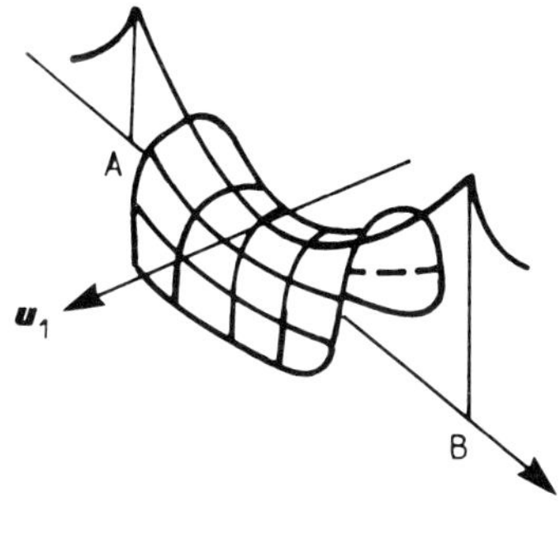

Figure 6.35 A two-dimensional illustration of the curvature measurement of the ED at the (3, −1) bond critical point of the AB bond.

in the Hessian matrix are positive and one is negative. At the cage critical point, of (3, +3) type, all the eigenvalues are positive. The Laplacian of ρ (the trace of the Hessian matrix) characterizes the total ED curvature, its sign being dependent on the relationship of λ_1, λ_2, and λ_3 values and signs. In the case when $|\lambda_1|+|\lambda_2|>\lambda_3$, the ED is concentrated at a critical point; in the opposite case the ED is depleted here.

Thus, the bond path character and the ED curvature at the critical points are of great interest for characterization of the chemical bond. The three-dimensional distribution of the Laplacian of ρ which characterizes both the ED curvature and ED concentration and depletion in each point of a bonded system is of the same significance. First, the Laplacian of ρ makes it possible to reveal the non-evident peculiarities of ED distribution such as shell structure of atoms (figure 2.18) and electron lone pairs (figure 2.19). Second, it points out the areas in a bonded system where the local potential energy is higher than the kinetic one. Finally, $\nabla^2\rho$ is connected with local electron energy, $H_e(\boldsymbol{r})$, by the simple expression (2.74) which links the topological (structural) properties with energetic characteristics of a bonded system.

In order to consider the topological ED properties, which can be associated with characteristics of the chemical bond, some specificity of the topological analysis of ED calculated from x-ray diffraction data should be discussed. The first question which needs an answer is, can the topological theory be applied to dynamic ED? The answer is positive, fortunately. Bader *et al* (1981) emphasized that molecular graphs for all nuclear configurations in the neighbourhood of the minimum-energy geometry are equivalent. Therefore, the same set of critical points describes static and dynamic systems. The quantitative estimations of topological characteristics depend, of course, on the character of nuclear thermal motion (Stephens and Becker 1983).

There are three different methods in the experimental ED topological analysis (Tsirelson 1994). One of them consists of a Fourier series for $\rho(\boldsymbol{r})$ differentiation over coordinates in order to arrive at $\nabla\rho(\boldsymbol{r})$ and $\nabla^2\rho(\boldsymbol{r})$ functions. Both the latter functions are also Fourier series which diverge, in general, independently of the number of terms in the series. Fortunately, as has been shown in subsection 3.6.1, differentiation by the Lanczos (1957) method allows one to obtain a mathematically correct expression (3.138) for $\nabla^2\rho$. Although the results remain dependent on the series truncation effect, in simple inorganic compounds, such as fluorite, CaF_2 (see figure 3.26), the correct sign of the function $\nabla^2\rho$, calculated in Fourier series form, is obtained in the internuclear space.

To study the negative influence of ED thermal smearing on topological characteristics Kapphahn *et al* (1988) have analysed the behaviour of ED and its Laplacian at different temperatures in the bond regions in promolecules of ethane, C_2H_6, ethylene, C_2H_4, and acetylene, C_2H_2. It proved that for rigid C–C bonds this influence is not crucial; in particular, signs of $\nabla^2\rho$ are the same, whereas for C–H bonds the topological characteristics change significantly due to the large thermal motion of H atoms relative to C atoms. Therefore, in the general case, an accurate low-temperature x-ray diffraction experiment is needed in order to obtain correct topological ED characteristics by the methods discussed above.

The second method consists of the calculation of $\nabla\rho(\boldsymbol{r})$ and $\nabla^2\rho(\boldsymbol{r})$ represented as a sum of the promolecule ED, $\tilde{\rho}$, in position space and the Fourier series for deformation ED in reciprocal space (Kapphahn *et al* 1988, 1989). Then the expression for the Laplacian of ρ is presented as

$$\nabla^2\rho(\boldsymbol{r}) = \nabla^2\tilde{\rho}(\boldsymbol{r}) + \nabla^2\,\delta\rho(\boldsymbol{r}). \tag{6.31}$$

The first term on the right-hand side of (6.31) is easily calculated using the tabulated atomic wave-functions of Clementi and Roetti (1974) or the Fourier transform of the atomic scattering amplitudes, expressed in parametric form (Schwarzenbach and Thong 1979). The promolecule ED is calculated as a static function and thermal smearing consequences are to a great extent rejected for the Laplacian of ρ.

The Fourier series for $\nabla^2\,\delta\rho$ is better converged than for $\nabla^2\rho$. The elimination of statistical errors in $\nabla^2\,\delta\rho$ can be performed in the same manner as in $\delta\rho$ (3.141). Besides, the calculations of Epstein and Swanton (1982) and Stephens and Becker (1983) have confirmed Stewart's (1977c) prediction that the experimental deformation ED, which follows only lattice modes and suffers less from thermal smearing, can be regarded as near the pseudostatic one. Moreover, $\nabla^2\,\delta\rho$ can be calculated for the static model deformation ED. Therefore, the use of the expression (6.31) instead of (3.138) in order to calculate the Laplacian of ρ is preferable.

The topological analysis of the static multipole ED is most attractive. The topological characteristics derived are to a large extent free from the effects of series termination, thermal smearing, and experimental statistical errors. Therefore, this approach, proposed by Kapphahn *et al* (1988) is now more often used (see e.g. Stewart 1991, Gatti *et al* 1992).

Kapphahn *et al* (1988), using the low-temperature x-ray diffraction data of van Nes and Vos (1978, 1979a, b) and van Nes (1978), have performed the topological analysis of ethane, ethylene, and acetylene. The multipole static ED was used in the case of C_2H_2 and C_2H_4, whereas in the study of C_2H_6 the expression (6.31) was used. These compounds crystallize only at 100 K and lower and only a small temperature interval exists between the melting point and the temperature of x-ray experiments. Therefore the thermal smearing can significantly distort the ED curvature in the compound discussed. In spite of this fact the critical points (3, −1) were found on C–C and C–H bonds in all molecules. The topological characteristics derived are presented in table 6.14 together with corresponding *ab initio* values for free molecules. The agreement is quite satisfactory for C–C bonds but is worse for C–H bonds. The latter demonstrates, among other things, that thermal deconvolution for light H atoms is only marginally achieved.

The α-oxalic acid dihydrate, $(COOH)\cdot 2H_2O$, crystallizes at room temperature and its ED topological characteristics should be much less dependent on the thermal effect at low temperature. This compound has been investigated at 125 K by Laidig and Frampton (1991). Satisfactory agreement of the experimental (multipole) and theoretical topological characteristics was achieved, in general (table 6.15). At the same time, unexpected features are the theoretical values of the Laplacian of ρ on C–O and C═O bonds which have a positive sign. Since the Hessian matrix eigenvectors λ_1 and λ_2 in both theory and experiment are negative, as expected, the difference in sign should be attributed to the inadequate description of the theoretical ED curvature along the bonds. Perhaps the STO-3G basis set used in theoretical calculations is not enough to correctly reproduce the ED curvature.

An important property of ρ at a bond critical point is the ratio λ_1/λ_2 of its negative curvatures along axes perpendicular to the bond path. The ellipticity

$$\varepsilon=\lambda_1/\lambda_2-1 \tag{6.32}$$

provides a link with and a quantitative generalization of the concept of σ and π character of a bond. If the axes associated with λ_1 and λ_2 are symmetrically equivalent (the C–C bond in the free ethane molecule), then $\lambda_1/\lambda_2=1$ and $\varepsilon=0$. The ED in the plane perpendicular to the C–C bond axis at its midpoint has axial symmetry. The density of a π orbital is not spatially separated from that of the σ orbitals (Coulson *et al* 1952) and all bonds, including multiple bonds, are topologically equivalent. They exhibit a single (3, −1) critical point; however, the contours in a display of ρ in the C═C internuclear section for ethylene are elliptical in shape (figure 6.36). Hence,

Table 6.14 Topological characteristics of bonds at (3, −1) critical points for ethane (85 K), ethylene (85 K), and acetylene (141 K) in crystals and for free molecules (Kapphahn *et al* 1988).

Bond	R_e (Å)	ρ (e Å^{-3})	$\nabla^2\rho$ (e Å^{-5})	λ_1 (e Å^{-5})	λ_2 (e Å^{-5})	λ_3 (e Å^{-5})	ε
C–C (crystal)	1.510	1.61	−16.13	−12.09	−9.25	5.21	0.31
C–C[a] (free molecule)	1.530	1.52	−9.94	−9.11	−9.11	8.28	0.00
C=C (crystal)	1.336	2.16	−16.71	−16.82	−14.17	14.28	0.19
C=C (free molecule)	1.322	2.25	−22.30	−16.12	−13.23	7.04	0.22
C≡C (crystal)	1.183	2.84	−31.24	−22.38	−22.36	13.24	0.00
C≡C (free molecule)	1.194	2.69	−30.46	−15.46	−15.46	0.46	0.00
C–H[b] (crystal) C_2H_6	0.960	1.55	−8.78	−13.01	−13.01	16.24	0.00
C–H (crystal) C_2H_4	1.071	1.40	−5.47	−13.22	−5.24	12.99	1.52
C–H (crystal) C_2H_2	1.043	1.42	−7.12	−10.26	−10.24	13.38	0.00
C–H (free molecule)	1.086	1.88	−23.71	−17.47	−17.47	11.07	0.00

[a]The values for free molecules calculated in 6-31G basis set by Bader *et al* (1983).
[b]The values for C–H bonds have been averaged.

Table 6.15 The critical point properties for theoretical and experimental ED distributions in α-oxalic acid dihydrate (Laidig and Frampton 1991). The first rows are results of topological analysis of ED with the multipole model adjusted to x-ray diffraction data; the second ones are results of topological analysis of theoretical (STO-3G) ED.

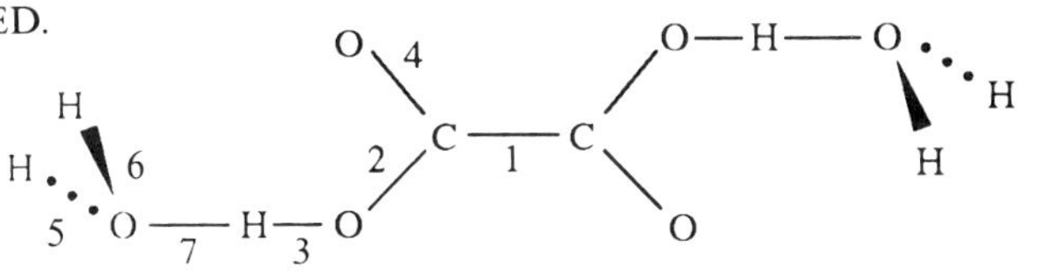

Bonds	ρ	$\nabla^2\rho$	λ_1	λ_2	ε
1	0.256	−0.498	−0.563	−0.427	0.319
	0.237	−0.566	−0.412	−0.378	0.089
2	0.369	−1.128	−0.932	−0.747	0.248
	0.298	+0.528	−0.586	−0.515	0.138
3	0.269	−0.776	−1.284	−1.240	0.035
	0.319	−3.083	−1.611	−1.596	0.010
4	0.375	−1.437	−1.749	−1.705	0.026
	0.346	+0.919	−0.751	−0.576	0.305
5	0.382	−1.558	−1.915	−1.853	0.033
	0.420	−2.990	−2.055	−1.953	0.035
6	0.427	−1.251	−1.172	−1.035	0.132
	0.354	−2.985	−2.253	−2.155	0.045
7	0.080	+0.542	−0.167	−0.147	0.136
	0.074	+0.307	−0.176	−0.172	0.024

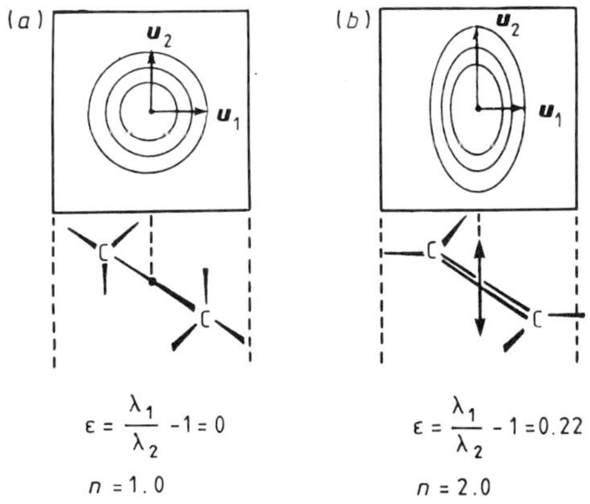

Figure 6.36 An illustration of the density distribution $\rho(\boldsymbol{r})$ around C–C single and C=C double bonds at the saddle point of the molecules ethane (*a*) and ethylene (*b*). The directions of steep and soft curvature of ρ_b are given by the eigenvectors $\boldsymbol{u}_1$ and $\boldsymbol{u}_2$. The latter is taken as the 'direction' of the bond ellipticity as indicated by the double-headed arrow (Cremer and Kraka 1984a).

λ_1 and λ_2 are not degenerate in ethylene, as they are in ethane, and their associated eigenvectors define a unique pair of orthogonal axes, perpendicular to the bond path: the minor axis along which the magnitude of the negative curvature of ρ is a maximum (λ_1) and the major axis along which its magnitude is a minimum (λ_2). In ethylene the major axis is perpendicular to the plane of the molecule and is coincident with the direction of the elongation of ED due to the π component in orbital theory. The ellipticity of the triple C≡C bond in the free acetylene molecule, however, is equal to zero, due to the presence of two π bonds.

Thus, the bond ellipticity provides a quantitative measure of the extent to which the ED is preferentially distributed in a particular plane containing the bond axis. That effect manifests itself as $\delta\rho$ peak elongation on the deformation ED maps (see figure 6.19). It should be kept in mind, that conjugative and hyperconjugative interactions and steric factors can result in ED distribution features which mimic the picture of a multiple chemical bond (see section 6.2). The crystal environment can lead also to the distortion of the axial symmetry in formally single and formally triple bonds. Such an effect is observed, for example, in the ethane crystal (see table 6.14) in which the experimental estimate of ellipticity of the C–C bond is different from zero. Kapphahn *et al* (1988) found that the eigenvector $\boldsymbol{u}_2$, which shows the direction of the slowest ED decrease, deviates only by 15° from the direction toward the nearest-neighbouring molecule in the crystal. Because the C atom thermal vibration tensor axes are oriented in other ways, such ED polarization can be treated as being a result of intermolecular interaction.

The carbon–carbon bonds exhibit substantial ellipticities in three-membered rings. The close proximity of bond and ring critical points reduces the magnitude of the negative curvature of the bond critical point which lies in the ring surface. For instance the carbon–carbon bonds of cyclopropane, C_3H_6, whose bond paths are noticeably curved outside the geometrical perimeter of the ring (see figure 2.16), also as a consequence of the proximity of the ring critical point, exhibit substantial ellipticities ($\varepsilon = 0.11$) with their major axes coincident with the ring surface. In four-membered and larger rings, the ring critical point is more distant from the bond critical points, therefore there is only a small or insignificant interaction between them. For example, in the free molecule of cyclobutane, C_4H_8, the value of ED at the ring critical point is only 0.53 e Å^{-3}, considerably less than the value of ED at the C–C bond critical point, 1.613 e Å^{-3} (the ellipticity of the perimeter bond is $\varepsilon = 0.02$). Bader *et al* (1983) concluded that substantial in-plane ellipticity is a property of three-membered rings only and their chemistry should in turn be unique among ring systems in exhibiting behaviour consistent with a 'π-like' charge distribution in the plane of the ring.

The critical point position relative to nuclei in some cases can be a qualitative characteristic of a chemical bond (Bader *et al* 1982, Cremer and Kraka 1984a). Indeed, the more electronegative bonded atom has the larger basin, therefore the boundary surface is closer to the neighbouring atom. Correspondingly, the critical point (3, −1) is shifted along a bond path toward this latter atom. The linear correlation between the differences of electronegativities of A (A = Be, B, C, and N) and H atoms and distances from the H nucleus to the bond critical point has been really observed in molecules of AH_n type; however, the generality of this characteristic should not be overestimated.

If valence angles in molecule or crystal structures are not consistent with the model of directed valence, the bond paths are not coincident with the internuclear axes between the bonded pairs of atoms. This is the case of strained or bent bonds (Coulson and Moffitt 1949). The bond paths are curved outward from the perimeter of a three- or four-membered ring or from the edges of a cage structure; therefore, the bond path length, L, should not be equated to internuclear distance, R_e (Bader *et al* 1982). The bond path length is the property of a bond. The values of L and their ratios to the corresponding R_e values for bonds in some highly strained molecules are listed in table 6.16. As a consequence of the curvature of a bond path, the ED is not distributed so that the forces of attraction exerted on the nuclei are maximal. This effect leads to a weakening of the bond in spite of the fact that strained bonds, in general, exhibit internuclear distances which are shorter than normal. The strain energy (see table 6.16) can be related to energy gain.

Bader *et al* (1982) and Cremer and Kraka (1984a) have found on the basis of quantum chemical calculations for hydrocarbons that the value of ED at the bond critical point position, $\boldsymbol{r}_b$, is correlated with the bond path length L. For the carbon–carbon bond this correlation is well described by a linear relationship

$$\rho(\boldsymbol{r}_b) = aL + b. \qquad (6.33)$$

The $\rho(\boldsymbol{r}_b)$ values can be also correlated with the Lewis bond order n through the relationship

$$n = \exp\left\{A[\rho(\boldsymbol{r}_b) - B]\right\}. \qquad (6.34)$$

The a, b, A, and B coefficients depend on the basis set used in quantum chemical calculations. They are tabulated for some bonds by Cremer and Kraka (1984a). Results of calculations using both relationships for hydrocarbons are shown in figures 6.37 and 6.38. Observation of these figures reveals that points connected with different compounds are clustered in four groups, each corresponding to formal (Lewis) bond order $n = 1$, 1.5, 2, and 3. The percentage variation in the value of $\rho(\boldsymbol{r}_b)$ is equal to the percentage variation in the bond lengths within each cluster of points. It can be concluded that the four kinds of carbon–carbon bond, resulting from the Lewis model,

Table 6.16 Bond path lengths (L), internuclear distances (R_e), and bond energies in some hydrocarbons. (Reprinted with permission from Bader *et al* (1982). © 1982 American Chemical Society.)

Molecule	L (Å)	L/R_e	(3, −1) critical point shift d (Å)	Strain energy[a] (kcal mol^{-1})
Cyclobutane	1.554	1.000	0.013	7.2
Cyclopropane	1.528	1.018	0.096	8.0
Bicyclo[1.1.1]pentane	1.550	1.004	0.067	7.3
	1.503 (bridge)	1.024	0.117	19.6
Bicyclo[1.1.1]butane	1.526	1.017	0.092	13.4
	1.524	1.021	0.106	—
Cyclopropene	1.300	1.018	0.092	—
Tetrahedrane	1.506	1.023	0.122	17.2

[a]Strain energy is calculated relative to the energy of the C–C bond of ethane as zero.

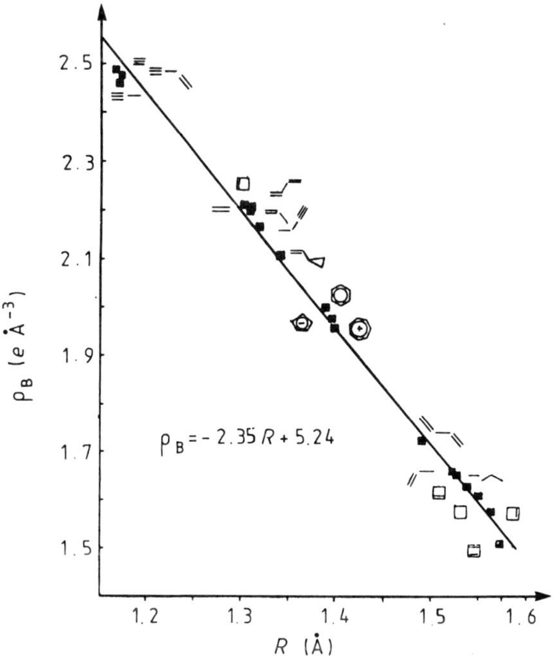

Figure 6.37 The relationship between the ED at the (3, −1) bond critical point position, $\boldsymbol{r}_b$, and the internuclear distance, R_e (HF STO-3G calculations) (Bader *et al* 1983, Cremer and Kraka 1984a).

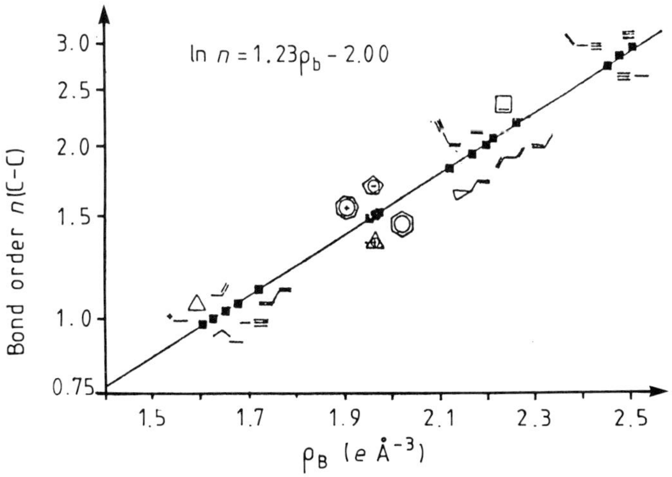

Figure 6.38 The correlation between the C–C bond order n and the electron density values ρ_b at the (3, −1) bond critical points (HF STO-3G calculations) (Bader *et al* 1983, Cremer and Kraka 1984a).

are reflected not only in the existence of a characteristic bond length, but also in values of ED at the bond critical point.

The dependences discussed are observed for C–C bonds in a wide range of neutral and charged, strained and unstrained systems. The polar covalent bonds of A–B type can be typified in a similar manner.

In spite of the fact that the existence of the bond path in a stable system is the necessary and sufficient condition for chemical bond formation, the sign of the Laplacian at a bond critical point, which depends on the relationship between the negative Hessian matrix eigenvalues λ_1 and λ_2 and positive eigenvalue λ_3, cannot be fixed by this condition. It depends on the character of atomic interaction, i.e. on the character of the chemical bond. It results in natural division of the atomic interactions into two types. The negative λ_1 and λ_2 values measure the degree of contraction of ED toward the bond critical point, perpendicular to the bond path. The positive λ_3 value measures the degree of ED contraction from the bond critical point toward each of the neighbouring nuclei. If negative curvatures dominate, the ED is locally concentrated in the region of the bond critical point and is shared by both nuclei. The local electronic potential energy in the critical point region is more than twice the excesses of the local kinetic energy. Such a picture is typical for a covalent bond. If the positive curvature dominates, ED is concentrated separately in each of the atomic basins (closed-shell interactions). The bond in this case is dominated by the relatively large positive contribution to the local kinetic energy of a system. The closed-shell interactions are usually connected with ionic and hydrogen bonds and van der Waals interactions. The properties of ED at bond critical points in different molecules, as obtained from quantum chemical calculation in the extended basis set at the near-HF level (Bader and Essen 1984), are listed in table 6.17. Some examples of the Laplacian of ρ distribution pictures in different chemical bonds are presented in figure 6.39. The negative $\nabla^2\rho$ values between nuclei in the case of shared interactions correspond to the low electronic potential energy region extending over both atomic basins. As can be seen from figure 6.40, a shift of the centre of negative $\nabla^2\rho$ is observed in the polar covalent bond. This shift correlates with bond polarity. Simultaneously, the volume of the negative $\nabla^2\rho$ region between nuclei is decreased. For closed-shell interactions the negative $\nabla^2\rho$ regions are atomic-like (figure 6.39), aside from small polarization effects. This results from the Pauli exclusion principle. Thus, the spatial regions where the local electronic potential energy dominates the kinetic energy are confined separately to each atom. That is reflected in the contraction of the ED toward each nuclei. The $\nabla^2\rho$ around the bond critical point is positive.

All this treatment was based on theoretical evidence. Now it is important to conclude to what extent it corresponds to the picture of the Laplacian of ρ, as reconstructed from x-ray diffraction data. The comparison with deformation ED maps should also be useful. Kapphahn *et al* (1988) calculated

Table 6.17 Characterization of atomic interactions via local properties of electron density (all quantities in atomic units). (Reprinted with permission from Bader and Essen (1984).)

Molecule and state	$\rho(r_b)$	$\nabla^2\rho(r_b)$	Eigenvalues of Hessian of $\rho(r_b)$: λ_1	λ_2	λ_3	$\lvert\lambda_1\rvert/\lambda_3$	Kinetic energy contributions: $G(r_b)_\perp$	$G(r)_\parallel$	$G_\perp/G_\parallel$	$G(r_b)/\rho(r_b)$
Shared interactions										
$H_2(^1\Sigma_g^+)$	0.2728	−1.3784	−0.9917	−0.9917	0.6049	1.64				0.062
$B_2(^3\Sigma_g^-)$	0.1250	−0.1983	−0.0998	−0.0998	0.0014	71.3	0.0223	0.0048	4.65	0.396
$N_2(^1\Sigma_g^+)$	0.7219	−3.0500	−1.9337	−1.9337	0.8175	2.37	0.3042	0.0170	17.89	0.866
$NO(^2\Pi)$	0.5933	−2.0353	−1.6460	−1.6460	1.2568	1.31	0.2366	0.0463	5.11	0.88
$NO^-(^3\Sigma^-)$	0.5755	−2.0851	−1.5883	−1.5883	1.0914	1.91	0.2284	0.0534	4.22	0.89
$O_2(^3\Sigma_g^-)$	0.5513	−1.0127	−1.4730	−1.4730	1.9333	0.76	0.2053	0.0721	2.85	0.88
Closed-shell interactions										
$He_2(^1\Sigma_g^+)$	0.0367	0.2501	−0.0774	−0.0774	0.4049	0.19	0.000	0.0540	0.000	1.47
$Ne_2(^1\Sigma_g^+)$	0.1314	1.3544	−0.3436	−0.3436	2.0417	0.17	0.0141	0.3024	0.047	2.52
$Ar_2(^1\Sigma_g^+)$	0.0957	0.4455	−0.1388	−0.1388	0.7231	0.20	0.0064	0.1144	0.056	1.33
$LiCl(^1\Sigma^+)$	0.0462	0.2657	−0.0725	−0.0725	0.4106	0.18	0.0033	0.0577	0.057	1.39
$NaCl(^1\Sigma^+)$	0.0358	0.2004	−0.0401	−0.0401	0.2806	0.14	0.0035	0.0396	0.099	1.30
$NaF(^1\Sigma^+)$	0.0543	0.4655	−0.0897	−0.0897	0.6449	0.14	0.0081	0.0890	0.090	1.94
$KF(^1\Sigma^+)$	0.0554	0.3132	−0.0717	−0.0717	0.4566	0.16	0.0067	0.0647	0.104	1.41
$MgO(^1\Sigma^+)$	0.0903	0.6506	−0.1331	−0.1331	0.9169	0.15	0.0170	0.1351	0.126	1.87
Hydrogen bond in $(H_2O)_2$	0.0198	0.0623	−0.0247	−0.0240	0.1110	0.223	0.0061	0.0147	0.415	0.806
Hydrogen bond in $(HF)_2$	0.0262	0.1198	−0.0406	−0.0360	0.1994	0.204	0.0013	0.0243	0.053	1.027

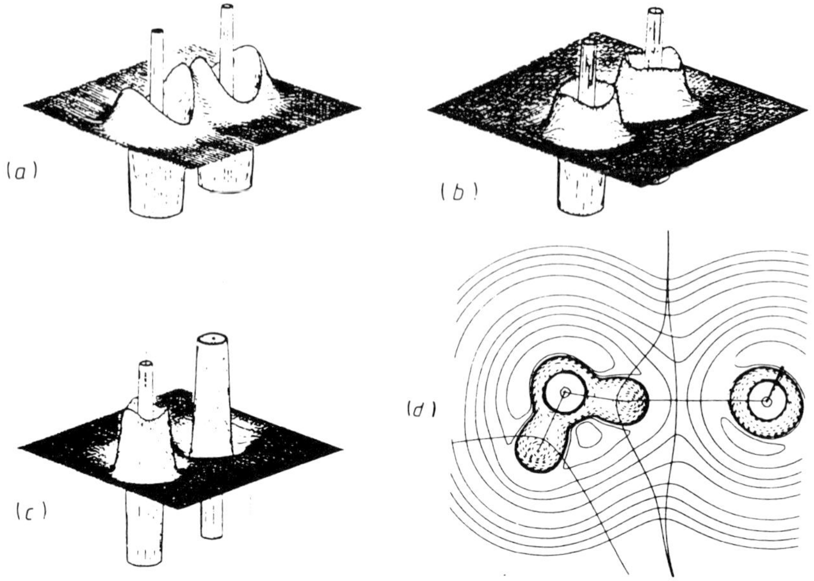

Figure 6.39 Contour maps of the Laplacian of ρ for molecules with shared and closed-shell interactions: (*a*) an N_2 molecule with a covalent bond; (*b*) an Ar_2 noble gas molecule; (*c*) a KF molecule with an ionic bond; (*d*) an $HO\ldots OH_2$ complex with a hydrogen bond. Maxima and broken lines indicate regions with an ED concentration.

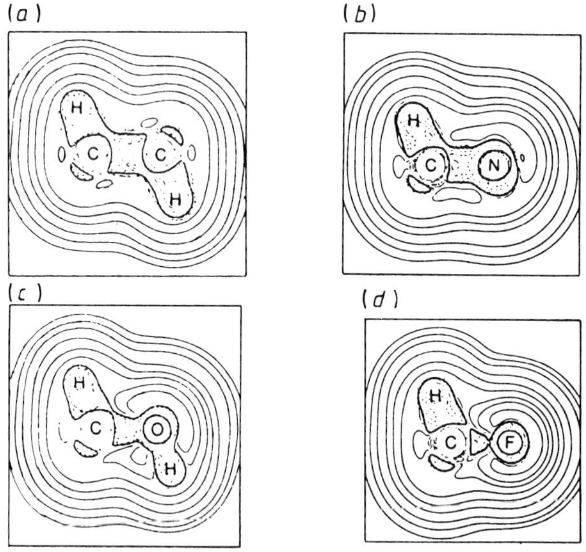

Figure 6.40 A contour line diagram of the Laplacian of ED in molecules with different polarity of the covalent chemical bond: (*a*) C_2H_6 (staggered); (*b*) CH_3NH_2 (staggered, HCN *trans*); (*c*) CH_3OH (staggered, HCOH *trans*); (*d*) CH_3F. Broken lines indicate regions with ED concentration (Cremer and Kraka 1984b).

$\nabla^2\rho$ maps for ethane, ethylene, and acetylene from x-ray diffraction data (figure 6.41). They found that the region of negative $\nabla^2\rho$ values between nuclei in the C–C bond really exists and it is decreased when the bond order is increased (at the same time the total ED in corresponding bond critical points increases—see table 6.14). The Laplacian of ρ in C–H bonds is nearly the same in all compounds. The positions of excessive $\delta\rho$ peaks on experimental deformation ED maps (figure 6.41(*b*), (*d*), (*f*)) coincide with negative $\nabla^2\rho$ regions; however volumes of these peaks do not correlate with bond order. For ethane, C_2H_6, a quite satisfactory agreement with $\nabla^2\rho$ distribution resulting from quantum chemical calculation is also observed (see figure 6.40(*a*)).

The Laplacian of ρ in the fluorite, CaF_2, compound with an ionic bond, calculated from x-ray diffraction data (figure 3.24), in accordance with theoretical arguments demonstrates the positive $\nabla^2\rho$ region in the middle of the Ca–F bond path, while negative regions of $\nabla^2\rho$ surround the Ca^{2+} and F^- ion positions. Strong asymmetry of the last $\nabla^2\rho$ regions is observed, in contrast with the analogous picture in polar chemical bonds of the separate molecules (figure 6.39(*c*) and figure 6.40).

Studies of the crystalline α-oxalic acid dihydrate (Laidig and Frampton 1991), the urea–phosphoric acid complex (Souhassou 1991), and urea (Zavodnik *et al* 1994) give examples of systems with intermolecular closed-shell interactions. They found that in H bonds of these crystals the Laplacian of ρ, reconstructed from x-ray diffraction data, at the (3, −1) critical points is positive. These results also agree with theoretical conclusions discussed.

The intermolecular interaction in the Cl_2 crystal gives another interesting example of closed-shell interaction. The crystal structure of solid chlorine is layered, with molecules in planes parallel to the (100) plane in the orthorhombic unit cell. Short intermolecular contacts of 3.28 Å, a value significantly less than twice the accepted van der Waals radius of 1.8 Å, are observed in the (100) plane. This indicates the presence of a specific intermolecular bonding in the Cl_2 crystal, the structure of which is not described in terms of a non-directional van der Waals potential. Indeed, Tsirelson *et al* (1995) found, experimentally and theoretically, that intermolecular interaction lines connect the lone-pair charge concentration on each Cl atom with a charge depletion at the end of the neighbouring molecule; each atom participates in two such interactions in the (100) plane. Additionally, the shape of the Cl atom in a crystal proved to be very far from spherical. Both these observations explained why the usual (spherical) van der Waals potential is not applicable to description of the solid Cl_2 structure.

The topological analysis allows one to prove some semiclassical models of the chemical bond, for example, the Gillespie (1992) concept of electron pairs. Simple inorganic molecules, such as polyhaloid compounds of ions

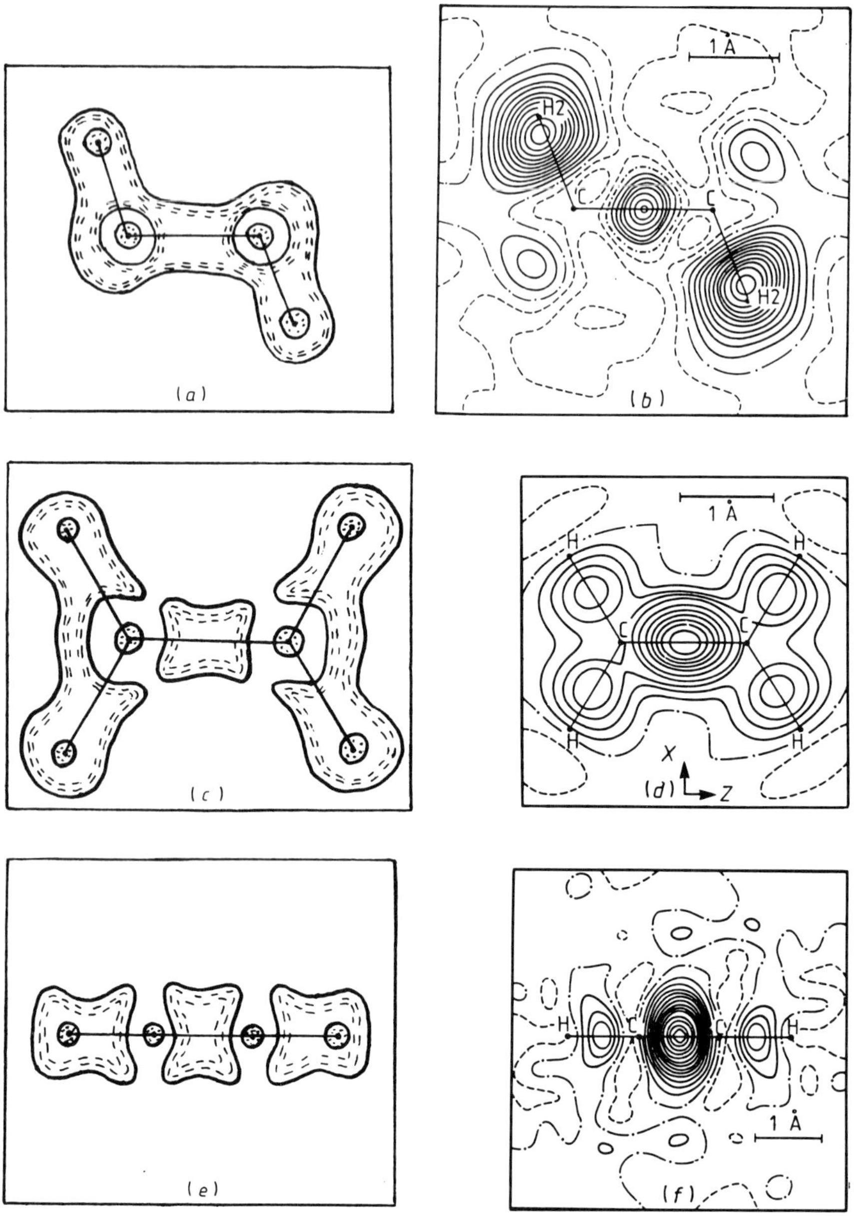

Figure 6.41 The Laplacian of ED (Kapphahn *et al* 1988) and deformation ED (van Nes 1978) in ordinary, double, and triple covalent carbon–carbon bonds: (*a*), (*b*) ethane; only carbon EDs in the $\delta\rho$ map are subtracted; (*c*), (*d*) ethylene; (*e*), (*f*) acetylene. Solid lines and broken lines correspond to positive and negative $\delta\rho$ values, respectively. Contours are at $0.05\,e\,\text{Å}^{-3}$. In regions where lines are dashed $\nabla^2\rho<0$. Around nuclei $\nabla^2\rho\ll0$.

ClF_3 and ClF_2^+, which contain so-called stereochemically active lone-pair electrons at the central atom, can be presented as an example. In order to describe the structure of such angle-like ions, the hypothesis that the repulsion of two valence electron pairs and two electronic lone pairs should be minimal is often used. This means that corresponding ED concentrations should be observed in corners of a tetrahedron. Antipin *et al* (1988) have found that on the experimental $\delta\rho$ map in tetrafluoroborate difluorocloronium, $(BF_4)^-(ClF_2)^+$, in addition to bond ED accumulations, there are two $\delta\rho$ maxima which form an angle of $\sim 150°$ with the Cl atom in the plane normal to the plane of cation $(ClF_2)^+$ (figure 6.42). Positions of these maxima can be, as usual, attributed to sites of localization of two lone pairs of the Cl atom. The maxima of electron pairs of Cl–F bonds and lone pairs of the Cl atom form a distorted tetrahedron (the valence F–Cl–F angle is 96.4°). Bader *et al* (1984) have shown that regions of local ED concentrations in the free cation $(ClF_2)^+$ indeed make a distorted tetrahedron; the angle formed by negative areas of $\nabla^2\rho$ on the bonds is 99° whereas the corresponding angle in the bisectral plane is $\sim 143°$. Hence, the simple model of electron pairs is justified for the compounds discussed.

The topological theory can also be linked to Fukui's (1971) frontier orbital model. As was noted by Bader and MacDougall (1985), in many molecules the regions of space where highest occupied MOs and lowest unoccupied MOs of the reactants are most localized coincide with the regions of ED concentration and depletion, respectively. It can be concluded that the Laplacian of ρ provides a bridge between the ED and orbital approaches to the understanding of chemical reactivity.

Thus, the experimental data concerning the Laplacian of ρ support the main conclusions concerned with the description of the chemical bond and intermolecular interaction based on Bader topological theory. They agree as well with the description of the chemical bond in terms of deformation ED, semiclassical, and orbital representations. It is important also that the

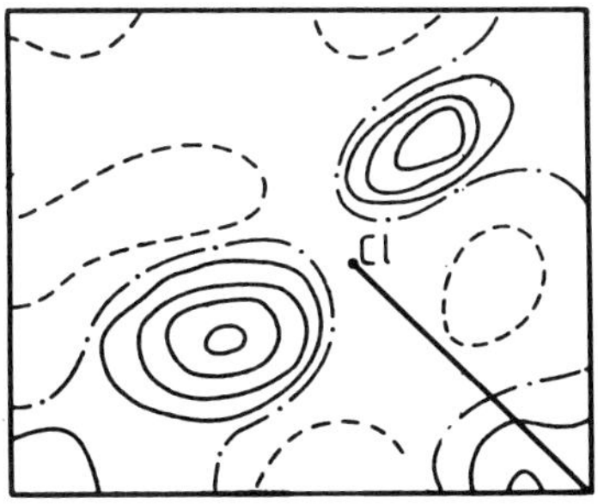

Figure 6.42 The deformation density of the $[BF_4^-][ClF_2^+]$ complex in the section which is normal to the plane of the ClF_2^+ cation (the direct line is the bisector of the F–Cl–F angle). The line intervals are 0.07 $e\,Å^{-3}$ (Antipin *et al* 1988).

topological analysis allows us to study the interatomic interactions in those cases when standard deformation ED does not reveal the chemical bond features. For example, the standard $\delta\rho$ maps of hydrogen peroxide, H_2O_2, experimental (Savariault and Lehmann 1980) and theoretical (Breitenstein *et al* 1983), have shown the negative $\delta\rho$ region between the bonded O atoms (figure 6.43(*a*) and (*b*)). Nevertheless, Cremer and Kraka (1984b) have found that the $\nabla^2\rho$ map (figure 6.43(*c*)) exhibits ED concentration along the O–O bond path. Therefore, the O–O bond can be considered as typically covalent. The main advantage of the topological approach to the chemical bond analysis is clearly seen in this example: there is no need to use either any reference state or any orbital model. At first glance, conclusions concerned with the chemical bond features, based on topological analysis, can be treated as nearly model free. Unfortunately, this is, sometimes, not the case and the influence of basis set on theory and the experimental errors can be very

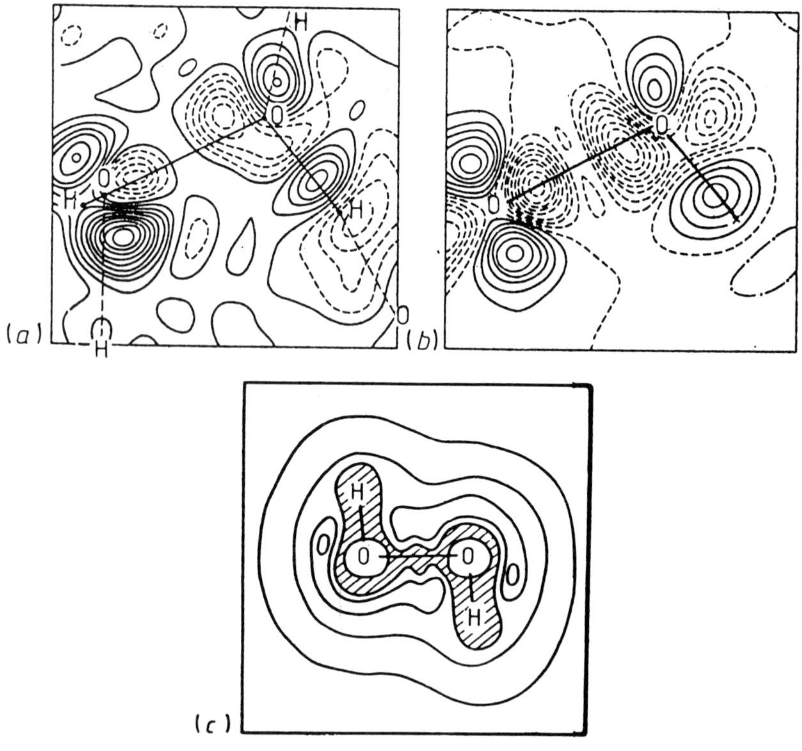

Figure 6.43 Experimental (*a*) and theoretical (*b*) deformation density and Laplacian of ρ (*c*) for hydrogen peroxide, H_2O_2. The line intervals are 0.1 $e\,\text{Å}^{-3}$. The shaded area on the $\nabla^2\rho$ map corresponds to $\nabla^2\rho < 0$. ((*a*) Reprinted with permission from Savariault and Lehmann 1980. (*b*) Breitenstein *et al* 1983. (*c*) Cremer and Kraka 1984b.)

important. For example, standard deformation ED is negative in the internuclear space of the F_2 molecule while hybrid deformation ED gives an excessive $\delta\rho$ peak in this region (figure 6.44). Which type of chemical bond should be assigned to this molecule? The early calculation of the Laplacian of the ED in the STO basis set near HF quality (Bader and Essen 1984) did not allow us to answer this question: the $\nabla^2\rho$ value in the midpoint of the F–F bond proved to be positive. Tsirelson (1994) has shown, however, that the use of a very large basis set containing diffuse f functions results in a negative value, -0.045 au, of $\nabla^2\rho$ (figure 6.44). The quality of calculation can also affect the position of critical points (Edgecombe and Boyd 1986) and volume size of pseudoatoms (Stefanov and Cioslowski 1995).

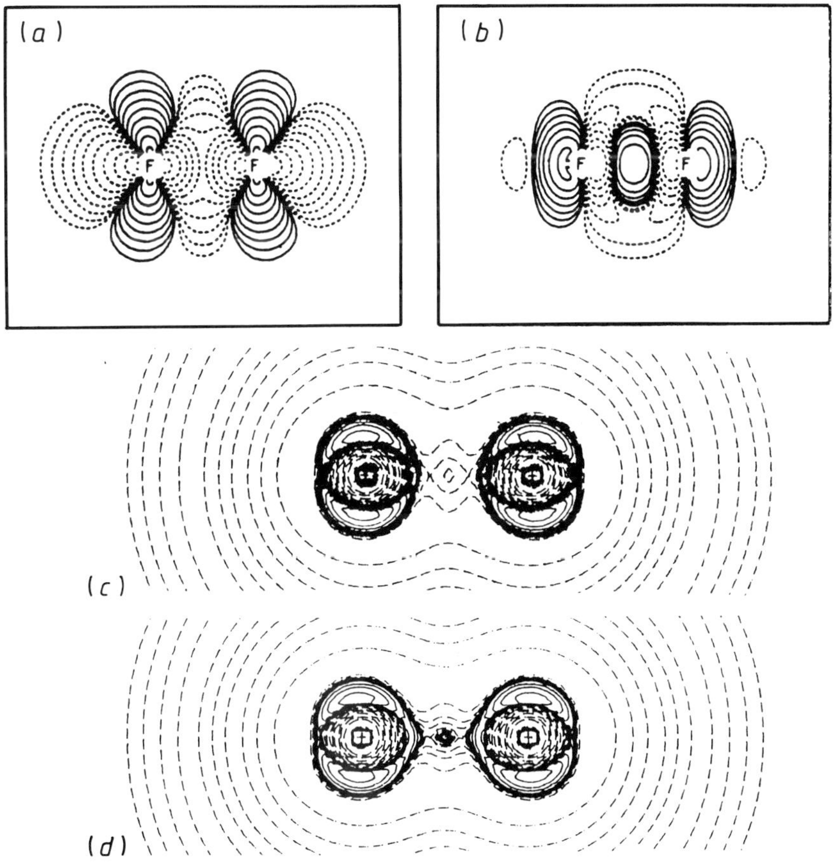

Figure 6.44 The F_2 molecule: (*a*) standard and (*b*) hybrid deformation densities; Laplacian of electron density: (*c*) 6–311++G(2d) and (*d*) 6–311++G(3df) basis sets (Tsirelson 1994). Solid lines indicate regions of ED concentration. ((*a*) and (*b*) Reprinted with permission from Kunze and Hall 1987.)

Cremer and Kraka (1984a) have used energetic considerations for characterization of the chemical bond. The bond formation is associated with a gain in molecular energy resulting from a complex interplay of changes in potential and kinetic energy. The analysis plotted in subsection 2.4.2 shows that the domination of the negative local electron potential energy in internuclear space, as takes place in the case of a covalent bond, gives evidence for the stabilizing energetic role of ED accumulation in this region: the local electron energy density $H_e(\mathbf{r})$ is negative here in this case; the relation $\frac{1}{4}\nabla^2\rho(\mathbf{r}) < |V(\mathbf{r})|$ can be fulfilled even when $\nabla^2\rho$ is positive. In a stable system, in which the bond is usually considered as covalent, the local electron energy density is negative at the bond critical points independent of the $\nabla^2\rho$ sign (see table 6.18). This allowed Cremer and Kraka (1984a) to formulate the characteristics of covalent chemical bonds. The covalent bond is formed if (a) the bond critical point (3, −1) between atoms exists (necessary condition) and (b) the local electron energy density H_e at the bond critical point is negative (sufficient condition). It should be added that the area around the (3, −1) critical point with $H_e(\mathbf{r}) < 0$ can be defined as the bonding region.

Table 6.18 Characterization of some covalent bonds by local properties of electron and energy densities (Cremer and Kraka 1984a).

Molecule	Bond	$\rho(\mathbf{r})$ (e Å^{-3})	$\nabla^2\rho(\mathbf{r}_b)$ (e Å^{-5})	$G(\mathbf{r}_b)$ (hartree)	$V(\mathbf{r}_b)$ (e Å^{-3})	$H_e(\mathbf{r}_b)$
H_2	H–H	1.692	−27.243	0.0	−1.910	−1.910
LiH	Li–H	0.230	3.448	0.236	−0.249	−0.013
BeH_2	Be–H	0.604	4.138	0.606	−0.890	−0.284
BH_3	B–H	1.251	−7.742	0.837	−2.221	−1.324
CH_4	C–H	1.846	−22.977	0.284	−2.349	−2.065
NH_3	N–H	2.312	−41.737	0.350	−3.621	−3.271
N_2H_4	N–N	2.195	−17.350	1.033	−3.405	−2.372
H_2O	O–H	2.475	−50.155	0.445	−4.399	−3.954
H_2O_2	O–O	2.249	−9.509	1.528	−3.721	−2.193
H_2OO	O–O	1.418	12.354	1.761	−2.665	−0.904
HF	H–F	2.420	−68.987	0.499	−5.830	−5.331
F_2	F–F	2.335	2.908	2.247	−4.292	−2.045
CH_3NH_2	C–N	1.866	−22.762	0.978	−3.549	−2.571
CH_3OH	C–O	1.775	−3.493	2.416	−5.081	−2.665
NH_2OH	N–O	2.155	−13.522	1.149	−3.360	−2.166
HOF	O–F	2.190	−1.047	1.829	−3.732	−1.903
CH_2NH	C=N	2.737	−9.515	4.170	−9.002	−4.832
N_2H_2	N=N	3.199	−30.818	2.072	−6.296	−4.224
$O_2(^3\Sigma_g)$	O=O	3.979	−26.304	3.543	−8.935	−5.392
HNO	N=O	3.574	−38.539	4.810	−12.318	−7.508

With these criteria, the chemical bond in the F_2 molecule, where $H_e(\mathbf{r}_b)<0$, is certainly covalent, in accordance with the chemical behaviour of this molecule. The simple lithium organic compounds where the character of the chemical bond was not discovered for a long time give another example of the usefulness of the local energy density consideration. The experimental $\delta\rho$ maps in the Li–C bonding region in phenylthiomethyllithium and methylthiomethyllithium complexes show ED accumulation near C atoms (Amstutz *et al* 1984) which may be treated as a manifestation of the covalent nature of this bond. At the same time, the Laplacian of ρ usually is positive along the Li–C bond in different compounds (Ritchie and Bachrach 1987). Cremer and Kraka (1984a) in analogous molecules found that in analogous fragments on the Li–C bond the local electron energy density $H_e(\mathbf{r})>0$. Therefore, the interaction of Li and C atoms should be considered as rather ionic while the $\delta\rho$ maximum shift from C toward Li atoms may be attributed to polarization effects.

Unfortunately, it is impossible, considering $H_e(\mathbf{r})$, to give an analogous definition of the ionic, hydrogen, and van der Waals bonds and to separate these bonds from each other. Positive values of the Laplacian of ρ and of local electron energy at the bond critical point are observed in these bonds. Therefore, all these bonds in the topological analysis framework are only unified as closed-shell interactions.

6.5 The nature of the chemical bond

As we have seen, the real picture of ED distribution in the molecules and crystals is strongly varied. The following question arises: what general physical factors result in keeping the atoms of a system together? The search for the answer to this question is known as the establishing of the nature of the chemical bond. There is extensive literature on this subject: Ruedenberg (1962), Feinberg and Ruedenberg (1971), Hirshfeld and Rzotkiewicz (1974), Wilson and Goddard (1972a, b), Bader (1975a, b, 1981), Cremer and Kraka (1984a), Bader and Essen (1984), Schwarz *et al* (1985), and Low and Hall (1990) have studied this matter from different points of view. Let us try to discuss this problem on the basis of ED distribution analysis of stable molecules and crystals.

There are two approaches to the analysis of the nature of the chemical bond. The first one originates from the work of Berlin (1951) who applied the Hellmann–Feynman theorem to chemical bond consideration. According to this theorem, the force exerted on any fixed nucleus, calculated from the ED (theoretical or reconstructed from the x-ray diffraction data), is classical electrostatic attractive. This force is connected by equations (2.88) and (2.89) with the energy which is needed to move the nucleus in the electrostatic field of electrons and other nuclei from one nuclear configuration to another. In principle, the binding energy can be determined in this way. The interpretation of the chemical bond in terms of the forces exerted on the nuclei is called the chemical binding approach. The forces acting on a nucleus in an equilibrium system from all other nuclei and electrons should be zero.

Berlin (1951) suggested for diatomics the value $f(\boldsymbol{r})$

$$f(\boldsymbol{r})=(Z_1/r_1)\cos\theta_1+(Z_2/r_2)\cos\theta_2 \tag{6.35}$$

which is a projection onto the internuclear axis of the total force acting at a point $\boldsymbol{r}$ on a nucleus with charge Z_i from the unit negative charge. Using (6.35), the force $\boldsymbol{F}(R)$ (R is the internuclear distance) acting on nuclei is presented as

$$\boldsymbol{F}(R)=\frac{Z_1Z_2}{R_2}-\frac{1}{2}\int f(\boldsymbol{r})\rho(\boldsymbol{r})\,\mathrm{d}\boldsymbol{r}. \tag{6.36}$$

At equilibrium $R=R_{eq}$ and the force, $\boldsymbol{F}(R_{eq})=0$. If one agrees that $\rho(\boldsymbol{r})>0$ (not taking into account that this function describes the distribution of the negative charge) then the contribution of the last term to the force $\boldsymbol{F}(R)$ at any point $\boldsymbol{r}$ depends on the sign of $f(\boldsymbol{r})$ at the same point. The negative charge in regions with $f(\boldsymbol{r})>0$ decreases the $\boldsymbol{F}$ value (binds the nuclei) whereas in regions with $f(\boldsymbol{r})<0$ it increases the $\boldsymbol{F}$ value. Berlin called regions where $f(\boldsymbol{r})>0$ the binding regions and those where $f(\boldsymbol{r})<0$ the antibinding regions. To stabilize a system it is necessary that the ED in the binding region at $R=R_{eq}$ create a force which would compensate the nuclear repulsions and the action upon them of the ED in antibinding regions.

Schrödinger perturbation theory shows that mutual polarization is taking place when atoms began to interact. This results in an ED shift toward a neighbouring atom and in induction of a dipole moment. Because chemical bond formation is often accompanied by ED accumulation in the internuclear space, Bamsai and Deb (1981) suggested that the requirement $\delta\rho>0$ in the binding region is necessary (but not sufficient) for the system to be stable. However, it becomes clear now that this is not the case (Bader 1981). For instance the Be_2 molecule is known to be unstable, whereas topographical features of its standard $\delta\rho$ map are the same as in C_2 and N_2 molecules. At the same time, the standard $\delta\rho$ map for the stable F_2 molecule has a negative $\delta\rho$ value in the binding region (see figure 6.44). Besides, it is now firmly established that even in diatomic molecules, and especially in crystals, $\delta\rho>0$ not only in the internuclear space: the excessive $\delta\rho$ peaks can be observed also in the regions behind nuclei (see figure 6.12). In other words, when forming a system of atoms, the electrons move into antibinding (according to Berlin) regions as well. Such ED redistribution in diatomics has been confirmed by gas electron diffraction (Fink *et al* 1979a, b). Thus, the formation of a stable system is provided by a very fine balance of forces acting on nuclei, and corresponding ED redistribution.

An analysis of ED redistribution during the chemical bond formation has been performed by Spackman and Maslen (1985). They have described the force acting on a nucleus in diatomics via the promolecule ED, $\tilde{\rho}$, and the deformation ED, $\delta\rho$:

$$F(R)=\frac{Z_1Z_2}{R^2}-\frac{1}{2}\int f(\boldsymbol{r})\tilde{\rho}(\boldsymbol{r})\,\mathrm{d}V-\frac{1}{2}\int f(\boldsymbol{r})\delta\rho(\boldsymbol{r})\,\mathrm{d}V. \tag{6.37}$$

$\tilde{\rho}$ is positive anywhere but $\delta\rho$ may have both positive and negative signs; therefore, the coincidence of regions with the same f and $\delta\rho$ signs should provide system stabilization. This is why the area of negative $\delta\rho$ in the antibonding regions behind the atomic nuclei, if they exist, should correspond to binding. Indeed, there are many cases when minima of $\delta\rho$ exist behind the nucleus on the continuation of the bond line (see e.g. figures 6.18(*a*), 6.22(*b*), and 6.24(*b*)).

Spackman and Maslen (1985) have considered the difference in the binding nature in N_2 and F_2 molecules on the basis of equation (6.37). There is a significant ED concentration in the first molecule along the internuclear line and on its continuation behind the nuclei. In the second molecule there are $\delta\rho$ minima between and behind nuclei (figure 6.44). This leads to the difference in corresponding integrands in the last term of the expression (6.37) (figure 6.45). In the N_2 molecule the binding region between nuclei, where $f(\boldsymbol{r})>0$ and $\delta\rho>0$, gives the main contribution to the positive value of the integral $\int f(\boldsymbol{r})\delta\rho(\boldsymbol{r})\,\mathrm{d}V$. In contrast, in the F_2 molecule the main contribution to binding is determined by the regions behind the nuclei and by the torus area around the bond line, where $f(\boldsymbol{r})<0$ and $\delta\rho(\boldsymbol{r})<0$. In both molecules, of great significance are the regions near nuclei where the $\delta\rho$ character is quite different. Thus, it becomes clear that in order to establish the chemical bond nature it is important to analyse the ED in the whole molecular space.

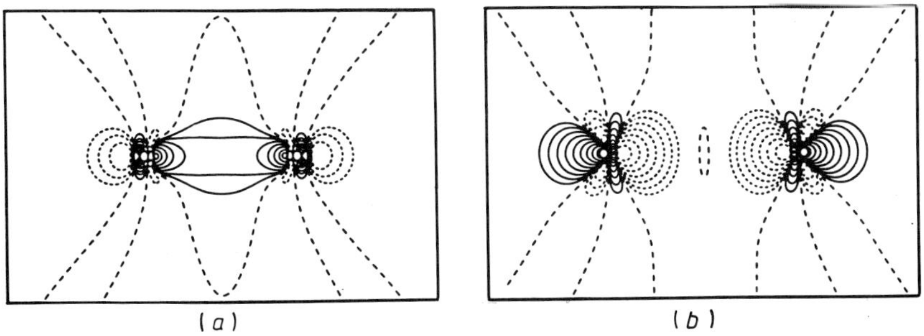

Figure 6.45 Contour maps of the force per unit charge multiplied by deformation ED, $f\ \delta\rho$: (*a*) the N_2 molecule; (*b*) the F_2 molecule. Contours differ by a successive factor of two, starting from $\pm 25.0\ e\ \text{Å}^{-3}$. Positive contours are solid, negative short dashed; the zero contour is long dashed (Spackman and Maslen 1985).

Silberbach (1991) noted that components $\tilde{\rho}$ and $\delta\rho$ in the expression (6.37) do not obey the variation principle. Therefore, the forces calculated may appear to have no physical meaning. It can be useful to calculate the function (6.35) for each molecular orbital separately: this will help in dividing them into binding, non-binding, and antibinding MOs (Bader 1981). The corresponding forces have, in this case, clear physical meaning. In general, however, the binding concept cannot explain the appearance or absence of the chemical bond exhaustively. Besides, Spackman and Maslen (1985) pointed out a noticeable drawback of the approach: if one postulates that $\rho(\boldsymbol{r})$ of a system is negative everywhere (not positive, as was supposed above), the binding and antibinding regions should be transposed, since charges of nuclei are always positive. This approach is also not applicable to metastable states, the energy of which is higher than the energy of the set of separate atoms: the forces exerted on some nuclei for such states can be zero. The force is zero also at the top of the chemical reaction energy barrier which divides the initial and final products. The more serious objection, however, is the other one. First, as noted in section 2.2, a static system of electric charges cannot be in equilibrium (Ernshow theorem). Second, in order to explain the nature of the chemical bond the local density consideration is not enough; the kinetic energy term related to wave-function derivatives should also be taken into consideration (Teller 1962). The electrostatic approach based on the Hellmann–Feynman theorem (2.62) is dependent on local potential energy only and, therefore, it does not allow one to draw any conclusions regarding the kinetic part of the total bond energy. The ED is distributed in such a way that the total energy of the system is minimal. Thus, we arrive at the necessity of considering another, energetic (bonding), approach in order to be able to analyse a chemical bond properly.

The energetic approach to the nature of the chemical bond has been considered from different points of view. The first careful analysis of a covalent bond was performed by Ruedenberg (1962) for the H_2 molecule and H_2^+ ion. His conclusions can be summarized as follows. If the interatomic distance decreases the constructive interference of atomic wave-function leads to the decrease of electron energy relative to the energy of the set of free atoms because of the decrease of the corresponding orbital gradients. The electronic charge, then, shifts from the regions near nuclei, where the potential is low, and into the interatomic space (see figure 6.11(*e*)) increasing the potential energy of a system. The ED increases in the vicinity of nuclei simultaneously. As a result, the kinetic energy of core electrons increases and the potential energy decreases. The total energy of a system is lower than that of the separate atoms: the total energy gain is because of the kinetic energy decrease during the chemical interaction. This conclusion is, however, in disagreement with the virial theorem, according to which the change in kinetic energy, on average, is positive and the change in potential energy is negative during system formation from separate atoms.

The Ruedenberg analysis, as later proved, cannot be transferred to the more complicated systems, because the local kinetic and potential energy distributions in these systems are more complicated. Studying this problem, Bader and Preston (1969) have presented the kinetic energy density as

$$T(\mathbf{r}) = -\tfrac{1}{4}\nabla^2\rho(\mathbf{r}) + G(\mathbf{r}). \tag{6.38}$$

Using the natural orbital expansion of the one-electron DM (2.22), $G(\mathbf{r})$ can be written in the form

$$G(\mathbf{r}) = \frac{1}{2}\sum_i \lambda_i \nabla\varphi_i^*(\mathbf{r})\nabla\varphi_i(\mathbf{r}) = \tfrac{1}{2}[\nabla\nabla\Gamma^1(\mathbf{r},\mathbf{r}')]|_{r=r'} = \frac{1}{8}\sum_i \frac{\nabla\rho_i(\mathbf{r})\nabla\rho_i(\mathbf{r})}{\rho(\mathbf{r})} \tag{6.39}$$

Here

$$\rho_i(\mathbf{r}) = \lambda_i \varphi_i(\mathbf{r})\varphi_i(\mathbf{r}) \tag{6.40}$$

where λ_i are the orbital occupation numbers. The integral of $\nabla^2\rho(\mathbf{r})$ over the whole space is zero (see subsection 2.4.1). Therefore the average $T(\mathbf{r})$ value, denoted as T, is equal to the average value of $G(\mathbf{r})$.

Bader and Preston introduced also the parallel

$$T_{\parallel} = \tfrac{1}{2}\langle(\partial\psi/\partial z)(\partial\psi/\partial z)\rangle = -\tfrac{1}{2}\langle\psi|\partial^2/\partial z^2|\psi\rangle \tag{6.41}$$

and perpendicular

$$T_{\perp} = \tfrac{1}{2}\langle(\partial\psi/\partial x)\partial\psi/\partial x\rangle + \langle(\partial\psi/\partial y)\partial\psi/\partial y\rangle$$
$$= -\tfrac{1}{2}\langle\psi|\partial^2/\partial x^2 + \partial^2/\partial y^2|\psi\rangle \tag{6.42}$$

components of the average kinetic energy. For a free spherical atom $T_{\parallel} = \frac{1}{2}T_{\perp}$ and $\delta = (T_{\perp} - T_{\parallel})/T = \frac{1}{3}$. In bonded systems the deviation of the deformation index δ from the atomic value $\frac{1}{3}$ can serve as a measure of changes of the kinetic energy components during the chemical bond formation. These characteristics are useful in the analysis of the bond energy components.

As can be seen from (6.39), for a one-electron system such as the H_2^+ ion the kinetic energy is totally determined by $\nabla\rho$. However, for a two-electron system such as the H_2 molecule, the natural orbital expansion results in the bonding orbital accounting for only 98% of total ED (Das and Wahl 1966). Thus, the kinetic energy of such systems is determined only approximately by $\nabla\rho(\mathbf{r})$. The component of $\nabla\rho$ which is parallel to the bond line is small in the H_2 molecule due to the ED concentration between nuclei. Therefore, the $G(\mathbf{r})$ value in this area is decreased moving from a nucleus to the bond centre (figure 6.46); the contribution to T of the gradient of ED, parallel to the bond line, is also decreased. The perpendicular component of ED gradient here has a much higher value than the parallel component because of the ED contraction to the bond line. Both components of $\nabla\rho$ give large contributions to $G(\mathbf{r})$ behind nuclei, because the ED here is increasing equally

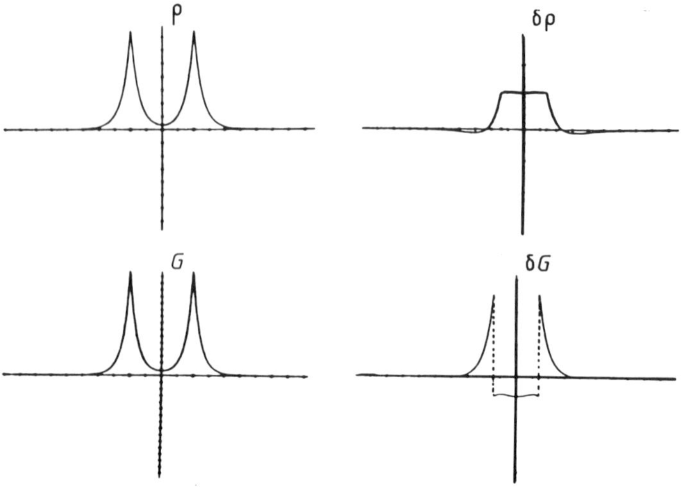

Figure 6.46 The H_2 molecule: profiles of electron density, $\rho(\mathbf{r})$, kinetic energy, $G(\mathbf{r})$, and corresponding difference functions referenced to atomic states. All values are given in au. (Reprinted with permission from Bader and Preston 1969.)

approaching the nucleus. The value of G between nuclei is less than the sum of the corresponding contributions of separate atoms.

The kinetic energy density component behaviour influences the average energy values as well. For example, in the H_2 molecule at equilibrium $T_{\|}(R_{eq}) < \frac{1}{2}T_{\perp}(R_{eq})$, whereas $T_{\|}(R_{eq}) < T_{\|}(\infty)$ and $T_{\perp}(R_e) > T_{\perp}(\infty)$. The deformation index $\delta = 0.4740$ is more than the atomic value $\frac{1}{3}$. It is remarkable that in the H_2 molecule the smallest $T_{\|}$ value is not at the equilibrium distance (1.4 au) but at the distance of 2.0 au. At equilibrium $T_{\|} = 0.3043$ au is still lower than the atomic value, 0.333 au; however, at $R < 2.0$ au $T_{\perp}$ is noticeably increased. Therefore, in a stable H_2 molecule the kinetic energy is increased due to the $G_{\perp}(\mathbf{r})$ increase. As the virial theorem states, the potential energy is decreased at the same time. Ruedenberg (1962) took into account only the kinetic energy component parallel to the bond line and his conclusion should be considered keeping this in mind.

In a non-stable system like He_2 ($R_{eq} = 2$ au) bonding and antibonding MOs are occupied by electrons. The deformation ED in the internuclear space is negative because of the Pauli exclusion principle. In this region in He_2 both parallel and perpendicular components of $G(\mathbf{r})$ and, consequently, the T, are larger in comparison to those of free atoms. The $T_{\|}$ component is increased more rapidly than $T_{\perp}$ as one atom approaches another. Due to this fact, the deformation index $\delta(R_{eq}) = 0.290$ is lower than the atomic value $\frac{1}{3}$. The kinetic energy density in the internuclear space of the He_2 molecule is systematically higher than that for free atoms. Thus, there are no areas in this molecule where the ED accumulation decreases the potential energy and simultaneously leads to low kinetic energy.

The N_2 molecule (electronic configuration $1\sigma^2 1\sigma^{*2} 2\sigma^2 2\sigma^{*2} 3\sigma^2 1\pi^4$) is an example of a many-electron system of general type where $G(\mathbf{r})$ cannot be even approximately connected with $\nabla\rho$. This is because the sum of orbital density gradients cannot be described as the gradient of the sum of orbital densities. The $G(\mathbf{r})$ distribution in the internuclear space is determined by both bonding and antibonding MOs occupied by electrons. The antibonding σ MOs have nodal surfaces in the internuclear distance centre. This leads to a noticable contribution here of their gradients to the kinetic energy density component parallel to the bond line. The π MOs give also a contribution to $G_{\parallel}(\mathbf{r})$. Thus, a softening of $\nabla\rho$ in such a system does not yet mean that the $G_{\parallel}(\mathbf{r})$ component is small.

The profile of the difference $\delta G(\mathbf{r}) = G_{N_2}(\mathbf{r}) - \sum_{at} G_N(\mathbf{r})$ (figure 6.47) shows that δG at the bond midpoint, where in the N_2 molecule the ED is accumulated, is larger in comparison to the separate atoms. In the vicinity of nuclei the kinetic energy density is negative both in the depletion ED area from the binding region side and in the accumulation ED regions behind a nucleus, i.e. in regions usually associated with electron lone pairs. Thus, the ED concentration along the diatomic molecule axis in the antibinding region makes the kinetic energy lower in relation to free atoms. It is interesting to note that such complicated kinetic energy density behaviour in the N_2 molecule results in a deformation index $\delta = 0.3331$, which does not differ, in fact, from the atomic value.

Preston (1969) calculated the $\delta G(\mathbf{r})$ maps for B_2, C_2, O_2, and F_2 molecules. The former two are similar to that of N_2, while the O_2 and F_2 exhibit areas in which $\delta G < 0$ in antibinding regions with the same characteristic features as for the N_2 molecule. The O_2 molecule has a region around the bond midpoint with $\delta G < 0$; the corresponding region for the F_2 molecule extends

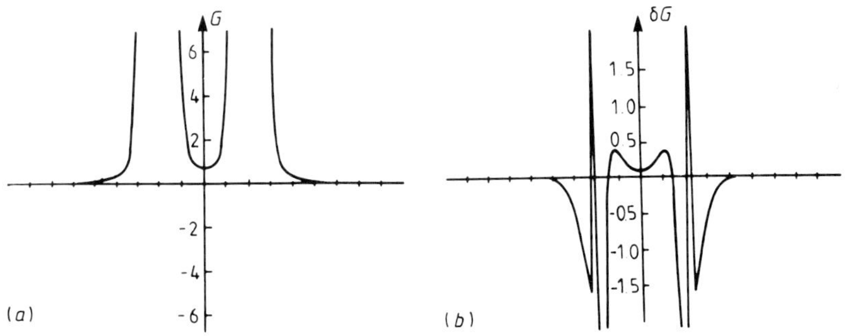

Figure 6.47 The N_2 molecule: kinetic energy density, $G(\mathbf{r})$ (*a*) and the difference between kinetic energy density in the molecule and in the promolecule, $\delta G(\mathbf{r})$ (*b*). All values are given in au. (Bader and Preston 1969).

almost throughout the binding region. The decrease in $\delta G(\mathbf{r})$ in the antibinding regions is greater than that in the binding regions in both molecules.

The $\delta G(\mathbf{r})$ distribution around the Li nucleus in LiH is nearly the same as in the Li_2 molecule: in the antibinding region $\delta G(\mathbf{r})$ is negative where $\delta\rho(\mathbf{r})$ is positive and vice versa in the binding region. This reflects the polarization of the Li atom core. In the binding region near the H nucleus $\delta G(\mathbf{r})<0$ and in the antibinding region $\delta G(\mathbf{r})>0$. This is like the $\delta G(\mathbf{r})$ distribution in the H_2 molecule. In the NH molecule the density of kinetic energy near the N nucleus is distributed as in the N_2 molecule and that near the H nucleus as in the H_2 molecule (figure 6.48). Generalizing, Bader (1981) concluded that since $G(\mathbf{r})$ near the H nucleus is defined mainly by the ED gradient, the $G(\mathbf{r})$ map analysis permits one to find what part of molecular space in hydrides can be successfully described by a single orbital wave-function.

We have seen that the kinetic energy distribution in an arbitrary many-electron system is different from that of the H_2 molecule. Thus, the H_2 molecule cannot be the probe for the chemical bond analysis.

The ED gradient softening along the bond line is not the most essential factor for bond formation. This conclusion results from analysis of constructive interference density maps for the H_2 molecule (figure 6.11(*e*)) and for some diatomic, triatomic, tetratomic and pentatomic hydrides (figure 6.15). Moreover, as a result of expressions (6.38) and (6.39), the kinetic energy density, $T(\mathbf{r})$, depends not only on the natural orbital gradients, but also on the Laplacian of ρ. The $\nabla^2\rho(\mathbf{r})$ and $G(\mathbf{r})$ signs may either coincide or differ at some point of space; the $G(\mathbf{r})$ sign depends on the sign of the $\nabla\varphi_i\cdot\nabla\varphi_1^*$ product. Favourable orbital contragradience is needed for a chemical bond to be formed, as was noted by Wilson and Goddard (1972a, b).

Bader and Essen (1984) calculated the components of $G_{\parallel}(r_b)$ and $G_{\perp}(r_b)$ at the bond critical points (3, −1) in some molecules with different types of chemical bond (table 6.17). The ratio of these values reflects the ratio of the

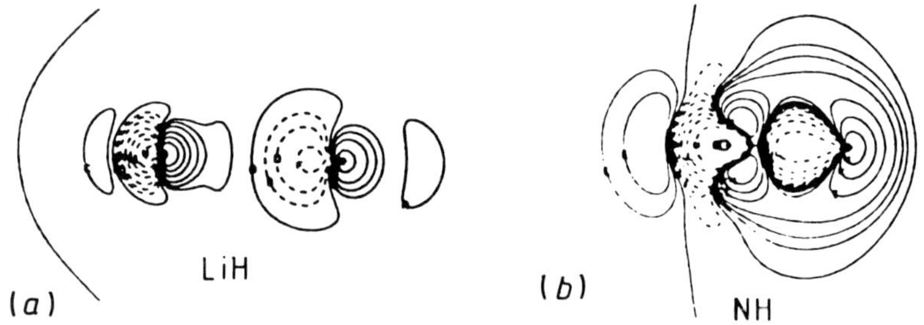

Figure 6.48 The contour maps of the difference between kinetic energy densities in the molecule and standard promolecule: (*a*) LiH; (*b*) NH. Solid contours point out the increase of $G(r)$; the first solid line is zero. The line intervals differ by a successive factor of two, starting from ± 0.002 au. (Bader and Preston 1969.)

ED curvature components at the bond point. If it proves to be that for shared interactions $G_{\|}(r_b) < G_{\perp}(r_b)$, the inverse is observed for closed-shell interactions. Besides, the kinetic energy density at critical point (3, −1) related to unit electron density, $G(\mathbf{r}_b)/\rho(\mathbf{r}_b)$, is more than unity for the closed-shell interactions and is less than unity for shared interactions. This means that, if the positive ED curvature is large because of ED contraction toward nuclei, and the $G(\mathbf{r}_b)/\rho(\mathbf{r}_b)$ value is also large enough, the value of its component parallel to the bond line is larger than the normal component. For shared interactions the inverse picture is observed, and the $G(\mathbf{r}_b)/\rho(\mathbf{r}_b)$ value is small. In the H_2 molecule, this ratio is abnormally small; hence, the H_2 molecule is atypical in this case also.

Let us look now at the F_2 molecule where the Laplacian of ρ in the internuclear space is only slightly negative: at the bond critical point $\nabla^2\rho(\mathbf{r}_b) = -0.045$, $\lambda_1 = \lambda_2 = -1.118$, $\lambda_3 = 2.190$ au. The ratio $G(\mathbf{r})/\rho(\mathbf{r}_b)$ is 0.91 close to N_2 and O_2 molecules; however, the ratio $G_{\perp}(\mathbf{r}_b)/G_{\|}(\mathbf{r}_b) = 0.074$ is considerably smaller than corresponding values in the case of a shared interaction. Thus, the chemical bond in the F_2 molecule has some specificity.

The analysis of local electron energy density $H_e(\mathbf{r})$ is important especially for the study of the chemical bond nature. According to (2.74) this characteristic is determined by the Laplacian of ρ and by the density of the electron potential energy $V(\mathbf{r})$. Hence, $H_e(\mathbf{r})$ unifies two approaches to chemical bond consideration: the first is electrostatic, operating with the electric field forces which depend on potential energy, and the second is energetic, operating with the kinetic energy of electrons which depends on the space distribution of both ED and wave-functions. The ED and its derivatives combine these approaches and this point needs more attention.

As we have seen, not only the internuclear space but also the vicinity of nuclei should be considered in order to analyse the cause of chemical bond formation. The ED near nuclei is responsible mainly for the forces exerted on nuclei. The overall picture of deformation ED reflects the changes in ED which result in the changes in energy discussed. Ruedenberg (1962) insisted on the point that the ED contraction near nuclei takes place when the chemical bond is formed, and it is responsible for the kinetic energy increase in these areas. However, Hirshfeld and Rzotkiewicz (1974) have found that such a contraction takes place only in the H_2 molecule. The density of atomic cores is, as a rule, polarized in the direction opposite to dipole-like deformation of valence ED. The net electrostatic effect of σ orbital forces in the first-row AH and A_2 molecules proved to be nearly zero. This conclusion was supported by Schwarz *et al* (1985) who found such ED distribution in the core σ MO of the N_2 molecule (figure 6.49). Hirshfeld and Rzotkiewicz noted also that the forces exerted on nuclei strongly depend on the number of electrons in interacting atoms and, correspondingly, on the electron principal, orbital and magnetic quantum numbers. These characteristics mainly influence the $\delta\rho$ distribution which is usually interpreted from the chemical bond point of

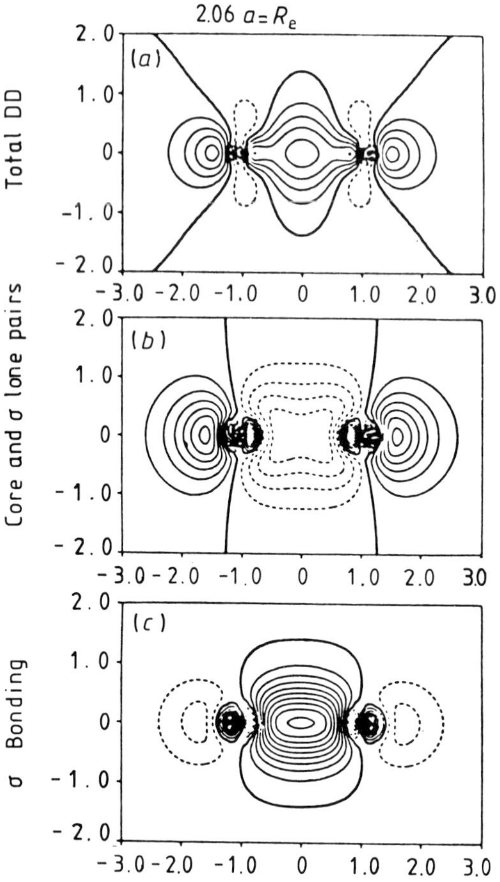

Figure 6.49 The N_2 molecule: (*a*) the standard electron deformation density; (*b*) the eight-electron core and lone-pair contribution to deformation density ($1\sigma_{g,u}$ and $2\sigma_{g,u}$ MO density minus 1s and 2s AO densities); (*c*) the two-electron σ bond contribution to the deformation density ($3\sigma_{g,u}$ MO density minus $2p_{y,x}$ AO densities). The MOs were calculated via natural orbital decomposition. The line intervals are 0.27 $e\,Å^{-3}$. (Reprinted with permission from Schwarz *et al* 1985.)

view. The constructive interference in the σ bond region will be lower for atoms having 3s and 3p electrons in the outer shell in comparison with first-row atoms. The reason for this is that the former have more nodes in their wave-functions. At the same time, the π interactions of the two types of atom should not differ noticeably. If a system has occupied π orbitals then they are mainly responsible for chemical bonding, because the influence of σ orbitals is mutually compensated to a great degree.

Unfortunately, as already noted in chapters 2 and 3, the ED near nuclei cannot be established with high accuracy either theoretically or experimentally. The consequences of the corresponding uncertainty have been analysed by Hirshfeld (1984b) in the ED investigation of tetrafluoroterephthalodinitrile, $C_8N_2F_4$. The force balance in the framework of the multipole model is not reached by taking into account a small accumulative $\delta\rho$ peak in the C–F bond (table 3.11). Therefore, dipole terms have been added to the crystal structural model with electron occupation parameters and exponential parameters to be varied. This has not changed either the R factor or $\delta\rho$ maps. The electric field at the nuclei sites remains the same as well, in the limit of estimated standard deviations. A large correlation has been observed, however, between additional dipole parameters and atomic coordinates, decreasing these parameter values. Analogous results have been obtained by Eisenstein and Hirshfeld (1983) by *ab initio* quantum chemical calculation of formylfluoride, HC(O)F. The $\delta\rho$ map in the C–F bond region was nearly the same as on the experimental map in the $C_8N_2F_4$ crystal. Hirshfeld stated that small deformations of the atomic core can be observed experimentally in the simultaneous investigation of the same crystal by accurate x-ray and neutron diffraction methods with an accuracy in nuclear localization $\sim 10^{-4}$ Å. The use of very low temperatures is necessary here for this purpose.

It should be noted that in a homonuclear diamond-like system the theoretical calculations (Zunger and Freeman 1977c) and x-ray experiment (see figure 2.11) revealed the ED contraction near nuclei. This was explained by interpenetration of wave-functions of neighbouring atoms. The same effect has been observed in different homonuclear crystals: Be (Tsirelson *et al* 1987a), Cu (Parini 1988), V (Tsirelson *et al* 1988b), and α-Fe (Ohba and Saito 1982). The accuracy of $\delta\rho$ determination near nuclei was in these studies at least several times higher than the effect measured, therefore the results can be considered as statistically reliable.

We may conclude that electrostatic effects can be considered as the precursor of chemical bond formation due to binding forces exerted on nuclei in a promolecule. However, the specific bond character is determined by the kind and number of interacting atoms and the Pauli principle effect. The chemical bond is formed as a result of constructive and destructive quantum mechanical interference of electron wave-functions of these atoms. The resulting ED distribution reflects the competition of all these factors and determines the energetical features of the chemical bond formed. Therefore, the characteristics of ED features can be used in chemical bond classification.

7

ELECTRON DENSITY AND CRYSTAL PROPERTIES

According to Hohenberg and Kohn (1964), the ED determines all properties of an atom, molecule, or crystal in the electronic ground state. However, the corresponding analytic dependences are successfully obtained only in rare cases. This is possible if the operator $\hat{A}$, which describes the property that we are interested in, is a one-electron operator (see (2.19)) and has local character. The expectation value of such an operator is calculated in the following way:

$$\langle \hat{A} \rangle = \int \hat{A}\rho(\boldsymbol{r})\,\mathrm{d}\boldsymbol{r}. \tag{7.1}$$

Some of the crystal property characteristics which can be calculated in this way are illustrated in figure 7.1.

However, the fundamental role of ED is not restricted to the ground state. The parameters of some models, which describe the properties dependent on excited states or properties which have non-local character, can sometimes be obtained from the ED. Figure 7.1 also gives examples of such characteristics. Methods for calculation of properties of both kinds from ED reconstructed from x-ray diffraction data will be presented below. An important problem which arises in this case is the comparison of calculated values with those that can be obtained by other experimental methods. Because determination of each of these quantities has its own specific features, these comparisons should be made with some caution.

7.1 Electrostatic effects

7.1.1 *Electrostatic potential in crystals*

The EP inside a crystal, $\varphi(\boldsymbol{r})$, introduced in subsection 2.5.1, is produced by the total charge density $\sigma(\boldsymbol{r})$ (2.77) and may be directly measured by electron diffraction (see section 4.4). The total charge density $\sigma(\boldsymbol{r})$ consists of both nuclear and electronic components. Correspondingly, the EP, which is connected to $\sigma(\boldsymbol{r})$ by the Poisson equation (2.79), also has two components, φ_n (2.80) and φ_e (2.81). Assuming that $\varphi(\infty)=0$, the EP can be described by

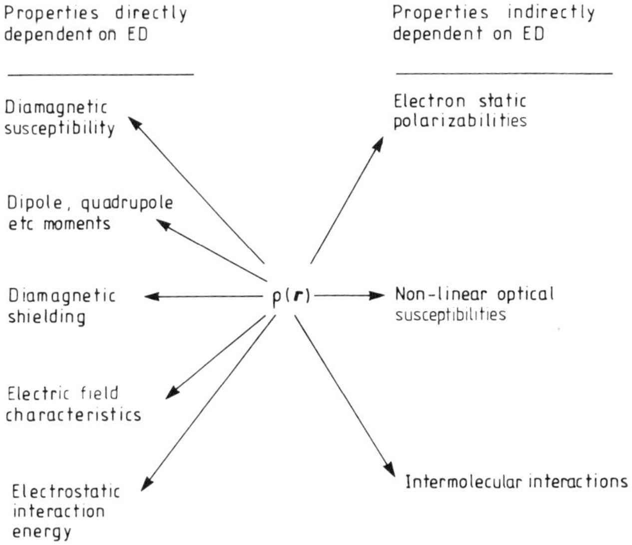

Figure 7.1 Some crystal property characteristics which can be calculated via ED.

the equation

$$\varphi(\boldsymbol{r}) = \int_{-\infty}^{\infty} \{\sigma(\boldsymbol{r}')/4\pi\varepsilon_0|\boldsymbol{r}-\boldsymbol{r}'|\}\,\mathrm{d}\boldsymbol{r}' = \sum_a Z_a/(4\pi\varepsilon_0|\boldsymbol{r}-\boldsymbol{R}_a|)$$

$$-\int_{-\infty}^{\infty} \{\rho(\boldsymbol{r}')/4\pi\varepsilon_0|\boldsymbol{r}-\boldsymbol{r}'|\}\,\mathrm{d}\boldsymbol{r}' \qquad (7.2)$$

where the radius vector $\boldsymbol{r}$ is drawn from a unique point, the same for all unit cells in the crystal. Another choice of origin for EP calculation can be made by setting the average electrostatic potential over a unit cell equal to zero:

$$\int_V \varphi(\boldsymbol{r})\,\mathrm{d}\boldsymbol{r} = 0. \qquad (7.3)$$

Such a choice of origin is more convenient for representation of the EP in the form of a Fourier series and is usually used in electron diffraction. It gives the same result as the Ewald method for evaluating Madelung sums (Harris 1976).

The EP is written as a Fourier transform of the total charge density $\sigma(\boldsymbol{r})$ in the following way:

$$\varphi(\boldsymbol{r}) = (8\pi^3\varepsilon_0)^{-1} \iint \sigma(\boldsymbol{r}) \exp(\mathrm{i}\boldsymbol{S}\cdot\boldsymbol{r}') S^{-2} \exp(-\mathrm{i}\boldsymbol{S}\cdot\boldsymbol{r})\,\mathrm{d}\boldsymbol{S}\,\mathrm{d}\boldsymbol{r}'$$

$$= -(8\pi^3\varepsilon_0)^{-1} \int F_t(\boldsymbol{S}) S^{-2} \exp(-\mathrm{i}\boldsymbol{S}\cdot\boldsymbol{r})\,\mathrm{d}\boldsymbol{S} \qquad (7.4)$$

where the relationship

$$(1/|\boldsymbol{r}-\boldsymbol{r}'|)=(2\pi^2)^{-1}\int S^{-2}\exp[-\mathrm{i}\boldsymbol{S}\cdot(\boldsymbol{r}-\boldsymbol{r}')]\,\mathrm{d}\boldsymbol{S} \tag{7.5}$$

has been used. The structure amplitudes, $F_t(\boldsymbol{S})$, entering (7.4) are the Fourier components of the density $\sigma(\boldsymbol{r})$. The nuclear part of $\sigma(\boldsymbol{r})$ can be easily calculated, accurately, while the electronic part of this density, $\rho(\boldsymbol{r})$, is determined by the kinematic structure amplitudes obtained from x-ray diffraction data. Therefore x-ray diffraction as well as electron diffraction can be applied to the study of inner EP in crystals.

The periodicity of the crystal lattice leads to the periodicity of the charge density, $\sigma(\boldsymbol{r})$, and the EP, $\varphi(\boldsymbol{r})$. We will denote the radius vector in an arbitrary cell by $\boldsymbol{R}_n+\boldsymbol{r}$, where $\boldsymbol{R}_n$ is a radius vector drawn from the general coordinate origin to the coordinate origin attached to the cell. Making the substitution $\boldsymbol{r}\rightarrow\boldsymbol{R}_n+\boldsymbol{r}$ in (7.5), the EP in the crystal can be written as a sum of the contributions of all unit cells:

$$\varphi(\boldsymbol{r})=-(8\pi\varepsilon_0)^{-1}\int F_t(\boldsymbol{S})S^{-2}\exp(-\mathrm{i}\boldsymbol{S}\cdot\boldsymbol{r})\sum_n\exp(\mathrm{i}\boldsymbol{S}\cdot\boldsymbol{R}_n)\,\mathrm{d}\boldsymbol{S}. \tag{7.6}$$

If the vector $\boldsymbol{S}$ tends towards a reciprocal lattice vector $\boldsymbol{H}$, as in single-crystal diffraction, and the number of cells is great enough, then the sum over cells will tend towards a sum of Dirac delta functions:

$$\sum_n\exp(-\mathrm{i}\boldsymbol{S}\cdot\boldsymbol{R}_n)=\frac{1}{V}\sum_{\boldsymbol{H}}\delta(\boldsymbol{S}-2\pi\boldsymbol{H}). \tag{7.7}$$

Substituting summation for integration in (7.6) and taking into account that $\mathrm{d}\boldsymbol{S}=(2\pi)^3\,\mathrm{d}\boldsymbol{H}$, we have

$$\varphi(\boldsymbol{r})=-(4\pi^2\varepsilon_0V)^{-1}\sum_{\boldsymbol{H}}F_t(\boldsymbol{H})H^{-2}\exp(-2\pi\mathrm{i}\boldsymbol{H}\cdot\boldsymbol{r}). \tag{7.8}$$

Sommer-Larsen *et al* (1990) showed that, strictly speaking, in a real finite crystal a term should be added to (7.8) which takes into account the distribution of charge over the surface of a crystal. Therefore the EP inside the crystal turns out to be dependent on its size and form. Deep inside the crystal, surface effects can be described by a constant term in the potential.

If $\boldsymbol{H}=0$, then $F_t(\boldsymbol{H})=0$ and the quantity $F_t(\boldsymbol{H})|\boldsymbol{H}|^{-2}$ is undefined. The situation can be clarified by returning to expression (7.6), writing the Fourier transform of the charge density $\sigma(\boldsymbol{r})$ as

$$F_t(\boldsymbol{S})=\int\sigma(\boldsymbol{r})\exp(\mathrm{i}\boldsymbol{S}\cdot\boldsymbol{r})\,\mathrm{d}\boldsymbol{r} \tag{7.9}$$

and expanding $\exp(\mathrm{i}\boldsymbol{S}\cdot\boldsymbol{r})$ into a series over small values of the vector $\boldsymbol{S}$

$$F_t(\boldsymbol{S}) = \int_V \sigma(\boldsymbol{r})\,\mathrm{d}\boldsymbol{r} + \mathrm{i}\int_V \sigma(\boldsymbol{r})(\boldsymbol{S}\cdot\boldsymbol{r})\,\mathrm{d}\boldsymbol{r} - \frac{1}{2}\int_V \sigma(\boldsymbol{r})(\boldsymbol{S}\cdot\boldsymbol{r})^2\,\mathrm{d}\boldsymbol{r} + \cdots. \tag{7.10}$$

Then, as shown by Avery *et al* (1984), EP (7.8) can be represented in the form

$$\varphi(\boldsymbol{r}) = -(4\pi^2\varepsilon_0 V)^{-1} \sum_{\boldsymbol{H}\neq 0} F_t(\boldsymbol{H})|\boldsymbol{H}|^{-2}\exp(-\mathrm{i}2\pi\boldsymbol{H}\cdot\boldsymbol{r}) + \varphi(0) \tag{7.11}$$

where the constant term $\varphi(0)$ is equal to

$$\varphi(0) = -(6\pi\varepsilon_0 V)^{-1}\int_V r^2\sigma(\boldsymbol{r})\,\mathrm{d}\boldsymbol{r}. \tag{7.12}$$

In obtaining this expression it was assumed that the unit cell is electroneutral and does not have a dipole or quadrupole moment. Spackman and Stewart (1981) showed that this can always be achieved by a suitable choice of periodically repeated fragments of charge density in a crystal; transfer to another cell simply entails changing the value of the constant term in (7.11). It should be noted that in non-centrosymmetric crystals the origin should be chosen very carefully, in order to satisfy the condition that the dipole moment of the unit cell is equal to zero, keeping in mind the condition (7.3). It is clear that in electrically neutral crystals the value of $\varphi(0)$ can be taken to be zero. Nevertheless, surface charge effects should be accounted for when EP reconstructed from x-ray diffraction data is compared with that obtained from other experimental methods.

There are two ways of calculating EP from x-ray diffraction data in which the influence of Fourier series termination and thermal ED smearing effects are minimal (Konobeevskij 1948, Sirota *et al* 1972, Moss 1982, Varnek *et al* 1981, Stewart 1979, 1982). The first uses the multipole model and the second consists of representing EP as a sum of procrystal potential and deformation EP. In the first method, the multipole model allows the representation of static EP through the pseudoatomic moments of ED, $\mu_{\alpha\beta\ldots\eta}$ (3.102), provided EP is given by the convergent series

$$\varphi(\boldsymbol{r}) = (4\pi\varepsilon_0)^{-1}\sum_\mu \left\{\frac{\mu_0^\mu}{r} + \sum \frac{\mu_\alpha^\mu r_\alpha}{r^3} + \sum_{\alpha\beta}\frac{(3r_\alpha r_\beta - \delta_{\alpha\beta})\mu_{\alpha\beta}^\mu}{2r^5}\right\} + \varphi(0). \tag{7.13}$$

The constant term was determined by Becker and Coppens (1990) as

$$\varphi(0) = -(6\pi\varepsilon_0)^{-1}\sum_\mu \left\{r_\mu \int \sigma_\mu(\boldsymbol{r})\,\mathrm{d}\boldsymbol{r} + 2r_\mu \int_V \boldsymbol{r}\sigma_\mu(\boldsymbol{r})\,\mathrm{d}\boldsymbol{r} + \int_V r^2\sigma_\mu(\boldsymbol{r})\,\mathrm{d}\boldsymbol{r}\right\}. \tag{7.14}$$

Here $\sigma_\mu(\boldsymbol{r})$ denotes the total charge density of pseudoatom μ and the summation over μ is within the unit cell. The analytical expressions relating $\varphi(0)$ to the parameters of multipole models can be easily obtained from (7.14).

The multipole model approach to EP calculation has mainly been applied to molecular crystals. The compounds imidazole, $C_3H_2N_4$ (Stewart 1982), alloxan, $C_4H_2N_2O_4$ (Swaminathan *et al* 1985), parabanic acid, $C_3H_2N_2O_3$ (He *et al* 1988), complex thiourea–parabanic acid, $CH_4N_2S \cdot C_3H_2N_2O_3$ (Weber and Craven 1987), cytosine and adenine (Eisenstein 1988), L-alanine (Destro *et al* 1988), deuterated cytosine monodeuterate, $C_4H_2D_3N_3O \cdot D_2O$ (Weber and Craven 1990), and some nucleic acid components (Klooster 1992) have been studied. All of these molecules are of interest from the biological point of view and analysis of the EP allows us to understand the mechanism of their mutual interactions and interactions with other molecules.

The investigation of alloxan (2,4,5,6,(1H,3H)-pyrimidinetetrone), gives an example of the application of the EP approach to the study of properties of such compounds. The crystal structure of alloxan (space group $P4_12_12$, $Z=4$) consists of a herringbone arrangement of nearly planar molecules. The latter lie on the twofold axes which are parallel to the [110] face diagonals of the unit cell. Fragments of alloxan crystal structure are presented in figure 7.2. The intermolecular contacts O═C···O involving the carbonyl group are shorter than the sum of the Van der Waals radii (3.1 Å). The shortest of these distances, C(5)···O(6), is only 2.73 Å at 123 K. The nature of such interactions in alloxan is not well understood as discussed by Burgi *et al* (1974). The hydrogen bonds exist in alloxan crystals as weak bifurcated interactions with long NH···O distances, 2.32 and 2.35 Å.

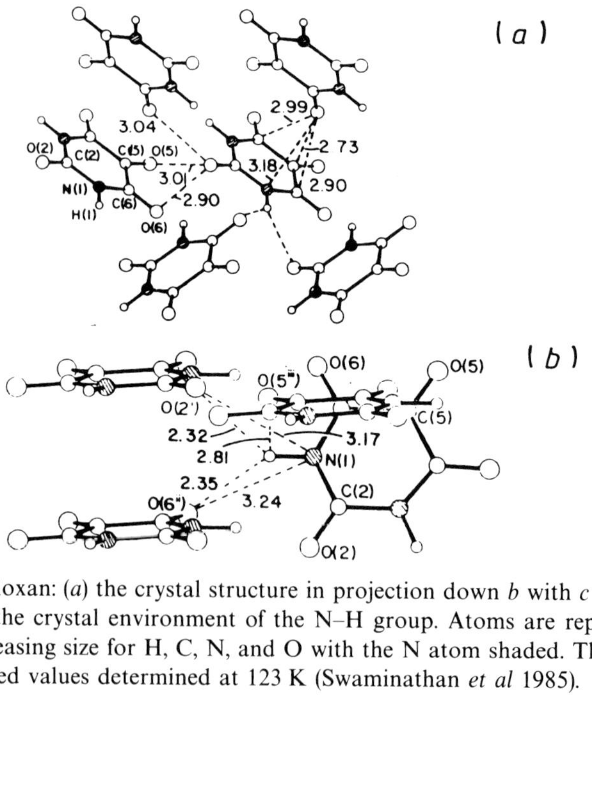

Figure 7.2 Alloxan: (*a*) the crystal structure in projection down *b* with *c* running up the page; (*b*) the crystal environment of the N–H group. Atoms are represented as circles of increasing size for H, C, N, and O with the N atom shaded. The distances (Å) are rounded values determined at 123 K (Swaminathan *et al* 1985).

The EP in alloxan was calculated by Swaminathan *et al* (1985) from static multipole model parameters derived from x-ray diffraction data at 123 K. The potential maps for an isolated molecule removed from the crystal and for small groups of molecules arranged in the same way as in the alloxan crystal are depicted in figure 7.3. In the isolated molecule the positive EP on either side of the molecular plane is more extensive at the C(5) atom than at the other carbonyl C atoms. The short intermolecular C···O distance at the C(5) atom appears to be derived primarily from deshielding of the carbonyl C atom nucleus. This allows the close approach of the weakly electronegative O(6) atoms (which have charge $-0.11\,e$ as follows from multipole model refinement) on either side of the molecular plane. When molecules are brought together, as in a crystal, saddle points of nearly zero potential are formed

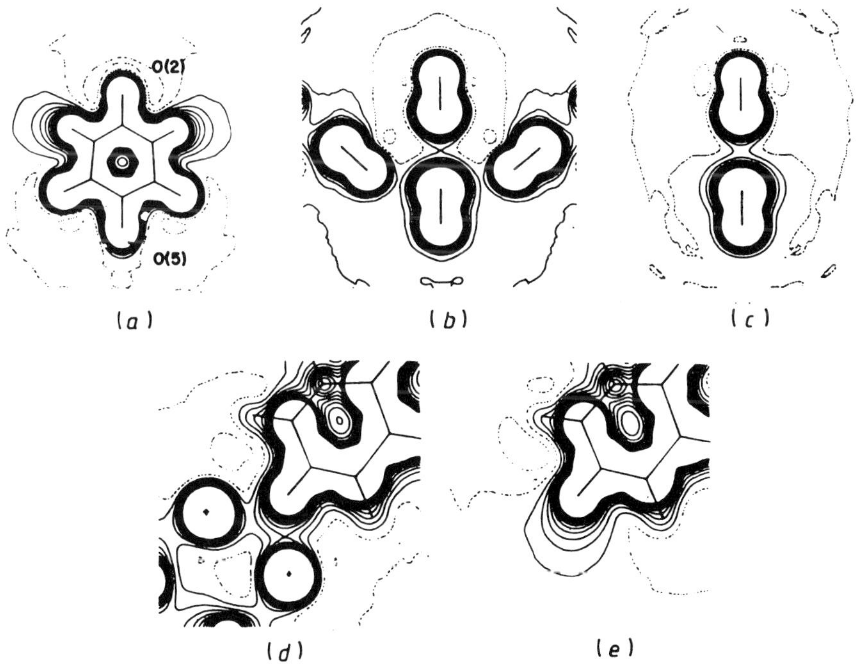

Figure 7.3 The EP for alloxan. (*a*) The EP in the plane of the isolated molecule. (*b*) The EP for a group of three molecules in the section normal to the molecular plane and passing through the axis O(2)C(2)C(5)O(5). Atoms O(6), which form the shortest C···O contact, are ± 0.1 Å from this plane. (*c*) The EP for the isolated central molecule in the same section. (*d*) The EP for the group of three molecules which form the H bond in the section of the best LS plane through atoms N(1), H(1), O(2^i), and O(6^{ii}) which are involved in the weak bifurcated H bond. (*e*) The EP for the isolated molecule in the same section. The contour intervals are $0.05\,e\,\text{Å}^{-1}$ corresponding to about one e.s.d.; solid lines correspond to positive (relative to mean inner potential) values of EP; the regions around nuclei with high positive values of EP are not shown (Swaminathan *et al* 1985).

along the lines C(5)···O(6) (figure 7.3(*b*)). The EP values around the O(6) atom in the isolated molecule are also nearly zero, therefore C···O interactions are only weakly attractive. Analogous electropositive EP bridges were observed in parabanic acid $C_3H_2N_2O_3$, between O atoms in the carbonyl group and carbon atoms from neighbouring molecules (corresponding distances are 2.75 and 2.94 Å).

The values of potential near O atoms in both molecules mentioned are almost the same, about -100 kJ mol^{-1}. This is not the case in urea, where the minimum in EP near the O atom is -440 kJ mol^{-1} (Stewart 1983). Thus, the carbonyl O atom in urea is more electronegative than the same atom in alloxan and parabanic acid. This conclusion supports the suggestion of Weber *et al* (1980), resulting from H bond geometry analysis, that parabanic acid is more effective as an H donor through its NH groups than as an acceptor through its carbonyl O atoms. For urea the opposite is true. Consequently, the strong H bond with N···O distance 2.66 Å, with parabanic acid as donor and urea as acceptor, is important for the existence of the 1:1 crystal complex involving these molecules. The EP in the 1:1 complex of thiourea–parabanic acid was studied at 298 K by Weber and Craven (1987). Unfortunately, some inconsistencies in the ED of thiourea in the complex and in the homomolecular crystal, noted by Kutoglu *et al* (1982) do not allow us to discuss now the results obtained.

The depletion of ED at the carbonyl C atom above and below the molecular plane and the EP picture observed are in agreement with the supposition of Craven and McMullan (1979) that, in alloxan (and in several other molecular crystals discussed by Burgi *et al* (1974)), the C···O distances could be regarded as normal, if a smaller van der Waals radius for C atoms were assumed. For this reason the value of 1.5 Å (the currently accepted value is 1.7 Å) was recommended by Swaminathan *et al* (1985) for carbonyl C atoms.

In isolated alloxan molecules (figure 7.3(*a*)) there are diffuse positive EP regions around the H atoms and negative regions around the O atoms. The most negative value of potential, $-100(33)$ kJ mol^{-1}, lies near the O(2) atom. In the crystal (figure 7.3(*d*)), the negative potential from O(6) and O(2) atoms is compensated by a positive contribution from the H atom and saddle points of weak positive potential are formed. These points are not equivalent because their EP values are different. Hence, attractive interactions in the bifurcated H bond are also not equivalent, in spite of the small difference in distances ($\Delta l = 0.03$ Å) and nearly symmetrical deformation ED. Calculation of EP allows one to reveal this non-equivalence.

Electropositive bridges of EP in H bonds were also found between pairs of molecules in parabanic acid. He *et al* (1988) supposed that such bridges may be a general property of short-range electrostatic internuclear interaction.

Another approach to the calculation of EP in crystals (Stewart 1982) consists of summation of the promolecule (procrystal) EP, $\tilde{\varphi}(\boldsymbol{r})$, and the deformation EP, $\delta\varphi(\boldsymbol{r})$. In this case total EP is presented as

$$\varphi(\boldsymbol{r}) = \tilde{\varphi}(\boldsymbol{r}) + \delta\varphi(\boldsymbol{r}) + \varphi(0) \tag{7.15}$$

where $\delta\varphi(\boldsymbol{r})$ can be calculated according to (2.92) as

$$\delta\varphi(\boldsymbol{r}) = (4\pi^2\varepsilon_0|e|V)^{-1} \sum_{\boldsymbol{H}\neq 0} M_H\{F(\boldsymbol{H}) - \tilde{F}(\boldsymbol{H})\}H^{-2}\exp(-2\pi i\boldsymbol{H}\cdot\boldsymbol{r}). \tag{7.16}$$

The constant term $\varphi(0)$ can be found from the condition (7.3). In order to obtain correct values for the EP, it is essential that the quantities $\tilde{F}(\boldsymbol{H})$, $\varphi(\boldsymbol{r})$, and $\varphi(0)$ are calculated using the same wave-functions. The series (7.16) is quickly converged and a resolution in the $\delta\varphi$ map of the order of 1% is attainable.

In this approach it is necessary to take thermal motion into account properly: either to calculate $\varphi(\boldsymbol{r})$ from static structure amplitudes (for example, using the multipole model), or to include thermal smearing in the terms $\varphi(\boldsymbol{r})$ and $\varphi(0)$. Fortunately, as was noted by Stewart (1977c) and later confirmed by Epstein and Swanton (1982), the deformation EP in the first approximation can be seen as characteristic of a static crystal. Therefore it is possible to directly use experimental dynamic structure amplitudes to calculate $\delta\varphi(\boldsymbol{r})$; however, the quantities $\varphi(\boldsymbol{r})$ and $\varphi(0)$ should be calculated for a static crystal. When low-temperature diffraction data are used, the uncertainty introduced into the final result by the above approximation is substantially reduced.

Total EP has been calculated using (7.15) for a few inorganic crystals only: Be (Spackman and Stewart 1981), stishovite, SiO_2 (Spackman *et al* 1987), phenakite, Be_2SiO_4 (Downs and Gibbs 1987), and boric acid, $B(OH)_3$ (Sommer-Larsen *et al* 1990). From analysis of the EP maps, calculated in the cited works, two main conclusions can be drawn: (a) the channels, running through the crystal structure (if they occur as in stishovite and phenakite), are characterized by relatively low values of inner potential; (b) analysis of the circular contours of EP around the positions of the atoms allows the approximate evaluation of their relative 'size' in the crystal. Probably this observation will allow a self-consistent set of crystal atomic radii to be worked out in the future, though such radii will have no more physical meaning than any other set of radii.

EP is a substantially flatter function than the ED. In order to distinguish details of the EP one can use the deformation electronic EP $\delta\varphi$ (7.16) (Vainstein 1960, Sirota *et al* 1972, Avery 1979, Stewart 1979) by analogy with the deformation ED. Deformation EP characterizes the electrostatic field, which, when being summed with the procrystal field (including the nuclei), provides the stability of the system. The sign and value of $\delta\varphi$ show how the electronic potential at any point of a crystal differs from that of the superposition of electronic potentials due to a system of non-interacting free atoms. The obvious advantage of the deformation potential is its independence from the constant term $\varphi(0)$. The $\delta\rho$ maps are negative in the regions where concentration of electron charge takes place, keeping in mind the sign of the electron.

Using the definitions and results of subsection 3.7.1 it is possible to arrive at an expression for the calculation of the variance of the deformation EP:

$$\sigma^2[\delta\varphi(\boldsymbol{r})]=(8\pi^4\varepsilon_0)^{-1}\left\{\eta\sum_{H>0}M_H\left[\frac{\sigma^2(|\delta F|)}{H^4}\cos^2\alpha+\frac{|\delta F|^2\sigma^2(\alpha)}{H^4}\sin^2\alpha\right]\right.$$
$$\left.+2\sigma^2(k)\left[\sum_{H>0}M_HC(\boldsymbol{H})\cos\alpha/|\boldsymbol{H}|^2\right]^2\right\}. \tag{7.17}$$

Everything which has been said regarding distribution of the mean square deviation $\sigma(\delta\rho)$ over a unit cell applies equally to $\sigma(\delta\varphi)$: these errors similarly increase at atomic positions, on closed symmetry elements, and can be reduced by the use of a filtration procedure. At the same time the relative error in deformation EP proves to be less than that of ED. The reason for this lies in the fact that the contribution of high-angle reflections (which are measured with lower statistical precision) to the EP is lower than in the case of ED.

The Fourier series (7.16) can be calculated either directly, using experimental kinematic structure amplitudes, or by substituting the latter by model structure amplitudes (calculated with multipole model parameters, for example). In the first case the deformation EP will be dynamic, in the latter case, dynamic or static depending on whether thermal parameters were used in structure factor calculation or not. The structure amplitudes for the procrystal, $\tilde{F}(\boldsymbol{H})$, must be calculated in the same manner.

The investigation of boric acid, $B(OH)_3$, at 105 K (Sommer-Larsen *et al* 1990) gives one the possibility of comparing the two methods of EP calculation discussed. The deformation EP maps calculated for all cases are presented in figure 7.4 together with the total EP calculated according to (7.15). The total EP distribution allows one to imagine the molecular packing in the crystal, however only deformation maps permit one to analyse the detailed EP features such as electronic structure of H bonds. The two types of map agree with each other fairly well. The most significant differences in static and dynamic maps are observed around the O atom centres: apparently the multipole model cannot describe the ED of $B(OH)_3$ adequately.

It is possible to conclude that the dynamic maps of deformation EP indeed give a good approximation to the static deformation EP in the areas of chemical bonding and intermolecular interaction.

The deformation EP obtained from x-ray diffraction data has been studied in a fairly broad range of compounds. We will survey several results obtained for crystals with a variety of types of chemical bond. First of all we will look at the features of the deformation EP in homoatomic crystals. In the covalent crystals diamond (Varnek *et al* 1982), silicon (Spackman and Stewart 1981, Varnek 1985), and germanium (Varnek 1985) a qualitative correspondence between $\delta\rho$ and $\delta\varphi$ was observed. The regions of ED accumulation in the

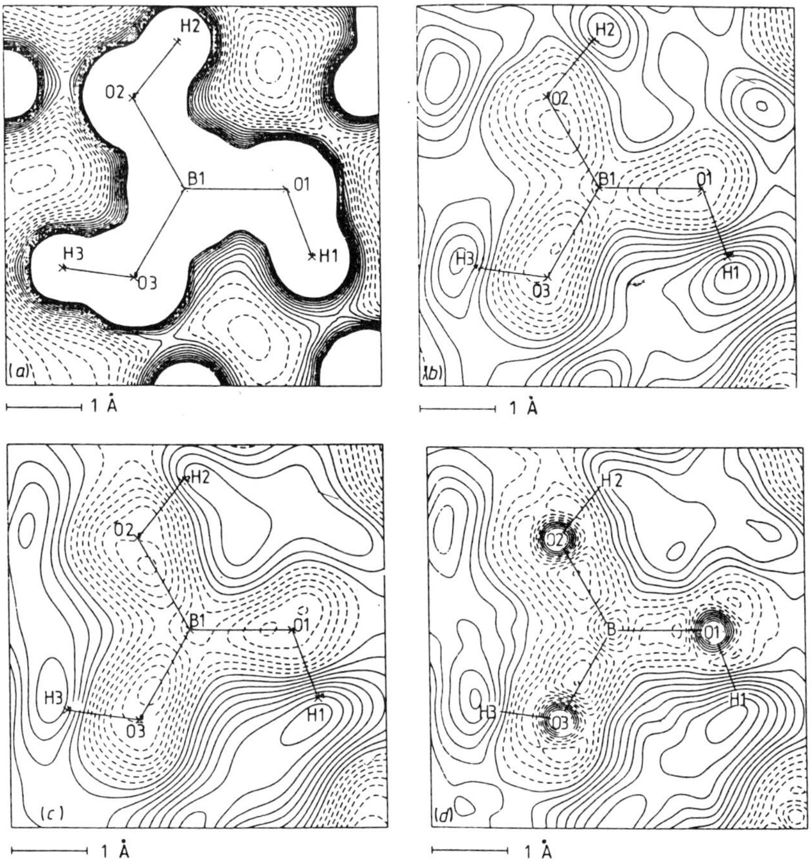

Figure 7.4 The EP in boric acid, $B(OH)_3$, section in the plane of the molecule: (*a*) the total EP (the lines around nuclei are omitted); (*b*) the dynamic deformation EP calculated via Fourier series; (*c*) the dynamic deformation EP calculated via Fourier series with multipole model structure amplitudes; (*d*) the static deformation EP calculated via Fourier series with multipole model structure amplitudes. The contour intervals are $0.05\,e\,\text{Å}^{-1}$; the negative contours are represented by broken lines (Sommer-Larsen *et al* 1990).

crystal correspond to negative values of $\delta\varphi$ (see appendix D): this signifies that in these regions the electronic part of the total EP in a crystal is larger in absolute value than the EP of the corresponding procrystal part. The regions of negative $\delta\varphi$ are spread along interatomic vectors and include the nuclear positions. In crystallographic holes the $\delta\varphi$ values are positive. The deformation EP distribution in the silicon crystal, which is presented in figure 7.5, illustrates the described features. In the V crystal (Tsirelson *et al* 1988b) with metallic bonds (see figure 3.35) the deformation EP distribution

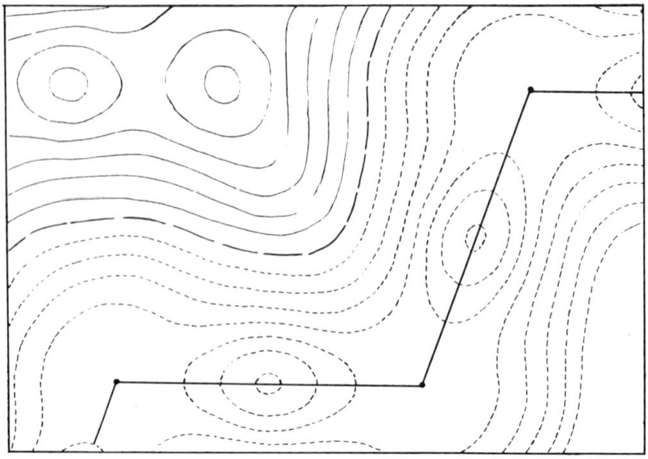

Figure 7.5 The deformation EP in the (110) plane of silicon. The line intervals are 0.05 $e\,Å^{-1}$; solid contours correspond to positive values of $\delta\varphi$ (Varnek 1985).

is quite different: between nearest nuclei, along the (110) direction where an excessive $\delta\rho$ region can be seen, the deformation EP is positive, whereas between next-nearest-neighbour nuclei along the (100) direction $\delta\varphi$ is negative. The $\delta\varphi$ values are also negative around nuclei.

Multicentre chemical bonds, which are characteristic of electron-deficient compounds, are formed in the Be crystal. Tsirelson *et al* (1987a) found that the ED is slightly increased around nuclei and decreased in the octahedral hole positions (figure 7.6(*a*)). The experimental $\delta\varphi$ map (figure 7.6(*b*)) shows that the regions of electron concentration are, at the same time, regions of negative deformation EP. The $\delta\varphi$ map obtained from the theoretical calculation of the Be crystal performed by Dovesi *et al* (1982) does not show the negative $\delta\varphi$ regions around nuclei (figure 7.6(*c*)). Perhaps this is connected with the inadequate behaviour near nuclei of the Gaussians which were used to approximate the basis functions (this problem was discussed in subsection 2.1.3).

The remarkable features in the deformation EP map in the Be crystal appear in the distribution of $\delta\varphi$ along the hexagonal z axis. An unbroken 'channel' of positive deformation EP is observed, running through the crystal and connecting the octahedral holes. Such a feature allows one to predict possible locations of impurity atoms and their diffusional movement.

The domination of the negative value of the electronic potential in homonuclear crystals near the nuclei is closely connected with the redistribution of the energy components when the atoms are placed in a chemical bond (see section 6.5). From expressions (2.86) and (2.87), derived from the DFT keeping in mind the non-negativity of the total EP of the procrystal, it follows that in homonuclear systems the value of $\delta\varphi$ should be

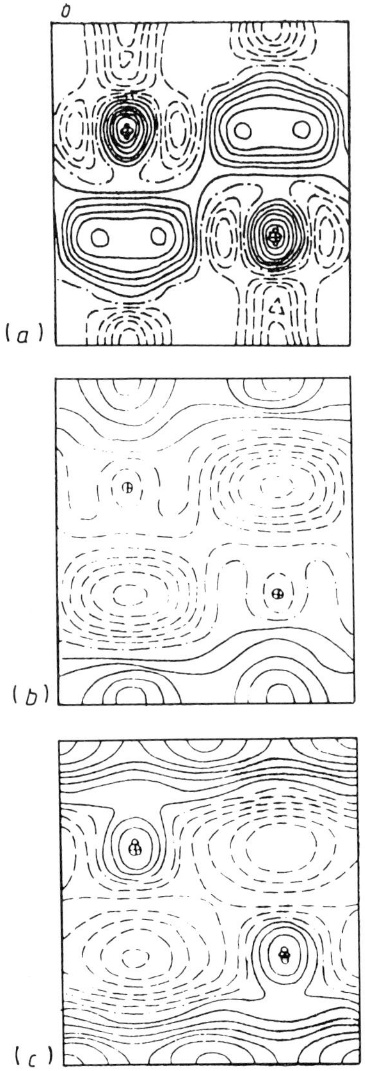

Figure 7.6 The deformation ED (*a*) and experimental (*b*) and theoretical (*c*) deformation EP in the (110) plane of the Be crystal. The line intervals are 0.015 $e\,Å^{-1}$ (*a*) and 0.01 $e\,Å^{-1}$ (*b*), (*c*) (Tsirelson *et al* 1987a).

negative at the nuclei in order for the system to be bonded. We note, however, that these expressions should be treated as qualitative only; attempts to evaluate the energy of atomization of diamond, silicon, beryllium, and vanadium on the basis of these equations quantitatively led to large discrepancies with thermochemical data (Bentley 1979, Varnek 1985). The only exception is diamond, for which a satisfactory agreement with experiment was obtained. As was noted by Bentley (1979), equation (2.87) does not satisfy the variational principle; therefore the lack of success of the above-mentioned calculations is not surprising.

A simple correlation between $\delta\rho$ and $\delta\varphi$ is observed in homoatomic crystals only. In multiatomic compounds the connections between these functions are more complicated, for example, in ruby, Al_2O_3:Cr^{3+} (Tsirelson *et al* 1985a); the $\delta\rho$ and $\delta\varphi$ maps are shown in figure 7.7. The sections displayed contain Al–O bonds and the centre of inversion of the structure. Quite noticeable in the $\delta\rho$ map is the 'bridge' of the deformation ED, which also appears in the deformation EP map. Regions of positive $\delta\varphi$ form quasicontinuous strips along the $\cdots$Al$\cdots$Al$\cdots$ lines. Regions of negative $\delta\varphi$ include the O atom sites and are elongated along the O––O$\cdots$ lines in planes parallel to (001). The minimum $\delta\varphi$ lies approximately 0.58 Å from the centre of the O atoms. Although the regions of negative $\delta\varphi$ coincide reasonably well with the positions of accumulative peaks on bond lines in $\delta\rho$, the $\delta\varphi$ values at these points differ substantially, while the $\delta\rho$ maxima are practically identical. This is explained by the fact that the EP at any point is created by charge which is distributed throughout the crystal, while the ED function has local character. This result supports the conclusion of Bader (1975a) that structural fragments are transferable only in so far as their electron distribution is the same in different compounds. Note that an important role in crystals is also played by the space disposition of structural units, leading to virial forces which act on the given fragment.

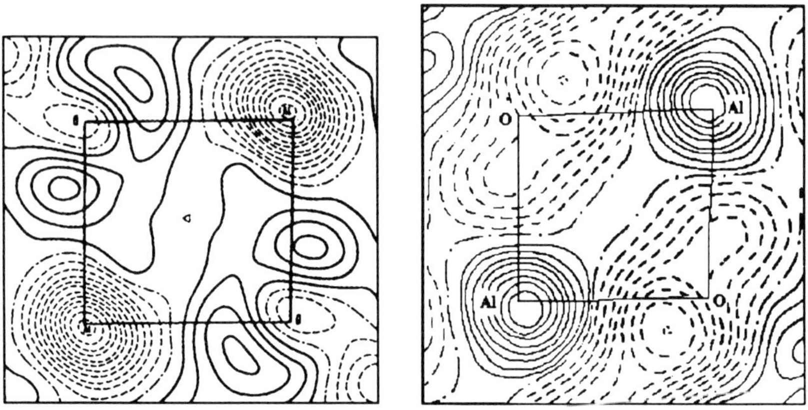

(*a*) (*b*)

Figure 7.7 The deformation ED (*a*) and deformation EP (*b*) in ruby α-Al_2O_3:Cr^{3+} in the plane containing all the chemical bonds. Impurity Cr atoms are included in the procrystal. The line intervals are 0.025 $e\,Å^{-1}$. (Reprinted with permission from Tsirelson *et al* 1985a.)

The general character of the ruby $\delta\varphi$ distribution is a consequence of layering of the corundum structure. It is connected with the anisotropy of ruby mechanical, electrical, and optical properties, although a simple link between these properties and ED is not obvious.

Besides the crystals considered above, the deformation EP has been studied in GaAs, GaP, and InP (Sirota *et al* 1972), fluorite, CaF_2 (Streltsov 1986), stishovite, SiO_2 (Spackman and Stewart 1984), phenakite, Be_2SiO_4 (Tsirelson *et al* 1986), lithium metaborate, $LiBO_2$ (Kirfel *et al* 1983), lithium formate deuterate (Tsirelson and Lobanov 1986), and spinels and garnets (Streltsov 1986, Lobanov *et al* 1988, 1989). In crystals of organic compounds the deformation EP has been described in imidazole, $C_3H_2N_4$ (Stewart 1982), tricyclo[4.4.1.0^{1,6}]undeca-2,4,7,9-tetraene-11,11-dicarbonitrile, $C_{13}H_8N_2$ (Bianchi *et al* 1984), and L-alanine, $C_3H_7NO_2$ (Destro *et al* 1988). All these studies contain interesting and important crystallochemical information. For example, the phenakite crystal is a good model for study of the characteristics of luminescent matrices such as Zn_2SiO_4 and the solid electrolytes such as Li_2GeO_4 and Li_2MoO_4. The properties of all these isostructural compounds can be connected with the hexagonal channels along the z axis in this structural type (space group $R\bar{3}$), the diameters of which are more than 5 Å. It is usually assumed that they provide a place of localization of the impurity

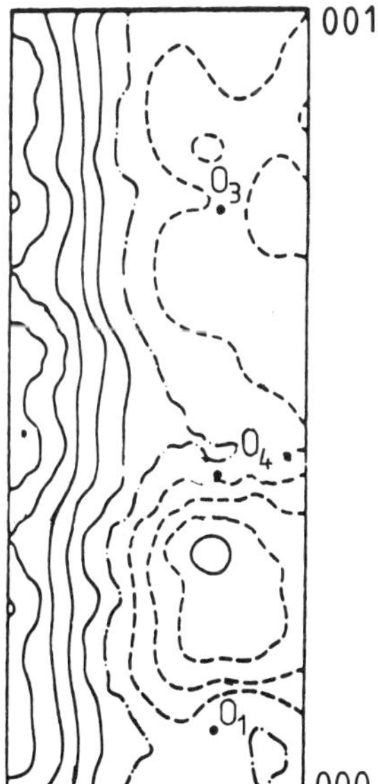

Figure 7.8 The deformation EP in the hexagonal channel along the z axis in phenakite, Be_2SiO_4, in the section through points (000) and (001) and the O3 atom. The line intervals are 0.07 $e\,Å^{-1}$ (Tsirelson *et al* 1986).

ions and of their transportation along the z axis. As found by Tsirelson *et al* (1986) the continuous region of excess deformation EP is spread along these channels throughout the crystal with three small local maxima at the levels of $z=0.25$, 0.50, and 0.75 (figure 7.8). Thus, in the phenakite crystal the conditions for easy diffusion of impurity ions along these channels have actual physical meaning. The local maxima in $\delta\varphi$ indicate the sites where the impurity ions can be trapped.

In summary, the directions in which analysis of the experimental EP is helpful can be outlined. One of these is concerned with the crystallographic aspect of the nature of various crystal properties. Another case, substantially less developed, may turn out to be helpful in solid state physics. For example, analysis of distribution of the EP along the internuclear vectors can give a foundation for the validity of calculation of nonlinear optical characteristics in the framework of the model of the linear anharmonic oscillator (see subsection 7.3.2). Consideration of the EP allows us to evaluate the importance of accounting for the non-spherical parts of the potential in electronic band structure calculations. Moreover, the EP, derived from x-ray diffraction data, can be directly included in the Hamiltonian and used in such calculations, at least for simple crystals (Avery *et al* 1981, Sommer-Larsen and Avery 1987). Of course, these possibilities do not exhaust all possible ways of using the experimentally derived distribution of the EP. First, it is also becoming possible to study qualitatively the (Su and Coppens 1995) interionic interactions in a crystal using experimentally measured quantities. Second, the atomic charges, which are fitted with the EP, can be obtained (Klooster 1992). They are transferable to other (analogous) molecules or crystals and result in forcing a field very close to the true field (Ghermani *et al* 1993). Finally, starting with the EP, one can easily arrive at other crystal-field characteristics using a minimum number of approximations. Some of these possibilities will be considered below.

7.1.2 *Electrostatics in intermolecular interactions*

The electrostatic interaction is a very important and, as a rule, predominant part of the atomic and molecular interactions in crystals (see subsection 2.5.2). The energy of electrostatic interaction of two non-penetrating systems, E_{es}, is described by expression (2.89). This expression is most applicable to molecular crystals where the ED can be presented approximately by superposition of non-overlapped EDs of separate molecules. Let us restrict ourselves to this case and consider the calculation of the electrostatic part of the energy of intermolecular interaction using the ED derived from the x-ray diffraction experiment.

In order to evaluate the contribution of different terms to the energy of total intermolecular interaction, W, the Rayleigh–Schrödinger perturbation theory is usually used (see e.g. Kaplan 1985, Buckingham 1978). The intermolecular interaction potential is considered as a perturbation. The first term of the perturbation theory describes the electrostatic interaction of neutral molecules without taking into account their mutual influence. The higher orders of perturbation theory give the polarization (induction and dispersion) interaction energies with exchange and charge transfer between molecules being taken into account. As a result, the total interaction energy is presented as a sum (2.90). The accurate and exhaustive application of quantum mechanical methods to the calculation of W can be achieved for a small number of atoms or small molecules only (see e.g. Zahradnik 1988). For large molecules and, especially, for crystals the mathematical difficulties do not allow one to obtain sufficiently accurate wave-functions for ground and excited states, which are needed for W calculations. Therefore semiempirical and, moreover, empirical methods are in use for such systems (Pullman 1978, Pertsin and Kitaigorodsky 1986). First of all, many-particle (non-additive) interactions in a system are approximated by sums of pair potentials derived either from quantum chemical calculation on isolated dimers under their different geometrical configurations or by effective pair potentials which are derived from empirical data. The empirical potentials effectively include the contribution of non-additive terms. It is assumed that the atomic interaction anisotropy is compensated on average and that these potentials are therefore isotropic.

Pair potentials are often (not obviously) expressed as a sum of atom–atom interactions. In this case the most general expression for description of the intermolecular interactions is

$$W=\frac{1}{2}\sum_{A\neq B}\sum_{\mu\nu}[-a_{AB\mu\nu}r_{ab\mu\nu}^{-6}+b_{AB\mu\nu}\exp(-\alpha_{AB\mu\nu}r_{ab\mu\nu})] \\ +E_{es}(A,B,\mu,\nu)+U_H. \qquad (7.18)$$

The indices A and B range over the molecules, μ and ν refer to atoms in these molecules. The first term inside the square brackets describes the dispersion attraction of particles, the second describes their repulsion, E_{es} is the electrostatic interaction energy and U_H is the hydrogen bond energy (if H bonds are present in the crystal). As can be seen from this expression, the electrostatic part of the interaction is separated from the short-range repulsion and dispersion parts, which are described by the Lennard-Jones potential. However, a serious problem is hidden here. The Lennard-Jones potential parameters, $a_{AB\mu\nu}$, $b_{AB\mu\nu}$, and $\alpha_{AB\mu\nu}$, can be derived from different experiments in the framework of a number of models, as described in detail by Kaplan (1985). Therefore their values are model dependent. In particular, the electrostatic interaction is represented in most of the models as the Coulomb interaction of point charges. Because of this, the part of the electrostatic energy caused by higher-order charge distribution moments is implicitly represented in the potential parameters which describe non-electrostatic interactions. This fact decreases the transferability of potentials, even in the case of homological series. In addition, the net atomic charge values turn out to be dependent on the procedure of their determination (section 6.2), which can lead to essentially different energies E_{es} when the point charge approximation is used (Hagler 1977).

The disadvantages mentioned have been partly avoided by Spackman (1986a) who used the Gordon and Kim (1972) model (see chapter 2) for obtaining the short-range repulsion potentials. This model is an example of the superpositional model in which the inhomogeneous electron gas approximation has been used for the energy density description. Expressing the repulsive part of the potential at low energies by a single exponent, Spackman has determined the parameters $b_{AB\mu\nu}$ and $\alpha_{AB\mu\nu}$ for 16 atoms usually found in molecular crystals. The corresponding potentials are more repulsive than any others (figure 7.9).

Adding to the new potential the dispersion term expressed via dynamical polarizabilities, Spackman arrived at a set of internally consistent atom–atom potentials which were successfully used in practical calculations (Spackman 1986b, 1987). However, their application assumes a more complete and explicit account of the electrostatic part of intermolecular interaction than the most usually used point charge approximation.

In an attempt to find an adequate method of calculating the energy E_{es}, Hirshfeld and Mirsky (1979) have compared three levels of E_{es} computation for crystals which consisted of the non-polar molecules C_2H_2, CO_2, and C_2N_2. In the simplest case each molecule was substituted by point quadrupoles, positioned at the centre of symmetry. On the next, intermediate, level each atom in a molecule was substituted by a point charge; the value of each was chosen to give the true molecular quadrupole moment. Finally,

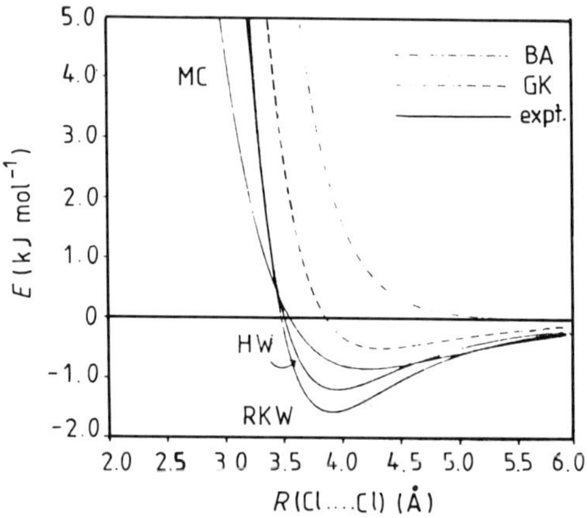

Figure 7.9 The potential of the Cl ··· Cl interaction. Theoretical potentials: BA, *ab initio* calculation of Bohm and Alrichs (1982); GK, Spackman's (1986a) calculation using the Gordon–Kim model. Experimental potentials: RKW, Reynolds *et al* (1974); HW, Hsu and Williams (1979); MC, Mirsky and Cohen (1978).

the ED of a molecule was partitioned into pseudoatomic fragments according to the Hirshfeld scheme and for each pseudoatom the net charge, dipole, and quadrupole moments were determined and placed at atomic nuclear positions. The HF wave-functions for free molecules were used in all these calculations. It was found that for molecules at distances of more than 15 Å all three models gave approximately the same values of E_{es}. However, at smaller distances the results of simpler models differ from those of the multipole model. The latter gives E_{es} values lower in absolute value than in the other cases and results in more precise values of the total intermolecular interaction energy, as calculated according to equation (7.18).

Because the multipole model gives the best approximation to the molecular ED it becomes clear how important an adequate description of the ED in the study of intermolecular interaction is. As has been shown by Spackman (1986b, 1987) even the use of the atomic point multipole moments, obtained by multipole expansion of theoretical ED of separate molecules, allows calculation of the energy, equilibrium geometry, force constants, and vibrational frequencies of molecular H-bonded complexes in very good agreement with experiment. However, not all possibilities of the method are used here in full measure: in particular, the contributions to the energy of the mutual molecular polarization and charge transfer are lost. As was pointed out by Tsirelson (1978), these drawbacks can be partly avoided in E_{es} calculation by the use of ED from x-ray diffraction experiments. Indeed, let us consider a molecular crystal without hydrogen bonds in which all molecules are separated by surfaces with ED values which do not exceed one e.s.d. of ED deformation. The assignment of some volume to a molecule approximately solves the problem of the description of intermolecular electron exchange. Being in the ground state, molecules in a crystal are under the influence of the electric field of their neighbours. The crystal many-electron wave-function can be presented in this case as an antisymmetrized product of wave-functions, describing each of the molecules in the crystal; each of these wave-functions, in turn, can be represented as a linear combination of single-determinant wave-functions. One of the latter describes the ground state of the non-disturbed (isolated) molecule and the others describe some excited states induced by the crystal field. The action of the crystal field leads to the self-consistent deformation of ED of each molecule which induces electric moments of different orders. As demonstrated by Krijn and Feil (1988a) using α-oxalic acid dihydrate as an example, the accuracy of the x-ray diffraction experiment is enough to fix corresponding deviations in ED (see figure 3.34). Table 7.1 shows that the dipole moments of molecules in crystals calculated using parameters of a multipole model derived from x-ray diffraction experiment have, as a rule, larger values than those for free molecules. This is a manifestation of the mutual attraction of the molecules in crystals.

Thus, the calculation of the energy of the electrostatic interaction from the experimental ED is equivalent to its calculation in the framework of

Table 7.1 Dipole moments μ (debye) (1 debye $= 3.335\,64 \times 10^{-30}$ C μ) of molecules in crystals and in solution or gas. (Reprinted with permission from Spackman *et al* (1988). © 1988 American Chemical Society.)

Molecule	$\lvert\mu\rvert$ Crystal	$\lvert\mu\rvert$ Solution or gas
Imidazole	4.8 (0.6)	4.0(1)
9-methyladenine	1.8 (1.0)	~3.0
Cytosine (in hydrate)	8.0 (1.4)	~7.0
Water (D_2O in cytosine monohydrate)	2.3 (0.3)	1.8
Urea	5.4 (0.5)	3.8–4.6
Thiourea (in crystal complex)	5.2 (1.8)	4.9

perturbation theory. The precision of calculation is limited by three factors: by the precision of the experimental ED data processing; by the adequacy of the separation of the ED asphericity and anisotropy of thermal vibrations; and by the choice of the volume of each molecule in a crystal. The first two factors are quite obvious. As far as the third factor is concerned, it is clear that in the case of noticeable molecular electron cloud overlap (for example in the presence of strong H bonds or short intermolecular contacts) results can be strongly distorted. As will be shown below, this is indeed the case.

Let us rewrite the expression (2.89) taking into account the above remarks:

$$E_{es} = \frac{1}{4\pi\varepsilon_0} \int_{\Omega'} \mathrm{d}\boldsymbol{r}' \int_{\Omega_0} \frac{\sigma(\boldsymbol{r})\sigma(\boldsymbol{r}')}{|\boldsymbol{r}-\boldsymbol{r}'|} \, \mathrm{d}\boldsymbol{r}. \tag{7.19}$$

Here $\sigma(\boldsymbol{r})$ is the total charge density in the crystal, Ω_0 is the volume of the initial molecule in a crystal, and Ω' is the space occupied by other molecules. It is assumed that mutual penetration of the molecular charge densities is negligible. The expression (7.19) differs from (2.89) in the fact that in the latter the charge densities represent the isolated non-disturbed systems, while charge densities in (7.19) are those of the molecules in the crystal.

We are mostly interested in the case when the electronic part of the total charge density is obtained from x-ray diffraction data. There are two ways of calculating the integral in (7.19). The first one (Tsirelson *et al* 1984a) is the direct numerical integration of (7.19), $\rho(\boldsymbol{r})$ being taken either in the form of a Fourier series (the series termination effect greatly influences the result in this case) or in parametric form, for example via the multipole model. The second way, tested by Moss and Feil (1981), is the use of pseudoatomic moments (3.102), point multipoles, the parameters of which are determined from a multipole model of ED. These moments describe the ED of pseudoatoms, not the molecular ED; however, outside the van der Waals radius they give nearly the same potential distribution as multipoles extended in space (figure 7.10). The pseudoatomic moments can be, of course, recalculated in molecular moments: the list of such molecular moments for some molecular crystals is given by Spackman (1992). If the space-extended multipole model describes the real ED adequately, then all the approaches mentioned arrive at nearly the same results.

For calculation of the integral (7.19) in practice, it is convenient to present the expression for $\sigma(r)$ in the form

$$\sigma(\boldsymbol{r}) = \sigma^0(\boldsymbol{r}) + \delta\rho(\boldsymbol{r}). \tag{7.20}$$

Here $\sigma^0(r)$ is the total (including nuclei) charge density of the procrystal from spherical atoms and $\delta\rho(r)$ is, as usual, the deformation ED. Applying any of the procedures described in section 6.2 for ED partitioning onto pseudoatoms, one can move further towards the atom–atom scheme and rewrite the expression (7.19) as

$$E_{es} = E_{p,es} + E_{pen} + \delta E_{es}. \tag{7.21}$$

Here

$$E_{p,es} = \frac{1}{4\pi\varepsilon_0} \sum_{A\mu < B\nu} \iint \frac{\sigma^0_{A\mu}(\boldsymbol{r})\sigma^0_{B\nu}(\boldsymbol{r}')}{|\boldsymbol{r} - \boldsymbol{r}'|} \, d\boldsymbol{r}' \, d\boldsymbol{r} \tag{7.22}$$

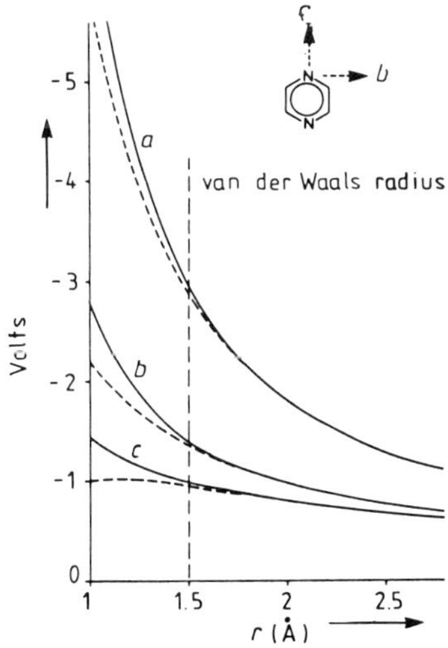

Figure 7.10 The EP due to the Hirshfeld deformation functions on nitrogen atoms of pyrazine, $C_4H_4N_2$. The potential is calculated in three perpendicular directions, two of which are indicated on the figure; the third direction, ***c***, is perpendicular to the plane of the molecule. The solid line gives the potential based on deformation densities up to fourth order, extended in space. The dashed line gives the potential due to point multipoles up to fourth order on the site of the N atom (Moss and Feil 1981).

is the electrostatic energy of interaction of spherical atoms in the procrystal,

$$E_{pen} = -\frac{1}{4\pi\varepsilon_0} \sum_{A\mu < B\nu} \int\int \left[\frac{\sigma^0_{A\mu}(\boldsymbol{r})\delta\rho_{B\nu}(\boldsymbol{r}')}{|\boldsymbol{r}-\boldsymbol{r}'|} + \frac{\delta\rho_{A\mu}(\boldsymbol{r})\sigma^0_{B\nu}(\boldsymbol{r}')}{|\boldsymbol{r}-\boldsymbol{r}'|} \right] \mathrm{d}\boldsymbol{r}'\, \mathrm{d}\boldsymbol{r} \tag{7.23}$$

is the electrostatic interaction component of the pseudoatomic deformation ED, with spherical parts of the penetrating field of only partly screened nuclei of other atoms, and

$$\delta E_{es} = \frac{1}{4\pi\varepsilon_0} \sum_{A\mu < B\nu} \int\int \frac{\delta\rho_{A\mu}(\boldsymbol{r})\delta\rho_{B\nu}(\boldsymbol{r}')}{|\boldsymbol{r}-\boldsymbol{r}'|} \mathrm{d}\boldsymbol{r}'\, \mathrm{d}\boldsymbol{r} \tag{7.24}$$

is the electrostatic energy of interaction deformation ED of pseudoatoms. The energy decomposition used develops the main line indicated in subsection 2.5.2 (see expression (2.95)), which was continued later in the analysis of EP in crystals.

As we have seen earlier, the electrostatic energy of atomic interactions in a procrystal, $E_{p,es}$, is always attractive at typical bond separations. When the interatomic distance increases its absolute value decreases rapidly. Epstein (1982) has found that $E_{p,es}$ at intermolecular distances, with an accuracy of about 1%, can be approximated by the Born–Mayer-type expression

$$E_{p,es} = -A \exp(-\alpha R). \tag{7.25}$$

The parameters A and α for several atoms are presented in table 7.2. It can be easily seen that at distances of about 5 Å the $E_{p,es}$ values become insignificantly small. The E_{pen} energy is usually repulsive. Its value varies, depending on the interacting atomic types and the orientation of interacting molecules. The E_{pen} values were calculated by Spackman (1986b) for some molecular pairs (the HF wave-functions for isolated molecules were

Table 7.2 Values of A and α (au) in the expression $E_{p,es} = -A \exp(-\alpha R)$ describing electrostatic interaction between some spherical atoms.

Atom–atom	Range of distances	A	α
N–H[a] ($\xi = 1.0$)	$3.2 \leqslant R \leqslant 5.6$	0.8994	1.588
N–H ($\xi = 1.24$)	$3.2 \leqslant R \leqslant 5.6$	1.708	2.080
N–N	$4.4 \leqslant R \leqslant 6.8$	43.21	1.898
O–O	$4.2 \leqslant R \leqslant 6.6$	53.74	2.129
F–F	$3.8 \leqslant R \leqslant 6.2$	59.32	2.294
Cl–Cl	$4.6 \leqslant R \leqslant 7.0$	206.1	1.832
Br–Br	$5.6 \leqslant R \leqslant 8.0$	420.0	1.767

[a]For N–H interaction two sets of parameters are given: one for exponential value $\xi = 1.0$ corresponding to the isolated atom wave-function; the second is for standard molecular value $\xi = 1.24$.

used). These values, together with other contributions to the interaction energy are listed in table 7.3. Analysing these data it may be noted that neglecting $E_{p,es}$ can sometimes lead to overestimation of the total energy: for example, for the pair CO–HF, the discrepancy is about 50%.

The electrostatic pseudoatomic deformation ED interaction is always attractive; the corresponding energy dominates in the electrostatic interaction (table 7.3). Because odd pseudoatomic moments are equal to zero for a spherical atom, the deformation ED mainly determines the values of point atomic multipoles. Though these values depend both on the multipole model used and on the ED partitioning scheme, Spackman (1986b) found that up to octupole level and higher, the dependence of calculated energies on the ED model is negligibly small. Formulae for calculation of the electrostatic energy of different point multipole interactions are given in table 7.4.

Using the Stevens (1979) multipole expansion parameters for the ED of the Cl_2 crystal, Tsirelson *et al* (1984a) calculated different energy contributions to E_{es}. The structure of the single-crystal Cl_2 (space group *Cmca*) consists of molecular layers parallel to the *bc* plane. Each molecule possesses a crystallographic inversion centre and lies in a mirror plane perpendicular to the *a* axis. There are short intermolecular contacts of 3.284 Å between molecules in *bc* layers: the doubled Van der Waals radius for Cl is 3.60 Å. The existence of such short contacts in the Cl_2 structure leads to the remarkable attraction between the atomic spherical parts: inside a sphere with radius 4.5 Å $E_{p,es}$ is approximately equal to $-45\ \mathrm{kJ\,mol^{-1}}$. The numerical integration of expressions (7.22)–(7.24) gave a rather large value, $E_{es} = -85\ \mathrm{kJ\,mol^{-1}}$, neglecting the small positive value of E_{pen}. The other contributions to intermolecular energy, calculated with Spackman's (1986a) parameters, were $E_{rep} = 31.7\ \mathrm{kJ\,mol^{-1}}$ and $E_{dis} = -15.6\ \mathrm{kJ\,mol^{-1}}$. Keeping in mind that $E_{p,es}$ is included in the repulsion potentials used, one obtains a value of total intermolecular energy of the Cl_2 crystal of $-24\ \mathrm{kJ\,mol^{-1}}$. Since the repulsion is certainly overestimated because of the too short intermolecular contacts in Cl_2, it can be concluded that the energy obtained is in satisfactory agreement with the experimentally measured sublimation heat (English and Venables 1974) of the Cl_2 crystal, extrapolated to 0 K and corrected for zero vibration energy: $|\Delta H_0| = 31.9\ \mathrm{kJ\,mol^{-1}}$. The small value of relaxation energy while the Cl_2 molecule transfers from the crystalline to the gas phase can be ignored because the distance 1.991(1) Å, within the limit of experimental accuracy, coincides with spectroscopy and with gas electron diffraction data.

Investigating the electrostatic interaction of two pyrazine, $C_4H_4N_2$, molecules described by point multipoles with parameters refined from x-ray diffraction data measured at 184 K, Moss and Feil (1981) found that, in order to obtain correct quantitative results, the experimental data need to be more precise than for mapping of deformation ED. Because of the low diffraction quality of the pyrazine single crystal, a small variation of the multipole model parameters led to the same R factor and to the same $\delta\rho$ maps; the latter were in good agreement with theoretical maps. However, values of electrostatic energies corresponding to these different sets of parameters were

Table 7.3 The decomposition of E_{tot} (kJ mol^{-1}). (Reprinted with permission from Spackman (1986b).)

Dimer	E_{rep}	E_{pen}	δE_{es}	E_{disp}	E_{tot}	E_{expt}
$(HF)_2$	8.1	1.9	−25.9	−2.4	−18.3	−19.1
$(HCl)_2$	3.1	0.9	−8.0	−2.3	−6.4	−9.5
HF–HCl	3.3	1.9	−12.2	−1.8	−8.9	
HCl–HF	6.5	0.8	−17.5	−2.7	−12.8	
N_2–HF	3.4	1.5	−9.3	−1.9	−6.4	−7.4
N_2–HCl	2.9	0.6	−5.8	−2.2	−4.5	
N_2–HCN	3.0	0.5	−5.7	−2.3	−4.5	
CO–HF	3.6	1.2	−10.8	−2.0	−8.0	
CO–HCl	3.0	0.5	−7.0	−2.3	−5.8	
CO–HCN	3.3	0.4	−7.7	−2.5	−6.5	
OC–HF	6.2	5.4	−21.1	−2.7	−12.2	−11.8
OC–HCl	5.0	2.0	−11.8	−2.9	−7.6	−6.8
OC–HCN	5.6	1.7	−11.6	−3.2	−7.5	
HCN–HF	14.7	7.7	−50.9	−4.4	−32.9	−26.1
HCN–HCl	11.9	3.4	−32.6	−4.8	−22.1	−14.4
$(HCN)_2$	13.2	2.8	−36.6	−5.5	−26.0	−18.4
C_2H_2–HF	6.6	3.9	−22.0	−3.4	−14.9	
C_2H_2–HCl	5.9	1.6	−13.7	−4.0	−10.2	−7.4
C_2H_2–HCN	6.7	1.4	−14.7	−4.4	−11.0	−7.7
CO_2–HF	4.5	1.5	−14.7	−2.3	−11.0	
CO_2–HCl	3.7	0.7	−9.6	−2.7	−7.8	
CO_2–HCN[a]	4.2	0.5	−10.6	−2.9	−8.8	
CO_2–HCl[b]	3.8	0.0	−6.2	−3.5	−6.0	
NCCN–HF	9.6	4.7	−29.6	−3.6	−18.8	
NCCN–HCl	7.8	2.0	−18.7	−4.0	−12.8	
HCCCN–HF	10.0	4.9	−36.2	−3.7	−25.0	
HCCCN–HCl	8.5	2.3	−23.8	−4.2	−17.2	
N_2O–HF	5.0	1.2	−14.3	−2.5	−10.7	
N_2O–HCl	4.3	0.5	−9.6	−3.0	−7.8	
SCO–HF	4.2	1.4	−14.3	−2.3	−10.9	
SCO–HCl	3.7	0.7	−9.5	−2.8	−7.9	
HF–Cl_2[c]	1.7	0.0	−3.5	−1.5	−3.3	
HF–Cl_2[d]	1.9	0.6	−4.6	−1.7	−3.8	

[a] Linear, hydrogen bonded.
[b] T shaped, 'anti' hydrogen bonded.
[c] 'Anti' hydrogen bonded.
[d] Hydrogen bonded.

spread significantly. Feil and Moss (1983) showed later that the use of point multipoles derived from sufficiently precise molecular wave-functions led, for the pyrazine pair, to an E_{es} which agrees well with the result of direct theoretical calculation. Even the orientational dependence was described satisfactorily, whereas the experimental data treatment gave only a qualitative picture.

Fortunately, in cases when the x-ray diffraction data are accurate enough, the interaction energetics of molecules in crystals can be successfully analysed with the help of the above approach. Spackman *et al* (1988) performed such an analysis for the crystals with structures illustrated in figure 7.11. Atomic point multipole moments were used, which were obtained from fitting of the model to the x-ray diffraction data up to octopoles on the atoms C, N, O, and S and up to quadrupoles on H atoms. Repulsive potential parameters were obtained from the Gordon–Kim electron gas model and dispersion parameters were obtained from dynamic polarizabilities. The term $E_{p,es}$ was neglected in energy calculation, because it is included in the Gordon–Kim model. For description of H bonds the interaction term connected with repulsion was omitted (it was assumed that repulsion energy is zero). The latter approximation resulted from the work of Buckingham and Fowler (1983), who successfully predicted the structure of Van der Waals complexes assuming that the H-bonded proton lies inside the Van der Waals sphere of the proton acceptor. This implies that the exchange repulsion between the proton and its acceptor on the H bond is very small. The results of the calculations of Spackman *et al* (1988) are included in table 7.5. In the first compound studied, imidazole, the crystal structure is characterized by chains of H-bonded molecules along the *c* axis, stacked such that each N(1) atom is superimposed on another at a distance of 3.42 Å (these atoms belong to molecules 1 and 3, figure 7.11(*a*)). The interaction along the *c* axis is attractive, with an energy value of $\pm 3.0 \pm 2.0$ kJ mol^{-1}. The H bond energy in the

Table 7.4 Expressions for the interaction energy between the various multipoles (Moss and Feil 1981). q, μ_α, $\theta_{\alpha\beta}$, $\Omega_{\alpha\beta\gamma}$, and $H_{\alpha\beta\gamma\eta}$ are net atomic charge and point dipole, quadrupole, octupole, and hexadecapole moments, respectively.

$$U_{m-m}{}^{a} = q^{(1)}q^{(2)}/4\pi\varepsilon_0 R$$
$$U_{m-d} = (R_\alpha/4\pi\varepsilon_0 R^3)[q^{(2)}\mu_\alpha^{(1)} - q^{(1)}\mu_\alpha^{(2)}]$$
$$U_{m-q} = (R_\alpha R_\beta/4\pi\varepsilon_0 R^5)[q^{(2)}\theta_{\alpha\beta}^{(1)} + q^{(1)}\theta_{\alpha\beta}^{(2)}]$$
$$U_{m-o} = (R_\alpha R_\beta R_\gamma/4\pi\varepsilon_0 R^7)[q^{(2)}\Omega_{\alpha\beta\gamma}^{(1)} - q^{(1)}\Omega_{\alpha\beta\gamma}^{(2)}]$$
$$U_{m-h} = (R_\alpha R_\beta R_\gamma R_\eta/4\pi\varepsilon_0 R^9)[q^{(2)}H_{\alpha\beta\gamma\eta}^{(1)} + q^{(1)}H_{\alpha\beta\gamma\eta}^{(2)}]$$
$$U_{d-d} = (1/4\pi\varepsilon_0 R^3)\mu_\alpha^{(1)}\mu_\alpha^{(2)} - (3/R^2)(\mu_\alpha^{(1)}R_\alpha)(\mu_\beta^{(2)}R_\beta)]$$
$$U_{d-q} = (5R_\alpha R_\beta R_\eta/4\pi\varepsilon_0 R^7)[\mu_\eta^{(1)}\theta_{\alpha\beta}^{(2)} - \mu_\eta^{(2)}\theta_{\alpha\beta}^{(1)}] + (2R_\alpha/4\pi\varepsilon_0 R^5)[\theta_{\alpha\beta}^{(1)}\mu_\beta^{(2)} - \theta_{\alpha\beta}^{(2)}\mu_\beta^{(1)}]$$
$$U_{d-o} = -(7R_\alpha R_\beta R_\gamma R_\eta/4\pi\varepsilon_0 R^9)[\mu_\eta^{(2)}\Omega_{\alpha\beta\gamma}^{(1)} + \mu_\eta^{(1)}\Omega_{\alpha\beta\gamma}^{(2)}] + (3R_\alpha R_\beta/4\pi\varepsilon_0 R^7)[\mu_\gamma^{(2)}\Omega_{\alpha\beta\gamma}^{(1)} + \mu_\gamma^{(1)}\Omega_{\alpha\beta\gamma}^{(2)}]$$
$$U_{q-q} = (35\theta_{\alpha\beta}^{(1)}\theta_{\gamma\eta}^{(2)}/12\pi\varepsilon_0 R^9)R_\alpha R_\beta R_\gamma R_\eta - (5\theta_{\alpha\beta}^{(2)}\theta_{\alpha\gamma}^{(1)}/3\pi\varepsilon_0 R^7)R_\beta R_\gamma + \theta_{\alpha\beta}^{(1)}\theta_{\alpha\beta}^{(2)}/6\pi\varepsilon_0 R^5$$

[a] Summation over repeated subscripts is implied.

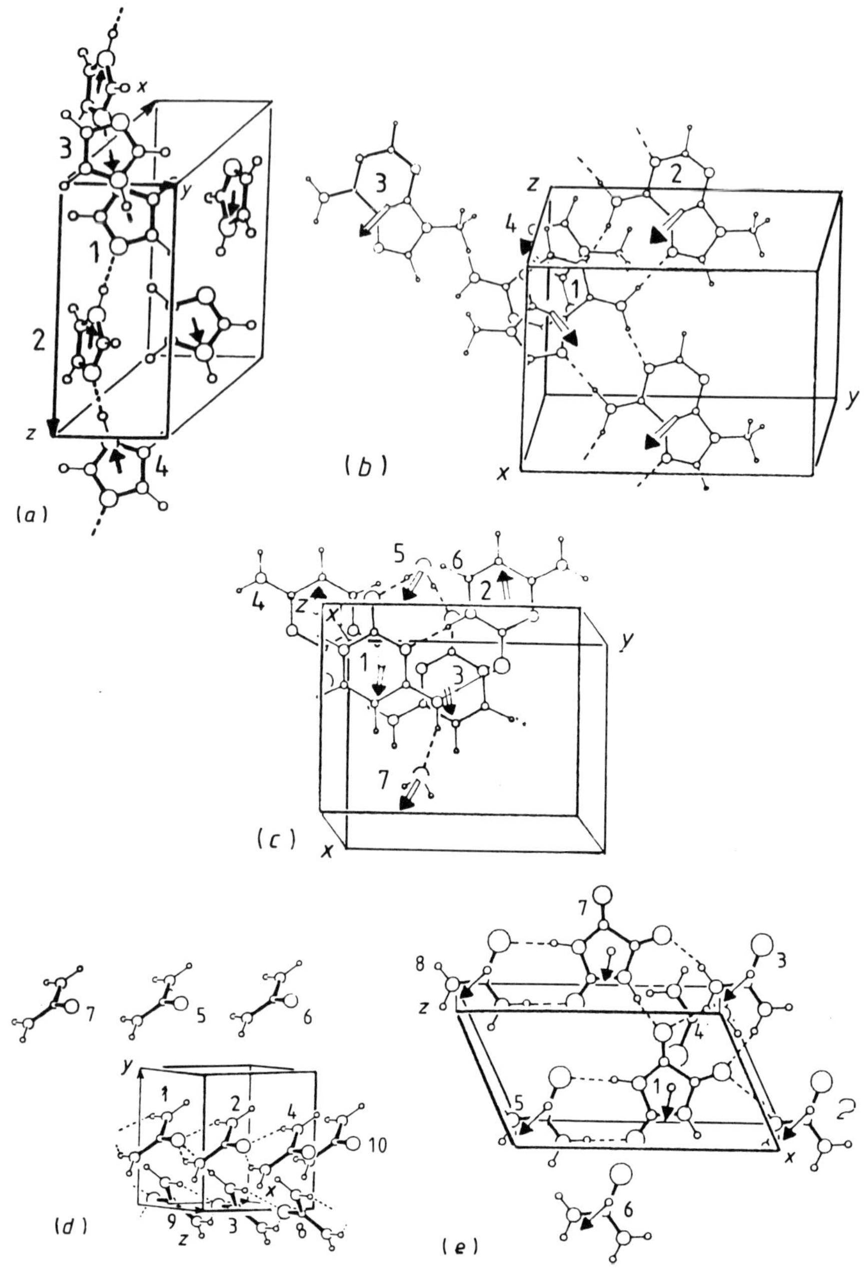

Figure 7.11 The crystal structures of some compounds studied by Spackman *et al* (1988): (*a*) imidazole; (*b*) 9-methyladenine; (*c*) cytosine monodeuterate; (*d*) urea, (*e*) 1:1 complex of thiourea with parabanic acid. Circles in order of decreasing size represent S, O, N, C, and H atoms. The direction of the dipole moment is indicated by arrows originating at the molecular centre of mass.

dimer having the crystal geometry is close to the value obtained from the trimerization energy by Luck (1976), and agrees to within one e.s.d. with the value obtained in chloroform solution, -16.7 ± 3.8 kJ mol^{-1}. The interaction energy between molecules 1 and 3, which have antiparallel dipole moments, is mostly electrostatic and dipolar in nature: the corresponding molecular dipole–dipole energy is -35.7 kJ mol^{-1}. The repulsive energy almost cancels the dispersion energy in this case.

The crystal structure of 9-methyladenine consists of sheets of molecules parallel to the (100) plane. Two N–H···NH bonds link adjacent molecules in ribbons along the *c* axis. The calculated energy for the H-bonded dimer of 9-methyladenine molecules, -16 ± 8 kJ mol^{-1}, is close to the value of $-17(3)$ kJ mol^{-1} obtained for self-association of 9-ethyladenine in solution (Kyogoku *et al* 1967). The total energy and the electrostatic energy of interaction of molecules 1 and 3 (figure 7.11(*b*)) is repulsive; this indicates that the C–H···N interaction should not be regarded as an H bond. It is interesting that the interaction energy for molecules 1 and 4 (figure 7.11(*b*)) with extensive molecular overlap is strongly repulsive. This results from the interaction of oppositely directed molecular dipoles.

In crystals of cytosine monodeuterate the cytosine molecules form H-bonded ribbons extending along the *b* axis. These ribbons stack along the *c* axis, with each ribbon linked to the immediate neighbour in the stack by water molecules H bonded to the carbonyl groups. The bonding of amino hydrogens in one stack to the water oxygen in a neighbouring stack results in formation of three-dimensional H bonding networks. The strongest interaction is between molecules 1 and 2 (figure 7.11(*c*)) which has two H bonds. The corresponding energy value is close to -57.4 kJ mol^{-1}, which was obtained by Kudritskaya and Danilov (1976) from theoretical calculation for the dimer in the same configuration. The cytosine molecules 1 and 3 and 1 and 4 (figure 7.11(*c*)), involving stacking interactions, have molecular dipole moments approximately parallel and antiparallel, correspondingly. The parallel arrangement of dipoles results in a strongly repulsive interaction whereas the antiparallel arrangement is attractive. The net interaction between layers is attractive due to H bonds of cytosine molecules with water molecules between adjacent layers. As was pointed out by Spackman *et al* (1988), relative strengths of these H bonds correlate well with H···O distances in the structure.

The crystal structure of urea consists of ribbons of doubly H-bonded molecules arranged in a head-to-tail fashion along the *c* axis. The plane of each ribbon is perpendicular to adjacent ribbons pointing in the opposite direction along the *c* axis. The urea dimer, involving doubly H-bonded molecules 1 and 2 (figure 7.11(*d*)) related head to tail by *c* translation, has the largest interaction energy. In the dimer with one H bond (molecules 1 and 3) the interaction energy is about one-half. This shows that all four H bonds at the urea O atom are approximately of the same strength.

Table 7.5 Energies of interaction (kJ mol^{-1}) of some organic molecules. Each energy is for a pair of molecules with a configuration as found in the crystal structure. Molecules are numbered as shown in the figures. Each pair involves molecule 1 which has fractional atomic coordinates (x, y, z) taken from the appropriate reference. The second molecule has atomic coordinates that can be derived by the space group symmetry operation which is given. E_{es} and $U = E_{rep} + E_{pen} + E_{disp}$ are as defined in the text. Negative values represent attractive interactions. E.s.d.s for E_{es} are given in parentheses. (Reprinted with permission from Spackman *et al* (1988). © American Chemical Society.)

Second molecule position	E_{es}	U	$W = E_{es} + U$	Comment
Imidazole (space group $P2_1/c$)				
2:$(x, \frac{1}{2} - y, \frac{1}{2} + z)$	−44(14)	14	−30	dimer with one H bond
3:$(-x, 1 - y, -z)$	−37(5)	−3	−40	partial overlap, adjacent ribbon
4:$(x, y, 1 + z)$	−3(2)	0	−3	second neighbours, same H-bonded ribbon
9-methyladenine (space group $P2_1/c$)				
2:$(x, \frac{1}{2} - y, \frac{1}{2} + z)$	−29(8)	13	−16	dimer with two H bonds
3:$(x, -\frac{1}{2} - y, \frac{1}{2} + z)$	1(4)	5	6	dimer with short C–H ··· N_3 distance
4:$(-x, -y, 1 - z)$	15(10)	6	21	dimer with extensive overlap
Cytosine monodeuterate (space group $P2_1/c$)				
2:$(-x, \frac{1}{2} + y, \frac{1}{2} - z)$	−96(27)	21	−75	cytosine dimer with two H bonds
3:$(x, \frac{1}{2} - y, \frac{1}{2} + z)$	42(16)	−5	37	cytosine dimer with partial overlap
4:$(-x, -y, -z)$	−18(15)	−2	−20	cytosine dimer with partial overlap
5:(x, y, z)	−42(20)	19	−23	cytosine with water as H bond donor
6:$(x, \frac{1}{2} - y, \frac{1}{2} + z)$	−24(16)	9	−15	cytosine with water as H bond donor
7:$(1 + x, y, z)$	−22(16)	18	−3	cytosine with water as H bond acceptor

Urea (space group $P\bar{4}2_1m$)				
2:(x, y, $1+z$)	−50(14)	7	−43	head-to-tail dimer with H bonds
3:(y, $-x$, $1-z$)	−32(10)	7	−25	side-by-side antiparallel dimer, one H bond
4:(x, y, $2+z$)	−5(3)	0	−5	see figure 7.11
5:(x, $1+y$, z)	4(2)	−2	2	
6:($1+x$, $1+y$, z)	7(2)	0	7	
7:($-1+x$, $1+y$, z)	4(1)	0	4	
8:(y, $-x$, $2-z$)	5(4)	0	5	
9:(y, $-x$, $-z$)	1(8)	6	7	
10:($1+x$, y, $1+z$)	−1(3)	0	−1	
1:1 complex of thiourea with parabanic acid (space group $P2_1/m$)				
2:($1+x$, y, z)	9(13)	2	11	heterodimer with one H bond
3:($1+x$, y, $1+z$)	−28(24)	9	−19	heterodimer with one H bond
4:($1-x$, $1-y$, $1-z$)	−5(13)	−1	−6	heterodimer with one H bond
5:(x, y, z)	−32(14)	14	−18	heterodimer with one H bond
6:(x, y, $-1+z$)	−1(9)	2	1	heterodimer with one H bond
7:(x, y, $1+z$)	−44(21)	8	−36	parabanic acid dimer with one H bond
Alloxan (space group $P2_12_12$)				
2:($\frac{1}{2}-y$, $\frac{1}{2}+x$, $\frac{1}{4}+z$)	0(2)	−2	−2	dimer with three short C···O distances
3:($\frac{1}{2}+y$, $-\frac{1}{2}-x$, $-\frac{1}{4}+z$)	−6(4)	1	−5	dimer with three short NH···O distances
Parabanic acid (space group $P2_1/n$)				
2:($\frac{1}{2}-x$, $\frac{1}{2}+y$, $\frac{1}{2}-z$)	10.9(4)	−12.2	−1.3(4)	dimer with 2.75 and 3.08 Å C···O distances
3:(x, y, $-1+z$)	5.4(1)	−6.4	−1.0(1)	dimer with 2.94 and 3.08 Å C···O distances

Electrostatic interactions in urea are long range: even the interaction energy of molecules 1 and 4 separated by a distance of twice the *c* period (9.732 Å) is attractive. For other pairs of molecules, including interactions with all next-nearest neighbours, the energies are, as a rule, repulsive. This is in agreement with the strongly dipolar nature of the urea molecule and the relative orientation of dipoles in the structure.

The urea crystal cohesive energy obtained from the model discussed is -66 ± 24 kJ mol^{-1}. This agrees satisfactorily with the observed sublimation energy at 298 K: 88 kJ mol^{-1} (Susuki *et al* 1956).

In all molecular crystals considered, the ED and potential characteristics, parameters of which do not contain electrostatic components, were used in calculations; one managed to obtain values of intermolecular interaction energy which are close to independent experimental data. An important circumstance was that the corresponding x-ray diffraction data were measured at low temperatures and the separation of thermal smearing and asphericity effects was carried out successfully enough. Besides, the molecular ED spatial separation is well pronounced in this case, which is one of the conditions for successful electrostatic energy calculation. The use of room-temperature x-ray data decreases the value of such calculations in the framework of the model used. An example of this is the crystal 1:1 complex of thiourea and parabanic acid. The energy values in table 7.5 calculated with multipole parameters seem physically reasonable. However, the significance of the calculated energies is marginal in terms of their e.s.d. Thus, a low-temperature x-ray experiment is an important condition for obtaining reliable energy characteristics with experimental ED.

Alloxan, the structure of which was described in section 7.1 (see figure 7.2), and parabanic acid, $C_3H_2N_2O_3$, (He *et al* 1988) are compounds for which assumption of good separation of the molecules in the crystal is invalid. In both crystals there are short C$\cdots$O intermolecular contacts: 2.73 Å in alloxan and 2.75 and 2.94 Å in parabanic acid. At such separations the sum of E_{rep} and E_{disp} gives a net repulsion. In order to take into account attractive interaction between these molecules in the crystals Spackman *et al* (1988) omitted corresponding terms in the C$\cdots$O potentials. However the energy values obtained were too small to explain high crystal densities and sublimation temperatures higher than 500 K. The interaction energy for alloxan molecules due to three close C$\cdots$O contacts is only -2 (2) kJ mol^{-1}. The H bond interactions are also weakly attractive. The electrostatic interaction energies for parabanic acid molecules forming such contacts are even repulsive. Thus, the approximation used is not suitable for crystals with close intermolecular distances.

It proved to be the case that the atomic multipole moments have good transferability within the group of identical compounds. This was found empirically by Berkovitch-Yellin and Leiserowitz (1980): using point multipoles for the acetamide molecule, calculated from low-temperature x-ray diffraction data for the complex acetamide–allenedicarboxylic acid, $C_2H_5NO{\cdot}C_5H_4O_2$, they have obtained a molecular dipole moment for formamide, CH_3NO, in good accordance with experiment. The use of the point multipoles from treatment of simple molecules allowed the study of packing features in the complicated crystals of primary amides and *cis* and *trans* secondary amides (Berkovitch-Yellin and Leiserowitz 1980), carboxylic acids (Berkovitch-Yellin and Leiserowitz 1982), and N-acylated amino acid (Berkovitch-Yellin *et al* 1983) and the analysis of the role of C–H···O and C–H···N interactions in these crystals (Berkovitch-Yellin and Leiserowitz 1984). The transferability of point multipoles was also used in the calculations of the habit of organic crystals with known structures and symmetry (Berkovitch-Yellin *et al* 1985, Berkovith-Yellin 1985, Boek *et al* 1991). The shape of crystals is determined by relative rates of deposition of molecules on their various faces. The growth rates are determined primarily by the strength of binding of the molecules which arrive at the various crystal surfaces, and this depends on the orientation of the molecules within the crystal relative to its various faces. The anisotropy of electrostatic atom–atom interactions, calculated with multipole–multipole terms, plays an important role in determining the binding values on different faces. Knowledge of this allows, from calculations, the prediction of the crystal morphology in good agreement with observations for crystals grown by sublimation. In this way the possibility for *ab initio* derivation of crystal morphology can be seen.

7.1.3 *The electric field gradient at nucleus sites and quadrupole interaction*

An important physical characteristic, depending on ED in the crystal and connected to diffraction and spectroscopy methods, is the electric field gradient (EFG) at the nuclear positions. If the charge density of the nucleus is non-uniform, and the nucleus is non-spherical (nuclear spin greater than $\frac{1}{2}$), then, in addition to Coulomb interactions of the point nucleus with its surroundings, effects arise resulting from the interaction of the nuclear charge density moments with the multipole moments of the ED relative to the nuclei. These effects lead to splitting of the nuclear energy levels (Flygare 1978). Because there is no nuclear electric dipole moment, the second non-zero term in the interaction energy expansion over the displacement from the origin plays a main role. This term arises from the interaction of the nuclear quadrupole moment, eQ, with the EFG, $\nabla \boldsymbol{E}_{\alpha\beta}$, created by all electrons and other nuclei of the crystal. The EFG is a second-order tensor for which $\Sigma_\alpha \nabla \boldsymbol{E}_{\alpha\alpha}=0$. Therefore, in the principal coordinate system is has only two independent components, which can be calculated from the equation

$$\nabla \boldsymbol{E}_{\alpha\beta} = -(4\pi\varepsilon_0)^{-1} \int_{-\infty}^{\infty} [(3r_\alpha r_\beta - \delta_{\alpha\beta}|\boldsymbol{r}|^2)/|\boldsymbol{r}|^5]\sigma(\boldsymbol{r})\,\mathrm{d}\boldsymbol{r}. \tag{7.26}$$

Here r_α is the projection of vector $\boldsymbol{r}$ onto the axes $\alpha = x, y, z$ in the local coordinate system, and $\sigma(\boldsymbol{r})$ is the total charge density in the system. In order to simplify the notation an asymmetry parameter is often used:

$$\eta=(\nabla E_{xx}-\nabla E_{yy})/\nabla E_{zz} \qquad |\nabla E_{zz}|\geqslant|\nabla E_{yy}|\geqslant|\nabla E_{xx}|. \tag{7.27}$$

It is easily seen that $0\leqslant\eta\leqslant 1$. In a cubic crystal field $\nabla E_{xx}=\nabla E_{yy}=\nabla E_{zz}$, and, as the trace of the EFG tensor is equal to zero, each of the EFG components should be zero. For an axial symmetric charge distribution around the z axis, $\nabla E_{xx}=\nabla E_{yy}=-\frac{1}{2}\nabla E_{zz}$ and $\eta=0$. The sign of EFG is often understood to be the sign of the largest component ∇E_{zz}.

The nuclear quadrupole moment eQ and the EFG $\nabla E_{\alpha\beta}$ enter as products of the quadrupole coupling constant e^2qQ ($eq=-\nabla E$), through which the energy of nuclear quadrupole transitions is expressed. These energies can be very precisely measured with the help of nuclear quadrupole (NQR) or nuclear magnetic (NMR) resonance and Mössbauer spectroscopy (MS); for molecules in the gas phase microwave spectroscopy can be used. In NQR, NMR, and microwave spectroscopy the transitions occur between the ground state energy sublevels; meanwhile MS deals with transitions between ground and excited states. However, knowledge of e^2qQ is not sufficient to define the value and sign of eQ and eq simultaneously. For many atoms the magnitude of eQ can be measured by the molecular beam method (Ramsey 1956) and its sign is determined by special methods, although by no means always (Thosar *et al* 1983). However for atomic states with short lifetimes, quadrupole moments cannot be reliably measured. This situation, in particular, is observed with the quadrupole moment of the ^{57}Fe nucleus excited to the 2.3×10^{-15} J (14.4 keV) level. In such cases difficulties arise in the analysis of resonance spectra regarding the choice of electronic structure parameters which would provide a reliable value of e^2qQ. This is especially the case for complicated spectra where the influence of sample non-stoichiometry as well as magnetic interactions is large. Therefore, independent determination of the magnitude and sign of EFG allows the magnitude and sign of eQ to be determined and makes interpretation of spectra simpler and more reliable. Knowledge of EFG is also necessary for study of cation distribution on non-equivalent structural sites, in studies of the influences of impurity atoms on crystal properties, for checking the validity of the proposed model of the nucleus, for nucleus quadrupole moment calculation and so on.

For a long time the EFG calculations at nuclei sites were performed on the basis of the static crystal ionic model (Sharma and Das 1964, Hafner and Raymond 1968). The lattice contribution to EFG, $eq_{\alpha\beta}^{lat}$, was calculated in the case of non-cubic symmetry when it is not equal to zero. It consists of a contribution from atomic charges and a contribution arising from dipole and higher-order deformations of the ligand electron shells. These contributions were taken into account with the help of point multipoles, which were assumed to lie outside the electron cloud of the atom under consideration. The contributions of partly occupied electron shells of the atom were added to $eq^{lat}_{\alpha\beta}$. These components of EFG were calculated for isolated ions in certain valence states.

Because the ionic model does not take all of the peculiarities of ED into account and distorts the value of electrostatic interaction of a nucleus with its neighbourhood, a correction coefficient is introduced into calculations: the Sternheimer screening factor, γ (Sternheimer 1950). This factor takes into account the quadrupole deformation of part of the electron cloud of the atom under consideration, which is not included in the calculation of $eq^{ion}_{\alpha\beta}$ because the ED of an atom is assumed to be spherical. In general, γ is a function of $\boldsymbol{r}$, $\gamma = \gamma(\boldsymbol{r})$, where $\boldsymbol{r}$ is the distance from the nucleus to external charges. For light atoms, γ is small and can be either positive or negative. For second-row atoms and for heavier atoms this function is positive when $r < 1$ Å (it has shielding character), and for $r > 1$ Å it is negative (and has antishielding character). The shielding function $\gamma(\boldsymbol{r})$ is usually calculated for atoms in certain valence states by the variational procedure, many-body perturbation method, coupled HF method etc (Thosar *et al* 1983, Guser *et al* 1995).

Unfortunately, even introduction of the γ correction factors does not allow one to describe (within the ionic model) all details of the crystal field for specific compounds. These details are often very important for calculation of EFG and their loss leads to contradictory results, a good example of which is the well known 'ferrous–ferric anomaly' (Sharma 1971). In ferrous compounds, evaluation of the quadrupole moment of a ^{57}Fe nucleus excited to 2.3×10^{-15} J from the experimental quadrupole interaction constant and EFG calculated in the ionic model led to a value of Q_{Fe} of about 0.18×10^{-28} m^2. The sign of Q in this excited state is positive (Thosar *et al* 1983). Analogous evaluations for ferric compounds gave $Q_{^{57}Fe} = (0.1\text{–}0.4) \times 10^{-28}$ m^2. However, the quadrupole moment of the ^{57}Fe nucleus in both types of compound has the same magnitude, and it is therefore clear that these discrepancies are connected with the method of EFG calculation. Indeed, a survey of molecular complexes consisting of a central iron atom and its ligand embedded in a crystal had been able to substantially narrow the range of $Q_{^{57}Fe}$ values. A more sophisticated crystal model led to a $Q_{^{57}Fe}$ value of $(0.18 \pm 0.02) \times 10^{-28}$ m^2. This value is close to the value $Q_{^{57}Fe} = 0.156 \times 10^{-28}$ m^2 derived on the basis of non-empirical EFG calculations in $FeCl_2$ and $FeBr_2$ molecules by Ellis *et al* (1983).

Nevertheless, all-electron calculations of clusters embedded in the crystal, carried out by the X_α method and by the UHF method for α-Al_2O_3 (Nagel 1985a), α-Fe_2O_3 (Nagel 1985b, Kelires and Das 1987), and Li_3N (Kelires *et al* 1987) did not allow complete elimination of the uncertainty in the EFG value connected with the defects in the crystal model. Usually non-empirically calculated clusters contain only a small number of atoms. The influence of the surroundings on these atoms is reduced to a simulation of the crystal field by a set of point multipoles. In ionic crystals, this approach turns out to be justified: for example, in α-Al_2O_3, it leads to a value of eq_{zz} at the Al nucleus equal to $-0.607\ 10^{21}\ V\,m^{-2}$ (Nagel 1985a). This agrees to within 8% with the experimental value $eq_{zz} = -0.655 \times 10^{21}\ V\,m^{-2}$, which follows from the quadrupole interaction constant measured by Pound (1950) using $Q_{^{27}Al} = 0.151 \times 10^{-28}\ m^2$; the signs of the EFG z components also coincide. However, for framework compounds with partly covalent multicentred chemical bonds, the fragments accessible to calculation do not allow for overlap of electron clouds with each other and with the atoms from the rest of the crystal. For example, from a calculation of the $Fe_2O_3^{12+}$ cluster, which simulates a crystal of α-Fe_2O_3 (Nagel 1985b), it follows that the iron d_{z^2} orbital, directed along the C_3 axis, is unoccupied by electrons. This contradicts the data of Goodenough (1971) and the results of Beri *et al* (1983). In the latter work, the overlaps of the orbitals of ions and local, non-local (including orthogonalization of orbitals), and long-term crystal effects affecting EFG at the nucleus were successively taken into account. At the same time, redistributions of ED when the crystal is formed from the ions Fe^{3+} and O^{2-} were not included in the calculation. The calculated quadrupole interaction constant was six times smaller than the experimental value. However, if one assumes that the iron atom d_{z^2} orbital in α-Fe_2O_3, which forms together with the other orbitals the a state, has a larger electron occupancy than the other d AOs, the discrepancy between theory and experiment is reduced. The validity of such an assumption is supported by the distribution of experimental deformation ED close to the cations in α-Fe_2O_3, which shows a high degree of filling by electrons of the a state (figure 7.12(a, b)). The $\delta\rho$ bridge found experimentally between the cations along the C_3 axis is not reproduced, however, in the calculation of the Fe_2O_3 fragment. Therefore the reason for the lack of success of Nagel's (1985b) EFG calculation in α-Fe_2O_3 by the cluster method appears to be the inadequacy of the crystal model used in the calculation.

The best results in EFG calculations were achieved by using the all-electron full-potential LAPW method with inclusion of exchange and correlation in the LDA (Blaha *et al* 1985, Schwarz *et al* 1990, 1994, Schwarz and Blaha 1991). This method, which explicitly takes into account all polarization effects and does not require the introduction of Sternheimer factors, was used to calculate EFG in the insulators Li_3O and Cu_2O, the molecular crystals Cl_2, Br_2, and I_2, and the high-temperature superconductor $YBa_2Cu_2O_7$. In all compounds satisfactory agreement between the calculated EFG and the values obtained from the quadrupole coupling constant was achieved (provided Q is reliably known).

X-ray diffraction, which gives an ED with a resolution of approximately 0.2 Å, allows one to calculate the electronic part of EFG, as it does by theoretical methods. Stewart (1972, 1977c) pointed this out. In this approach it is highly significant that the EFG at the nucleus depends not on the complete ED, but on its quadrupole deformation relative to this nucleus.

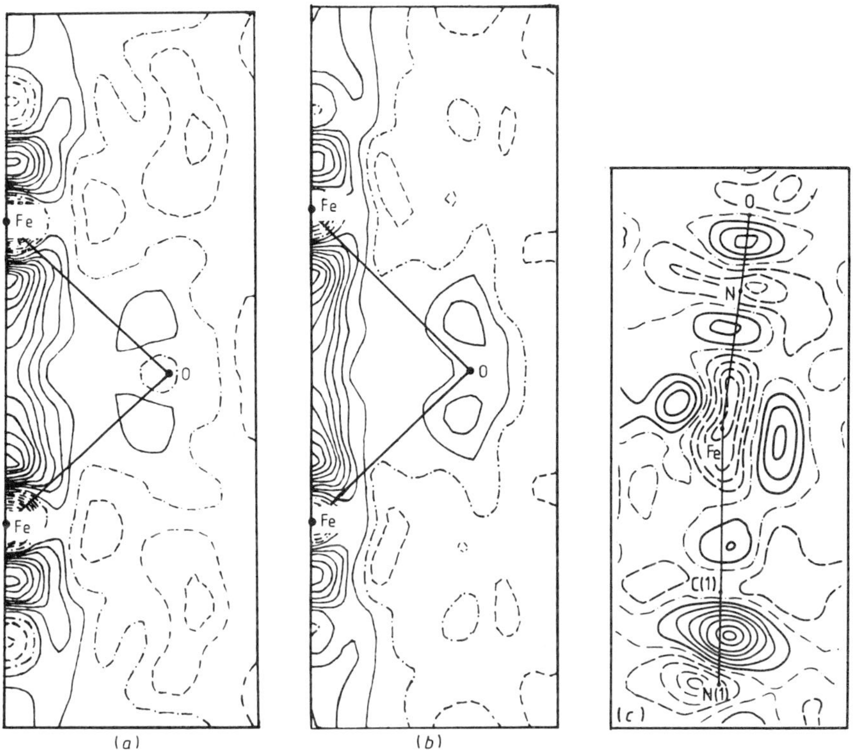

Figure 7.12 The deformation ED in haematite, α-Fe_2O_3 (*a*) at 295 K, (*b*) at 153 K, in the section through the *c* axis and Fe and O atoms, and (*c*) in sodium nitroprusside $Na_2[Fe(CN_5)NO]\cdot 2H_2O$ at 153 K in the section through the CN–Fe–NO plane (the crystallographic *m* plane). The line intervals are 0.1 $e\,Å^{-3}$ (Antipin *et al* 1987, Tsirelson *et al* 1988a, 1989).

The quadrupole component of the radial probability density for an electron, located at distance r from the nucleus, takes the form $R_n^2 r^2 \sim r^{2n} \exp(-2\xi r)$ if we approximate the radial part of the wave-function by a single STO (C.1). The distances from the nucleus to the maxima of this function, $r_{max} = n/\xi$, for different subshells of several free atoms are tabulated in table 7.6. As can be seen from these data, the quadrupole deformation of the 1s subshell lies outside the limit of resolution of the x-ray diffraction method for atoms from the first and second rows of the periodic table; this is also true for the 1s and (2s, 2p) subshells of the 3d elements. Bentley and Stewart (1974) found that in molecules which consist of first-row atoms the EFG induced by 1s electrons is up to 0.02 au, which is rather small in comparison with total EFGs, where the magnitudes lie between 0.40 and 2.87 au. It should also be noted that information about the quadrupole deformation of the core electron subshells is not lost completely. Bentley and Stewart (1974) calculated, for several molecules consisting of first-row elements, projection coefficients

$$p = \int_0^\infty \rho_{quadr}^{core}(r)\rho_{quadr}^{val}(r)r^2\,\mathrm{d}r$$

$$\times\left(\int_0^\infty [\rho_{quadr}^{core}(r)]^2 r^2\,\mathrm{d}r \int [\rho_{quadr}^{val}(r)]^2 r^2\,\mathrm{d}r\right)^{-1} \tag{7.28}$$

which serve as the measure of linear dependence between ρ_{quadr}^{core} and ρ_{quadr}^{val}. These coefficients, listed in table 7.7, show that a substantial part of the quadrupole core deformation is represented by a valence quadrupole term in LS analysis. Stewart (1977c), using the N_2 molecule as an example, showed that the error in EFG at the nuclei due to insufficient resolution in the x-ray diffraction data constituted 10–15%.

At first sight it seems that the accuracy of determination of ED in the neighbourhood of the nuclei would be a major source of error in EFG. However, near the nuclei, where errors in determination of ED are maximum, the deformation ED has a dipole character (see figure 6.49). In the region where quadrupole deformation takes place (see table 7.6), the experimental error in $\delta\rho$ is substantially less and, in any case, is obscured by the uncertainty in EFG connected with resolution.

As usual, two approaches to calculation of EFG using x-ray diffraction data exist. One consists of using a Fourier series with the structure amplitudes as coefficients, in which spirit it is close to the non-empirical theoretical method of EFG calculation. The other method is based on the ED

Table 7.6 Positions of radial ED maxima for electronic subshells of some atoms.

	N		P			Fe					
	1s	2s,2p	1s	2s,2p	3s,3p	1s	2s	2p	3s	3p	3d
r_{max} (10^{-10} m)	0.08	0.54	0.04	0.20	0.84	0.02	0.11	0.10	0.35	0.37	0.43

Table 7.7 Core versus valence density function projection coefficients p for some atoms in diatomic molecules.

	B		C		N		O	
	BH	BF	CH	CO	NH	N_2	OH	CO
p	0.752	0.632	0.613	0.564	0.581	0.632	0.624	0.599

reconstructed within the multipole model with fixed electron cores. This requires, generally speaking, the introduction of the Sternheimer screening factors, although in a way that slightly differs from the standard free-atom method, as follows from the analysis presented above. Before looking at these approaches, it is necessary to assess a problem common to both: the problem of the influence of the atomic thermal motion on the EFG. This influence depends, first of all, on the mutual orientation of the principal axes of the thermal vibration tensor and the EFG tensor, and can be highly significant, as seen from $\delta\rho$ maps for α-Fe_2O_3 at different temperatures. Therefore, it is highly desirable to use x-ray diffraction data collected at as low a temperature as possible in order to obtain undistorted components of EFG. The multipole model allows us to calculate the quasistatic EFG provided that the thermal motion effects and the asphericity of the pseudoatom EDs are well separated. It should be emphasized that in resonance methods, which have characteristic times that are different from those in x-ray diffraction, atomic thermal motion appears in a different way. So, in NQR, an atom 'feels' the time-averaged value of EFG, the magnitude of which depends on the degree of correlation in the thermal motion of neighbouring atoms relative to the atom being considered. This leads to the conclusion that the EFG at the nuclear position, and consequently the quadrupole coupling constant, depend in NQR upon temperature. Lowering the temperature usually increases the resonance frequency of NQR (Hedvig 1975). In MS a more complicated dependence is observed (Thosar *et al* 1983). Here it is very important how the electronic populations of levels vacant at 0 K change with temperature. It is clear that the EFG which appears from spectroscopy methods will, generally, not be equal to the EFG at the time-averaged position of the nucleus which is calculated from dynamic experimental ED. However, the difference between them will decrease as the temperature of MS, NQR, NMR, and x-ray experiments is reduced.

We will now examine the calculation of EFG by application of Fourier series, as was suggested by Schwarzenbach and Thong (1979). If one represents the total charge density $\sigma(\boldsymbol{r})$ in (7.26) in the form of (7.20), then the EFG can be written in the following way:

$$eq_{\alpha\beta}=eq^{0}_{\alpha\beta}+\delta(eq_{\alpha\beta}) \qquad \alpha, \beta=x, y, z. \tag{7.29}$$

Here $q^{0}_{\alpha\beta}$ is the EFG arising from the distribution of electrons and nuclei in the procrystal; $\delta(eq_{\alpha\beta})$ is the EFG arising from the deformation ED. The static term $eq^{0}_{\alpha\beta}$ is easy to calculate using the tables of Clementi and Roetti (1974) when the crystal structure is known. The magnitude of $\delta(eq_{\alpha\beta})$ can be calculated by substituting in (7.26) the deformation ED for the total charge density. Then, in terms of the structure amplitudes,

$$\delta(eq_{\alpha\beta})=-\frac{e}{3\varepsilon_0 V}\sum_{\boldsymbol{H}}\left\{\frac{3H_\alpha H_\beta-\delta_{\alpha\beta}|\boldsymbol{H}|^2}{|\boldsymbol{H}|^2}\right\}\{F(\boldsymbol{H})-\tilde{F}(\boldsymbol{H})\}\exp(-2\pi\mathrm{i}\boldsymbol{H}\cdot\boldsymbol{r}). \tag{7.30}$$

Stewart (1977c) and Epstein and Swanton (1982) showed that the deformation ED contribution to the EFG, calculated in accordance with (7.30), to a first approximation can be considered as characteristic of the static crystal. However, it is better to use static values $F_{mod}(\boldsymbol{H})$ instead of $F(\boldsymbol{H})$ calculated within the framework of the multipole model (Lewis *et al* 1982). Therefore, having calculated the static term $q^0_{\alpha\beta}$ using the time-averaged nuclear positions (this results from penetration terms from non-local spherical charge densities of neutral atoms), we can compare the total EFG (7.29) with spectroscopic data.

In order to reduce the effect of random error, when summing the series (7.30), it is advisable to use a regularization filtering procedure (see section 3.6). In the crystallographic coordinate system the expression for the component $\delta(eq_{\alpha\beta})$ including filtering factor M has the form

$$\delta(eq_{\alpha\beta}) = -\frac{e}{\varepsilon_0 V}\sum_H M_H \left\{\frac{H_\alpha H_\beta}{|\boldsymbol{H}|^2} - (\boldsymbol{a}_\alpha \cdot \boldsymbol{a}^*_\beta)/3\right\} \times [F(\boldsymbol{H}) - \tilde{F}(\boldsymbol{H})]\exp(-2\pi \mathrm{i}\boldsymbol{H}\cdot\boldsymbol{r}) \tag{7.31}$$

where $\boldsymbol{a}_i$ and $\boldsymbol{a}^*_i$ are direct and reciprocal lattice vectors, $H_i = h, k, l$. The filtering factor has a form similar to (3.137). The expression for the variance in EFG is obtained as described in subsection 3.7.1. It has the form

$$\sigma^2[\delta(eq_{\alpha\beta})] = \left(\frac{e}{\varepsilon_0 V}\right)^2 \sum_{H>0} M_H \left\{\frac{H_\alpha H_\beta}{|\boldsymbol{H}|^2} - \frac{\boldsymbol{a}_\alpha \cdot \boldsymbol{a}^*_\beta}{3}\right\}^2 \{\sigma^2[\delta_A(\boldsymbol{H})]\cos^2(2\pi \mathrm{i}\boldsymbol{H}\cdot\boldsymbol{r}) + \sigma^2[\delta_B(\boldsymbol{H})]\sin^2(2\pi \mathrm{i}\boldsymbol{H}\cdot\boldsymbol{r})\} \tag{7.32}$$

$$F(\boldsymbol{H}) - \tilde{F}(\boldsymbol{H}) = \delta_A(\boldsymbol{H}) + \mathrm{i}\delta_B(\boldsymbol{H})$$

(it is assumed that reflections have been measured over a reciprocal space hemisphere).

Expressions (7.31) and (7.32) are easily transformed to the principal coordinate system connected with the atom under consideration. When the EFG is calculated in reciprocal space, the symmetry and long-range order are automatically taken into account, while in position space, in order to ensure that the integral (7.26) converges properly, it is necessary to include in the calculation a large number of coordination spheres around the given nucleus.

It is clear from the above expressions that the accuracy of determination of EFG depends on the precision and completeness of the x-ray reflection measurement and on the correctness of determination of the scale factor. It is important to emphasize that exclusion, or insufficient precision, of some of the high-angle reflections along a few directions in reciprocal space can influence the calculation of any component of EFG in a drastic way, breaking the condition $\Sigma q_{\alpha\alpha} = 0$. Therefore this condition should be controlled in the calculation process.

Let us examine a calculation of EFG at the ^{57}Fe nuclei in haematite, α-Fe_2O_3, and sodium nitroprusside (SNP), $Na_2[Fe(CN)_5NO]\cdot 2H_2O$ (Tsirelson *et al* 1987b). Results of low-temperature, accurate x-ray diffraction experiments at 153 K served as source data for the calculation. The atomic thermal motion in these crystals hardly influences the quadrupole coupling constants; they remain unchanged in magnitude and sign in the temperature range from 100 to 295 K (Khramov and Polosin 1983, Hauser *et al* 1977). This means that the EFG is constant in magnitude and sign in these compounds. The position symmetry of the iron ions in haematite (D_{3d}) is such that the hexagonal unit cell c axis and main axis z of the EFG tensor coincide. The components of EFG at the Fe nucleus are connected by the relationship $eq_{xx} = eq_{yy} = -\frac{1}{2}eq_{zz}$, that is, the EFG is completely characterized by a single component eq_{zz}. In the SNP crystal (space group *Pnnm*) the iron ion occupies a special position, whose point symmetry can be approximately described as C_{4v}. Therefore it can be considered that $eq_{xx} \cong eq_{yy}$, and here, as in haematite, the EFG is characterized by a single component in the coordinate system of the principal axes: as Grant *et al* (1969) have found the asymmetry parameter for EFG in SNP $\eta \simeq 0.01$. Then it is justifiable to state

$$\alpha = (eq_{zz})_{Fe_2O_3}/(eq_{zz})_{SNP} \cong \Delta_{Fe_2O_3}/\Delta_{SNP}. \tag{7.33}$$

Here $\Delta_{Fe_2O_3}$ and Δ_{SNP} are the values experimentally measured by MS of quadrupole splitting of the spectral lines; for the ^{57}Fe nucleus in both compounds, the value of α is easily obtained from MS data, which are readily available over a wide range of temperatures. Therefore equation (7.33) makes it possible to directly evaluate the accuracy of EFG calculations on the basis of x-ray diffraction data. The crystals considered have quite different structures; therefore, it is hardly to be expected that the values obtained for eq_{zz} will differ from the true value by exactly the same factor, which is connected with the loss of part of the EFG due to insufficient resolution. The disagreement in values of α obtained from x-ray and spectroscopy data can be, consequently, a measure of the accuracy of the x-ray diffraction determination of EFG.

The $\delta(eq_{\alpha\beta})$ component of EFG at the Fe atom nucleus with coordinates (0; 0; 0.355 20) in α-Fe_2O_3 was calculated by Tsirelson *et al* (1987b) according to formula (7.31). In order to exclude the influence of the filtering procedure on the result, calculations were carried out using different values for β. It was found that, when filter parameter β (see (3.137)) lay in the interval $1 \leqslant \beta \leqslant 2$, the value of $\delta(eq_{zz})$ remained unchanged within the limits of one estimated standard deviation. At the same time, filtration made the error one-half the size. The result obtained with $\beta = 1.7$ was recognized as being the most reliable, as the condition $\delta(eq_{zz}) = \delta(eq_{yy}) = -\frac{1}{2}\delta(eq_{zz})$ was nearly exactly fulfilled. In this case $\delta(eq_{zz}) = (0.33 \pm 0.05) \times 10^{22}$ V m^{-2}. The contribution to the EFG of the procrystal constructed from spherical atoms, eq^0_{zz}, was an order smaller than the standard deviation in (eq_{zz}).

At 295 K the value of the component $\delta(eq_{zz})$ of the Fe nucleus in haematite was found to be twice that at 153 K; the sign of EFG at both temperatures was the same. Such a distortion in EFG is primarily due to the third-order anharmonicity in Fe atom thermal vibrations. One of the components of the tensor which describes the thermal vibrations is oriented along the c axis (that is, along the z axis of the local coordinate system): it induces a significant additional EFG at the iron nucleus at room temperature. This influence of anharmonicity is manifested distinctly upon analysis of the x-ray haematite harmonic atomic thermal vibration parameters at both temperatures. When the temperature is lowered from 295 to 153 K, the component of the thermal vibration tensor directed along the c axis of the unit cell is more markedly reduced than the component oriented in the basal plane. For the Fe atom, these changes are, respectively, 3.9-fold and threefold, and for the O atom, 2.7- and 2.4-fold. A similar situation occurs for MS data. In this way, the relative orientations of the EFG tensor and the tensors describing thermal vibration become factors influencing the result of EFG calculation.

In the SNP crystal the largest component of EFG, eq_{zz}, corresponds to the Fe–NO direction in the ab plane, while one of the remaining EFG components lies along the crystallographic c axis. This significantly simplifies the calculation. A calculation using expression (7.31) gave a value of $\delta(eq_{zz})=(1.17\pm0.14)\times10^{22}\ \mathrm{V\,m^{-2}}$ on the ^{57}Fe nucleus. We note that the large value of $\delta(eq_{zz})$ obtained agrees excellently with the character of the deformation ED around the Fe nucleus in SNP (figure 7.12(c)). The magnitude of the eq^{0}_{zz} term turned out to be negligibly small. Thus, in both compounds, EFG is almost completely determined by the deformation ED. Thong and Schwarzenbach (1979) observed the same situation in the study of the low-quartz structure of $AlPO_4$. In Li_3N, however, the contribution of the procrystal to the EFG at the Li nuclei is significant (Lewis and Schwarzenbach 1981b).

The α value (7.33) derived from low-temperature accurate x-ray diffraction experiments for α-Fe_2O_3 and SNP proved to be equal to 3.55 ± 0.11. The specially undertaken measurement of the quadrupole splitting Δ (Tsirelson *et al* 1987b) in a very pure sample of α-Fe_2O_3 at 295 K gave the result $\Delta_{Fe_2O_3}=3.69(15)\times10^{-27}$ J. Hauser *et al* (1977) at 100 K found $\Delta_{SNP}=13.187(3)\times10^{-27}$ J. This leads to the value of $\alpha_{MS}=3.58\pm0.14$. Thus, the EFG values provided by different methods coincide with each other in the limit of one e.s.d. with a real accuracy of 12–15%. This agrees well with Stewart's (1977c) predictions.

The value of Δ is equal to $\frac{1}{2}e^2q_{zz}Q$ in the crystals considered. Therefore, it is possible to determine the unknown value and sign of the quadrupole moment of the ^{57}Fe nucleus from the values of x-ray EFG and experimental MS values of Δ. The combination of values mentioned above leads to $Q_{^{57}Fe}=(0.14\pm0.02)\times10^{-28}\ \mathrm{m^2}$. This is in remarkable agreement with $Q_{^{57}Fe}=0.156\times10^{-28}\ \mathrm{m^2}$ from the theoretical non-empirical quantum mechanical calculation (Ellis *et al* 1983) and with $Q_{^{57}Fe}=(0.16\pm0.02)\times10^{-28}$ m^2 from the direct calculation of nuclear structure (Bolotin *et al* 1978).

There is a wide spread in the quadrupole splitting values in haematite in the literature, perhaps due to the non-controlled stoichiometry of samples measured. The range of this spread is $(3.23$–$3.69)\times10^{-27}$ J. If one takes the mean value, 3.46×10^{-27} J, and the value $eq_{zz}=0.33\times10^{-28}$ V m^{-2} one can obtain the result $Q_{^{57}Fe}=0.13\times10^{-28}$ m^2. Thus, the value $Q_{^{57}Fe}=0.14\times10^{-28}$ m^2 seems to be reasonably independent of the quality of sample measured.

An interesting case exists when the EFG value at the nuclear position is nearly entirely determined by its atomic surroundings. This is the case for the EFG on the hydrogen nucleus. The electron shell of the H atom is significantly displaced toward the neighbouring atom along a chemical bond line. Because this atom has no core electrons the x-ray EFG value in this case is not limited by the experimental resolution. Tegenfeldt and Hermansson (1985) used this fact and calculated, by the method of Schwarzenbach and Thong, the EFG at the hydrogen nucleus in crystals of $LiOH\cdot2H_2O$ (at 295 K), $LiNO_3\cdot3H_2O$ (120 K), and $NaHC_2O_4\cdot H_2O$ (120 K). The x-ray diffraction data were collected at $0.94\leqslant(\sin\theta/\lambda)_{max}\leqslant1.05$ Å; however, the authors used the extrapolation properties of the multipole model, with the help of which the static structure amplitudes $F(\boldsymbol{H})$ in (7.31) were calculated. It was found that the contributions $(eq)^0$ to the EFG are positive in all cases: as a rule they exceed the negative contributions to the EFG arising from the deformation ED fourfold to eightfold. The value of the quadrupole coupling constant obtained versus H bond (with oxygen as acceptor) distance is shown in figure 7.13. A correlation curve obtained by Berglund *et al* (1978) and summarizing a large number of quadrupole coupling constants determined by NMR is also plotted in figure 7.13. It can be noted that the x-ray values of e^2qQ lie, as a rule, higher than the correlation curve obtained from spectroscopic data. This is connected with the manifestation of the thermal motion effect in spectroscopy. Tegenfeldt and Hermansson found that taking into account the effect of thermal motion displaces the x-ray values of e^2qQ in the direction of their spectroscopic analogues. Another reason for the disagreement is the uncertainty in determination of the H atom positions: an error in the O–H bond length of 0.01 Å leads to a shift in the value of the quadrupole coupling constant of 45 kHz. Taking this into account, the agreement between the results of different methods can be considered completely satisfactory.

An important practical problem of the approach under consideration is the influence of extinction, if it is large, on the values of the EFG components obtained. Thus, in crystals of $AlPO_4$, Li_3N, and α-Al_2O_3, where extinction strongly influences the values of the structure amplitudes, agreement of EFG components calculated from x-ray and NQR data is unachievable. Moreover, the signs of these components turn out to depend on the algorithm used to calculate extinction correction. Besides, the results of the calculation depend

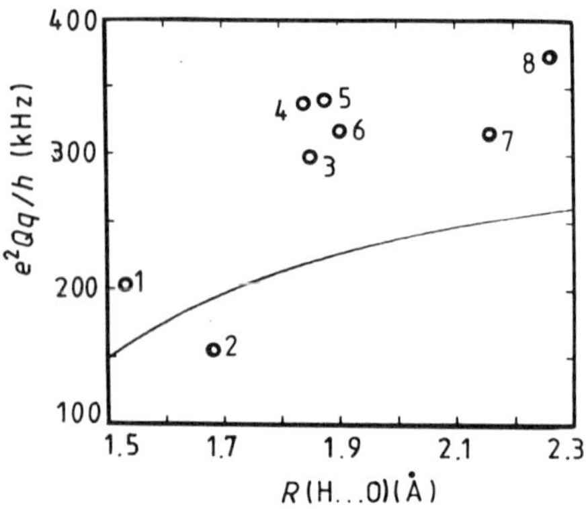

Figure 7.13 Quadrupole coupling constants for deuterium calculated from x-ray diffraction data by Tegenfeldt and Hermansson (1985) versus H···O H bond length: 1, H(oxalate) in $NaHC_2O_4 \cdot H_2O$; 2, $H(H_2O)$ in $LiOH \cdot H_2O$; 3, $H2(H_2O)$ in $LiNO_3 \cdot 3H_2O$; 4, $H2(H_2O)$ in $NaHC_2O_4 \cdot H_2O$; 5, $H1(H_2O)$ in $NaHC_2O_4 \cdot H_2O$; 6, $H1(H_2O)$ in $LiNO_3 \cdot 3H_2O$; 7, $H3(H_2O)$ in $LiNO_3 \cdot 3H_2O$; 8, $H(OH^-)$ in $LiOH \cdot H_2O$. The solid line is the correlation curve from NMR measurements (experimental points deviate by less than 10 kHz from the line).

strongly on the completeness of the Fourier series. Restori and Schwarzenbach (1986) found that in cuprite, Cu_2O, where three-eighths of the intensities are too weak to be measured with an x-ray tube, the magnitude and sign of EFG at the Cu nucleus were undetermined. Later, however, Kirfel and Eichhorn (1990), using synchrotron radiation, measured all of these weak reflections. The result was impressive: the value of the eq_{zz} component of EFG at the Cu nucleus is $1.35(6) \times 10^{22}$ V m^{-2} when thermal motion is considered in the harmonic approximation and $1.40(6) \times 10^{22}$ V m^{-2} in the anharmonic approximation. This agrees excellently with the value of 1.34×10^{22} V m^{-2} which follows from the results of NQR measurements of Krueger and Meyer-Berkhout (1952). It should be noted that the e.s.d. cited does not reflect the real accuracy of these results: differences in the methods of synchrotron radiation scattering measurement lead to a spread in the value of EFG of about 22%. Unfortunately in α-Al_2O_3 it proved impossible to achieve the same outcome with synchrotron radiation as in Cu_2O.

Another approach to the calculation of EFG consists of the use of the multipole model (Stewart 1972, Cromer *et al* 1976, Su and Coppens 1992). In this case the components of EFG at nucleus μ are defined by the relationship

$$(eq_{\alpha\beta})_\mu = \frac{1}{4\pi\varepsilon_0}\left\{ \sum_{\nu \neq \mu} Z_\nu R_{\nu\mu}^{-5}[3(R_{\nu\mu})_\alpha (R_{\nu\mu})_\beta - \delta_{\alpha\beta} R_{\nu\mu}^2] - \sum_\nu \int \rho_\nu(\mathbf{r}_\nu) r_\mu^{-5}[3(r_\mu)_\alpha (r_\mu)_\beta - \delta_{\alpha\beta} r_\mu^2]\, d\mathbf{r}_\nu \right\} \tag{7.34}$$

The first term in this relationship is the contribution to the EFG from all the other nuclei, the second term is the contribution from the electronic parts of all the pseudoatoms (including the atom under consideration) approximated using the multipole model, Z_ν is the nuclear charge, $R_{\nu\mu}$ is the length of the internuclear vector $\boldsymbol{R}_{\nu\mu}$ starting at point μ, the projection of which onto Cartesian coordinate axes is written as $(R_{\nu\mu})_\alpha$, $\boldsymbol{r}_\mu$ is the distance from nucleus μ to the point with ED $\rho_\nu(r_\nu)$. The local contribution to EFG from the ED of 'its own' pseudoatom is calculated using simple analytical formulae. For example, the z component of EFG in the Hansen and Coppens (1978) multipole model has the form

$$(eq)^{val}_{zz} = -(e/4\pi\varepsilon_0)[6 \times 3^{1/2}\alpha^3/5n_2(n_2+1)(n_2+2)]C_{20} \tag{7.35}$$

where n_2 is the power index in the quadrupole component of the pseudoatom ED radial part, the values α and C_{20} are model parameters determined by LS (it is assumed that α in (7.35) is measured in units of m^{-1}). Formulae for calculation of the components of EFG arising from the contributions of neighbouring pseudoatoms (including the nuclei) are obtained using the results of Pitzer *et al* (1962). Epstein and Swanton (1982) obtained these formulae for the Stewart multipole model (3.88), while Su and Coppens (1992) and Tsirelson and Belyaev (1992) obtained them for the Hansen–Coppens model (3.91).

It is clear that EFG found via multipole model parameters is model dependent. In particular, when the model is optimized there are always negative correlation dependences between pseudoatomic quadrupole electronic and thermal parameters, which influence the value of the EFG. Stewart (1976) showed that the values of the corresponding correlation coefficients are reduced not only when the length of the structure amplitude set is increased, but also when the atomic displacements are decreased. This is an additional argument in favour of using low-temperature x-ray diffraction data.

The example of L-alanine, $C_2H_7NO_2$ (Destro *et al* 1989), allows us to discuss the possibilities and difficulties in using the multipole model for EFG calculation. The x-ray diffraction experiment for this compound was carried out at 23 K ($\sin\theta/\lambda \leqslant 1.08$ Å^{-1}). The Stewart multipole model was used to reconstruct the static ED. Because the experimental temperature was very low it was reasonable to suppose that temperature influence on EFG value would be minimal. Moreover, the maximal EFG tensor components eq_{zz} of N, O1, O2, and C3 atoms in L-alanine are oriented approximately normal to the maximum axes of harmonic atomic thermal vibration ellipsoids. The EFG tensor orientation for one of the atoms, the N atom, is presented in figure 7.14. The highest correlation between thermal and quadrupole multipole parameters was found for the oxygen pseudoatoms; however, even in this case the values of correlation coefficients were moderate (about -0.7). Thus, it may be concluded that satisfactory quasistatic electronic parameters of pseudoatoms in L-alanine were obtained.

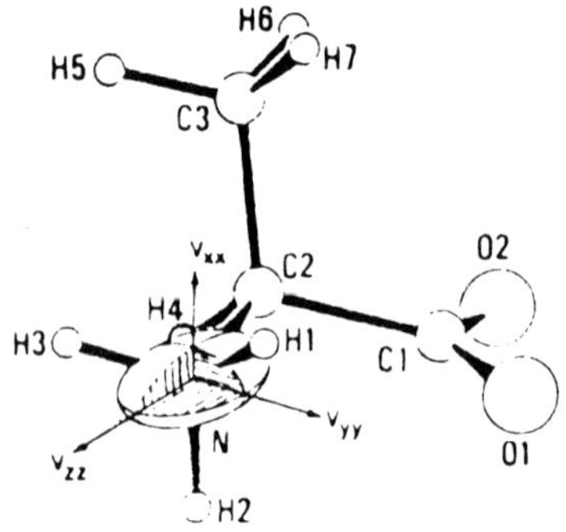

Figure 7.14 A schematic representation of the EFG tensor at the N atom in an L-alanine single crystal. The lengths of the axes of the ellipsoids are proportional to the corresponding principal values $eq_{\alpha\alpha}$ shifted to make all three values positive (Destro *et al* 1989).

The variation of the number of terms in the multipole model of L-alanine influenced the value of EFG tensor components but not their signs or asymmetry parameters η. All of the $eq_{\alpha\beta}$ values led to quadrupole interaction constants satisfactorily agreeing with those from NQR data. Table 7.8 presents the results of calculation of the EFG at the nucleus position of the N atom. It should be noted that small quadrupole electronic occupancy parameters of the N atom gave the main contribution to the EFG. Accounting for the neighbour hydrogen atom contributions led to overestimation of EFG at the nuclear position of the nitrogen atom. The reason is perhaps the simplest approximation accepted for the hydrogen atoms: single STOs with an exponential parameter $\alpha = 2.48$ were used. Therefore, the contribution of H atoms to the EFG at the N nucleus position was restricted by only a penetration term, which is obviously not the case.

The principal components of EFG at nuclei of non-hydrogen atoms in L-alanine calculated according to model I from table 7.8 are presented in table 7.9 for two cases. One of them concerns a crystal, the second a single molecule which is 'removed' from the crystal. Comparing the two corresponding sets of results for each pseudoatom, one can see the following. Signs of all components in the 'molecule' and in the crystal are the same (excluding the C2 pseudoatom where the zero eq_{xx} value makes the eq_{yy} and eq_{zz} values easily interchangeable). The values of δ angles which characterize the mutual disorientation of tensors show that principal axes of EFG in the two cases are nearly parallel; the exceptions are pseudoatoms C2 and C3. The discrepancy in EFG values does not exceed two e.s.d. Thus the differences in the EFG values and in orientation of their tensors for the majority of the non-hydrogen pseudoatoms of L-alanine are not negligible but, at the same time, not too large.

For oxygen atoms the quadrupole coupling constants appear to be equal

Table 7.8 The quadrupole coupling constant of the N atom in L-alanine as calculated from x-ray diffraction data at 23 K with variation of the multipole model. $Q_{^{14}N} = 0.0174(2) \times 10^{-28}$ m^2 according to Ensllin *et al* (1974).

Model		e^2qQ (J $\times 10^{27}$)	η
I.	Multipoles up to octupole level on non-hydrogen pseudoatoms; H atoms are described by spherical functions	1.59(13)	0.3(1)
II.	As I but multipoles with smallest populations are omitted (the octupoles on O1 and O2, the quadrupoles on N, the dipoles and quadrupoles on C3 pseudoatoms)	0.86(13)	0.3(1)
III.	As II plus the five quadrupole parameters for the N pseudoatom	1.66(13)	0.3(1)
IV.	As III plus the dipole and quadrupole parameters of the C3 pseudoatom	1.59(13)	0.3(1)
V.	As I but with the quadrupoles on the N pseudoatom removed	0.86(13)	0.3(1)
VI.	Contribution from N pseudoatom only	0.73(13)	0.7(1)
NQR measurements (77 K)[a]		0.798	0.26

[a] Hunt's (1974) measurement for a polycrystalline sample.

$-4.51(33) \times 10^{-27}$ J (O1 atom) and $-5.37(33) \times 10^{-27}$ J (O2 atom) with the quadrupole moment $Q_{^{17}O} = -0.0269 \times 10^{-28}$ m^2. There are no direct data on the same values from spectroscopic measurements for L-alanine; however the values of Destro *et al* are in good agreement (in the limit of one e.s.d.) with that for carboxylic acid and some acid salts, including negatively charged O=C–O groups.

One should emphasize that the extensions of basis sets in the non-empirical MO SCF calculations of the quadrupole coupling constants and asymmetry parameters in L-alanine does not increase the precision of results and does not change their signs. What is surprising is that the best computational result (4% deviation from spectroscopic data) has been achieved for the N atom with the minimal STO-3G basis. Thus, we may conclude that the quantum mechanical EFG calculations are, at present, not especially reliable

Table 7.9 Values of EFGs, $eq_{\alpha\alpha}$, on the N nucleus in L-alanine calculated from parameters of the multipole model at the octupole level and from SCF MO calculations (Destro *et al* 1989). $eq_{\alpha\alpha}$ is in units of 10^{14} esu cm^{-3}. Calculations assume quadrupole moments $Q_N = 0.0174 \times 10^{-28}$ and $Q_0 = -0.0265 \times 10^{-28}$ m^2.

		Multipole model					SCF MO calculations			
							STO-3G		6-31G**	
Atom	α	$eq_{\alpha\alpha}$ (crystal)	η	$eq_{\alpha\alpha}$ ('molecule')	η	δ (°)	$eq_{\alpha\alpha}$	η	$eq_{\alpha\alpha}$	η
O1	x	−3(2)	0.85(12)	−5(2)	0.68(12)	7	1.121	0.96	−10.777	0.52
	y	−33(3)		−28(3)		3	58.111		−33.797	
	z	36(3)		33.3(3)		7	−59.232		44.574	
O2	x	−12(2)	0.42(9)	−10(2)	0.52(9)	9	−0.654	0.98	−11.052	0.55
	y	−30(3)		−32(3)		9	−61.083		−38.161	
	z	42(3)		42(3)		0	61.737		49.213	
N	x	−7(2)	0.29(14)	−9(2)	0.16(12)	14	−3.553	0.23	−0.606	0.78
	y	−12(2)		−13(2)		12	−5.635		−4.937	
	z	19(2)		22(2)		5	9.188		5.543	
C1	x	2(1)	0.69(16)	1(1)	0.83(15)	5	1.632	0.57	3.852	0.52
	y	11(1)		13(1)		1	5.877		12.344	
	z	−13(1)		−14(1)		4	−7.509		−16.196	
C2	x	0(1)	1.0(9)	0(1)	0.98(9)	2	2.430	0.74	5.239	0.40
	y	−23(1)		23(1)		89	16.185		12.138	
	z	23(1)		−24(1)		91	−18.615		−17.377	
C3	x	−8(1)	0.18(9)	−9(1)	0.18(9)	32	−2.222	0.82	0.736	0.50
	y	−12(1)		−12(1)		31	−2.242		2.231	
	z	20(1)		21(1)		5	2.464		−2.967	

and the use of the x-ray diffraction method is of importance in interpretation of spectroscopic measurements.

All of the results discussed were achieved without taking the Sternheimer factor into account. As can be seen from the discussion above, for molecules consisting of the first-row elements, the uncertainties in results arising because of partial loss of information owing to unsatisfactory resolution are not higher than statistical error in EFG determination (though the latter always gives a reduced error estimation). The data of Bentley and Stewart (1974) and Chandler and Spackman (1982) show that for the N atom in the NH molecule the corresponding screening Sternheimer factor is $R=\gamma(\text{core})\cong 0.1$. Thus, the real information lost because of the lack of resolution is about 10%. In the case of hexamethylenetetramine, $C_6H_{12}N_6$, Parini *et al* (1985) also came to the conclusion that ignoring the Sternheimer factor in multipole calculation of EFG in organic molecular crystals is quite acceptable. Investigating the stability of the results with the number of terms in multipole expansion regarding the EFG values at N nuclear positions (see table 3.15) they also found a nearly constant gradient value on increasing the expansion length from quadrupole to octupole. In all cases, the maximal discrepancy between the calculated values of quadrupole coupling constant of the ^{14}N nucleus and the spectroscopic results in $C_6H_{12}N_6$ was about 10%.

Epstein and Swanton (1982) for imidazole and Destro *et al* (1989) for L-alanine compared the two approaches to EFG calculation mentioned above: those based on difference Fourier series and based on the multipole model. Unfortunately, we conclude that up to now it has been impossible to obtain coinciding results for EFG values from these two approaches for both substances mentioned. The reason is not yet known. Further attempts should be made in order to provide quantitative reproducibility in EFG calculation for any system. However even now the results of x-ray diffraction experiments are useful in spectroscopic data analysis both from a quantitative point of view and especially in determination of EFG constant signs; the latter appear always to be correctly determined from x-ray diffraction.

Finishing this section, we touch on the interpretation of Mössbauer quadrupole splitting spectra for iron containing compounds with the multipole model. The core electrons in the Fe atom make a notable contribution to the EFG in contrast to light atoms. Because the core electronic subshell in the multipole model is presented in a spherical approximation its contribution to the EFG can be given in the form $(eq_{zz})^{val}\Rightarrow(1-R)(eQ_{zz})^{val}$ where R is the screening Sternheimer factor. As far as the antiscreening correction, γ_∞, is concerned, the successive introduction of different contributions to EFG from all multipoles of neighbouring pseudoatoms, as described by Epstein and Swanton (1982) and Su and Coppens (1992), allows us to neglect it. However, if the influence of the neighbourhood is considered in a point charge approximation, the introduction of screening correction γ_∞ becomes necessary.

The first iron containing compound investigated from the EFG point of view was pyrite, FeS_2 (Stevens *et al* 1980). The iron ion, Fe^{2+}, in this compound is located in the centre of a trigonally distorted octahedron made of S atoms (the symmetry site is $\bar{3}$). The disposition of ligands is such that their contribution to the EFG at the iron nucleus is negligible. The x-ray diffraction experiment was carried out at room temperature; nevertheless the resolution achieved was high ($\sin\theta/\lambda \leqslant 1.46$ Å^{-1}). The value of the single independent component of EFG at the iron nucleus, caused by the local quadrupole term, was equal to $-(0.51 \pm 0.30) \times 10^{22}$ V m^{-2}. In the original work, Stevens *et al* (1980) used the obsolete value $R = 0.22$ and $Q_{^{57}\mathrm{Fe}} = 0.2 \times 10^{-28}$ m^2 in order to calculate the quadrupole coupling constant. Nevertheless they arrived at a result which agrees with MS data satisfactorily. If a more accurate value $R = 0.12$ is used (Sternheimer 1972, Sen and Narasimhan 1977) and a value $Q_{^{57}\mathrm{Fe}} = 0.14 \times 10^{-28}$ m^2, which is more realistic, then the data of Stevens *et al* lead to the value $(1 - R)e^2qQ = -(10.1 \pm 5.9) \times 10^{-27}$ J. This is in excellent agreement with the experimental value of the quadrupole coupling constant, $|(9.74 \pm 0.10)| \times 10^{-27}$ J, reported by Finklea *et al* (1976).

Unfortunately, further investigations of iron porphyrines and phtalocyanines carried out at temperatures of 100–120 K, displayed a discrepancy in EFG data. In some compounds, such as (*meso*-tetraphenylporphinato)iron (III) methoxide, $C_{45}H_{31}N_4OFe$ (Lecomte *et al* 1983), and iron (II) phthalocyanine, $C_{32}H_{16}N_8Fe$ (Coppens and Li 1984), agreement to within 10% was achieved between the calculated quadrupole coupling constants and the MS data. In other compounds, (*meso*-tetraphenylporphinato)iron (II), $C_{44}H_{28}N_4Fe$ (Tanaka *et al* 1986, Li *et al* 1990), iron (II) tetraphenylporphyrine bis(tetrahydrofuran), $C_{52}H_{44}N_4O_2Fe$ (Lecomte *et al* 1986), and bis(pyridine)-(*meso*-tetraphenylporphinato)iron (II), $C_{54}H_{38}N_6Fe$ (Li *et al* 1988), the values of the quadrupole coupling constant calculated from x-ray data were a long way from experimental values (although their signs were usually correct). At first sight, the lack of success of these calculations might lead to a pessimistic conclusion concerning the prospects of using the multipole model for interpretation of quadrupole splitting of Mössbauer spectra, a situation which is analogous to that in non-empirical calculation of the EFG (Delley 1991). However, the analysis of Tsirelson and Belyaev (1992) showed that such a conclusion would be premature. Firstly, in the investigations cited, in order to simplify calculations for the iron pseudoatom, an approximate higher symmetry was used, D_{4h} (or C_{4v}), although in real crystals the distortion from this symmetry was significant and was noticeable in deformation ED maps. Secondly, contributions from the neighbouring atoms to EFG at the iron nucleus were treated in different ways for different compounds. In some compounds, only the influence of point atoms was considered, and in others multipoles of higher orders were taken into account. In all cases, the Sternheimer antiscreening correction factor γ_∞ was introduced, although, as has already been noted, its use is only justified when the environment is described as 'external' point ions. Both circumstances noticeably influence the value of EFG at the nucleus. Therefore properly and uniformly performed calculations of the quadrupole coupling constants in the above-mentioned compounds, taking into account the real ED distribution, would allow a more objective judgement of the possibilities of the multipole model for EFG calculation at nuclei in low-symmetry positions.

7.2 Diamagnetic susceptibility

Diamagnetic susceptibility, χ_d, which for isotropic substances is a coefficient of the proportionality between the magnetization of the sample and the external magnetic field applied, is a characteristic which is directly dependent on the ground state ED (Flygare 1978). Therefore it is natural to try to determine this characteristic using x-ray diffraction data. Sirota (1962) and Sirota and Sheleg (1963), with this goal in mind, used, as an approximation, spherical Gaussian functions adjusted to experimental ED. They obtained, for diamond and a series of binary compounds, results close to those obtained by direct magnetic measurements. Weiss (1966) pointed out the possibility of evaluating the diamagnetic susceptibility of inert gases from small-angle x-ray scattering data. Ivanov-Smolenskii *et al* (1983) suggested a more general solution to the problem, which we will follow here.

For a system possessing spherical or axial symmetry and subject to Larmor's theorem, the diamagnetic susceptibility can be calculated according to the classical Langevin formula (Ashcroft and Mermin 1976)

$$\chi_d^z = -\frac{\mu_0 e^2}{4m}\left\langle \Psi_0 \middle| \sum_i^N x_i^2 + y_i^2 \middle| \Psi_0 \right\rangle. \tag{7.36}$$

Here e and m are the electronic charge and mass, μ_0 is the magnetic constant, and x_i and y_i are the projections of the electron radius vectors onto the corresponding coordinate axes. In (7.36) the summation takes place over all electrons and it is assumed that the magnetic field is directed along the z axis; other components of the diamagnetic susceptibility are obtained by cyclical replacement of the x, y, z indices. From this point we will not write the z index in χ_d^z. Transferring to the ED, as has been done in (1.10), expression (7.36) can be rewritten using the electron density of the μth pseudoatom, ρ_μ:

$$\chi_d^z = -\frac{\mu_0 e^2}{4m}\int (x_\mu^2 + y_\mu^2)\rho_\mu(\mathbf{r})\,\mathrm{d}V. \tag{7.37}$$

We place a local coordinate system in the centre of the μth pseudoatom and represent the crystal as a combination of non-overlapping pseudoatoms. Let the ED be presented as a Fourier series. Then the diamagnetic susceptibility of a single pseudoatom of type μ is written in the following way:

$$\chi_d^z = -\frac{\mu_0 e^2}{4mV} \sum_{hkl} F(hkl) J_\mu(hkl) \tag{7.38}$$

where

$$J_\mu(hkl) = \int_{\Omega_\mu} (x_\mu^2 + y_\mu^2) \exp[-2\pi i(hx + ky + lz)] \, dV. \tag{7.39}$$

Here Ω_μ is the volume belonging to the μth pseudoatom. In this way, through multiplication of (7.38) by Avogadro's number, N_A, the molar diamagnetic susceptibility for ionic crystals can be calculated, within the superpositional model, which includes ions with spherically symmetric deformed electron clouds (spherical pseudoatoms). The problem is reduced to the calculation of the integral J_μ (7.39) and summing of the contributions to χ_d over pseudoatoms. The accuracy of the calculation will be dependent on the adequacy of the choice of volume of the μth pseudoatom, on the character of the ion deformation (in other words, on the features of the chemical bonds in a crystal), and on the precision of determination of the structure amplitudes $F(hkl)$. The problem of choice of the Ω_μ volume was discussed in detail in chapter 6.

Ivanov-Smolenskii *et al* (1983) used the approach described to calculate the diamagnetic susceptibility of crystals with structural type NaCl. The EDs of ions in these crystals are sufficiently well separated in space, while the basic part of their electron cloud deformation is spherically symmetric. Comparison of electron charges of the pseudoatoms in the LiF crystal, found by integration over a variety of possible choices for the atomic volumes, showed (table 7.10) that these values weakly depend on the concrete method of ED partitioning onto pseudoatoms. Therefore, in calculations of the diamagnetic susceptibility using x-ray data for crystals with NaCl-type structure, the simplest scheme was chosen. The volume of each pseudoatom was represented as a Wigner–Seitz cell, having the form of a cube with side length $a/2$ (a is the length of the cubic unit cell). The ions are situated at the centres of the cells. In this case the integration in (7.39) can be carried out

Table 7.10 Electronic charges of ions in an LiF crystal obtained by integration of electron density over different ionic volumes.

Ion	Wigner–Seitz cell	Modified Wigner–Seitz cell	Sphere with radius chosen at ED minimum
Li^+	2.29(5) *e*	2.28(5) *e*	2.05(5) *e*
F^-	9.71(5) *e*	9.75(5) *e*	9.70(9) *e*

analytically and the resulting average molar diamagnetic susceptibility of the crystals, taking into account their symmetry, is determined by the relationship

$$\chi_d^z = -\frac{\mu_0 e^2 N_A a^2}{4mV}\left[(n/96) + \sum_{h=2,4,\ldots} (-1)^{h/2} F(h00)/2\pi^2 h^2\right] \tag{7.40}$$

which contains the structure amplitudes with corresponding signs (n is the number of electrons in the unit cell). In this equation a fast-converging Fourier series is present, which includes only reflections with positive even values of h, as the remaining terms in the series mutually cancel due to opposite signs of the cation and anion contributions. The error in χ_d, because of errors in the experimental structure amplitudes, can be evaluated by the approximate formula

$$\sigma(\chi_d) = \frac{\mu_0 e^2 N_A a^2}{8\pi^2 m}\left[\sum_{h=2,4,\ldots} \frac{\sigma^2\{|F(h00)|\}}{h^4}\right]^{1/2}. \tag{7.41}$$

The influence of Fourier series termination in (7.40) on the result is illustrated in figure 7.15. The magnitude of the error associated with this effect is several times smaller than the root mean square error, $\sigma(\chi_d)$, calculated according to (7.41). The uncertainty resulting from arbitrariness in the choice of volumes Ω_m is, of course, not taken into account in (7.41).

The results of calculations of the molar diamagnetic susceptibility for NaCl-type crystals are presented in table 7.11. These values satisfactorily agree with magnetic measurements, although their errors, without question, are underestimated: a more realistic evaluation of the error in χ_d is 5–15%.

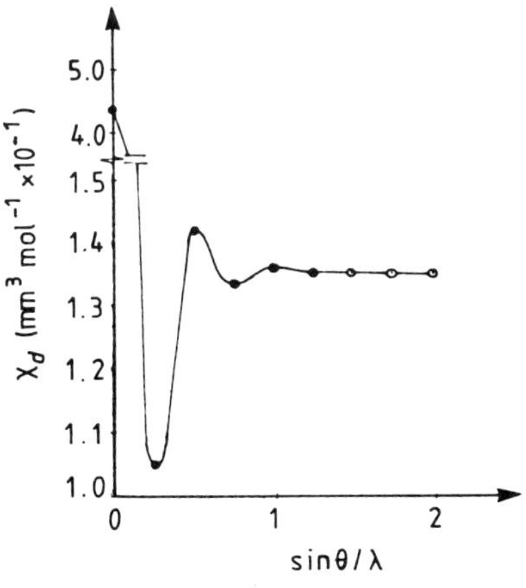

Figure 7.15 The convergence of diamagnetic susceptibility values calculated with x-ray structure amplitudes for the LiF crystal (Ivanov-Smolenskii *et al* 1983). Open circles correspond to theoretical values of structure amplitudes.

Table 7.11 Molar diamagnetic susceptibilities χ_d ($mm^3\ mol^{-1}$) and molar paramagnetic (Van Vleck) susceptibilities χ_p ($mm^3\ mol^{-1}$) calculated from x-ray structure amplitudes (Ivanov-Smolenskii *et al* 1983).

	$-\chi_d$		χ_p	
Crystal	X-ray	Magnetic measurements[a]	X-ray	Experimental
LiF	0.14(1)	0.131	0.009(1)	0.005
NaF	0.24(1)	{0.193 0.244[b]	0.042(2)	0
NaCl	0.45(2)	0.382	0.06(2)	0.001
KBr	0.67(2)	0.682	0.05(2)	
MgO	0.21(1)	0.231	0.08(1)	0.103
CaO	0.37(1)	0.356	0.18(1)	0.168
SrO	0.45(2)	0.496	0.01(2)	0.056
BaO	0.76(5)	0.716	0.39(5)	0.34

[a] Dorfman 1961.
[b] Kikoin 1976.

At the same time, the value of χ_d calculated from x-ray data is free from 'hard to control' errors, which arise when the diamagnetic part of the experimentally measured total magnetic susceptibility is determined. Moreover, these values can be used in some cases for evaluation of the paramagnetic (Van Vleck) susceptibility with the help of the relationship $\chi_p = \chi - \chi_d$. Values of χ_p calculated in this way and evaluated by other methods are presented in table 7.11 too. We note that the value of χ_p gradually increases along the row MgO, CaO, BaO; this reflects an increase of the mutual deformation ED of the ions in these crystals. The magnitude of χ_p for SrO falls away from this tendency, as noted by Dorfman (1961), because of insufficient accuracy in magnetic measurements for this crystal.

The results of these calculations using x-ray data are helpful for choosing the most reliable values of diamagnetic susceptibility from the different values presented in the literature. For example, the magnitude of $\chi_d = -0.244\ mm^3\ mol^{-1}$, found for the NaF crystal by Kikoin (1976), lies closer to the x-ray value and may be considered more exact than the value, $-0.193\ mm^3\ mol^{-1}$, of Dorfman (1961).

7.3 Optical characteristics

We considered earlier the crystal characteristics which depend on the ground state electron density only. In order to evaluate optical characteristics, the knowledge of the ground state is not enough: optical properties deal with electron excitations and the entire set of electron wave-functions or corresponding transition densities needs to be known to calculate the optical characteristics. However, the possibility of approximate calculation of optical (linear and nonlinear) characteristics exists, on the basis of ground state ED only; the approximations used are reasonable and the results achieved are quite satisfactory. The semiempirical character of these approximations needs justification every time they are applied to new classes of crystals.

7.3.1 *Electronic low-frequency (static) polarizability*

The polarizability is the coefficient of proportionality between an atomic dipole moment and local electric field: $p=\alpha\varepsilon_0\mathscr{E}_{loc}$. This is the tensor value, components of which are described as

$$\alpha_{\beta\beta}(\omega)=8\pi e^2\sum_{n\neq 0}\frac{\omega_{n0}|\langle\Psi_0|\beta|\Psi_n\rangle|^2}{\hbar(\omega_{n0}^2-\omega^2)}\qquad \beta=x,y,z. \tag{7.42}$$

The Ψ_i are the wave-functions of ground and excited electronic states of the atom, ω_{n0} are the frequencies connected with energies of electron transitions between these states, and ω is the electric field frequency. When $\omega\ll\omega_{n0}$ (the low-frequency region of the optical range of electromagnetic waves) the expression (7.42) can be rewritten in the isotropic case as

$$\alpha_{\beta\beta}=\alpha_{\beta\beta}(0)\approx 8\pi e^2\sum_{n\neq 0}\frac{|\langle\Psi_0|\beta|\Psi_n\rangle|^2}{\hbar\omega_{n0}}. \tag{7.43}$$

Using then the Thomas–Kuhn sum rule (Loudon 1973)

$$\frac{2m}{\hbar}\sum_n\omega_{n0}|\langle\Psi_0|\beta|\Psi_n\rangle|^2=k \tag{7.44}$$

(k is the number of electrons in an atom), replacing ω_{nm} in (7.43) and (7.44) by its mean value and taking into account the closure of the set of eigenfunctions of electronic Hamiltonian, $\{\Psi_n\}$, one can arrive at the expression

$$\alpha\cong(16\pi me^2/9\hbar^2k)(\langle\Psi_0|r^2|\Psi_0\rangle)^2. \tag{7.45}$$

This relation, including the wave-function of only the ground state, is valid for spherical particles. It has been derived starting with different suppositions by Unsöld (1927), Kirkwood (1932), Buckingham (1937), Goscinski and Delhalle (1989), and others, and is known as the Kirkwood approximation. It is most suitable for dielectrics with a wide energy gap, consisting of near-spherical particles, and provides good agreement with experiment (Dorfman 1961) in spite of the fact that the contribution to the polarizability is given mostly by the valence electrons, whereas the total electron number k is presented in the relation.

Ivanov-Smolenskii *et al* (1983, 1984) used this approximation to evaluate low-frequency electronic polarizabilities of ionic crystals with NaCl-type structure by using the ED reconstructed from the x-ray structure amplitudes. Their result for the mean polarizability of single ion μ with the notations of the previous chapters has form

$$\alpha_\mu = \frac{4\pi m e^2}{\hbar^2 k V^2}\left[\sum_{hkl} F(hkl) J_\mu(hkl)\right]^2. \tag{7.46}$$

The validity of the Kirkwood approximation was analysed first by calculations of the polarizability of separate ions according to (7.46) in crystals with NaCl-type structure. The separation of ED to pseudoatomic fragments was performed using a modified Wigner–Seitz scheme. It was found that $\alpha_F = 3.80 \times 10^{-30}$ m^3 in LiF and 6.76×10^{-30} m^3 in NaF crystals whereas $\alpha_{Li} = 0.44 \times 10^{-30}$ m^3 and $\alpha_{Na} = 1.84 \times 10^{-30}$ m^3. The α_F(LiF) value agrees well with the result of Fouler and Madden (1983): 3.68×10^{-30} m^3. The value of anion polarizability in NaF is near to the mean value for different crystals, $\alpha_F = 6.24 \times 10^{-30}$ m^3, indicated by Wilson and Curtis (1970). Let us note that the α_F value calculated from x-ray diffraction data is quite different from the free-ion F^- value in its ground state (3.04×10^{-30} m^3). As far as the cations are concerned, their x-ray polarizabilities are approximately three times higher than from the theoretical calculation for free ions ($\alpha_{Li} = 0.11 \times 10^{-30}$ m^3, $\alpha_{Na} = 0.56 \times 10^{-30}$ m^3) and two to three times higher than empirical values, $\alpha_{Li} = 0.11 \times 10^{-30}$ m^3 and $\alpha_{Na} = 0.66 \times 10^{-30}$ m^3, from Pearson *et al* (1984). The formula unit polarizabilities proved to be $\alpha_{LiF} = 4.24 \times 10^{-30}$ m^3 and $\alpha_{NaF} = 8.6 \times 10^{-30}$ m^3; these values agree with the averaged experimental values within a 13% limit (table 7.12), being somewhat higher. The same trend is observed for MgO also.

Table 7.12 Electronic low-frequency polarizabilities of formulae units α (10^{-30} m^3) calculated from x-ray structure amplitudes in the Kirkwood approximation (Ivanov-Smolenskii *et al* 1983).

Crystal	X-ray	Optical measurements[a]
LiF	4.1(1)	3.7
NaF	7.5(1)	7.3
NaCl	18.8(1.6)	14.0[b]
KBr	21.7(1.6)	22.0
MgO	5.9(1)	7.2
CaO	12.1(1)	11.8
SrO	15.6(1)	14.0
BaO	23.9(3.5)	21.2

[a] Dorfman 1961.
[b] Kikoin 1976.

The small discrepancy of x-ray and optical measurement results observed can be attributed to the overestimation of cation ED volumes chosen according to the modified Wigner–Seitz scheme. However, this kind of separation can be accepted as satisfactory because this discrepancy is not too large. The same conclusion was arrived at by Coppens and Guru Row (1978) who studied the atomic charges in the framework of the same scheme.

Choosing the pseudoatom volumes in NaCl-type crystals as cubic Wigner–Seitz cells with the size $a/2$ and placing pseudoatoms into the cell centres, the mean electronic low-frequency polarizability of molecular units can be written as

$$\alpha = \frac{4\pi m e^2 a^4}{\hbar^2 m}\left[\frac{n}{96} + \sum_{h=2,4,\ldots} \frac{(-1)^{h/2} F(h00)}{2\pi^2 h^2}\right]^2 \tag{7.47}$$

(a is the cubic unit cell period; m is the number of electrons in formula unit and n is the number of electrons in the unit cell). The standard deviation for α can be calculated according to a formula such as (7.41). As can be seen from table 7.12 the values arrived at according to (7.47) are in good agreement with the optical measurement data. Using the Klausius–Mossotti equation, the x-ray α values can be recalculated as dielectric permeabilities. The values of the latter lie between static, $\varepsilon(0)$, and high-frequency, $\varepsilon(\infty)$, dielectric permeabilities presented by Ashcroft and Mermin (1976). Thus, the x-ray α values allow us to satisfactorily evaluate the dielectric properties of NaCl-type crystals in the optical range of electromagnetic waves.

7.3.2 *Nonlinear optical susceptibilities*

Nonlinear optical effects appear when high-intensity electromagnetic radiation (for example, optical laser radiation) with an electric field strength comparable to the inner crystal field strength ($\sim 10^{10}$–10^{11} V m^{-1}) passes through a substance. In this case, the classical Maxwell linear relation between field and media polarization is broken down. The decomposition of the substance polarization vector, $\boldsymbol{P}$, over the external electric field, $\boldsymbol{E}$, is possible because of the small values of nonlinear effects; the decomposition coefficients, $\chi^{(i)}$, are connected to the substance's optical properties:

$$P_i = \varepsilon_0\left[\sum_j \frac{\partial P_i}{\partial E_j} E_j + \sum_{jk} \frac{\partial^2 P_i}{\partial E_j\, \partial E_k} E_j E_k + \cdots\right]$$

$$= \varepsilon_0\left[\sum_j \chi_{ij} E_j + \sum_{jk} \chi_{ijk} E_j E_k + \cdots\right] \qquad i = x, y, z. \tag{7.48}$$

In particular, the quadratic susceptibility, $\chi^{(2)}_{ijk}$, is responsible for the second-harmonic generation, linear electrooptical effect, and parametric light generation (Zernike and Midwinter 1973). The $\chi^{(2)}_{ijk}$ tensor components, as well as higher-order susceptibilities, can be measured experimentally.

However, in order to investigate the nature of nonlinear effects it is necessary to have a model which will provide a connection between the nonlinear properties observed and the molecule or crystal structure features. In the optical range of electromagnetic waves the nonlinear effects arise mainly because of nonlinear electronic polarization of a medium. Thereforc the model parameters should be dependent on the ED distribution. Let us consider a model which uses parameters calculated from the ED, and permits us to evaluate and to predict the optical properties of complicated heterodesmic compounds.

The quantum mechanical methods of calculation of optical susceptibilities were reviewed by Korolkova *et al* (1985). Their analysis can be summarized as follows. Perturbation theory can be strictly applied for susceptibility calculations; the external electromagnetic wave is the small perturbant. However, for complicated heterodesmic crystals it is very difficult to calculate, from first principles, accurate enough wave-functions of the excited electronic states. Therefore, a number of simplifications are introduced. In particular, when calculating, according to perturbation theory, the corrections to the probe wave-function, the differences in energy states in the denominator are often replaced by a single effective value, the latter being the mean width of the energy gap, E_g. Additionally, the matrix elements of the operators of the electron transitions from the occupied level into the unoccupied ones are calculated in simplified form, using hybrid wave-functions. Sometimes, the wave-function parameters are optimized by a variation procedure. Unfortunately, even such strong simplifications allow us to study only simple binary compounds with σ bonds, and moreover with unsatisfactory accuracy.

Some semiempirical schemes have therefore been suggested. Harrison (1980) presented a crystal as a set of dipoles created by valence electrons weakly bonded with cores and placed in internuclear regions. These electrons polarize in the external field. The value of polarization is determined from the minimum-energy condition for a 'free' dipole's energy, adding to this the energy of their interaction with the external field. The decomposition of polarization over the field and summation over the bonds permit a relation to be obtained which provides satisfactory results for nonlinear susceptibilities of binary compounds. The positive role of introducing ED into a computational scheme can be seen even in this simple approach. A more impressive result has been achieved by connecting the nonlinear effects with an anharmonic oscillator which describes the vibrations of valence electrons in the external field (Bloembergen 1965). Depending on the chemical bond character, the optical susceptibilities have different natures. In ionic crystals where ions are clearly separated in space, the nonlinear effects can be assumed to be mainly due to anharmonic vibrations of valence electrons relative to the cores. In covalent crystals one should consider the vibrations of ED concentrated in the interatomic space, e.g. the vibrations of bond charge electrons. Two methods of calculation have been accordingly suggested: the method of Coulomb anharmonicity (Meisner 1975) and the method of bond charges (Levine 1969, 1973a, b). Both methods have been successfully applied to nonlinear optical characteristic calculations for crystals with single heterodesmic bonds which were inaccessible for such treatment earlier. However, both methods contain some free parameters determined from dielectric data or from theoretical and crystallographic considerations. These circumstances decrease the value of the methods. At the same time, these model parameters can be obtained from x-ray diffraction experiments. In order to realize this approach it is necessary to modify these two methods. This was done by Tsirelson *et al* (1984b) and Korolkova *et al* (1985). They have also extended the bond charge method to heterodesmic crystals with multiple bonds in order to investigate materials which can be used in technology. Let us consider both approaches and find the point where the experimental parameters determined by the x-ray diffraction method can be introduced into the models.

Let us begin with the bond charge model and consider, first, a homodesmic binary dielectric crystal with a wide energy gap, following Phillips (1970) and Van Vechten (1969). We present a crystal as a set of atomic cores merged in an inhomogeneous electron gas. Because the atomic cores are screened by valence electrons, their potential in the internuclear space is weak and the electrons can be considered to be nearly free. This picture corresponds to the pseudopotential method. All valence electrons take part in dielectric screening, therefore the latter is weakly sensitive to the band structure details. Penn (1962) suggested the use, for the dielectric crystal description, of a two-band isotropic model characterized by average gap width E_g. The latter in the nearly-free-electron model is connected with low-frequency dielectric permeability $\varepsilon(0)$ by the relation

$$\varepsilon(0)=1+(\hbar\omega_p/E_g)^2[1-E_g/4E_F+\tfrac{1}{3}(E_g/4E_F)^2]D. \tag{7.49}$$

Here both plasma frequency for valence electron vibrations, $\omega_p=(\langle\rho_{val}\rangle e^2/m)^{1/2}$, and the Fermi energy

$$E_F=(\hbar^2/2m)[3\pi^2\langle\rho_{val}\rangle]^{2/3} \tag{7.50}$$

are determined by the mean valence electron density $\langle\rho_{val}\rangle$. D is the correction factor related to the core d states; its value is close to unity. According to Phillips (1970), the total pseudopotential can be divided in heteropolar crystals into ionic and covalent parts. Accordingly, E_g can also be presented as a sum of heteropolar (ionic), C, and homopolar (covalent), E_h, contributions, $E_g^2=E_h^2+C^2$. It is assumed that all local field effects are included in E_g. The difference in screening of ions in this model is neglected and the same isotropic screening function for both bonded atoms

$$t=\exp(-k_s d/2) \tag{7.51}$$

is used. Here the Thomas–Fermi screening wave number is

$$k_s=[16m(3\pi^2\langle\rho_{val}\rangle)^{1/3}]^{1/2}(e/\hbar) \tag{7.52}$$

and d is the interionic distance. The method of calculation of E_h and C for crystals with heteropolar σ bonds was given by Levine (1973a, b). He has taken into account that E_h and C values depend on the ED distribution in the internuclear space. The ionic contribution, connected to electrons displaced along a bond path, is described by $C = e^2 b(z_\alpha/r_\alpha - z_\beta/r_\beta)t$. This depends on the difference of the screening Columb potentials of atomic cores at the point $r_\alpha = d - r_\beta$ and their charges z_α, and on the factor b, which reflects the average coordination numbers of α and β atoms and is the analogue of the local field factor. The covalent contribution is $E_h \sim d^{-s}$ (Van Vechten 1969). Thus, the Phillips–Van Vechten model is related to the valence ED, geometrical structure and electronic characteristics of atoms forming a crystal.

The generalization of this model to heterodesmic crystals with single bonds has been performed by Levine (1973a, b). He suggested that in compounds with covalent bonds only those electrons which are mainly localized in the interatomic space and are weakly bonded to atomic cores will be displaced in an external field $\boldsymbol{E}$. Therefore, the polarization of the substance can be mostly attributed to the electron bond charges. The contributions to the optical susceptibility from each atomic pair were assumed to be additive. The justification of this suggestion has been checked by experimental optical polarizability data analysis and comparison with calculations. Denoting the bond charge values by q^μ, it is possible to write a phenomenological expression which connects them with the macroscopic crystal susceptibility per bond, χ_b^μ:

$$q^\mu \, \Delta \boldsymbol{r}_\alpha = \chi_\beta^\mu \boldsymbol{E}. \tag{7.53}$$

Here $\Delta \boldsymbol{r}_\alpha$ is the displacement of the centre of charge of the bond in the external field ($\Delta r_\alpha = -\Delta r_\beta$). Nonlinear polarization of a medium is then the consequence of anharmonic vibrations of bond charges. Levine suggested that mainly the bond charge shift along the bond line should be considered. Later, Pietsch *et al* (1985), while studying GaAs, experimentally showed that the displacement of ED, connected with forbidden reflections, along the internuclear vector is indeed much larger than the displacements in other directions (figure 7.16). The bond charge approximated by Gaussians, in the electric field, 40 kV cm^{-1}, is displaced by $\sim 0.12(3) \times 10^{-13}$ m along the bond path in GaAs. In this way, in order to describe bond charge movement it is enough to use the linear anharmonic oscillator equation and to connect the terms in their solution, responsible for frequency doubling, to longitudinal quadratic bond polarizabilities (hyperpolarizabilities), β^μ. The latter, in its turn, is connected to change in the electric field of the linear bond polarizabilities, χ_b^μ, by the relation

$$\beta^\mu = \Delta \chi_b^\mu / 4E. \tag{7.54}$$

This can be rewritten as a function of change in the field of components E_h and C as follows:

$$\beta^\mu = -[\chi_b^\mu / 4(E_g^\mu)^2][\Delta(E^\mu)^2 + \Delta(C^\mu)^2]/E. \tag{7.55}$$

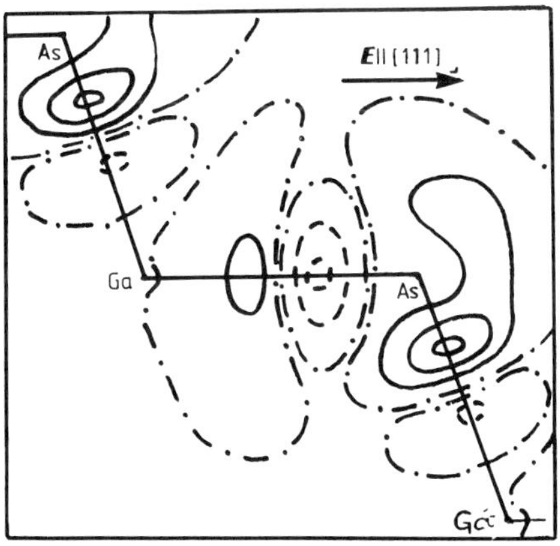

Figure 7.16 The section of difference function $\Delta\rho(\boldsymbol{r}) = \rho^{el}(\boldsymbol{r}) - \rho(\boldsymbol{r})$ in the (110) plane of the GaAs crystal: $\rho^{el}(\boldsymbol{r})$ is the ED calculated for forbidden reflections (222), (444), ..., measured in the external electric field of 40×10^5 V m^{-1}; $\rho(\boldsymbol{r})$ is the analogous value measured without an electric field. The line intervals are 0.006 e Å^{-3}. (Pietsch *et al* 1985.)

Differentiating E_h^μ and C^μ over the bond charge coordinates, r_α, and using (7.54), the dependence of homopolar and heteropolar contributions to E_g on the field can be found and β^μ can be calculated. The μ bond hyperpolarizability proves to be inversely related to the bond charge value q^μ.

The final result of Levine (1973a) for A_nB_m-type crystals ($m > n$) can be written in terms of macroscopic nonlinear susceptibility tensor components $d_{ijk} = \frac{1}{2}\chi_{ijk}^{(2)}$ as

$$d_{ijk} = \sum_\mu G_{ijk}^\mu N_b^\mu [\beta^\mu(C) + \beta^\mu(E_h)]. \tag{7.56}$$

(This value is convenient for comparison of calculated and experimental data.) Here N_b^μ is the number of μ bonds in a unit volume, G_{ijk}^μ is a geometrical factor which reduces the differently orientated μ bonds to a single coordinate system; G_{ijk}^μ is equal to the product of direction cosines for each μ-type bond in the unit cell. The explicit forms for the hyperpolarizability components are as follows:

$$\beta^\mu(C) = \frac{b^\mu t^\mu (Z_\alpha^\mu + (m/n) Z_\beta^\mu) C^\mu (\chi_\beta^\mu)|e|}{2\pi (E_g^\mu)^2 (d^\mu)^2 q^\mu} \tag{7.57}$$

$$\beta^\mu(E_h) = \frac{s(2s-1)(\chi_b^\mu)^2 f_c^\mu R^\mu \varepsilon_0}{q^\mu d^\mu} \left[\frac{r_0^\mu}{r_0^\mu - \bar{r}_c^\mu}\right]^2 \tag{7.58}$$

$$R^\mu = (r_\alpha^\mu - r_\beta^\mu)/(r_\alpha^\mu + r_\beta^\mu) \qquad f_c^\mu = (E_h^\mu / E_g^\mu)^2. \tag{7.59}$$

Here $r_0^\mu = d/2$; $\bar{r}_c^\mu$ is the mean core radius of the bonded atoms.

It can be concluded that some parameters of the charge bond model can be found by means of tabulated atomic characteristics. For example, the charges Z_α and radii $(r_c^\mu)_\alpha$ of atomic cores for s and p elements were determined by Boyd (1977a) by means of analysis of atomic radial ED distributions. Korolkova *et al* (1985), considering 20 binary crystals, found that the mean value of $(\bar{r}_c^\mu)_\alpha$ calculated with Boyd's data is equal to 0.328 r_0^μ. This is very close to the empirically found value, 0.35 r_0^μ (Levine 1973a). For d elements, the closest in spirit to the above approach are the sizes of 'empty' cores in the Ashcroft pseudopotentials (Harrison 1980). More significantly, the remaining model parameters can be obtained by means of x-ray diffraction experiments. Reconstructing ρ_{val} from the x-ray data can enable calculation of the plasma frequency of bond charge vibrations, ω_p, and the screening factor, t^μ. The coordinates of the centre of gravity of bond charge r_α^μ can be set equal to the position of maxima on the deformation ED maps. The bond charges can be determined from the experimental ED as well. Being a conceptual phenomenological notion, the bond charge cannot be determined in a unique fashion. Tsirelson *et al* (1984b) have analysed the different methods of bond charge determination. For binary compounds, Phillips (1970) used the equation $q^\mu = (2e/\varepsilon)$, where ε is the permittivity. Levine (1973a) calculated bond charges according to the equation

$$q^\mu/e = n_{val}^\mu[(1/\varepsilon) + (f_c/3)]. \tag{7.60}$$

One of the terms of this sum is connected with the overlap of spherical parts of atomic ED. This overlap is taken to be proportional to the covalence index of the bond, f_c. The second term in (7.60), in contrast to Phillips, describes the number of covalent electrons n_{val}^μ participating in a given bond. The coefficient of one-third in (7.60) was found by Levine from the requirement of agreement with experiment of the linear optical susceptibilities χ_b, which were calculated in wurzite and zincblende crystals with bond charges (7.60). Pietsch (1980, 1981, 1982a) obtained the bond charges from x-ray experiment, approximating the valence ED in the region of the chemical bond by Gaussians (see subsection 3.5.3). The values of the bond charges according to Pietsch agree reasonably with those according to Phillips, while they are systematically smaller than the bond charges of Levine (Tsirelson *et al* 1984b). The cause of this lies in the fact that Pietsch did not take into account the contribution from the spherical part of the atomic electron clouds to the bond charge. It, nevertheless, contributes also to the optical susceptibility.

In order to take into account all contributions to the bond charge, Tsirelson *et al* (1984b) suggested integrating the experimental valence ED over the volume Ω_μ assigned to the bond charge:

$$q^\mu = \int_{\Omega_\mu} \rho_{val}(\boldsymbol{r})\,\mathrm{d}V. \tag{7.61}$$

Though it is impossible to define Ω_μ in a unique way, Cohen (1981) noted that, because of the rapid decrease in ρ_{val} in directions away from the bond path, the values of q^{μ} determined from a variety of methods of choosing the volume are close to one another.

Berlinite, $AlPO_4$, demonstrates how accurately bond charges evaluated from experimental ED allow calculation of nonlinear optical characteristics. This crystal has α-quartz structure (space group $P3_121$) and displays birefringence, which is a condition for the appearance of synchronous nonlinear interactions (Zernike and Midwinter 1973). The berlinite structure is built from PO_4 and AlO_4 tetrahedra having common apices; the difference between Al–O and P–O bond lengths is approximately 1%. To calculate the experimental ED in $AlPO_4$, Korolkova *et al* (1985) used x-ray structure amplitudes measured by Thong and Schwarzenbach (1979). The electronic charges have been evaluated by integration of the valence ED over cubes with edges equal to the bond length d and parallelepipeds with a base size obtained by starting with a volume bounded by the value $\rho_{val} \cong 0$. The charges obtained differed by no more than 20%. Such accuracy of bond charge determination is enough for evaluation of optical properties. The values of q^{μ} used in the calculation after subtraction of the charges connected with the atom cores were as follows: $q(\text{Al–O}_1) = 0.72\ e$, $q(\text{Al–O}_2) = 0.69\ e$, $q(\text{P–O}_1) = 1.33\ e$, and $q(\text{P–O}_2) = 1.30\ e$. The positions of the centre of charge of the bonds in $AlPO_4$ were determined from deformation ED maps. As bond charges of all types in $AlPO_4$ only weakly depend on the bond length, their average values were used. The results of the berlinite optical characteristic calculations are displayed in table 7.13. Because of the difference in chemical bond character, reflected in the value of f_c (7.59), the linear susceptibility of the Al–O bond is 1.3 times smaller, while the hyperpolarizability β is three times larger than

Table 7.13 Calculated optical characteristics of the $AlPO_4$ crystal.

Characteristics	Bonds: Al–O		Bonds: P–O
f_c	0.37		0.54
χ_b^{μ} (10^{-30} m^3)	12.13		16.55
$\beta^{\mu}(C)$ (10^{-42} m^4 V^{-1})	54.45		84.1
$\beta^{\mu}(E_h)$ (10^{-42} m^4 V^{-1})	−119.42		−104.7
β (10^{-42} m^4 V^{-1})	−65.0		−20.6
n_{calc}		1.60	
$n_{exp} = (n_o + n_e)/2$		1.524	
Korolkova *et al* (1985)		−0.40	
d_{111} (10^{-12} m V^{-1}); Levine (1973a)		−0.33	
Experiment (Levine 1973a)		−0.41(3)	
$\lambda = 1.058\ \mu$m			

for bonds P–O. This is connected with the fact that, in the series of increasing bond lengths, P–O, Si–O, and Al–O, the potential in the region of the chemical bond becomes smaller and the value of E_g reduces. This should lead to larger hyperpolarizability values for longer bonds, which is observed in $AlPO_4$. In general, the bond hyperpolarizabilities are relatively small in $AlPO_4$ because of the partial cancellation of the ionic and covalent contributions.

The calculated values of the refractive indices and the independent component of the tensor d_{111} agree well with the experimental values, supporting the suitability of the method used for bond charge evaluation in calculations of optical characteristics. The closeness of the results to the data of Levine's (1973a) calculation where the values ε and q^{μ} were determined from dielectric measurements is also noticeable.

Korolkova *et al* (1985) showed that when evaluating nonlinear optical properties in a series of isostructural crystals it is sufficient to have bond charge values for one member only. Bond charges in the remaining crystals can be determined by using the relation $\chi^{\mu} \sim q^{\mu}/(E_g)^2$. Supposing that the local field factors in isostructural compounds are identical, it is not difficult to obtain a relationship for calculation of bond charges from already known values q^{μ}:

$$q^x = q^{\mu}(E_g^x/E_g^{\mu})^2(\chi^x/\chi^{\mu}). \tag{7.62}$$

A parameter b compensates the deficiency of the Thomas–Fermi description of atomic core screening by valence electrons in the ionic contribution to the average width of the energy gap. This depends on the average coordination number of the atom whose bonds are being considered, N_c, and on the degree of filling of the unit cell by bonds of μ type. Therefore

$$b^x = b^{\mu}(N^{\mu}/N_b^x)(d^{\mu}/d^x)^3. \tag{7.63}$$

Korolkova *et al* (1985) have estimated the optical properties of the crystals $FePO_4$, $GaPO_4$, and $AlAsO_4$, which are isostructural to $AlPO_4$. The geometrical factors in all these crystals were taken as being equal to those of $AlPO_4$. Crystal radii were used for evaluation of bond lengths. The parameter R^{μ}, which characterizes the displacement of the bond charge, was calculated with the help of covalent radii taking into account their dependence on the atomic coordination number. The bond parameters were calculated according to the above scheme, and through them the characteristics of the optical properties of the crystals in the series being considered were determined. The results of the calculation are presented in tables 7.14 and 7.15 and in figure 7.17. The agreement of the calculated refractive indices with the measured ones confirms that the approach discussed gives, at this stage, a completely satisfactory prediction. This provides a basis for the analysis of contributions of different bonds to the optical susceptibilities. As is apparent from table 7.14, a very large contribution to the linear susceptibility can be expected from the Fe–O and As–O bonds. Then, in

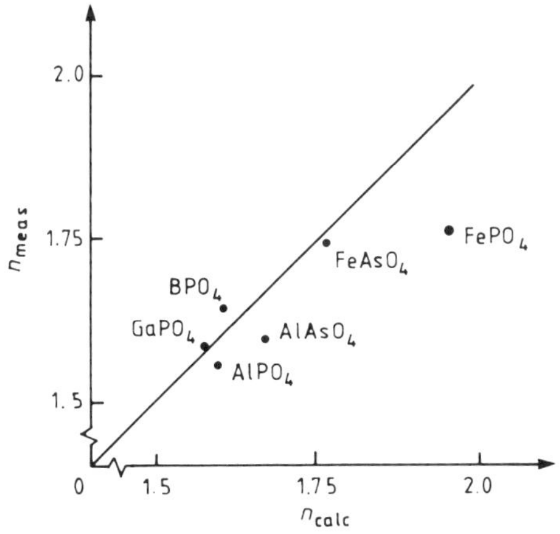

Figure 7.17 Calculated and measured refractive indices for crystals isostructural to $AlPO_4$ (Korolkova *et al* 1985).

accordance with the Miller rule (Zernike and Midwinter 1973) the hyperpolarizability of these bonds should be maximal. This is indeed the case: the values of β(Fe–O) and β(As–O) are roughly an order greater (in absolute value) than the hyperpolarizabilities of the other bonds. The macroscopic nonlinear characteristics calculated for $FeAsO_4$ are larger than they are in the phosphates. In this way the prediction leads to the conclusion that, among compounds of this series, this crystal has the best potential prospect as a material for construction of quantum electronics devices, especially as the natural frequencies of lattice vibration of this crystal lie lower than the working frequency of $1.8 \times 10^{-15}\, c^{-1}$ of the widespread neodymium garnet laser (Korolkova *et al* 1985).

Korolkova *et al* (1986) adapted the approach described to find the structural fragments responsible for nonlinear properties of the crystal $KTiOPO_4$ (KTP) and its structural analogues $RbTiOPO_4$, $KAlFPO_4$, $KFeFPO_4$, $KAsOPO_4$, etc described by Stucky *et al* (1989). The structure of KTP (space group $Pn2_1a$) is formed by alternated distorted TiO_6 octahedra and PO_4 tetrahedra (figure 7.18). At room temperature potassium ions are partially disordered along the 2_1 axis (Belokoneva *et al* 1990). It was assumed that the main contribution to the nonlinear optical properties of KTP is due to the Ti–O and P–O bonds, and the contribution of the ionic K–O bonds was neglected. The latter bonds are characterized by low valence ED (Voloshina *et al* 1985) and a high degree of ionicity, leading to a small value of $\chi^{K-O}/(E_g^{K-O})^2$. At the same time, the volume occupied by the potassium polyhedra in the KTP structure is sufficiently large and it was subtracted from the whole unit cell volume in the calculation of the volumes assigned to the Ti–O and P–O bonds.

Table 7.14 Calculated linear and quadratic polarizability parameters for some α-quartz-type crystals (Korolkova *et al* 1985).

Bond	b	χ_b (10^{-30} m^3)	E_g (10^{-19} J)	$q/\lvert e\rvert$	$\beta(C)$ (10^{-42} m^4 V^{-1})	$\beta(E_h)$ (10^{-42} m^4 V^{-1})	β (10^{-42} m^4 V^{-1})
B–O	0.515	9.65	33.97	0.9	23.2	0	23.2
Fe–O	0.633	40.48	18.70	1.41	440	−805	−365
Ga–O	0.729	12.14	25.89	0.80	120	−176	−56
As–O	0.633	22.69	23.92	1.11	174	−440	−266

Table 7.15 Calculated components of nonlinear optical susceptibility tensor d_{ijk} (10^{-12} m V^{-1}) for α-quartz-type crystals and the contributions of separate bonds to them (Korolkova *et al* 1985).

Compound	d_{11}	
$AlPO_4$	−0.40	(Al–O: −0.30, P–O: −0.1)
BPO_4	0.014	(B–O: 0.131, P–O: −0.116)
$GaPO_4$	−0.32	(Ga–O: −0.234, P–O: −0.087)
$FePO_4$	−1.52	(Fe–O: −1.43, P–O: −0.082)
$AlAsO_4$	−1.53	(Al–O: −0.26, As–O: −1.27)
$BAsO_4$	−1.54	(B–O: 0.12, As–O: −1.66)
$GaAsO_4$	−1.47	(Ga–O: −0.218, As–O: −1.25)
$FeAsO_4$	−2.51	(Fe–O: −1.34, As–O: −1.17)

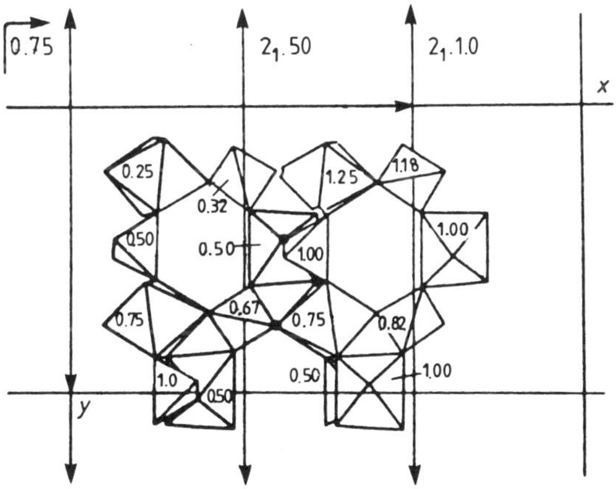

Figure 7.18 The crystal structure of KTP.

The following division of KTP into structural fragments was accepted: $KO_{3/2}\cdot TiO_{3/2}\cdot PO_2$, in determination of the bond volumes, average coordination numbers, and other characteristics. The bond charge values were determined by integration of the valence ED reconstructed from x-ray diffraction data. As is seen from the ρ_{val} maps, shown in figure 7.19, a larger transfer of ED to the oxygen atoms takes place in the more ionic Ti–O bonds. The charges of these bonds were determined by integration of ρ_{val} over a cube with side length $d-(r_c^{Ti}+r_c^{O})$: with this choice of volume practically all electronic charge in the interatomic space is successfully included in the consideration. For the more covalent P–O bonds the bond volume was approximated by a parallelepiped with boundaries chosen to be as close as

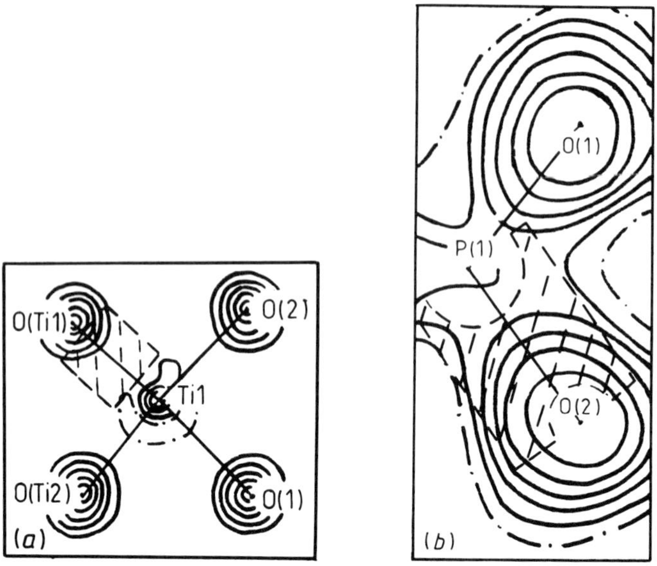

Figure 7.19 Valence ED in Ti–O (*a*) and P–O (*b*) bonds of the KTP crystal. The volumes assigned to bond charges are shown (Korolkova *et al* 1986).

possible to the surface with $\rho_{val} \approx 0$. The values obtained are within the range of 1.97–2.55 *e* for the P–O bond and 0.34–0.69 *e* for the Ti–O bond; a tendency for the bond charge to be reduced for longer bonds was observed. For calculation of the optical properties the average values of bond charges, 2.22 and 0.55 *e*, were used. For all P–O and Ti–O bonds the average bond lengths were also used: $d_0 = 1.538$ Å and $d_0 = 1.995$ Å, respectively. It was found that the positions of charges of the P–O and Ti–O bonds lie approximately halfway between the nuclei; the calculation of r_α according to Levine leads to the same conclusion. Consequently, $R^\mu \approx 0$ and $\beta \approx \beta(C)$.

The geometrical factors for the P–O bonds were taken to be identical while for the Ti–O bonds each factor was calculated separately. The hyperpolarizabilities of the Ti–O bonds were determined according to the formula $\beta = \beta_0(d/d_0)^\sigma$ (Levine 1974). The parameter σ depends on the bond length, degree of screening of the core electrons by the valence electrons, and degree of ionicity of the bond: for the Ti–O bond $\sigma = 2.36$.

It was found that the Ti–O bond contribution to the linear polarizability is 1.4 times larger than the P–O bond contribution. The calculated refractive index was $n = 2.04$; this is close to the experimental value of 1.80. The relationship between Ti–O and P–O contributions for average hyperpolarizabilities is already 8:1. The components of the nonlinear susceptibility tensor for KTP are listed in table 7.16. They agree well with the experimental

Table 7.16 Components of nonlinear optical susceptibilities d_{ijk} (10^{-12} m V^{-1}) of KTP-type crystals.

		d_{211}	d_{222}	d_{233}
Calculation	$KTiOPO_4$	5.2	19.2	5.0
Experiment (Zumsteg *et al* 1976)	$KTiOPO_4$	6.5–7.7	13.7	5.0–6.1
Prognosis	$KAlFPO_4$	0.19	2.35	1.14
Prognosis	$KFeFPO_4$	1.29	9.60	1.86
Prognosis	$KTiOAsO_4$	5.20	20.10	4.60

results. Therefore, it can be concluded that the nonlinear properties of the KTP crystal arise mainly from the high polarizability of the Ti–O bonds, from differences in their lengths, and from their mutual orientation. In particular, the strong distortion of the octahedra in the KTP structure destroys the mutual compensation of contributions from oppositely directed bonds and promotes the increase of quadratic susceptibility. It is sometimes suggested that the dominant contribution to the nonlinear properties in KTP results from the shortest Ti–O bond, 1.72 Å (Hansen *et al* 1991). However, expansion of expression (7.56) into a series in small displacements Δd from the average value of d_0

$$d_{ijk}(\mathrm{TiO_6}) = \sum_{\mu} G^{\mu}_{ijk}\beta^{\mu}_{0}\left(1+\sigma\frac{\Delta d^{\mu}}{d_0}\right) \tag{7.64}$$

and taking into account the fact that the Ti–O bond lengths in KTP lie within the range 1.723–2.150 Å shows that contributions of other bonds are significant as well. Therefore the explanation of the nature of KTP optical nonlinearity given by Korolkova *et al* (1986) is more correct. This, by the way, also explains the lack of success of the calculation of KTP nonlinear optical characteristics carried out by Zumsteg *et al* (1976), which took into account only the shortest Ti–O bond.

The large nonlinear optical characteristics of KTP and its high radiation damage threshold, which is connected with the rigid framework of PO_4 tetrahedra, suggests that among other crystals possessing the same structure there may be prospective optical materials. Korolkova *et al* (1986) have done *a priori* qualitative evaluation of nonlinear properties for four compounds: $RbTiOPO_4$, $KFeFPO_4$, $KAlFPO_4$, and $KTiOAsO_4$. Replacement of the K^+ ion by Rb^+ only increases the unit cell dimensions by 1.4%, while the Rb–O bond, like the K–O bond, exhibits a substantially ionic character and its contribution to the nonlinear properties can be neglected. Therefore the nonlinear susceptibility in $KTiOPO_4$ and $RbTiOPO_4$ should not differ. This is in agreement with the experimental data of Zumsteg *et al* (1976). The same characteristics were calculated for other crystals, as was described above for crystals isostructural to $AlPO_4$. The results are listed in table 7.16. They show that $KAlFPO_4$ and $KFeFPO_4$ crystals have worse nonlinear characteristics than KTP. The distortion of the octahedra because of the introduction of F atoms in the place of one of the O atoms in the formula unit compensates the significantly lower value of hyperpolarizability of the Fe–F bonds: $\beta(\text{Fe–F}) = 1.7 \times 10^{-42}\ \text{m}^4\,\text{V}^{-1}$ where $\beta(\text{Ti–O}) = 48.2 \times 10^{-42}\ \text{m}^4\,\text{V}^{-1}$. This is caused by an increase in the number of polarized 3d electrons in the transition ion. The Fe–O bonds are almost equal in length; their hyperpolarizabilities are relatively small: $\beta(\text{Fe–O}) \cong 6.7 \times 10^{-42}\ \text{m}^4\,\text{V}^{-1}$. All of this taken together leads to the small values of the nonlinear optical characteristics in these crystals. In addition, absorption of radiation due to features of the Fe^{3+} optical spectrum makes the $KFeFPO_4$ crystal unsuitable for use in nonlinear optics. The $KTiOAsO_4$ crystal, judging by the prediction, is just as suitable as KTP for optical use. The experimental data of El-Brahimi and Durand (1986) support this assumption.

The valuable conclusion reached by Korolkova *et al* (1986) is as follows. In order to increase the nonlinear optical characteristics of octahedral structural units, it is necessary to introduce into them 3d and 4d elements with a small number of valence d electrons. These elements will form easily polarizable bonds. In combination with distortion of the octahedra, this is the necessary condition for creation of nonlinear optical properties.

Tsirelson *et al* (1984b) generalized the bond charge method, making it suitable for dielectrics with multiple bonds, and used it for study of the optical properties of formates. The ED distribution in one of these, lithium formate deuterate (LFD), $HCOOLi \cdot D_2O$ (space group $Pna2_1$) was experimentally studied in detail by Ozerov *et al* (1981): these data were used for calculation of the bond charges. Because of the low valence ED, the contribution of the Li–O bonds was neglected and only the C$\cdots$O bond, assumed to be identical, and the C–H and D–O bonds were considered. The influence of H bonds did not require special attention, as they affect the position of the centre of charge of the $O–D_{1,2}$ bonds. The O atom lone pairs are strongly delocalized in this compound and therefore their contribution to screening is practically isotropic.

The bond charges in LFD, calculated by integration of ρ_{val}, were as follows: $q_{C-O} = 2.28\,e$, $q_{C-H} = 0.90\,e$, $q_{D-O} = 0.90\,e$. The calculated refractive index was 1.25, which was close to the experimental value $n = 1.45$, measured at a wavelength of 1.06 μm (Singh *et al* 1970). The calculation showed that the hyperpolarizability of the C$\cdots$O bond was three times larger than the roughly equal hyperpolarizabilities of the remaining bonds, including bonds of the water molecule. Covalent components make the main contribution to all β values. This is connected with the asymmetrical disposition of bond charges along the line joining the nuclei, which is manifested in R^{μ} (7.59). The

Table 7.17 Components of the nonlinear optical susceptibility tensors, d_{ijk}, for the formate crystal family (10^{-12} m V^{-1}) (Tsirelson *et al* 1984b).

Compound	Tensor component	Bond charge model	Experiment $\|d_{ijk}\|$ $\lambda = 1.06$ μm
$HCOOLi \cdot D_2O^a$	d_{322}	−1.13	1.27(9)
	d_{311}	0.12	0.11(2)
	d_{333}	1.81	1.86(11)
$Ba(HCOO)_2$	d_{123}	−0.08	0.12
$Sr(HCOO)_2$	d_{123}	−0.88	0.80
$Sr(HCOO)_2 \cdot 2H_2O$	d_{123}	−0.13	0.50
$Y(Ho,Er)(HCOO)_3 \cdot 2H_2O$	d_{123}	0.14	0.17–0.25

[a]Experiment gives for $LiCOOH \cdot D_2O$ $d_{311}d_{333} < 0$ (Singh *et al* 1970).

calculated tensor d_{ijk} (table 7.17) agrees excellently with experimentally measured component values.

Tsirelson *et al* (1984b), having analysed the results of a variety of experimental studies of ED distribution in formate crystals, concluded that the $HCOO^-$ ion preserves all the basic details of deformation ED in all compounds. In the independent-bond approximation, nonlinear optical properties are determined, in addition to the geometry, by the hyperpolarizabilities of the C$\overset{...}{-}$O and C–H bonds was well as the O–H bonds if water is present in the crystal. Therefore, using bond parameters obtained for the LFD crystal, these authors evaluated nonlinear characteristics of other representatives of this family. Changes in the hyperpolarizabilities were accounted for by changes in the bond charges, which in turn were considered to be connected with changes in the interatomic distances and, consequently, in the bond volumes. In the absence of data about the positions of hydrogen atoms in several formates, the bond geometrical factor was calculated for C_{2v} ion symmetry at a C–H separation of 1.072 Å, which is correct for LFD.

The tensor d_{ijk} values obtained for the formates are listed in table 7.17. As for the phosphates, the prediction of nonlinear optical properties in a series of crystals having approximately identical, fairly complicated heterodesmic fragments gives positive results and may be recommended for practical searches for new nonlinear materials.

The success of these predictions permits one to analyse the nature of nonlinear optical properties in the formates. All bonds in $HCOOLi \cdot D_2O$ give a significant contribution to the d_{ijk} tensor components. However, if the d_{311} component is mainly determined by the contributions of the C$\overset{...}{-}$O and

C–H bonds, then the values of the two remaining components arise from the O–D bond, the contribution of which is two to three times larger than the contribution of the formate ion. This explains the lack of success of the theoretical calculation of Dewar *et al* (1976) for $HCOOLi \cdot H_2O$, where the contribution of water to the nonlinear optical properties was ignored. As a result, the signs of d_{ijk} components contradicted the experimental data. In another hydrate, $Sr(HCOO)_2 \cdot 2H_2O$, the influence of the water molecules on the nonlinear properties is extremely small because of features of their spatial arrangement in the unit cell. The main role here is played by the O⋯C⋯O fragment in the formate ion. In waterless strontium and barium formates the following feature is observed. In $Sr(HCOO)_2$ quadratic hyperpolarizability is connected mainly with the O⋯C⋯O fragment, whereas in $Ba(HCOO)_2$ the C–H bond gives a contribution to d_{123} rougly the same at this fragment, but opposite in sign. As a consequence, the latter crystal has a small d_{123} value.

All of the crystals considered are characterized by the presence of covalent chemical bonds, accompanied by localization of some of the valence electrons approximately halfway between nuclei. It is this that permits the use of the bond charge model for calculation of nonlinear optical properties. Crystals with ionic bonds are characterized by significant displacement of electrons toward the more electronegative atoms. In substantially ionic compounds this displacement becomes so large that explanation of nonlinear optical effects by anharmonic motion of the bond charges relative to several positions in internuclear space is no longer satisfactory. In such crystals, in place of bond charge vibrations, it is necessary to consider anharmonic motion of the valence electron cloud relative to the atomic cores. The corresponding method was suggested by Meisner (1975) and the following basic considerations were proposed.

Within the frame of the superpositional model of the crystal, it was assumed that dipole deformation of the electron cloud of the ions in a laser wave field, identical for all electronic displacements, results in polarization of the media. Presenting the dipole moment of every ion as a series in the field strength, it is possible to connect the expansion coefficients with the nonlinear optical susceptibility tensor components of the corresponding orders. For this it was taken into account that nonlinear displacements of part of the ionic electron cloud depend on electronic interactions and on interaction of valence electrons with atomic cores. In a first approximation, only the Coulomb (linear) part of this interaction may be considered, assuming that the forces connected with electron exchange and correlation remain unchanged and give no contribution to the generation of higher optical harmonics. This approximation is justified only for crystals with a small covalency of bonds. The forces causing anharmonic vibrations of the valence electron cloud of the ions arise from the action of the local electric field $\mathscr{E}_{loc}$. The latter is composed of the external field $\boldsymbol{E}$, Lorenz field $\boldsymbol{E}_L$, and ionic field outside the Lorenz cavity $\mathscr{E}'$ (the latter is accounted for by introducing the

Lorenz factor β). For a given ion it is convenient to expand the field $\boldsymbol{E}_L$ in a series over small displacements Δx_k of neighbouring atom clouds in the Lorenz cavity. The expansion coefficients, a_k, depend only on the crystal structure and can be easily calculated (see Born and Huang 1954). Then $\mathscr{E}_{loc}$ in the one-dimensional case can be given by

$$\mathscr{E}^i_{loc} = E + \sum_k [a^0_{ik} + g_{ik}(\beta, a^{(1)}_{ik})\Delta x_k + a^{(2)}_{ik}\Delta x^2_k + \cdots]Q_k \tag{7.65}$$

where Q_k is the electron shell charge displaced in the field. The summation is spread over all ions in a crystal. Placing this expression into the anharmonic oscillator equation, it is possible to obtain a relationship which describes nonlinear effects of different orders. The linear part of the electron cloud displacement, Δx_{lin}, can be expressed through the static electronic polarizability α, if all terms higher than linear are discarded in expansion (7.65). Then

$$\mathscr{E}^i_{loc} \approx E + \sum_k g^i_k \alpha_k (\mathscr{E}^k_{loc}) \tag{7.66}$$

and

$$\Delta x^i_{lin} = \alpha_i \mathscr{E}^i_{loc}/Q_i = (\alpha_i A_i/Q_i)E(\omega). \tag{7.67}$$

The parameters A_i, which depend on the structural coefficients g^i_k (Achmanov *et al* 1977) can be found by solving the system (7.66). In this case the local field, acting on the ion in the lattice, is sufficiently accurately taken into account. Then, choosing from the expansion (7.65) the term containing, for example, Δx^3_k, and using (7.67), we obtain

$$\Delta x^i_{nonlin}(3\omega) = \frac{\alpha_i(3\omega)}{Q_i} E^3(\omega) \sum_k a^{(3)}_{ik} \alpha^3_k A^3_k Q^{-2}_k. \tag{7.68}$$

Comparing this expression with the expansion

$$P(3\omega) = \eta E^3(\omega) = \sum_m N_m \, \Delta x^{(m)}_{nonlin}(3\omega) Q_m \tag{7.69}$$

(N_m is the number of ions of the mth type in the unit volume), we obtain an expression for cubic susceptibility:

$$\eta = \sum_i N_i \alpha_i(3\omega) \sum_k a^{(3)}_{ik} \alpha^3_k(\omega) A^3_k Q^{-2}_k. \tag{7.70}$$

This is one of the possible ways to write the result of Meisner (1975). From (7.70) it is seen that, apart from the structural coefficients, a_i, the method of Coulomb anharmonicity contains free parameters—the polarizability α_i and the quantities Q_i. Meisner determined their values from refractive indices according to the Phillips and Van Vechten dielectric theory of the chemical

Table 7.18 The dielectric permeability and optical cubic susceptibility η_{1111} (10^{-22} $m^2 V^{-2}$) of crystals with NaCl-type structure (Korolkova *et al* 1986).

		NaF	NaCl	LiF	MgO
$1-f_c$		0.95	0.94	0.91	0.84
ε	calc	2.4	3.0	2.1	2.5
	exp	1.7	2.3	1.9	3.0
η_{1111}	calc	0.50	6.6	0.31	0.44
	exp	0.49	2.4	0.47	4.7

bonds. Korolkova *et al* (1986) suggested that these parameters be determined from experimental ED and showed how this could be done in practice using a crystal of NaCl type.

Q_i is the charge of the valence electron cloud of the ith ion which is displaced in the field. The value of Q_i is equal to $Q_i=(p_{at}+p_{eff})/\sigma$. Here p_{at} and p_{eff} are the numbers of valence electrons of the atom and the effective ionic charge, respectively, and σ is a screening constant. In the homopolar approximation, accepted in the Phillips and Van Vechten dielectric theory, the screening is considered to be the same for all ions and equal to the value of the low-frequency dielectric permittivity, ε (Penn 1962). Calculation of the screening constant by the Clausius–Mossotti formula via the low-frequency electron polarizabilities, the determination of which from x-ray diffraction data is described above (see table 7.12), does not present any difficulty.

In order to calculate the polarization of the formula unit α, formula (7.70) was modified, using the following result of Achmanov *et al* (1977). The product $N_i\alpha_i A_i$ has the meaning of the contribution to the polarization of the crystal from one type of ion. Therefore the substitution $N_i\alpha_i A_i \rightarrow (\alpha/2)A_i N_0$ is valid. Introducing the average ionic charge displaced in the field $\bar{Q}=(Q_i+Q_k)/2$ and remembering that in crystals of NaCl type $A_i=A_k=(\varepsilon+2)/3$, one obtains

$$\eta=\frac{\alpha(3\omega)\alpha^3(\omega)}{2^4\bar{Q}^2}A^3N_0\sum_{ik}a_{ik}^{(3)}. \tag{7.71}$$

This formula is already suitable for practical calculations. With its help Korolkova *et al* (1986) calculated, ignoring dispersion, the cubic susceptibilities for a series of simple ionic crystals of NaCl structure (table 7.18). The use of the values of α calculated from x-ray diffraction data in the anharmonic Coulomb model led to a sufficiently good agreement with measurements. Attention should be paid to the fact that as the bond covalency, f_c, decreases the agreement between calculation and experiment improves in accordance with the initial assumptions of this method.

EPILOGUE

We have sincerely tried to throw light onto the different aspects of the ED concept in this book, beginning from its role in theory and finishing with the crystal properties. During our work on the book many new results have appeared; some of these have been included in the text and the rest need further analysis and generalization. It is not possible to include all aspects of ED concepts in one book. Moreover, they should not be seen from a single point of view. Meanwhile, we will be glad if our work stimulates the development of this very interesting field of physics and chemistry to which the efforts of many scientists have been applied. The story, in which electron density plays the main role, is far from complete.

APPENDIX A

THE SYSTEMS OF UNITS

The SI system of units, which we tried to use in this book in all possible cases, is not convenient for problems dealing with electronic properties of matter. The basic system of units in quantum chemistry is the atomic unit (au). In this system the Bohr radius, a_0, the Planck constant, $\hbar$, the electron charge, e, and the electron mass, m, are all equal to unity: $a_0 = \hbar = e = m = 1$. Due to this fact, the coefficients in the electronic Schrödinger equation are essentially simplified, and the numerical values of physical quantities do not contain any factors of 10^n type. The parameters of wave-functions in various data collections (see e.g. Clementi and Roetti 1974, Huzinaga 1984) are presented in the atomic system of units. The so-called electronic units system has been elaborated in crystallography. In this system, the unit of length is the ångstrom: 1 Å $= 10^{-10}$ m. The distances between the nuclei in molecules and crystals are of the order of 1 Å, and measurement of these distances in ångstroms allows one to delete a numerical factor 10^{-m} that is present in other systems of units. A feature of this system is that the ED is measured in units of the number of electrons in a cubic ångstrom, e Å^{-3}, i.e. the electron charge value and sign are not taken into account.

All listed units are used in this book, and table A.1 contains conversion factors which interrelate the units. The data presented can be supplemented by the facts that the energy is frequently measured in electron volts (1 au $= 27.2116$ eV $= 4.359\,81 \times 10^{-18}$ J) and that in electrostatic units (esu) the charge of an electron is $4.803\,242 \times 10^{-10}$ esu.

Table A.1 Atomic versus electronic and SI units for some physical quantities.

Quantity	Atomic units	Electronic units	International system of units (SI)
Charge of electron, e	1	1 e	1.6022×10^{-19} C
Mass of electron, m	1	9.1095×10^{-31} kg	9.1095×10^{-31} kg
Action, $\hbar$	1	1.0546×10^{-14} kg Å^2 s^{-1}	1.0546×10^{-34} J s
Length, a_0	1	5.2918×10^{-1} Å	5.2918×10^{-11} m
Inverse length, a_0^{-1}	1	1.8897 Å^{-1}	1.8897×10^{10} m^{-1}
Volume, a_0^3	1	1.4818×10^{-1} Å^3	1.4818×10^{29} m^3
Electron density, $\rho(\boldsymbol{r})$	1	6.7483 e Å^{-3}	1.0812×10^{12} C m^{-3}
Laplacian of the electron density, $\nabla^2\rho(\boldsymbol{r})$	1	2.4099×10^{1} e Å^{-5}	3.8611×10^{32} C m^{-5}
Force, F	1	8.2388×10^{2} kg Å s^{-2}	8.2388×10^{-8} N
Energy, E	1	4.3598×10^{2} kg Å^2 s^{-2}	4.3598×10^{-18} J
Electrostatic potential, φ	1	1.8897 e Å^{-1}	2.7212×10^{1} V
Electric field, $\boldsymbol{E}$	1	3.5711 e Å^{-2}	5.1423×10^{11} V m^{-1}
Electric field gradient $\nabla\boldsymbol{E}$	1	6.7483 e Å^{-3}	9.7174×10^{21} V m^{-2}
Electric dipole moment, μ[a]	1	5.2918×10^{-1} e Å	8.4784×10^{-30} C m
Electric quadrupole moment, Q[b]	1	2.8003×10^{-1} e Å^2	4.4866×10^{-40} C m^2

[a] 1 debye (D) = 0.3934 au = 0.2082 e Å = 3.3556×10^{-30} C m.
[b] 1 buckingham (B) = 0.7435 au = 0.2082 e Å^2 = 3.3556×10^{-40} C m^2.

APPENDIX B

VIBRATING ATOMS IN CRYSTALS AS QUANTUM OSCILLATORS

The vibration of an atom in a crystal can be considered as vibration of a quantum particle of mass μ in a potential of the general form $V = V(\boldsymbol{r})$. This problem, called the quantum oscillator problem, has a wide scope of applications. Let us expand the potential into a series over powers of displacements, $\boldsymbol{r}$, with respect to the equilibrium position

$$V(\boldsymbol{r}) = V_0 + \frac{1}{2}\sum_{ij}\left(\frac{\partial^2 V}{\partial x_i\,\partial x_j}\right)_0 x_i x_j + \frac{1}{6}\sum_{ijk}\left(\frac{\partial^3 V}{\mathrm{d}x_i\,\mathrm{d}x_j\,\mathrm{d}x_k}\right) x_i x_j x_k$$

$$= V_0 + \frac{1}{2}\sum V_{ij} x_i x_j + \frac{1}{6}\sum V_{ijk} x_i x_j x_k + \cdots \qquad x_i = x, y, z. \tag{B.1}$$

Constant V_0 is the potential energy of a particle at equilibrium; by appropriate choice of the origin it can be chosen as zero. Depending on the number of terms retained in expansion (B.1), two important cases can be distinguished.

(i) *The quantum harmonic oscillator*. In some problems the displacement of a particle from the equilibrium position can be considered to be small and the restoring force to be linearly dependent on the displacement. Then it is sufficient to retain only the quadratic term in (B.1) and, considering for simplicity one-dimensional motion, one can write the harmonic oscillator Hamiltonian as follows:

$$\hat{H}_0 = -(\hbar^2/2m)\partial^2/\partial x^2 + \tfrac{1}{2}\mu\omega^2\hat{x}^2 \tag{B.2}$$

(the vibration frequency is $\omega^2 = V_{ii}/\mu$). The Schrödinger equation for a quantum linear harmonic oscillator takes the form

$$-(\hbar^2/2m)\partial^2\psi(x)/\partial x^2 + (\tfrac{1}{2}\mu\omega^2 x^2 - E)\psi(x). \tag{B.3}$$

As is known (Messiah 1966), this equation can be transformed to the differential Hermite equation which can be solved only when the energy levels of a quantum oscillator have discrete values:

$$E_a = (a + \tfrac{1}{2})\hbar\omega \qquad a = 0, 1, 2, \ldots \tag{B.4}$$

where a is the quantum number. The energy of the lowest level at $a=0$ is the zero-point vibration energy. The eigenfunctions of equation (B.3) have the form

$$\psi_a(\xi)=[(2^a a!)^{-1}(\mu\omega/\hbar\pi)^{1/2}]^{1/2}\exp(-\xi^2/2)H_a(\xi) \tag{B.5}$$

where $H_a(\xi)$ is the Hermite polynomial of ath order which can be obtained by the formula

$$H_a(\xi)=(-1)^a\exp(\xi)^2(d^a/d\xi^a)\exp(-\xi^2) \tag{B.6}$$

where $\xi=(\mu\omega/\hbar)^{1/2}x$.

The three-dimensional harmonic oscillator can be considered as a set of three linear oscillators with potential energy $\frac{1}{2}\mu\omega^2(x^2+y^2+z^2)$. The energy of such an oscillator is $E_a=\hbar\omega(a_x+a_y+a_z+\frac{3}{2})$, $a_x, a_y, a_z=0, 1, 2,\ldots$ and their wave-function is a product of three components, each of which is determined by the expression (B.5): $\psi_{a_xa_ya_z}=\psi_{a_x}(\xi_x)\psi_{a_y}(\xi_y)\psi_{a_z}(\xi_z)$. If all three oscillators are equivalent (the isotropic oscillator), then one can obtain the same integer a by combining various quantum numbers a_x, a_y, a_z; i.e. the levels of a three-dimensional oscillator are degenerate. In the general case, the degeneracy is equal to $(a+1)(a+2)/2$. Therefore, in summing up over the states one must introduce the weight factors that take into account the state degeneracy multiplicities.

The mean value of a particle displacement in any state is zero, since in the integral

$$\langle\xi\rangle_a=\int_{-\infty}^{\infty}\psi_a^2(\xi)\xi\,\mathrm{d}\xi=0 \tag{B.7}$$

an odd function appears. For this reason, the mean square displacement

$$\langle\xi^2\rangle_a=\int_{-\infty}^{\infty}\psi_a^*(\xi)\xi^2\psi_a(\xi)\,\mathrm{d}\xi=0 \tag{B.8}$$

is usually considered. Using the recurrent relations existing between the Hermite polynomials of various orders and the orthonormalization properties of $\psi_a(\xi)$ functions, one can arrive at the relations $\langle\xi^2\rangle_a=a+\frac{1}{2}$ or $\langle x^2\rangle_a=(a+\frac{1}{2})\hbar/\mu\omega$. As a result, the expression for energy (B.4) can easily be transformed into the form $E_a=\mu\omega^2\langle x^2\rangle_a$.

(ii) *The quantum anharmonic oscillator*. A great number of crystal properties (heat conduction, superionic conductivity, and phase transitions, etc) are related to atom vibration caused by forces which are nonlinearly dependent upon displacements. The values of displacements can no longer be considered to be small in this case and one must retain in expansion (B.1) the terms higher than quadratic. This means that the anharmonicity must be taken into account. If the anharmonicity is small and the oscillator is isolated, the problem of determination of the eigenfunctions and eigenvalues of the Schrödinger equation for an anharmonic oscillator can be solved within the stationary perturbation theory framework. The Hamiltonian is written in this case as

$$\hat{H}=\hat{H}_0+\hat{H}'. \tag{B.9}$$

Here H_0 is the harmonic oscillator Hamiltonian (B.2) and the anharmonicity is taken into account in the one-dimensional case as a perturbation described by the operator

$$\hat{H}'=\beta\hat{x}^3+\gamma\hat{x}^4+\cdots. \tag{B.10}$$

Wave-functions, χ_a, and stationary state energies, ε_a, of an anharmonic oscillator are determined, as usual in the perturbation theory, in terms of known E_a energy values (B.4) and of harmonic oscillator wave-functions ψ_a (B.5). For a small perturbation it is sufficient to use the first-order approximation for wave-functions

$$\chi_a=\psi_a+\sum_{b\neq a}\frac{H'_{ab}}{E_a-E_b}\psi_b \tag{B.11}$$

and the second-order approximation for energies

$$\varepsilon_a=E_a+H'_{aa}+\sum_{b\neq a}\frac{|H'_{ab}|^2}{E_a-E_b}. \tag{B.12}$$

For an isolated oscillator, the contribution of a cubic term βx^3 to the matrix element $H'_{aa}=\beta\int_{-\infty}^{\infty}\psi_a^* x^3\psi_a\,dx$ is zero. The contributions of the fourth-order term, γx^4, and higher-order terms to the matrix element $|H'_{ab}|^2$ can be neglected as compared with the cubic term contribution. Taking these facts into account, one can write the approximate expression for an anharmonic oscillator energy as

$$\varepsilon_a\cong\hbar\omega(a+\tfrac{1}{2})-(15\beta^2/\hbar\omega)(h/\mu\omega)(a^2+a+\tfrac{11}{30})+\tfrac{3}{2}\gamma(\hbar/\mu\omega)$$
$$\times(a^2+a+\tfrac{1}{2}). \tag{B.13}$$

The presence of terms of odd power in the perturbation series expression (B.10) results in the fact that for high negative values of displacement the potential energy of the anharmonic oscillator can exceed the total energy (i.e. it can become higher than the depth of the potential well), and the stationarity of vibration can be violated. However, for small β values this situation is less probable and the solutions found from the perturbation theory coincide with stationary ones (Reisland 1973).

The vibration of atoms in a crystal

The set of vibrating bonded atoms in a crystal can be considered as a system of oscillators; each atom is a part of a macroscopic system and it is at thermodynamic equilibrium with its surroundings. The state of this atom is a mixture of 'pure' states and is completely described by the Hamiltonian $\hat{H}$ and temperature T. However, the interaction with the environment does not allow one to describe the behaviour of an atom in the crystal by a wave-function and requires the application of a statistical approach. In quantum mechanics this is achieved by means of a statistical operator (or density matrix)

$$\hat{\rho}=\exp(-\beta\hat{H})/\mathrm{Tr}[\exp(-\beta\hat{H})] \tag{B.14}$$

(here Tr denotes a sum over all possible 'pure' states). The mean value of physical quantity A, at a state described by statistical operator $\hat{\rho}$, is calculated as follows:

$$\langle A\rangle=\mathrm{Tr}(\hat{\rho}\hat{A}). \tag{B.15}$$

This results, in particular, in the fact that the position of a vibrating atom in a crystal can be described statistically only. For example, the PDF for isotropic harmonic displacements of such an atom from the equilibrium position is described by the relation

$$p(u)=[(\mu\omega/\pi\hbar)\tanh(\tfrac{1}{2}\beta h\omega)]^{1/2}\exp[-(\mu\omega/\hbar)\tanh(\tfrac{1}{2}\beta h\omega)u^2]. \tag{B.16}$$

This is the Gauss PDF with zero mean value and with the variance

$$\sigma_u^2=(\hbar/2\mu\omega)\coth(\tfrac{1}{2}\beta\hbar\omega) \tag{B.17}$$

$\beta=1/kT$, where k is the Boltzmann constant. The corresponding mean energy and its variance are given by

$$\langle E\rangle=(\tfrac{1}{2}+1/[\exp(\beta\hbar\omega)-1])\hbar\omega \tag{B.18}$$

$$\sigma_E^2=(\hbar\omega/2)^2[\coth^2(\beta\hbar\omega/2)-1]. \tag{B.19}$$

The first term in (B.18) presents the contribution of zero vibrations. At low temperatures ($T\to 0$)

$$p(u)\approx(\mu\omega/\pi\hbar)^{1/2}\exp(-(\mu\omega/\hbar)u^2) \tag{B.20}$$

$$\langle E\rangle\approx\hbar\omega/2 \qquad \sigma_E^2\approx 0. \tag{B.21}$$

Comparing these expressions with (B.4) and (B.5) one can see that they correspond to the ground state of the quantum oscillator. At high temperatures ($\beta\gg\hbar\omega$) the classical expressions

$$p(u)=(\mu\omega/2\pi\beta)^{1/2}\exp[-\mu\omega\beta u/2] \tag{B.22}$$

$$\langle E\rangle\approx kT \qquad \sigma_E^2\approx(kT)^2 \tag{B.23}$$

are valid. Thus, in the harmonic approximation at $T>0$ the state of an atom in a crystal is a mixture of the ground and excited states. The weights of these states depend on the temperature: the statistical weight of each state is given by

$$W_a=\exp(-E_a\beta)\Big/\sum_n \exp(-E_n\beta) \tag{B.24}$$

where E_a is the energy of the 'pure' state a.

The interatomic interactions in a crystal violate the harmonic model of atomic vibration. Considering a crystal to be a set of anharmonically vibrating atoms, the expression for a statistical operator in the spirit of perturbation theory should be presented as $\hat{\rho}=\hat{\rho}^0\hat{U}$. Here $\hat{\rho}^0$ describes the state of a system of harmonic oscillators, and operator $\hat{U}$ depends on the Hamiltonian of a perturbed system (B.9). Expanding $\hat{U}$ into the series in powers of $\hat{H}'$, one can show (Reisland 1973) that, in the second-order perturbation theory, the elements of the statistical operator matrix are as follows:

$$\rho_{nm}=N^{-1}(\beta)\left[\rho_n^0\delta_{nm}+H'_{nm}\frac{\rho_n^0-\rho_m^0}{E_n-E_m}+\sum_k\frac{H'_{nk}H'_{km}}{E_k-E_m}\right.$$
$$\left.\times\left(\frac{\rho_n^0-\rho_k^0}{E_n-E_k}-\frac{\rho_n^0-\rho_m^0}{E_n-E_m}\right)\right] \tag{B.25}$$

$$N(\beta)=1-\beta\sum_n H'_{nn}\rho_n^0-\beta\sum_{m\neq n}H'_{nm}H'_{mn}\frac{\rho_n^0-\rho_m^0}{E_n-E_m} \tag{B.26}$$

$$\rho_n^0=e^{-\beta E_n}/\mathrm{Tr}[\exp(-\beta\hat{H}_0)]. \tag{B.27}$$

In calculating the diagonal elements of the statistical operator matrix the uncertainty in energy denominators can be found as follows:

$$\lim_{E_n\to E_m}\frac{\exp(-\beta E_n)-\exp(-\beta E_m)}{E_n-E_m}=-\beta\exp(-\beta E_m). \tag{B.28}$$

Specifying the explicit form of the perturbation Hamiltonian (B.10), and using expressions (B.25) and (B.26), one can obtain the mean value of any oscillating physical quantity averaged over the whole anharmonic crystal.

APPENDIX C

ATOMIC ORBITALS AND THEIR ANALYTICAL APPROXIMATIONS

The spinless wave-functions describing the states of electrons in position space of atoms (AOs) can be presented in spherical coordinates (r, θ, φ) in the form

$$\Phi_{nlm} = N(n, l) R_{nl}(r) Y_{lm}(\theta, \varphi) \tag{C.1}$$

where n, l and m are quantum numbers. Such a form of description arises from the solution of the Schrödinger equation for a one-electron hydrogen-like atom with nuclear potential $|e|z(4\pi\varepsilon_0 r)^{-1}$ (Bethe and Salpeter 1957). The angular dependence of solutions is rigidly fixed by a set of spherical harmonics $Y_{lm}(\theta, \varphi)$; the explicit form of some of them is given in table 2.1. The radical function $R_{nl}(r)$ for a hydrogen-like atom obeys the differential Laguerre equation. This function is presented as follows:

$$R_{nl}(r) = -N(2zr/na_0)^l \exp(-zr/na_0) L_{n+1}^{2l+1}(2zr/na_0) \tag{C.2}$$

with

$$N(n, l) = [(2z/na_0)^3 (n-l-1)!/2n[(n+l)!]^3]^{1/2}. \tag{C.3}$$

Here $l = 0, 1, 2, 3, \ldots, n \geqslant l+1$. $a_0 = 4\pi\varepsilon_0\hbar^2/e^2 m$ is the Bohr radius, L_{n+1}^{2l+1} is the associated Laguerre polynomial (Arfken 1985). Since the Laguerre polynomials, with various quantum numbers n and l, are mutually orthogonal the radial functions R_{nl} are also orthogonal.

The use of accurate functions (C.2) in quantum chemistry calculations as basis sets is inconvenient because these functions include polynomials of various powers of r as multipliers. This makes the calculation of molecular integrals very complicated. Therefore, the accurate orbitals R_{nl} are approximated by simpler functions. The requirements for the choice of their form can be reduced to two general statements. First, these functions must approximate AOs in the best way, not violating their properties related to the quantum nature of electrons. Second, these functions should allow one to evaluate the integrals in calculating characteristics and properties of many-electron systems in the simplest manner.

The calculations of many-electron systems are most frequently carried out using the following analytical approximations of atomic wave-functions (the atomic system of units is used here).

(i) *Slater-type orbitals (STOs)*. These functions have the form

$$\Phi_{nlm}=(2\xi)^{n+1/2}[(2n)!]^{-1/2}r^{n-1}\exp(-\xi r)Y_{lm}(\theta,\varphi). \tag{C.4}$$

They correspond to the central force field with the potential

$$V(r)=-\xi n/r+[n(n-1)-l(l+1)]/2r^2 \tag{C.5}$$

i.e. for $l=-1$ the STO transfers into the AO of a hydrogen-like atom with $\xi=z/n$. Even a small number of STOs approximates well the HF AO. The STO of 1s type satisfies the asymptotic Kato condition (2.51) and correctly reproduces the electron behaviour near nuclei. The multiplier, which exponentially decreases with distance, allows one to describe correctly ED in the atomic valence shell. At the same time, the STOs with $l>1$ are nodeless, and the STOs with the same l but with different n are not mutually orthogonal.

The well known Slater rules (see e.g. Flygare 1978) can be used for the approximate evaluation of ξ values when a single STO is used as the AO. Clementi and Roetti (1974) have obtained ξ values corresponding to the atomic energy minimum for atoms up to $Z\leqslant 54$. They have also presented more accurate HF AOs approximated by STO sets. Hehre *et al* (1969) have averaged the data of quantum chemical calculations of simple molecules, carried out with optimization of exponential multipliers, and proposed a set of 'standard values' of ξ for light atoms (table C.1). These parameters provide a good initial approximation for a multipole analysis of organic crystals. The main disadvantage of STOs is the fact that in calculations of quantum mechanical integrals the functions centred at different atoms cannot be accurately reduced to a single coordinate system (Silverstone 1968, Rae 1978, Avery 1978). This circumstance necessitated the search for other types of basis function.

(ii) *Gaussian-type orbitals (GTOs)*. These functions, corresponding to the potential

$$V(r)=2\alpha^2r^2+[n(n-1)-l(l+1)]/2r^2 \tag{C.6}$$

are used either in the form of spherical functions

$$\Phi_{nlm}=\{[2(n-1)!/(2n-1)] \times[(2\alpha)^{2n+1}/\pi]^{1/2}\}^{1/2}r^{n-1}\exp(-\alpha r^2)Y_{lm}(\theta,\varphi) \tag{C.7}$$

or in the form of Cartesian functions

$$g_{pqk}=\{2^{2(p+q+k)+3/2}\alpha^{p+q+k+3/2}/2/(2p-1)!!(2q-1)!!(2k-1)!!\pi^{3/2}\}^{1/2} \times x^py^qz^k\exp(-\alpha r^2) \tag{C.8}$$

Table C.1 Standard molecular exponents for K, L, and M shells of atoms with $Z < 18$. (Reprinted with permission from Hehre *et al* 1969.)

Atom	K	L	M
H	1.24		
He	1.69		
Li	2.69	0.80	
Be	3.68	1.15	
B	4.68	1.50	
C	5.67	1.72	
N	6.67	1.95	
O	7.66	2.25	
F	8.65	2.55	
Ne	9.64	2.88	
Na	10.61	3.48	1.75
Mg	11.59	3.90	1.70
Al	12.56	4.36	1.70
Si	13.53	4.83	1.75
P	14.50	5.31	1.90
S	15.47	5.79	2.05
Cl	16.43	6.26	2.10
Ar	17.40	6.74	2.30

$((2p-1)!!=1\times3\times5\cdots(2p-1))$. The latter are real functions. The pre-exponential multipliers in (C.8), which depend on non-negative integers p, q, and k, describe the angular dependences of these functions. For $p+q+k=0$ we obtain a single 1s orbital, for $p=q=k=1$, three orbitals of 2p type, for $p+q+k=2$, six orbitals of 3d type (five of them are linearly independent), etc. The GTOs with the same value of quantum number l are non-orthogonal, like STOs.

The main advantage of GTOs, which facilitates the calculation of integrals, proceeds from the property that the product of two Gaussians centred at points $\boldsymbol{A}(A_x, A_y, A_z)$ and $\boldsymbol{B}(B_x, B_y, B_z)$ is equivalent to a third one centred at point $\boldsymbol{P}=(\alpha_1\boldsymbol{A}+\alpha_2\boldsymbol{B})/(\alpha_1+\alpha_2)$ that lies on line AB (Shavitt 1962). For example, for the GTO of the sth type

$$\exp(-\alpha_1 r_A^2)\exp(-\alpha_2 r_B^2) = \exp[-\alpha_1\alpha_2 r_{AB}^2/(\alpha_1+\alpha_2)]\exp-(\alpha_1+\alpha_2)r_p^2]; \qquad \text{(C.9)}$$

here $r_{AB}^2=(A_x-B_x)^2+(A_y-B_y)^2+(A_z-B_z)^2$. The integrals with s-type functions can be calculated especially easily. Then, differentiating their analytical expressions by using the relations of type

$$X_A\exp(-\alpha r_A^2)=(1/2\alpha)(\delta/\delta A_x)\exp(-\alpha r_A^2) \qquad \text{(C.10)}$$

they can be transformed into the integrals over p, d, etc functions.

As Klahn and Bingel (1977) have shown, another important circumstance is that the variational calculations with more and more extended Gaussian bases lead to results which converge to accurate values. In other words, that basis set is complete in the energy space.

In practical computations, either the Gaussian approximations to AOs are applied or the GTOs, whose parameters are found by optimizing the energy functional, are directly used as basis functions. In the first case the brief notation STO-NG is used for designating the minimal basis set, where $N = 2$–6 is the number of Gaussians used in the STO expansion. The double-zeta (DZ) basis set, in which the inner orbitals are approximated by four GTOs and three and one GTOs approximate the inner and outer regions of the valence AO, is designated 4-31G. The STO-4-31G* basis set is double zeta, extended by adding polarization functions of d type for all atoms of the second and third periods. If, in addition, the basis set includes polarization p AOs of hydrogen atoms, then it is designated as the 4-31G** basis set. The diffuse functions with small α values are denoted by + and ++, respectively.

If the calculation is carried out directly with the Gaussian basis set, the latter is designated by (NsMpLd), where N, M, and L indicate the number of Gaussians used of s, p, and d types, respectively. In order to lower the number of calculated integrals, the so-called 'contracted' sets of GTOs are often used. These sets represent fixed linear combinations of Gaussians of the same symmetry, centred at the same nucleus.

Primorac *et al* (1989) have proposed the expression of hydrogen-like AOs in terms of Hermitian polynomials and have tabulated parameters of short expansions of these functions. In such a form hydrogen-like AOs are suitable for calculations as well.

The convenience of calculations is achieved at the expense of the necessity of using a greater number of GTOs than STOs for correct description of AOs. This is due to the fact that the GTOs inadequately describe the behaviour of electrons both near atomic nuclei, and far from them, not satisfying conditions (2.51) and (2.53) (figure C.1). Besides, for d, f, etc AOs the number of Cartesian GTOs with corresponding symmetry exceeds the number of linearly independent functions. This can lead to instability of the computational procedures because nearly linear dependences arise.

The work of Baerends *et al* (1985) allows one to compare the results of various calculations of CO molecule characteristics obtained with different basis sets (table C.2). The data obtained show that, as the basis set is extended, the results confidently approach the exact numerical calculation data obtained by the HFS method. At the same time, the results for the dipole moment and dissociation energy differ from the experimental data since a single-determinant approximation is evidently insufficient for their accurate calculation.

Table C.2 Spectroscopic properties of CO as a function of basis set size, including basis-free results (Baerends *et al* 1985): R_e is the equilibrium distance, ω_e is the harmonic vibration frequency, D_e is the dissociation energy, μ_0 is the dipole moment, and μ_1 is the derivative of the dipole moment.

Basis	R_e (bohr)	ω_e (cm^{-1})	D_e (eV)	μ_0[a] (debye)	μ_1[a] (D $bohr^{-1}$)	Electric field gradient on nuclei: C	Electric field gradient on nuclei: O
DZ	2.175	2003	10.196	+0.259	1.67	0.745	0.637
DZD	2.137	2179	11.817	−0.076	1.68	0.819	0.661
DZDF	2.136	2183	11.914	−0.072	1.69	0.845	0.700
TZ	2182	1985	10.329	−0.060	1.76	0.921	0.606
TZD	2.139	2154	11.785	−0.230	1.78	0.985	0.676
TZDD	2.135	2152	11.808	−0.218	1.73	0.983	0.693
TZDF	2.132	2170	11.911	−0.197	1.76	1.001	0.730
TZDDF	2.132	2167	11.929	−0.220	1.75	1.002	0.730
QZ	2.177	2012	10.502	−0.104	1.65	0.858	0.571
QZDDF	2.132	2173	11.966	−0.243	1.71	0.980	0.744
Gaussian basis set	2.132	2160	11.98	−0.24	1.65		
Numerical HFS	2.128	2174	12.084	−0.241[b]	1.717[b]	0.9691(5)[c]	0.7559(15)[c]
	2.13	2170	12.0	−0.24	1.73		
Experiment	2.132	2170	11.2	−0.122	1.66		

[a] At 2.132 bohr.
[b] Using the theoretical $R_e = 2.127\,77$ au.
[c] At 2.130 bohr, in a 105×145 grid.

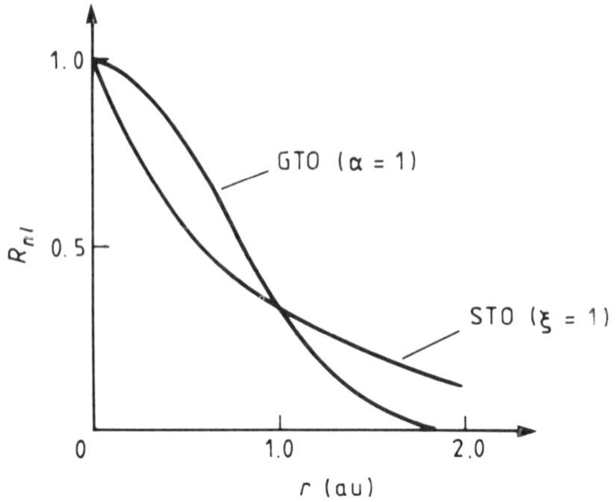

Figure C.1 The radial dependences of STOs and GTOs.

(iii) *Improvements to the basis sets.* Bishop (1964) and Steiner and Sykes (1972) have shown that the addition into the Gaussian basis set of functions of type

$$\psi=\begin{cases} r^l(\rho-r)^t & r\leqslant\rho \\ 0 & r\geqslant\rho \end{cases} \tag{C.11}$$

eliminates GTO defects caused by violation of the Kato (1957) cusp condition (2.51). Approximate cusps on nuclei can also be provided by using non-local operator identities for Dirac's delta function (Challacombe and Cioslowky 1994). The undesirable linear dependence effect can be weakened by using either piecewise polynomial functions (Gazquez and Silverstone 1977) or even-tempered basis sets, such as proposed by Raffenetti (1973) and Feller and Ruedenberg (1979)

$$\Phi_{klm}=N_{nl}\exp(-\xi_k r^p)r^l Y_{lm}(\theta,\varphi). \tag{C.12}$$

Here $p=1$ (STO) or 2 (GTO), $\xi_k=\gamma\beta^k$, $k=1, 2,\ldots, N$. Such a form of ξ_k appears due to the fact that the independent optimization of exponents in the HF self-consistent calculations yields a nearly linear dependence $\ln(\xi_k)=f(k)$. For each spherical harmonic $Y_{lm}(\theta,\varphi)$ the even-tempered functions have the same power of r. Thus, the basis set consists of 1s, 2p, 3d, etc functions, which is suitable for composing linear combinations of primitive functions. Besides, these functions are linearly independent for $\beta>1$.

The basis set extension is often preferable to the optimization of exponential multipliers. Silver *et al* (1978) have proposed the so-called universal basis sets, which are moderately large, flexible enough, and can be transferred from one system into another. By choosing basis set parameters from the consideration of small molecules these sets can be used in calculations of more complicated systems.

APPENDIX D

ELECTROSTATIC POTENTIAL DISTRIBUTION IN ATOMS

The presentation of molecules and crystals in the form of a superposition of spherical atoms (promolecules or procrystals), plus an addition related to the change of its ED caused by chemical bonding, is used extensively in this book. The EP of a procrystal is the sum of the potentials of the atoms, $\varphi_{at}(\mathbf{r})$, and the behaviour of members of this sum determines, in many respects, the main features of the potential distribution in a real crystal. Varnek (1985), Sommer-Larsen *et al* (1990), and Su and Coppens (1992) have obtained analytical expressions for the atomic EP in terms of orbital contributions. The STO and GTO expansions of AOs were considered. Let us examine the radial distributions of the orbital contributions of the total φ_{at} for atoms ranging from H to Ca, calculated by Varnek (1985) with HF wave-functions expanded over 4–8 STOs, depending on the AO type (Clementi and Roetti 1974). Corresponding curves are given in figure D.1. They are smooth and everywhere positive with several points of inflection and asymptotic behaviour of $1/r$ type at infinity. A zero derivative of the orbital contributions at a nucleus corresponds to a zero of electrostatic forces acting on a nucleus from the electrons (in accordance with the Hellmann–Feynman theorem). At large distances the potential of distributed charge associated with AOs, in accordance with the Gauss theorem, becomes close to the point charge potential, which is equal in magnitude to the orbital population situated on the nucleus. This explains the form of asymptotic behaviour at $r \to \infty$. One should also point out that the closer the orbital is to a nucleus, the more rapidly the potential curve coincides with a hyperbola. Figure D.1 shows that electrons of the 1s shell of elements of periods II–III practically completely screen the nuclear charges for $r \geqslant 2$ au. A similar degree of screening for other orbitals is achieved at higher r values. The relation between the orbital contraction ratio and its screening effect explains the mutual position of the potential curves at small and medium distances from a nucleus. Since the orbital contraction ratio increases with atomic number, the curves corresponding to the orbital contribution with greater atomic number lie higher.

The orbital contributions to the EP reflect the features of the distribution

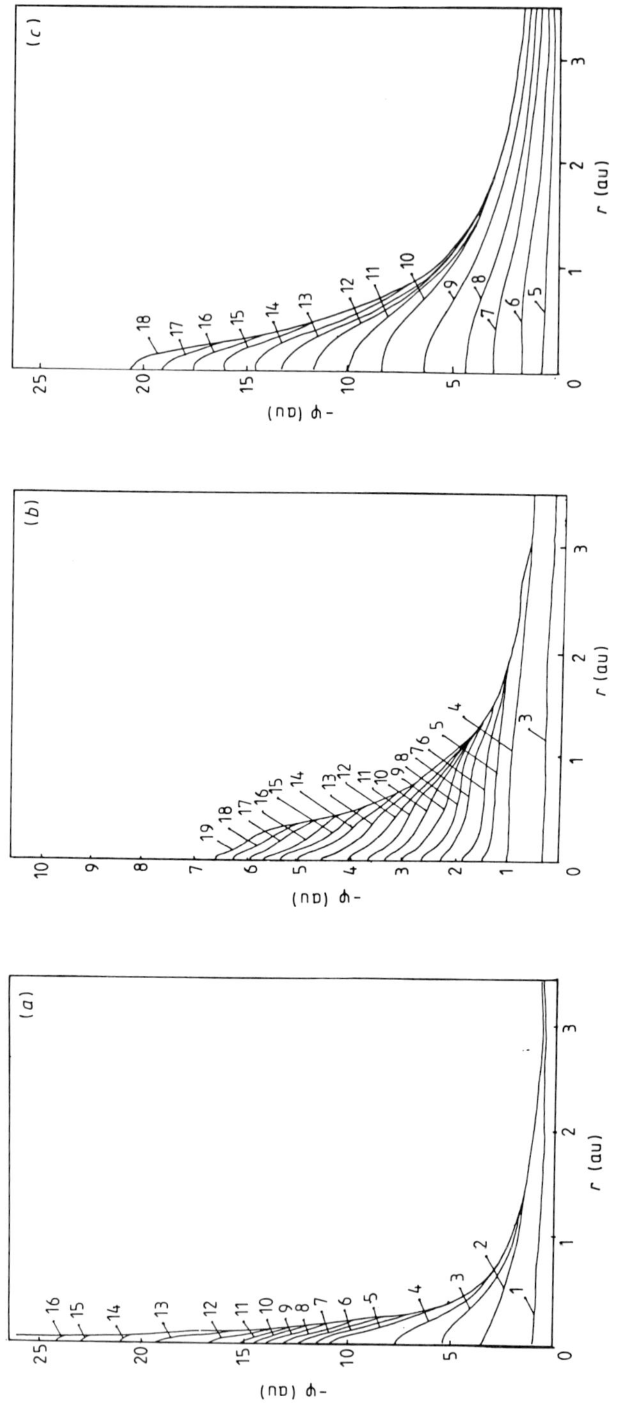

Figure D.1 The orbital radial distributions of the EPs for atoms from H to Ag: (*a*) 1s orbitals; (*b*) 2s orbitals; (*c*) 2p orbitals; (*d*) 3s orbitals; and (*e*) 3p orbitals. The numbers correspond to atomic numbers (Varnek 1985).

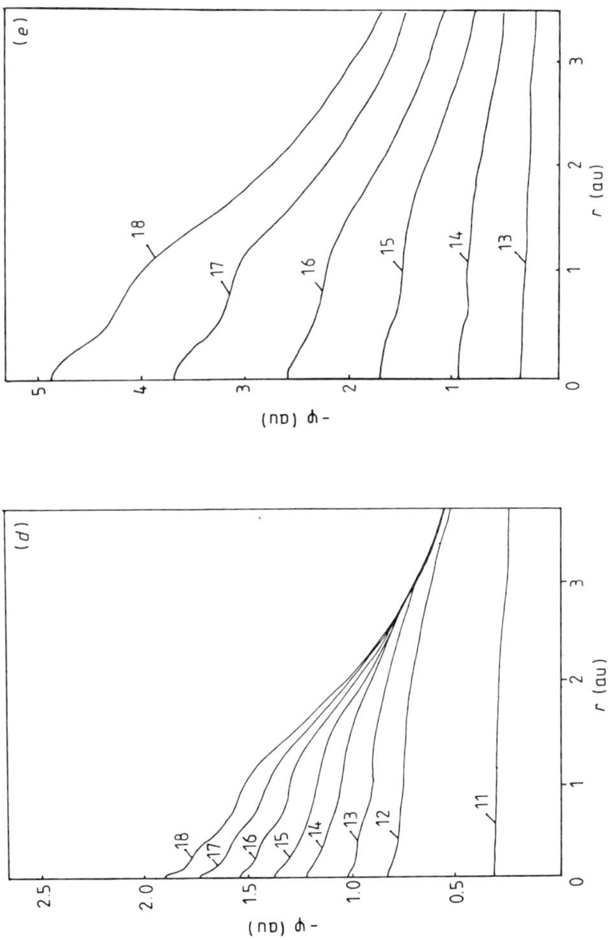

Figure D.1 *Continued*

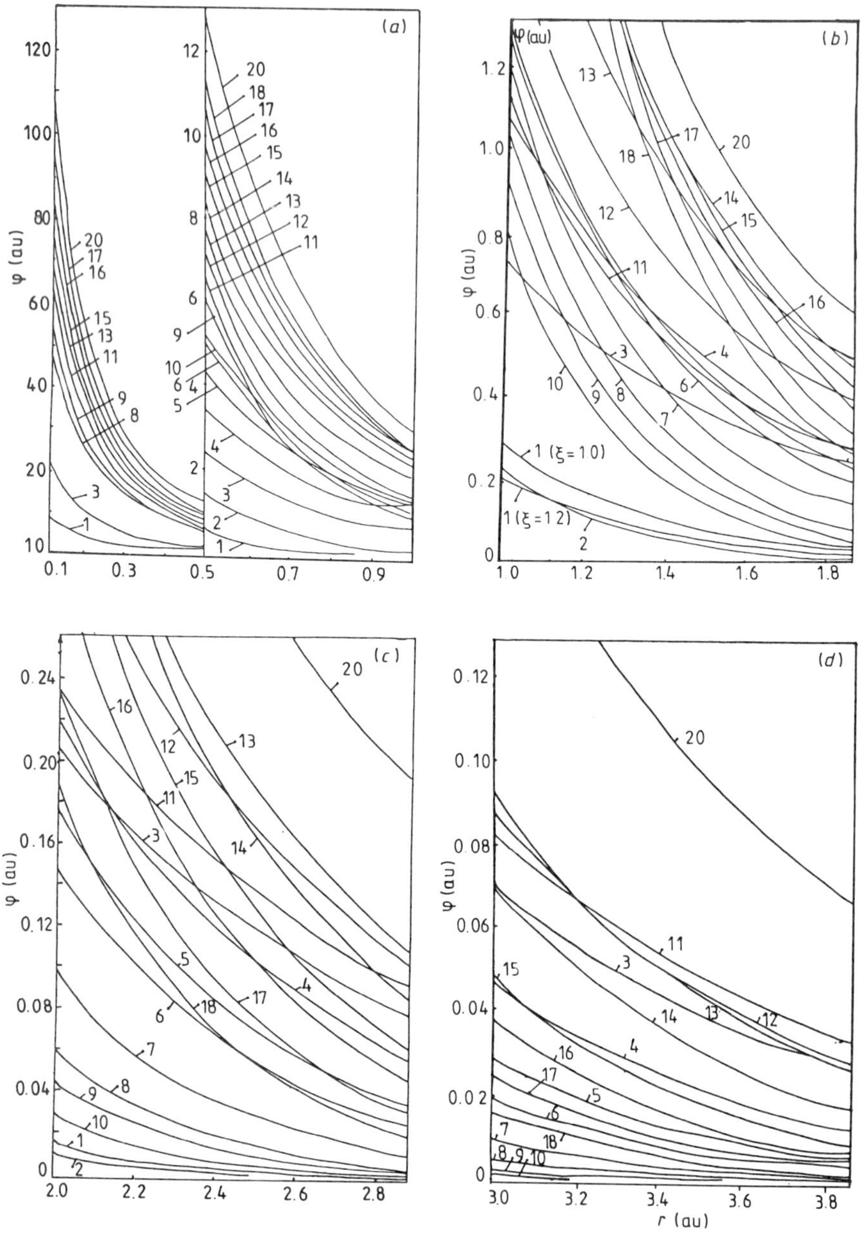

Figure D.2 The radial distribution of the EP of free atoms from H to Ca: (*a*) $0.1 \leqslant r < 1.0$ au; (*b*) $1.0 < r < 1.9$ au; (*c*) $2.0 < r < 2.9$ au; (*d*) $3.0 < r < 3.9$ au. The numbers correspond to atomic numbers (Varnek 1985).

of ED described by the given AO. In particular, the number of inflection points in each curve is equal to the number of nodes of the radial distribution function of the orbital electron density, $D_{AO}(r)=4\pi r^2\rho_{AO}(r)$.

The radial EP distributions of free atoms ranging from H to Ca are presented in figure D.2. The plot of the atomic potential is seen to be a smooth curve with a singularity at $r=0$ and an asymptotic approach to zero as r grows. At small distances the potential grows with increasing atomic number for elements of the same period and at large r values the opposite dependence is found.

Within the intervals from 0.5 to 2.15 au for atoms of period II, $\varphi_{at}(r)$ curves of each atom intersect all the rest of the curves for atoms of 'their own' period. The plots of the potential of some atoms of different periods intersect, the $\varphi_{at}(r)$ curve for the Na atom intersecting with the plot for Be, B, C, and N atoms at two points. One should note, however, that the atomic potential curves of elements related to the same group of the periodic system do not intersect at all. The plots for elements with a greater atomic number lie higher.

The mutual intersection of curves may be explained as follows. At small r values the potential is mainly determined by the field of an unscreened nucleus. Therefore, the $\varphi(r)$ value in this region increases with the atomic number. For large r values the potential is determined by a screened nuclear charge and by the valence shell electrons. As follows from the analysis of orbital contributions, as the atomic number grows within the same period, the ED contraction ratio increases and, as a result, the degree of nuclear screening by electrons also grows. The increase of an electron shell screening effect leads to the fact that $r \geqslant 2.14$ au for period II and at $r \geqslant 3.98$ au for period III the potential decreases as the atomic number grows.

The potential curves for atoms of the same group do not intersect each other because, for the same valence shell configuration, the ED of atoms with a large nuclear charge is more diffuse and, hence, only screens the nucleus more weakly. The validity of this statement can be confirmed by comparing the curves for a free (with STO exponent value $\xi=1.0$) and a 'compressed' ($\xi=1.2$) hydrogen atom. It is seen from figure D.2 that the curves are always higher for an atom with a more diffuse electron shell.

APPENDIX E

CREATION AND ANNIHILATION OPERATORS

In some problems, related to the transitions of quantum particles (quasiparticles) from one vibrational state into another or to the 'radiation–substance' interaction, it can be convenient to use the operators of creation and annihilation of particles, $\hat{a}^+$ and $\hat{a}$. These real dimensionless operators can be considered with a single linear harmonic oscillator as example. Its Hamiltonian (B.2) can be written in the form

$$\hat{H}_0 = \hat{p}^2/2m + \tfrac{1}{2}m\omega^2\hat{x}^2 \tag{E.1}$$

where $\hat{p}$ is the momentum operator; $\hat{p}^2 = -\hbar^2(\partial^2/\partial x^2)$. One can further introduce the linear transformations (Reisland 1973)

$$\begin{aligned} \hat{x} &= (\hbar/2\mu\omega)^{1/2}(\hat{a}^+ + \hat{a}) \\ \hat{p} &= \mathrm{i}(\mu\omega\hbar/2)^{1/2}(\hat{a}^+ - \hat{a}). \end{aligned} \tag{E.2}$$

The pairs $\hat{x}$ and $\hat{p}$ are quantum mechanical conjugated variables, for which the commutator

$$\hat{x}\hat{p} - \hat{p}\hat{x} = \mathrm{i}\hbar \tag{E.3}$$

is valid. Substituting (E.2) into (E.3), one can easily see that for new operators $\hat{a}^+$ and $\hat{a}$ a similar commutator has the form

$$\hat{a}\hat{a}^+ - \hat{a}^+\hat{a} = 1. \tag{E.4}$$

Substituting (E.2) into (E.1), one obtains

$$\hat{H}_0 = (\hbar\omega/2)(\hat{a}^+\hat{a} + \hat{a}\hat{a}^+) \tag{E.5}$$

and then, using (E.4), one gets

$$\hat{H}_0 = \hbar\omega(\hat{a}^+\hat{a} + \tfrac{1}{2}) = \hbar\omega(\hat{n} + \tfrac{1}{2}). \tag{E.6}$$

Here $n = \hat{a}^+\hat{a}$ is the particle number operator whose eigenvalues are zero and positive integers; they determine the eigenvalues of $\hat{H}_0$. We designate the eigenstates of Hamiltonian $\hat{H}_0$ as $|n\rangle$ and write down the Schrödinger equation for a harmonic oscillator:

$$\hat{H}_0|n\rangle = \hbar\omega(\hat{a}^+\hat{a} + \tfrac{1}{2})|n\rangle = E_n|n\rangle. \tag{E.7}$$

Multiplying both sides of this expression by $\hat{a}^+$ on the left and using (E.4) one obtains

$$\hbar\omega(\hat{a}^+\hat{a}a^+ - \hat{a} + \tfrac{1}{2}\hat{a}^+)|n\rangle = E_n\hat{a}^+|n\rangle \tag{E.8}$$

and, further,

$$\hbar\omega(\hat{a}^+\hat{a} + \tfrac{1}{2})\hat{a}^+)|n\rangle = \hat{H}_0\hat{a}^+|n\rangle = (E_n + \hbar\omega)\hat{a}^+|n\rangle. \tag{E.9}$$

This expression is the equation for energy eigenvalues: the state $a^+|n\rangle$ is the energy eigenstate with eigenvalue $(E+\hbar\omega)$. Determining the new eigenstate and eigenvalue as

$$|n+1\rangle = \hat{a}^+|n\rangle \tag{E.10}$$

$$E_{n+1} + E_n + \hbar\omega \tag{E.11}$$

we rewrite (E.9) in the form

$$\hat{H}_0|n+1\rangle = E_{n+1}|n+1\rangle. \tag{E.12}$$

If we repeat our considerations by multiplying both sides of expression (E.7) by a on the left then we can arrive at the equation

$$\hat{H}_0\hat{a}|n\rangle = (E_n - \hbar\omega)\hat{a}|n\rangle \tag{E.13}$$

which can easily be transformed, by using relations $|n-1\rangle = a|n\rangle$ and $E_{n-1} = E_n - \hbar\omega$, to the form

$$\hat{H}_0|n-1\rangle - E_{n-1}|n-1\rangle. \tag{E.14}$$

Thus, the action of operator $\hat{a}^+$ on the state vector $|n\rangle$ transfers the latter in the $|n+1\rangle$ state vector with increasing oscillation energy by quantum $\hbar\omega$ (exciting a new vibrational mode). The action of operator $\hat{a}$ corresponds to the $|n\rangle \rightarrow |n-1\rangle$ transition with simultaneous energy decrease by the value of $\hbar\omega$. If one acts with operator H_0 (E.6) on the ground state vector $|0\rangle$, then, taking into account that $\hat{a}|0\rangle = 0$ (since the oscillator energy cannot be negative), one obtains

$$\hat{H}_0|0\rangle = \tfrac{1}{2}\hbar\omega|0\rangle = E_0|0\rangle \tag{E.15}$$

where E_0 is the zero vibration energy. Then, taking into account (E.11), we can see that the expression for eigenvalues of harmonic oscillator energy has the standard form, which we have obtained earlier in a different way, namely

$$E_n = \hbar\omega(n + \tfrac{1}{2}). \tag{E.16}$$

As we have seen, the introduction of operators $\hat{a}^+$ and $\hat{a}$ allowed us to give a simple physical interpretation of quantum harmonic oscillator transitions from one vibrational state into another. At the same time, these operators are not Hermitian, i.e. they do not describe the quantities observed.

So far we have ignored the question concerning the normalization of state vector $|n\rangle$. If one takes this normalization into account, then expression (E.10) should be rewritten as

$$\hat{a}^{+}|n\rangle = (n+1)^{1/2}|n+1\rangle. \tag{E.17}$$

Accordingly, for operator $\hat{a}$ the relation

$$\hat{a}|n\rangle = n^{1/2}|n-1\rangle \tag{E.18}$$

is valid.

The $|n\rangle$ states are eigenstates of two operators: $\hat{H}_0$ and $\hat{n}$. Acting with $\hat{n}$ on $|n\rangle$, one obtains

$$\hat{n}|n\rangle = \hat{a}^{+}\hat{a}|m\rangle = n|n\rangle. \tag{E.19}$$

The eigenvalue of operator $\hat{n}$ in this expression shows what number of energy quanta $\hbar\omega$ exist above the ground state for an oscillator.

The arbitrary excited state $|n\rangle$ of an oscillator can be expressed in terms of the ground state $|0\rangle$ by the following formula (Loudon 1973):

$$|n\rangle = (n!)^{-1/2}(\hat{a}^{+})^{n}|0\rangle. \tag{E.20}$$

Consider now the action of creation and annihilation operators as applied to the change of the electromagnetic field state during scattering. The x-ray radiation field in some volume Ω_0 can be considered to be a set of quantum oscillators, each of which corresponds to some certain field mode. The modes are characterized by wave-vectors $\boldsymbol{k}_l$ and polarization state vectors $\boldsymbol{e}(\boldsymbol{k}_l, \mu)$. The number of photons with wave-vector $\boldsymbol{k}_l$ in volume Ω_0 is determined by the eigenvalue of operator $\hat{n}(\boldsymbol{k}_l, \mu)$, and the state of the $\boldsymbol{k}_l$th mode by the eigenvector of this operator $|n_{\boldsymbol{k}_l}\rangle$. The state vector of the total field is written as follows:

$$|n_{\boldsymbol{k}_0}, n_{\boldsymbol{k}_1}, n_{\boldsymbol{k}_2}, \ldots\rangle.$$

Let us now relate the creation and annihilation in scattering of photons or of energy quanta of modes, $\hbar\omega(\boldsymbol{k}_l, \mu)$, to the action of operators $\hat{a}^{+}(\boldsymbol{k}_l, \mu)$ and $\hat{a}(\boldsymbol{k}_l, \mu)$ for which relations (E.17) and (E.18) are satisfied. These operators, each of which acts on its own mode $(\boldsymbol{k}_l, \mu)$ only, are related to the operator of a vector potential of the electromagnetic field (Loudon 1973) as follows:

$$\hat{A}(\boldsymbol{r}, t) = \left(\frac{\hbar}{2\varepsilon_0\Omega_0}\right)^{1/2} \sum_{\boldsymbol{k}_l, \mu} \omega_l^{1/2}\{\hat{a}(\boldsymbol{k}_l, \mu)\exp[\mathrm{i}(\boldsymbol{k}_l\cdot\boldsymbol{r} - \omega_l t)]$$
$$+ \hat{a}^{+}(\boldsymbol{k}_l, \mu)\exp[-\mathrm{i}(\boldsymbol{k}_l\cdot\boldsymbol{r} - \omega_l t)]\boldsymbol{e}(\boldsymbol{k}_l, \mu). \tag{E.21}$$

The vector potential enters as a square in the perturbation operator $\hat{H}'$ of a system. The action of $\hat{A}^2(\boldsymbol{r}, t)$ on the field state vector results in the emission or absorption of two photons or in the annihilation of a photon, moving in the $\boldsymbol{k}_0$ direction, and in the appearance of a photon moving in the $\boldsymbol{k}_1$ direction (the scattering). Let us assume for simplicity that the radiation field consists of only one photon: $|1_{\boldsymbol{k}_0}, 0, 0, \ldots\rangle$. The term from $\hat{A}^2(\boldsymbol{r}, t)$ that is responsible for scattering has the form:

$$\hat{A}_{\mu\nu}(\boldsymbol{k}_0, \boldsymbol{k}_1) = (\hbar/2\varepsilon_0\Omega_0)(\omega_0\omega_1)^{-1/2}\{\hat{a}(\boldsymbol{k}_0, \mu)\hat{a}^+(\boldsymbol{k}_1, \nu)\exp[i(\boldsymbol{k}_0\cdot\boldsymbol{r} - \boldsymbol{k}_1\cdot\boldsymbol{r})] \times \exp[i(\omega_0 - \omega_1)t] + \hat{a}^+(\boldsymbol{k}_0, \mu)\exp[i(\boldsymbol{k}_0\cdot\boldsymbol{r} - \boldsymbol{k}_1\cdot\boldsymbol{r})] \times \exp[-i(\omega_0 - \omega_1)t]\} \times [\boldsymbol{e}(\boldsymbol{k}_0, \mu)\cdot\boldsymbol{e}(\boldsymbol{k}_1, \nu)] \tag{E.22}$$

and to describe the scattering one must calculate the matrix element

$$[A_{\mu\nu}(\boldsymbol{k}_0, \boldsymbol{k}_1)]^2_{np} = \langle\varphi_n(r), 0_{k_0}, 1_{k_1}, \ldots|(\hat{A}_{\mu\nu}(\boldsymbol{k}_0, \boldsymbol{k}_1))|\varphi_p(r), 1_{k_0}, 0_{k_1}, \ldots\rangle. \tag{E.23}$$

Here $\varphi_p(\boldsymbol{r})$ and $\varphi_n(\boldsymbol{r})$ are the wave-functions of a particle which scatters a photon, before and after scattering. Substituting (E.22) into (E.23) and separating wave-functions of the scattering particle and of the field, one obtains

$$[A_{\mu\nu}(\boldsymbol{k}_0, \boldsymbol{k}_1)]^2_{np} = (\hbar/2\varepsilon_0\Omega_0)(\omega_0\omega_1)^{-1/2}\langle n|\exp[i(\boldsymbol{k}_0\cdot\boldsymbol{r} - \boldsymbol{k}_1\cdot\boldsymbol{r})]|p\rangle \times \exp[-i(\omega_0 - \omega_1)t]\{\langle 0_{k_0}, 1_{k_1}|\hat{a}(\boldsymbol{k}_0, \mu)a^+(\boldsymbol{k}_1, \nu)|1_{k_0}, 0_{k_1}\rangle + \langle 0_{k_0}, 1_{k_1}|\hat{a}^+(\boldsymbol{k}_1, \nu)a(\boldsymbol{k}_0, \mu)|1_{k_0}, 0_{k_1}\rangle\} \times [\boldsymbol{e}(\boldsymbol{k}_0, \mu)\cdot\boldsymbol{e}(\boldsymbol{k}_1, \nu)]. \tag{E.24}$$

Acting with operators $\hat{a}^+$ and $\hat{a}$, which act here as photon creation and annihilation operators, on the radiation field state vectors, one can easily see that the expression in curly brackets is equal to two in this case. Then the matrix element (E.23) in question takes the form

$$[A_{\mu\nu}(\boldsymbol{k}_0, \boldsymbol{k}_1)]^2_{np} = (\hbar/\varepsilon_0\Omega_0)(\omega_0\omega_1)^{-1/2}\langle n|\exp[i(\boldsymbol{k}_0\cdot\boldsymbol{r} - \boldsymbol{k}_1\cdot\boldsymbol{r})]|p\rangle \times \exp[-i(\omega_0 - \omega_1)t][\boldsymbol{e}(\boldsymbol{k}_0, \mu)\cdot\boldsymbol{e}(\boldsymbol{k}_1, \nu)] \tag{E.25}$$

i.e. it depends on the time and on the state of the scattering matter; the term related to field polarization is insignificant here, because the incident radiation can also be polarized in this case. As can be seen from (E.25), the state of scattering matter can be judged by analysing the scattering data.

APPENDIX F

RECIPROCAL SPACE

The interpretation of diffraction from a crystal as a reflection of the radiation from a set of crystallographic planes is widely used. The orientation of each of these planes may be fixed by a set of points on it or by the direction of a vector normal to this plane. The specification of a plane's orientation by a normal vector can be made as follows. The interplanar distances, $d(hkl)$, where h, k, and l are Miller indices, are specified for each crystallographic plane. The reciprocal values $H(hkl)=d^{-1}(hkl)$ are calculated and the lengths $H(hkl)$ are put from the origin along the normal to the crystallographic planes (hkl). The ends of these vectors (equal to $\boldsymbol{n}H$) create a regular set of points. The set is called the reciprocal lattice in which each $\boldsymbol{H}$ vector corresponds to a definite orientation of an atomic (crystallographic) plane in the position space. It can be shown that the ordered system of points in position space leads to an ordered system of points in reciprocal (Fourier) space. This construction is carried out in figure F.1 for a two-dimensional orthorhombic lattice case.

However, this procedure is not needed in practice, because the periods and angles of reciprocal and position space lattices are connected in a simple way. The relations for vectors $\boldsymbol{a}$ and $\boldsymbol{a}^*$ are

$$\boldsymbol{a}^*=[\boldsymbol{bc}]/\boldsymbol{a}\cdot[\boldsymbol{bc}] \qquad \boldsymbol{a}=[\boldsymbol{b}^*\boldsymbol{c}^*]/\boldsymbol{a}^*\cdot[\boldsymbol{b}^*\boldsymbol{c}^*] \tag{F.1}$$

(symbol * refers to reciprocal space); the same expressions for $\boldsymbol{b}$ and $\boldsymbol{b}^*$ and $\boldsymbol{c}$ and $\boldsymbol{c}^*$ vectors are valid. Angles α, β, and γ and α^*, β^*, and γ^* can be determined using equations (F.1).

An example of the position space and reciprocal lattices of a monoclinic crystal is given in figure F.2.

The notions of wave-vector and momentum space have much in common with reciprocal space; they are widely used in solid state physics. In these cases the additional multiplier 2π appears sometimes in defining the reciprocal vectors. However, Born and Huang (1954) have nevertheless used in their textbook the crystallographic definition.

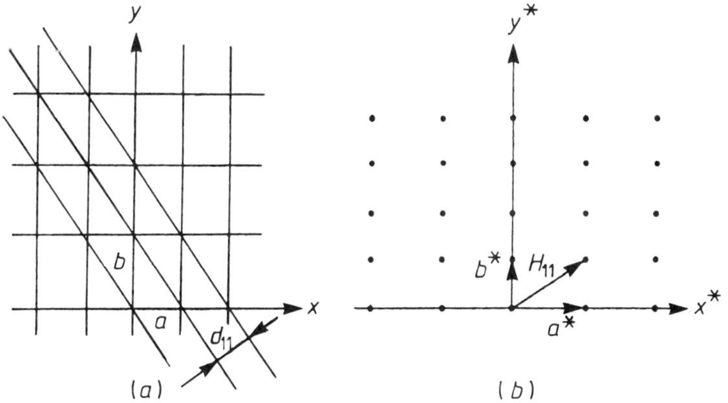

Figure F.1 A two-dimensional crystal and reciprocal lattices: (*a*) atomic planes in the orthorhombic lattice; (*b*) the reciprocal lattice.

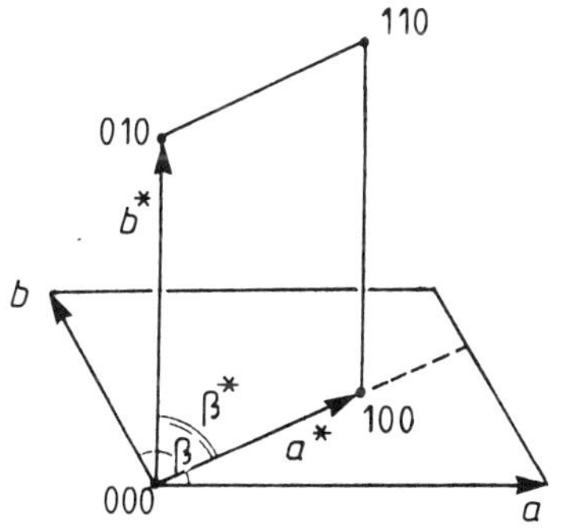

Figure F.2 The mutual disposition of crystal and reciprocal monoclinic lattice axes. The c and c^* axes are directed upward perpendicular to the drawing plane. Black circles are the reciprocal lattice points.

The reciprocal lattice vector can be expressed via reciprocal lattice periods as

$$\boldsymbol{H}(hkl) = h\boldsymbol{a}^* + k\boldsymbol{b}^* + l\boldsymbol{c}^*. \tag{F.2}$$

The $\boldsymbol{H}$ vectors are measured in m^{-1} or in $Å^{-1}$ or in au.

If the monochromatic incident beam of x-rays (or neutrons or electrons) with wave-vector $\boldsymbol{k}_0$ ($k_0/2\pi = 1/\lambda$) is directed along a certain crystal direction, then the Bragg equation for scattering by a crystallographic plane (hkl), $\lambda = 2d(hkl)\sin\theta$, may be rewritten in terms of the reciprocal lattice in the form

$$\sin\theta = \lambda(1/d(hkl))/2 = H(hkl)/2. \tag{F.3}$$

This equation has a simple geometrical interpretation (figure F.3). The

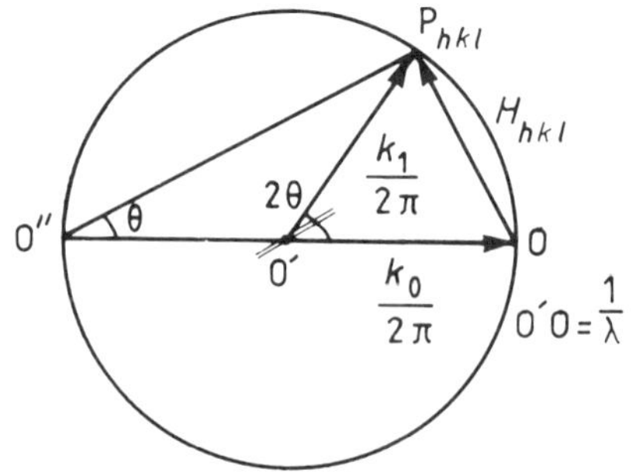

Figure F.3 Bragg reflection in reciprocal space. P is a reciprocal lattice point in the reflecting position; $\boldsymbol{H}_{hkl}=\boldsymbol{S}/2\pi$. Reflecting planes are also shown.

ratio of segment $|\boldsymbol{OP}|$ to the diameter of a sphere with radius $1/\lambda$ (the reflection sphere or Ewald sphere) is

$$\sin\theta=\tfrac{1}{2}\mathrm{OP}/\mathrm{OO}'=\mathrm{OP}/(2/\lambda)=(\mathrm{OP}/2)\lambda. \tag{F.4}$$

The comparison of the latter two equations shows that they are identical provided OP is equal to $H(hkl)$. This fact is used in the following Ewald construction. The Ewald sphere touches the zero point, O, of a reciprocal lattice (the origin). If the point (hkl) falls on the Ewald sphere, then the equations (F.3) and (F.4) become identical, the Bragg condition is satisfied, and the reflection takes place. Moreover, the O′P gives the reflected beam direction coinciding with wave-vector $\boldsymbol{k}/2\pi$; O″P characterizes the reflecting plane (hkl) orientation which is perpendicular to the reciprocal lattice point vector $\boldsymbol{H}(hkl)$. The condition of coherent elastic scattering may now be rewritten in vector form as

$$\boldsymbol{k}_1/2\pi-\boldsymbol{k}_0/2\pi=\boldsymbol{S}/2\pi. \tag{F.5}$$

In the case of inelastic or incoherent scattering this is not fulfilled.

The Ewald construction permits one to easily find the reflected beam direction with respect to the crystal orientation.

In the ideal case, the reciprocal lattice points are dimensionless and the reflection sphere is infinitely thin. When a crystal is arbitrarily oriented, a point can be found on the reflection sphere only accidentally. In order to 'catch' the reflection, it is necessary, taking into account the initial orientation, to turn the crystal around a certain set of axes (see figure 3.3). This procedure is identical to turning the $\boldsymbol{k}_0$ vector around the origin. All crystal rotations in position space have their analogues in the reflection sphere rotation around the origin. The three-dimensional construction for this case is presented in figure F.4. All reciprocal lattice points inside the sphere of radius $2/\lambda$ can, in principle, be placed on the reflection sphere S, and the reflections be observed and recorded by the detector.

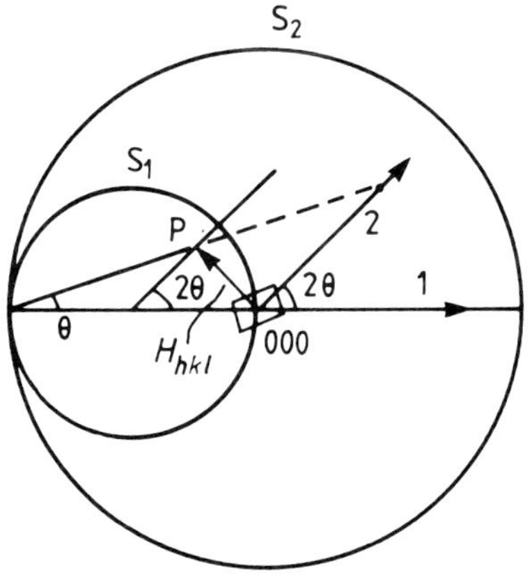

Figure F.4 The three-dimensional construction in reciprocal space. 1 is the incident beam; 2 is the reflected beam; P is the reflecting reciprocal lattice point on the Ewald sphere.

The characteristics of the four-circular x-ray diffractometer construction and the wavelength used restrict the dimension and shape of reciprocal space regions which can be examined diffractometrically.

If other reciprocal lattice points appear simultaneously on the Ewald sphere, so-called simultaneous reflections take place. The influence of these parasitic reflections on the main reflection is as follows. The x-ray beam, once reflected by a system of crystallographic planes, is also in the Bragg condition for another system; therefore, the second reflection can take place. This results in intensity redistribution between the rays and can either reinforce or weaken the reflection measured. In any case the value of intensity measured is distorted.

The simplest way to avoid simultaneous reflections is the experimental one. This can be done by rotation of the crystal examined around the scattering vector. The rotation, indeed, removes all the reciprocal points from the reflection sphere except for (*hkl*) (see figure F.4). This means that single scattering will take place provided another reciprocal point does not get onto the reflecting sphere. The procedure described can easily be performed with the four-circle x-ray goniometer.

The effect of simultaneous reflections can be used positively for the experimental measurement of reflection phases. Owing to interference, the intensity of radiation scattered in the vicinity of the reflection position depends on the ratio of reflection phases and changes significantly from point to point in reciprocal space. The interference analysis permits one to establish the reflection phase. This analysis is based on the Ewald theory of multiple dynamical diffraction (Ewald 1917) which uses Maxwell's equations for a medium with periodic complex dielectric constant, ε. The crystal wave field is a sum of planar waves with different wave-vectors. Solving Maxwell's equations when the Bragg condition is fulfilled, one can find a relation between structure amplitudes of incident, F_0, and scattered, F_i, waves. This approach was realized by Post (1983) and it is widely used nowadays (see e.g. Chang 1987). As a rule, the consideration is limited to centrosymmetric crystals and twofold diffraction. When the centre of symmetry is on the origin, the structure amplitudes have either 0 or π phases. In this case the sign of triple products turns out to be invariant with respect to the origin position in the unit cell. The structure amplitudes of incident and two reflected (F_1 and F_2) waves are related by the cubic equation

$$\left(F_0+\frac{2\varepsilon}{\gamma}\right)^3-\left(\sum_{i=1}^{3} F_i^2\right)\left(F_0+\frac{2\varepsilon}{\gamma}\right)+2F_H F_P F_{P-H}=0 \qquad \text{(F.6)}$$

the sum of non-zero roots of which is zero. The signs of roots are determined by the product $F_1F_2F_{2-1}$. In centrosymmetric structures only two variants of sign distributions, $(-++)$ and $(+--)$, are possible. They result in different scattering intensity distribution in the vicinity of the reciprocal lattice point responsible for the twofold scattering.

The experimental application of the phase measuring method described is rather cumbersome. Nevertheless, a significant number of signs can be determined by this method, not only for monatomic but for more complicated inorganic compounds as well (Gong and Post 1983). The same procedure can also be applied to forbidden reflections, which are especially important for determination of ED in internuclear crystal space and for refinement of anharmonic parameters of thermal atomic vibrations.

APPENDIX G

THERMAL DIFFUSE SCATTERING OF X-RAYS AND NEUTRONS

TDS is an inelastic scattering caused by atomic thermal movements in crystals. Separating the elastic, I_0, the inelastic, I_k, components (see equation (3.32)) and neglecting the anharmonic cross-term in (3.28) the total scattered intensity, I, can be presented as a sum

$$I(\boldsymbol{S}) = I_0(\boldsymbol{S}) + \sum_{k=0} I_k(\boldsymbol{S}). \tag{G.1}$$

I_k ($k = 1, 2, \ldots$) is the intensity of k phonon contributions to TDS, each of which can be written as

$$I_k(\boldsymbol{S}) = N^2 \sum_{\mu\mu'} f_\mu^*(\boldsymbol{S}) f_{\mu'}(\boldsymbol{S}) T_\mu(\boldsymbol{S}) T_{\mu'}^*(\boldsymbol{S}) \exp[i2\pi \boldsymbol{H}\cdot(\boldsymbol{r}_{\mu'} - \boldsymbol{r}_\mu)] I_{\mu\mu'}^{(k)}. \tag{G.2}$$

Here $\boldsymbol{S}$ is the scattering vector and $\boldsymbol{H}$ is the vector of the reciprocal lattice point, around which scattering is investigated. Let us restrict ourselves to one- and two-phonon scattering contributions which predominate over all others near the reciprocal lattice point. In the perturbation theory approximation, which takes up to second-order term contributions into account, the $I_{\mu\mu'}^{(1)}$ value can be presented in the form (Tsarkov and Tsirelson 1990)

$$I_{\mu\mu'}^{(1)} = I_{\mu\mu'}^a + I_{\mu\mu'}^b + I_{\mu\mu'}^c + I_{\mu\mu'}^d + I_{\mu\mu'}^e + I_{\mu\mu'}^f + I_{\mu\mu'}^g. \tag{G.3}$$

$I_{\mu\mu}^0$ here is the contribution of the zero order over perturbation (quasiharmonic). The other terms describe the anharmonic effects: $I_{\mu\mu}^b$ is the first-order term whereas the $I_{\mu\mu}^c$–$I_{\mu\mu}^g$ terms are second-order ones. In the high-temperature limit ($T > \theta_D$)

$$I_{\mu\mu'}^a = \frac{kTN}{s} \sum_\alpha \frac{(\boldsymbol{S}\cdot\boldsymbol{e}_\alpha^\mu)(\boldsymbol{S}\cdot\boldsymbol{e}_\alpha^{\mu'*})}{(m_\mu m_{\mu'})^{1/2}\omega_\alpha^2} \qquad \boldsymbol{q}_\alpha = \boldsymbol{q} - 2\pi\boldsymbol{H} \tag{G.4}$$

$$I^b_{\mu\mu'} = -\mathrm{i}\frac{(kT)^2}{2s^2}\sum_{\alpha\beta\gamma}\left\{\frac{(\boldsymbol{S}\cdot\boldsymbol{e}^{\mu *}_{\alpha})(\boldsymbol{S}\cdot\boldsymbol{e}^{\mu *}_{\beta})(\boldsymbol{S}\cdot\boldsymbol{e}^{\mu' *}_{\gamma})}{m_\mu(m_{\mu'})^{1/2}(\omega_\alpha\omega_\beta\omega_\gamma)^2}\Phi_{\alpha\beta\gamma}\exp(-\mathrm{i}2\pi\boldsymbol{H}_1\cdot\boldsymbol{r}_\mu)\right.$$
$$\left.-\frac{(\boldsymbol{S}\cdot\boldsymbol{e}^{\mu'}_{\alpha})(\boldsymbol{S}\cdot\boldsymbol{e}^{\mu'}_{\beta})(\boldsymbol{S}\cdot\boldsymbol{e}^{\mu'}_{\gamma})}{m_{\mu'}(m_{\mu})^{1/2}(\omega_\alpha\omega_\beta\omega_\gamma)^2}\Phi^*_{\alpha\beta\gamma}\exp(\mathrm{i}2\pi\boldsymbol{H}_1\cdot\boldsymbol{r}_{\mu'})\right\}\qquad \boldsymbol{q}_\gamma=\boldsymbol{q}-2\pi\boldsymbol{H}\quad \text{(G.5)}$$

$$I^c_{\mu\mu'} = -\frac{(kT)^2}{4s^2}\sum_{\alpha\beta\gamma}\left\{\frac{(\boldsymbol{S}\cdot\boldsymbol{e}^{\mu *}_{\alpha})(\boldsymbol{S}\cdot\boldsymbol{e}^{\mu' *}_{\beta})}{(m_\mu m_{\mu'})^{1/2}(\omega_\alpha\omega_\beta\omega_\gamma)^2}\Phi_{-\alpha\beta\gamma-\gamma}\right.$$
$$\left.+\frac{(\boldsymbol{S}\cdot\boldsymbol{e}^{\mu' *}_{\alpha})(\boldsymbol{S}\cdot\boldsymbol{e}^{\mu}_{\beta})}{(m_\mu m_{\mu'})^{1/2}(\omega_\alpha\omega_\beta\omega_\gamma)^2}\Phi^*_{-\alpha\beta\gamma-\gamma}\right\}\qquad \boldsymbol{q}_\alpha=\boldsymbol{q}-2\pi\boldsymbol{H}\quad \text{(G.6)}$$

$$I^d_{\mu\mu'} = \frac{(kT)^2}{2s^2}\sum_{\alpha\beta\gamma}\frac{(\boldsymbol{S}\cdot\boldsymbol{e}^{\mu}_{\gamma})(\boldsymbol{S}\cdot\boldsymbol{e}^{\mu' *}_{\gamma})|\Phi_{\alpha\beta\gamma}|^2}{(m_\mu m_{\mu'})^{1/2}\omega_\gamma(\omega_\alpha\omega_\beta\omega_\gamma^2)^2}\qquad \boldsymbol{q}_\gamma=\boldsymbol{q}-2\pi\boldsymbol{H}\quad \text{(G.7)}$$

$$I^e_{\mu\mu'} = \frac{(kT)^2}{4s^2}\sum_{\alpha\beta\gamma\delta}\left\{\frac{(\boldsymbol{S}\cdot\boldsymbol{e}^{\mu *}_{\alpha})(\boldsymbol{S}\cdot\boldsymbol{e}^{\mu' *}_{\beta})}{(m_\mu m_{\mu'})^{1/2}(\omega_\alpha\omega_\beta\omega_\gamma\omega_\delta)^2}\Phi_{-\alpha\beta\gamma}\Phi_{-\gamma\delta-\delta}\right.$$
$$\left.+\frac{(\boldsymbol{S}\cdot\boldsymbol{e}^{\mu' *}_{\alpha})(\boldsymbol{S}\cdot\boldsymbol{e}^{\mu}_{\beta})}{(m_\mu m_{\mu'})^{1/2}(\omega_\alpha\omega_\beta\omega_\gamma\omega_\delta)^2}\Phi^*_{-\alpha\beta\gamma}\Phi^*_{-\gamma\delta-\delta}\right\}\qquad \boldsymbol{q}_\alpha=\boldsymbol{q}-2\pi\boldsymbol{H}\quad \text{(G.8)}$$

$$I^f_{\mu\mu'} = \frac{(kT)^3}{6s^3N}\sum_{\alpha\beta\gamma\delta}\left\{\frac{(\boldsymbol{S}\cdot\boldsymbol{e}^{\mu' *}_{\alpha})(\boldsymbol{S}\cdot\boldsymbol{e}^{\mu *}_{\beta})(\boldsymbol{S}\cdot\boldsymbol{e}^{\mu *}_{\gamma})(\boldsymbol{S}\cdot\boldsymbol{e}^{\mu *}_{\delta})}{m_{\mu'}^{1/2}m_\mu^{3/2}(\omega_\alpha\omega_\beta\omega_\gamma\omega_\delta)^2}\Phi_{\alpha\beta\gamma\delta}\exp(-\mathrm{i}2\pi\boldsymbol{H}_2\cdot\boldsymbol{r}_\mu)\right.$$
$$\left.+\frac{(\boldsymbol{S}\cdot\boldsymbol{e}^{\mu}_{\alpha})(\boldsymbol{S}\cdot\boldsymbol{e}^{\mu'}_{\beta})(\boldsymbol{S}\cdot\boldsymbol{e}^{\mu'}_{\gamma})(\boldsymbol{S}\cdot\boldsymbol{e}^{\mu'}_{\delta})}{m_\mu^{1/2}m_{\mu'}^{3/2}(\omega_\alpha\omega_\beta\omega_\gamma\omega_\delta)^2}\Phi^*_{\alpha\beta\gamma\delta}\exp(\mathrm{i}2\pi\boldsymbol{H}_2\cdot\boldsymbol{r}_{\mu'})\right\}$$
$$\boldsymbol{q}_\alpha=\boldsymbol{q}-2\pi\boldsymbol{H}\quad \text{(G.9)}$$

$$I^g_{\mu\mu'} = -\frac{(kT)^3}{2s^3N}\sum\left\{\frac{(\boldsymbol{S}\cdot\boldsymbol{e}^{\mu' *}_{\beta})(\boldsymbol{S}\cdot\boldsymbol{e}^{\mu *}_{\gamma})(\boldsymbol{S}\cdot\boldsymbol{e}^{\mu *}_{\beta'})(\boldsymbol{S}\cdot\boldsymbol{e}^{\mu *}_{\gamma'})}{m_{\mu'}^{1/2}m_\mu^{3/2}(\omega_\alpha\omega_\beta\omega_\gamma\omega_{\beta'}\omega_{\gamma'})^2}\Phi_{\alpha\beta\gamma}\Phi_{-\alpha\beta'\gamma}\right.$$
$$\times\exp(-\mathrm{i}2\pi\boldsymbol{H}_3\cdot\boldsymbol{r}_\mu)+\frac{(\boldsymbol{S}\cdot\boldsymbol{e}^{\mu}_{\beta})(\boldsymbol{S}\cdot\boldsymbol{e}^{\mu'}_{\gamma})(\boldsymbol{S}\cdot\boldsymbol{e}^{\mu'}_{\beta'})(\boldsymbol{S}\cdot\boldsymbol{e}^{\mu'}_{\gamma'})}{m_\mu^{1/2}m_{\mu'}^{3/2}(\omega_\alpha\omega_\beta\omega_\gamma\omega_{\beta'}\omega_{\gamma'})^2}\Phi_{\alpha\beta\gamma}\Phi_{-\alpha\beta'\gamma'}$$
$$\left.\times\exp(\mathrm{i}2\pi\boldsymbol{H}_3\cdot\boldsymbol{r}_{\mu'})\right\}\qquad \boldsymbol{q}_\beta=\boldsymbol{q}-2\pi\boldsymbol{H}.\quad \text{(G.10)}$$

It can be seen here that the term $I^a_{\mu\mu}$ determines the intensity of inelastic quasiharmonic scattering on one phonon with wave-vector $\boldsymbol{q}_\alpha$ (or $-\boldsymbol{q}_\alpha$), related to the μ or μ' atom vibrations. The terms $I^b_{\mu\mu}$–$I^g_{\mu\mu}$ describe two competing processes: scattering with phonon creation, which then decomposes into two or several phonons because of phonon–phonon interaction, and scattering with annihilation of a phonon, which was created before the scattering act due to several-phonon interaction.

It is acceptable to consider the two-phonon contributions to TDS as harmonic, because the anharmonic additions are much less than anharmonic additions to one-phonon interactions. Then, in the high-temperature limit

$$I^{(2)}_{\mu\mu'}=\frac{(kT)^2}{2(s)^2}\sum_{\alpha\beta}\frac{(S\cdot e^{\mu}_{\alpha})(S\cdot e^{\mu'}_{\alpha}{}^{*})(S\cdot e^{\mu}_{\beta})(S\cdot e^{\mu'}_{\beta}{}^{*})}{m_{\mu}m_{\mu'}\omega^2_{\alpha}\omega^2_{\beta}}\qquad \boldsymbol{q}_{\alpha}+\boldsymbol{q}_{\beta}=\boldsymbol{S}-2\pi\boldsymbol{H}. \tag{G.11}$$

Let us consider the possible corrections to the calculation for TDS which use additional experimental information. The intensity measured is an integral quantity which is obtained after integration over all $\boldsymbol{S}$ values in the reciprocal lattice volume scanned. Let us neglect the frequency shift caused by inelastic scattering (it is two orders of magnitude lower than the x-ray spectrum line width) and introduce the vector $2\pi\boldsymbol{H}$ instead of $\boldsymbol{S}$ into equations (G.2) and (G.11). The summation over $\boldsymbol{S}$ will be replaced by summing over all values of the phonon mode wave-vectors. The integral intensities T can be written from (G.1) in the form

$$T(\boldsymbol{H})=T_0(\boldsymbol{H})(1+\alpha_1+\alpha_2). \tag{G.12}$$

Here T_0 is the elastic part of the total intensity, α_1 and $\alpha_2=\alpha^{qh}$ are corrections for one- and two-phonon TDS; they are determined as the ratios of inelastic integral intensities to elastic ones. In order to find the correction α_1, the anharmonic effect being taken into account, the $I^{(k)}_{\mu\mu'}$ (G.2) values should be determined for all directions near the reciprocal point and integrated over the whole volume scanned. This is difficult to do even in the framework of simplified models and for simple lattices. Therefore, following Leibfried and Ludwig (1961), anharmonic parameters $\Phi_{\alpha\beta\gamma}$ and $\Phi_{\alpha\beta\gamma\delta}$ from (3.21) are averaged over all polarization directions and over all vibrational normal mode wave-vectors. Then the integral analogue of $I^{(1)}_{\mu\mu'}$ intensity (G.3), which includes an anharmonic contribution, can be presented in the form (Kashiwase 1965)

$$T^{(1)}_{\mu\mu'}=(1+\alpha_T)T^{(a)}_{\mu\mu'} \tag{G.13}$$

where $T^{(a)}_{\mu\mu'}$ corresponds to $I^{(a)}_{\mu\mu'}$ (G.4). The correction α_T, which takes the anharmonic crystal properties into account, is determined as

$$\alpha_T=-\tfrac{3}{2}skTQ. \tag{G.14}$$

The Q and P values depend on the coupling parameters (3.17) and on the harmonic and anharmonic crystal models used (Leibfried and Ludwig 1961). Their calculations, and consequently the α_T evaluation from first principles, are very complicated even for simple crystals. The alternative approach, based on the experimental temperature dependences of specific heat capacity, elastic constants, thermal expansion, etc, is preferable. So, the specific heat capacity of the anharmonic crystal, C_p, in the low- and high-temperature limits is given by expressions

$$C_p(T)=3sNk[1+\tfrac{3}{2}skT(sP-\tfrac{1}{2}Q+\tfrac{2}{3}P')]\qquad (T>\theta_D) \tag{G.15}$$

$$C_p(T)=3sNk(4\pi^4/5)(T/\theta_{eff})^3\qquad (T\to 0). \tag{G.16}$$

In this expression $P' = 3\tilde{\varkappa}\gamma_G^2/\tilde{V}$, $\theta_{eff} = \tilde{\theta}_D[1 - \frac{27}{64}sk\tilde{\theta}_D(sP - \frac{2}{3}Q + \frac{8}{9}P')]$, $\tilde{\varkappa}$ is the compressivity, $\tilde{V}$ is the unit cell volume, γ_G is the Grüneisen parameter; the tilde marks the harmonic values. The expressions (G.15) and (G.16) permit one to find Q and P values from the properties measured at different temperatures and, then, to calculate the anharmonic correction α_T.

The penalty for the use of the Leibfried–Ludwig approximation is the information lost on the anharmonical TDS anisotropy with scattering vector $\boldsymbol{S}$, which appears in addition to the TDS anisotropy in the harmonic approximation. Moreover, this approximation has been initially used in description of anharmonic contributions to heat capacity, whereas the effective Hamiltonian, used in TDS theory, strictly speaking, differs from that used in heat capacity theory. However, the model uncertainty in α_T caused by this approximation is less than the uncertainty in the experimental data mentioned, therefore it cannot be taken into account.

The approximation (G.13) permits one to write the one-phonon TDS correction in the form

$$\alpha_1 = (1 + \alpha_T)\alpha_1^{qh} \tag{G.17}$$

Now the quasiharmonic values α_1 and α_2 should be determined. These are the only values which are usually used in practice. To do this, some extra experimental information should be used. It is necessary to take into account that near the reflecting crystal position both the vibration mode wave-vectors $\boldsymbol{q}_\alpha$ approach zero and the TDS scattering increases nonlinearly as $\boldsymbol{q}_\alpha$ approaches the reciprocal lattice point (see figure 3.5). The TDS here is determined mainly by acoustic phonons, the one-phonon contribution being dominant. The optical mode contribution practically does not depends on $\boldsymbol{q}_\alpha$ (Willis 1969) and is eliminated at the stage of background treatment. Therefore, the summation in (G.4) and (G.11) can be restricted to three acoustic modes. It is important that the $\boldsymbol{e}_\mu m_\mu^{-1/2}$ value for all atoms is identical in this case. The next simplification is that the phonon frequencies can be considered to be proportional to their wave-vector values (Debye approximation). It may be argued that this approximation is too rough. However, Helmholdt *et al* (1983) have shown that the implicit account of ω_α dispersion only weakly influences the result. Considering the x-ray beam to be planar and monochromatic, the expression for TDS correction can be written as follows (Stevens 1974, Sakata and Harada 1976)

$$\alpha_1^{qh} = \frac{kT}{(2\pi)^3} \int J_1(\boldsymbol{q}_\alpha)\, \mathrm{d}\boldsymbol{q}_\alpha \tag{G.18}$$

$$J_1(\boldsymbol{q}_\alpha) = \sum_{ij} h_i h_j (A_{ij}^{-1})/q_\alpha^2. \tag{G.19}$$

Here h_i are the orthogonal coordinates of the vector $\boldsymbol{H}$ in the reflecting position; A_{ij} is the Christoffel tensor, the elements of which can be determined from the experimental elastic constants (Nye 1962). The integration in (G.18) is over the reciprocal space volume scanned.

The Wooster (1962) approximation is convenient to use when calculating two-phonon TDS corrections. It takes into account that the two-phonon TDS intensity is maximal when the phonon wave-vectors $\boldsymbol{q}_\alpha$ and $\boldsymbol{q}_\beta$ in (G.11) are collinear. Then the expression for α_2 may be presented (Stevens 1974) as

$$\alpha_2 = \frac{(kT)^2}{(2\pi)^6} \int J_2(\boldsymbol{q}_\alpha)\, \mathrm{d}\boldsymbol{q}_\alpha \tag{G.20}$$

$$J_2(\boldsymbol{q}_\alpha) \cong (\pi^3/2) q_\alpha^2 [J_1(\boldsymbol{q}_\alpha)]^2. \tag{G.21}$$

The expressions presented permit a simple correction for α_1–α_2 splitting.

It is necessary to note that in the high-temperature limit the elastic constants measured at $T>0$ correspond to the quasiharmonic approximation. This is why the correction α_1^{qh} is considered as quasiharmonic. To calculate α_2 in the approximation chosen, the harmonic elastic constants, determined by extrapolation of their temperature dependence to 0 K, should be used. However, for practical α_2 calculation it is more convenient to use the elastic constants measured at the temperature of the x-ray diffraction experiment. This permits one to calculate α_1^{qh} and α_2 uniformly.

Let us consider now the specificity of description of neutron TDS in the anharmonic approximation. The shift of the radiation energy at one-phonon x-ray TDS, which is equal to the energy of created or annihilated phonons, is less than 0.1 eV, which in its turn is less than the spectral linewidth (1–5 eV). The energy of thermal neutrons is 10^{-3}–10^{-1} eV. In inelastic coherent neutron scattering, the change in energy

$$E = \hbar\omega = (\hbar^2/2m_N)(k_0^2 - k_1^2) \tag{G.22}$$

can be higher than the spectral interval. Therefore the neutron scattering should be described by the double differential scattering cross-section, $\partial^2\sigma/\partial\Omega\,\partial E$. As a result the total coherent neutron scattering intensity is determined by the expressions (Willis 1970)

$$I(\boldsymbol{S}) = \int \frac{\partial^2\sigma(\boldsymbol{S})}{\partial\Omega\,\partial E}\, \mathrm{d}E \tag{G.23}$$

$$\frac{\partial^2\sigma(\boldsymbol{S})}{\partial\Omega\,\partial E} = \frac{1}{2\pi}\frac{k}{k_0} \sum_{\substack{nn'\\ \mu\mu'}} b_\mu^* b_{\mu'} \mathrm{e}^{\mathrm{i}\boldsymbol{S}\cdot(\boldsymbol{R}_{n'\mu'} - \boldsymbol{R}_{n\mu})} \int_{-\infty}^{+\infty} \mathrm{d}t\, \mathrm{e}^{\mathrm{i}[\omega - (E - E_0)/\hbar]t} \times \langle \mathrm{e}^{\mathrm{i}S[\boldsymbol{u}_{\mu'}(t) - \boldsymbol{u}_\mu(0)]} \rangle. \tag{G.24}$$

Here b_μ is the μth atom coherent neutron scattering amplitude; E_0 and E are the crystal energies before and after scattering. After separating the intensity into elastic and TDS parts $I(\boldsymbol{S})$ can be written as

$$I(\boldsymbol{S})=\int_{-\infty}^{+\infty}\left\{\frac{\partial^2\sigma_0(\boldsymbol{S})}{\partial\Omega\,\partial E}+\frac{\partial^2\sigma_1(\boldsymbol{S})}{\partial\Omega\,\partial E}+\cdots\right\}\mathrm{d}E=I_0(\boldsymbol{S})+\sum_{k\neq 0}I_k(\boldsymbol{S}) \qquad \text{(G.25)}$$

which formally coincides with expression (G.1) substituting b_μ and $b_{\mu'}$ for f_μ and $f_{\mu'}$.

In the harmonic approximation for one-phonon TDS scattering

$$\frac{\partial^2\sigma_1(\boldsymbol{S})}{\partial\Omega\,\partial E}=\frac{k}{k_0}\frac{1}{2N\hbar}\sum_\alpha\frac{1}{\omega_\alpha}\left|\sum_\mu\frac{\boldsymbol{S}\cdot\boldsymbol{e}_\alpha^\mu}{\sqrt{m_\mu}}b_\mu T_\mu(\boldsymbol{S})\mathrm{e}^{\mathrm{i}(\boldsymbol{S}-\boldsymbol{q}_\alpha)\cdot\boldsymbol{r}_\mu}\right|^2$$

$$\times\frac{\delta(\omega-\omega_\alpha)-\delta(\omega+\omega_\alpha)}{1-\mathrm{e}^{-\beta\hbar\omega}}L(\boldsymbol{S}-\boldsymbol{q}_\alpha) \qquad \text{(G.26)}$$

$L(\boldsymbol{S}-\boldsymbol{q}_\alpha)$ is the Laue interference function. The δ function provides the energy conservation law fulfilment $\hbar\omega=\pm\hbar\omega_\alpha$; the signs + or − correspond to phonon creation or annihilation. As for x-rays, the maximal contribution to neutron scattering near the Bragg position gives the long-wave acoustical phonons. However, the scattering surface is now not the Ewald sphere but a more complicated surface which is different for the cases of phonon creation and annihilation. Therefore, the two cases of $v_{sound}\gtrless v_N$ should be considered separately. Willis (1970) showed that when $v_N<v_{sound}$, the neutron TDS is complicated but depends on scattering angle slightly and therefore can be merged into the background. The situation is more complicated when $v_{sound}\approx v_N$; therefore, one should avoid this case in structure analysis. Finally, when $v_N>v_{sound}$ the TDS is independent of v_N and the corresponding x-ray formula can be used here for correction calculations.

The anharmonic effects change the one-phonon scattering neutron cross-section (Maradudin and Ambegaokar 1964). First, the difference of δ functions in (G.26) is transformed into a Lorentz function which is further distorted by the interference of one- and many-phonon processes. The summing over α in (G.26) becomes a non-trivial problem due to this fact and has still not been worked out completely. However, it is clear that the difference in x-ray and neutron scattering leads to different anharmonicity manifestations in x-ray and neutron structure analysis. When anharmonicity is small, then the TDS corrections can be realized in both methods uniformly. If this is not the case, the thermal atomic parameters determined by x-rays and neutrons will differ from each other. This has been observed (Coppens 1978, Blessing 1995) and this circumstance should be taken into account in the combined x-ray and neutron diffraction investigations.

One should note, finally, that the anharmonic contributions to the temperature factor do not depend on the type of radiation. They are similar both for neutrons and x-rays.

APPENDIX H

STATISTICS IN X-RAY STRUCTURE ANALYSIS

Let us consider the main statistical descriptors used in modern x-ray structure analysis. The terminology and meaning of all quantities and functions are in accordance with the recommendations of the Subcommittee of the International Union of Crystallography (Schwarzenbach *et al* 1989). For details one is referred to the works of Hamilton (1964), Bevington (1969), Taylor (1982), and Wilson (1992).

The measured x-ray intensities can be treated as random variables x. The function $p(x)$ of the random variable x, such that the probability of finding x between a and b is given by

$$P(a \leqslant x \leqslant b) = \int_a^b p(x)\,\mathrm{d}x \qquad \text{(H.1)}$$

is called the probability density function (PDF). If $\boldsymbol{x} = (x_1, x_2, \ldots, x_n)$ is an ordered set of n random variables, the multivariate PDF is the function $p_m(\boldsymbol{x})$ such that

$$P(a_1 \leqslant x_1 \leqslant b_1; \ldots; a_n \leqslant x_n \leqslant b_n) = \int_{a_1}^{b_1} \cdots \int_{a_n}^{b_n} p_m(\boldsymbol{x})\,\mathrm{d}^n\boldsymbol{x}. \qquad \text{(H.2)}$$

The PDF of an element x_i of the set $\boldsymbol{x}$, $\tilde{p}(x_i)$, irrespective of the values of any other element, is defined by

$$\tilde{p}(x_i) = \int p_m(\boldsymbol{x})\,\mathrm{d}^{n-1}\boldsymbol{x} \qquad \text{(H.3)}$$

and called the marginal PDF.

The expected value (expectation) of the nth power of the random variable x is called the nth moment of the PDF, $p(x)$:

$$E(x^n) = \int_{-\infty}^{+\infty} x^n p(x)\,\mathrm{d}x. \qquad \text{(H.4)}$$

The first moment or mean is usually denoted by $E(x) = \mu$. The second moment about the mean, $E[(x-\mu)^2]$, is the variance of $p(x)$ and is denoted by σ^2.

The square root of the variance of a PDF, σ, is called the standard deviation. In a space of n dimensions, the mean of the multivariate PDF, $p_m(\boldsymbol{x})$, is an n vector $\boldsymbol{E}(\boldsymbol{x})$ with elements $\mu_i = E(x_i)$, equal to the mean of the marginal PDF of x_i. The $n \times n$ variance–covariance matrix (tensor of rank two in n dimensions) is defined by the second moments about the mean:

$$V_{ij} = E[(x_i - \mu_i)(x_j - \mu_j)] = \int (x_i - \mu_i)(x_j - \mu_j) p_m(\boldsymbol{x}) \, \mathrm{d}^n(\boldsymbol{x}). \tag{H.5}$$

For $i = j$ this is equal to the variance of the marginal PDF of x_i, $V_{ii} = \sigma_i^2$. For $i \neq j$

$$V_{ij} = \mathrm{cov}(x_i, x_j) = \sigma(x_i)\sigma(x_j)\rho_{ij} \tag{H.6}$$

where ρ_{ij} are correlation coefficients. The correlation matrix is defined by

$$\rho_{ij} = V_{ij}/(\sigma_i \sigma_j) \tag{H.7}$$

where $-1 \leqslant \rho_{ij} \leqslant 1$ and $\rho_{ii} = 1$.

The average of a set of values $\{x_i\}$, $1 \leqslant i \leqslant n$, is defined by $\bar{x} = (\Sigma_i x_i)/n$. If the $\{x_i\}$ are a sample (a finite set) of n independent observations of a single quantity x distributed according to a PDF $p(x)$ with mean μ and variance σ^2, then $\bar{x}$ is a minimum-variance unbiased estimate of μ, and

$$s^2 = \left[\sum_i (x_i - \bar{x})^2 \right] \Big/ (n-1) \tag{H.8}$$

is an unbiased estimate of σ^2. The variance of the PDF of $\bar{x}$ is σ^2/n and an unbiased estimate is obtained from s^2/n. These estimates do not require a complete knowledge of $p(x)$ to be available. Using a set of weights $\{w_i\}$ the weighted average is defined by $\bar{x}_w = \Sigma_i w_i x_i / \Sigma_i w_i$. If the $\{w_i\}$ do not depend on the $\{x_i\}$, then $\bar{x}_w$ is an unbiased estimate of μ. The weighted average is used if the n observations are drawn from the set of measurements with identical mean but different variances.

A mathematical expression leading from the observations to an estimate of the values of physical quantities is called the estimator. The estimator is constructed using the mathematical formulation of some accepted model. The latter include the physical quantities as parameters, values of which are estimated from observations. Any estimates of parameter values are random variables. The expected value, E, of function $f(x)$ of a random variable x whose PDF is $p(x)$ is defined as

$$E[f(x)] = \int_{-\infty}^{\infty} f(x) p(x) \, \mathrm{d}x. \tag{H.9}$$

The variance of a function of many random variables x_i, $f(x_1, x_2, \ldots)$ may be estimated by a first-order Taylor expansion provided that errors of x are independent, none of the errors are predominant, and the expected value of the sum of the errors is zero. Then

$$\sigma^2(f) = \sum_{ij} \frac{\partial f}{\partial x_i} \frac{\partial f}{\partial x_j} \operatorname{cov}(x_i, x_j). \tag{H.10}$$

In this way the ED distribution and ED characteristics error analyses were performed in section 3.7.

An error having an expected value of zero is called a random error. In diffraction methods there are statistical fluctuations in photon counts. Other effects such as irreproducible play in diffractometer settings and short-term fluctuations of temperature, pressure, and voltage contribute, at least in part, to the random errors. These terms can be reduced at the expense of increased measuring time. If the model and the method of refinement are perfect, errors of estimated parameters are also random.

An estimator of a statistical quantity is biased if the expected value of the quantity is not equal to the true value. In x-ray structure analysis bias is usually identified with a systematic error. In statistics, it is sometimes restricted to a particular type of systematic error arising from the mathematical model applied to the observations (Hamilton 1964). Any nonlinear operation or model can result in a bias, for example taking the square root of an intensity when computing kinematic structure amplitudes (see subsection 3.3.5). The bias is then due to the fact that the expected value of a function $f(x)$ is in general not a simple function of the expected value of x:

$$E[f(x)] \neq f[E(x)]. \tag{H.11}$$

The equality will hold for any distribution of x if $f(x)$ is a linear function of x. Biases in this restricted sense are proportional to the variances due to random errors in the observations. In order to estimate the error of the experimental result, the estimated standard deviation (e.s.d.) of a PDF is used. Methods used for obtaining e.s.d. of diffraction intensities take into account quantum counting statistics, the variations of periodically measured reference reflections, and the scatter among symmetry-equivalent reflections (see section 3.2). Such methods are part of the model, and not of the observations.

If F_j are the experimental structure amplitudes and $F_{model,j}$ are the corresponding calculated model values of n quantities, the differences $d_j = |F_j| - |F_{model,j}|$ are called deviates. The definition holds for any given choice of model parameters. The deviance between the calculated and observed quantities is

$$D_0 = \sum_{j=1}^{n} d_j^2. \tag{H.12}$$

The weighted deviance is

$$D_w = \sum_{j=1}^{n}\sum_{k=1}^{n} w_{jk} d_j d_k = \hat{d}^{\mathrm{T}} \hat{W} \hat{d} \tag{H.13}$$

where $\hat{d}$ is an n vector of deviates. The positive definite $(n \times n)$ weight matrix $\hat{W}$ may be written as a product $\hat{W} = \hat{B}^{\mathrm{T}}\hat{B}$ and equation (H.13) then becomes $D_w = \hat{d}^{\mathrm{T}}\hat{B}^{\mathrm{T}}\hat{B}\hat{d}$, where $\hat{B}\hat{d}$ is the vector of weighted deviates. If correlation terms of $\hat{W}$ are assumed to be negligible, $\hat{W}$ becomes a diagonal matrix, and the quantity commonly refined in LS is obtained:

$$D_w = \sum_{j=1}^{n} w_j d_j^2. \tag{H.14}$$

If the weights are reciprocals of the variances of the observed quantities, $w_j = \sigma_j^{-2}$, or more generally $W = \hat{V}^{-1}$, where $\hat{V}$ is the variance–covariance matrix of the observations, the weighed deviance is often called the scaled deviance.

The PDF of the quantity x with standard deviation σ about the mean μ

$$p(x) = \sigma^{-1}(2\pi)^{-1/2} \exp\{-\tfrac{1}{2}[(x-\mu)/\sigma]^2\} \tag{H.15}$$

is called the normal or Gaussian PDF. The normal PDF can originate from addition of a large number of small independent errors, each with its own non-normal distribution, but occurrence of a normal PDF does not imply this underlying structure. The standard normal deviate $z = (x-\mu)/\sigma$ has a normal distribution with zero mean and unit standard deviation, when x is distributed according to (H.13).

In order to measure the extent to which values $F_{model,j}$ of a set of n structure amplitudes approach the experimental values F_j the goodness-of-fit test, S, is used. For an LS refinement with weight matrix $\hat{W}$, it is defined as

$$S = [\hat{d}^{\mathrm{T}}\hat{W}\hat{d}/E(\hat{d}^{\mathrm{T}}\hat{W}\hat{d})]^{1/2}. \tag{H.16}$$

The deviation of S^2 from unity is a measure of the validity of the model used and of the estimate of the variance–covariance matrix $\hat{V}$ of the observations used to calculate $E(\hat{d}^{\mathrm{T}}\hat{W}\hat{d})$. If, and only if, the weight matrix $\hat{W}$ in the refinement is chosen to be the inverse of $\hat{V}$, $\hat{W} = \hat{V}^{-1}$, then $E(\hat{d}^{\mathrm{T}}\hat{W}\hat{d}) = n - m$, regardless of the form of the PDF; m is the number of variables in the model, and $n-m$ is the number of degrees of freedom. For a diagonal matrix $\hat{V}$, the value of S^2 then becomes

$$S^2 = (n - m^{-1}) \sum_{j=1}^{n} \sigma_j^{-2} d_j^2. \tag{H.17}$$

Further, for the $\hat{W} = \hat{V}^{-1}$ weights, if the deviates are normally distributed (i.e. with Gaussian multivariate PDF) the value of S^2 that will be exceeded in $100\alpha\%$ of replications is given by

$$(S^2)_\alpha = \chi^2_{n-m,\alpha}/(n-m) \tag{H.18}$$

where $\chi^2_{n-m,\alpha}$ is the $100\alpha\%$ point of the χ^2 distribution.

The discrepancy factor (the mean normalized deviate)

$$R = \sum_j d_j \left(\sum_j |F| \right)^{-1} \tag{H.19}$$

and weighted discrepancy factor (the normalized weighted deviance)

$$R_w = \left[D_w \Big/ \sum_j w_j |F_j|^2 \right]^{1/2} \tag{H.20}$$

are often used as a measure of the agreement between model and experimental structure amplitudes.

Along with the point statistical descriptors S, R, and R_w others can also be used for assessment of x-ray diffraction data quality and adequacy of the model used. So, a normal probability plot (Abrahams and Keve 1971) is often used. This is a graphical procedure in which the differences between two independent sets of measurements, or those between experiment and model, are analysed in terms of a normal PDF. The ordered experimental

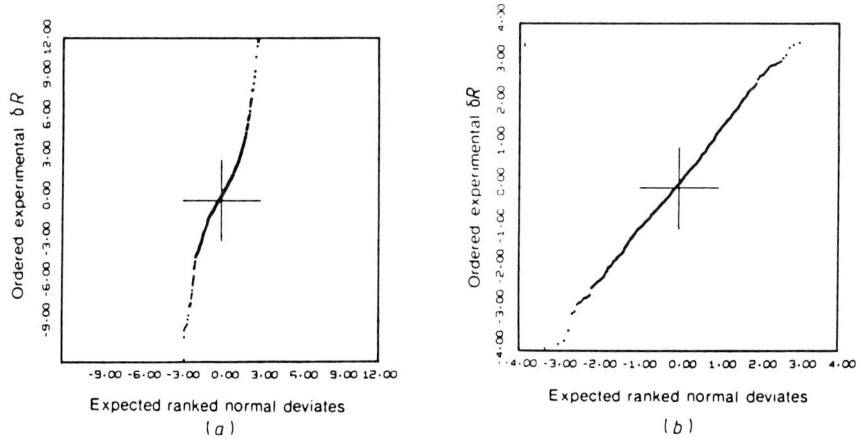

Figure H.1 Normal probability plots for ammonium perchlorate, NH_4ClO_4. (*a*) A δR plot calculated for the rigid superpositional model refinement. The slope of the LS line through the points is 2.36 and the y intercept is 0.22. The points for nine reflections fall outside the plot range. The plotted curve is far from the ideal shape, a straight line with slope 1.00. (*b*) A δR plot following the multipole model refinement. The slope of the LS line through the points is 1.22 and the y intercept is 0.12. The point for one reflection falls outside the plot range. Note the different plot ranges in (*a*) and (*b*).

normal deviates $\delta R_i = d_i/\sigma_i$ are plotted against ordered standard normal deviates. A resulting normal probability that is linear, with zero intercept and unit slope, shows that the experimental deviates are normally distributed. The presence of bias can be revealed in such a way. For example, a δR plot calculated after the ammonium perchlorate, NH_4ClO_4, crystal structure refinement (Lundgren 1979) showed that only the multipole model results in the absence of significant systematical error (figure H.1).

The correlation in experimental data resulting from uncorrected systematical errors can be revealed by Durbin–Watson (1950a, b) statistics. It is defined by

$$d = \sum_{j=2}^{N} (d_j - d_{j-1})^2 \left(\sum_{j=1}^{N} d_j^2 \right)^{-1} \tag{H.21}$$

where $d_j = |F_j| - |F_{model,j}|$ is the deviate. d takes values $0 < d < 4$. For no serial correlation a d value close to two is expected. With positive serial correlation, adjacent deviates tend to have the same sign and d becomes less than two, whereas with negative serial correlation (alternating signs of deviates) d takes values larger than two. If deviates d_j are ordered with $\sin\theta/\lambda$, the error due to inadequate corrections for extinction can be recognized at small scattering angle. The improper description of the atomic thermal vibration can be established in such a way at high values of $\sin\theta/\lambda$. Tables of values for testing d are given by Durbin and Watson (1950a, b).

APPENDIX I

FOURIER TRANSFORMATIONS OF ATOMIC ORBITAL PRODUCTS

Many problems of ED analysis with x-ray diffraction data require the calculation of the Fourier transform of atomic orbital products—orbital or partial scattering amplitudes.

$$f_{\mu\nu}(H) = \int \phi_\mu^*(\boldsymbol{r})\phi_\nu(\boldsymbol{r}) \exp(2\pi i \boldsymbol{H}\cdot\boldsymbol{r}) \, dV. \tag{I.1}$$

This matter has been studied in detail by McWeeny (1951, 1953), Weiss and Freeman (1959), Tavard *et al* (1964), Stewart *et al* (1965), Stewart (1969), Chandler and Spackman (1978), Rae (1978), Guidatti *et al* (1979), Avery and Ormen (1979), Tanaka (1988a), and others. We will consider here only those results which permit the calculation of $f_{\mu\nu}$ in the analytical form.

The method suggested by Weiss and Freeman (1959) and Freeman and Watson (1961a, b) is the most convenient for the calculation of the d and f orbital scattering amplitudes in fields of different symmetry. It deals with the orbitals, ϕ_μ, related to the same centre and uses the expansion of the exponent into a series (Arfken 1985)

$$\exp(2\pi i \boldsymbol{H}\cdot\boldsymbol{r}) = 4\pi \sum_{n=0}^{\infty} \sum_{m=-n}^{\infty} i^n j_n(2\pi H r) \mathscr{I}_n^m(\theta)\Phi_m^*(\varphi) \mathscr{I}_n^m(\beta)\Phi_m(\gamma). \tag{I.2}$$

Here $(2\pi H, \beta, \gamma)$ and (r, θ, φ) are the spherical coordinates of points which are given by vectors $2\pi\boldsymbol{H}$ and $\boldsymbol{r}$ in the Cartesian coordinate system with z as the quantization axis (figure I.1); $j_n(2\pi H r)$ are spherical Bessel functions; $\mathscr{I}_n^m$ are normalized associated Legendre polynomials; $\Phi(\varphi) = (1/\sqrt{2\pi}) \exp(im\varphi)$. Writing the wave-functions ϕ_μ in the spherical coordinate system as a product of an arbitrary radial part, $R(r)$, and an angular part, $\mathscr{I}_{l_\mu}^{m_\mu}(\theta)\Phi_\mu$, taking into account (I.2) and the fact that $dV = r^2 \sin\theta \, dr \, d\theta \, d\varphi$, one can arrive at

$$f_{\mu\nu} = 4\pi \sum_{n=0}^{\infty} \sum_{m=-n}^{\infty} i^n \left[\int_0^\infty R_\mu^*(r) R_\nu(r) j_n(2\pi H r) r^2 \, dr \right] \left[\int_0^\pi \mathscr{I}_{l_\mu}^{*m_\mu}(\theta) \mathscr{I}_{l_\nu}^{m_\mu}(\theta) \mathscr{I}_n^m(\theta) \right.$$
$$\left. \times \sin(\theta) \, d\theta \right] \left[\int_0^{2\pi} \Phi_{m_\mu}^*(\varphi)\Phi_{n_\nu}(\varphi)\Phi_m^*(\varphi) \, d\varphi \right] \mathscr{I}_n^m(\beta)\Phi_m(\gamma). \tag{I.3}$$

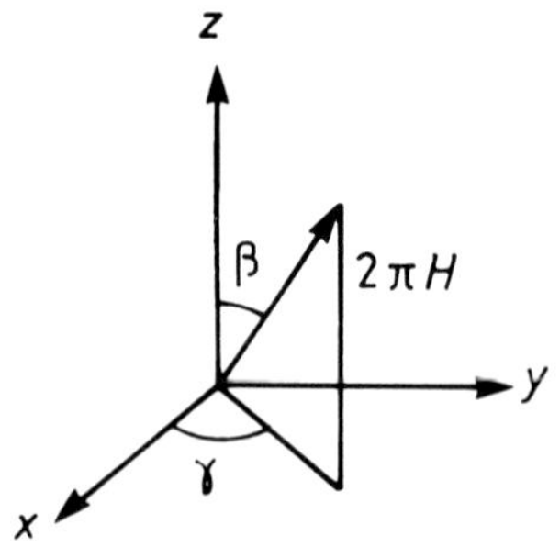

Figure I.1 The relation between the spherical and the Cartesian coordinate system for calculation of the Fourier transform of the AO product.

Table I.1 Condon–Shortley coefficients $C^n(l_\nu m_\nu, l_\mu m_\mu)$ for $l_\nu = l_\mu = 2$.

m_μ	m_ν	C^0	C^2	C^4	m_μ	m_ν	C^0	C^2	C^4
± 2	± 2	1	$-\frac{2}{7}$	$\frac{1}{21}$	0	0	1	$\frac{2}{7}$	$\frac{6}{21}$
± 2	± 1	0	$-\sqrt{6/7}$	$\sqrt{5/21}$	± 2	∓ 2	0	0	$\sqrt{70/21}$
± 2	0	0	$-\frac{2}{7}$	$-\sqrt{15/21}$	± 2	∓ 1	0	0	$\sqrt{35/21}$
± 1	± 1	1	$\frac{1}{7}$	$-\frac{4}{21}$	± 1	∓ 1	0	$-\sqrt{6/27}$	$-\sqrt{40/21}$
± 1	0	0	$-\frac{1}{7}$	$-\sqrt{30/21}$					

Integration over the angle φ gives

$$\int_0^{2\pi} \Phi^*_{m_\mu}(\varphi)\Phi_{n_\nu}(\varphi)\Phi^*_m(\varphi)\,d\varphi = (2\pi)^{-3/2}\int_0^{2\pi} \exp[i(m_\mu - m_\nu - m)\varphi]\,d\varphi$$

$$= \begin{cases} (2\pi)^{-1/2} & \text{if } m = m_\nu - m_\mu \\ 0 & \text{if } m \neq m_\nu - m_\mu. \end{cases} \qquad \text{(I.4)}$$

As a result, only the terms with $m = m_\nu - m_\mu$ are retained in the sum over m in (I.3). Integration over angle θ leads to the Condon–Shortley (1959) coefficients $C^n(l_\mu m_\mu, l_\nu m_\nu)$, namely

$$\int_0^{\pi} \mathscr{I}^{*m_\mu}_{l_\mu}(\theta)\mathscr{I}^{m_\nu}_{l_\nu}(\theta)\mathscr{I}^{m}_{n}(\theta)\sin(\theta)\,d\theta = ((2n+1)/2)^{1/2} C^n(l_\mu m_m, l_\nu m_\nu). \qquad \text{(I.5)}$$

The C^n coefficients for $l_\mu = l_\nu = 2$, i.e. for the d subshell, are presented in table I.1. We will restrict ourselves to this case only. It is necessary to note that the Condon–Shortley coefficients in table I.1 correspond to equation (I.2) in which the conjugated function $\Phi^*_m(\varphi)$ is present. These coefficients, for non-conjugated $\Phi_m(\varphi)$ functions, will be opposite in sign for odd $m_\nu - m_\mu$ values, i.e. $C^n(l_\mu m_\mu, l_\nu m_\nu) = (-1)^{m_\nu - m_\mu} C^n(l_\nu m_\nu, l_\mu m_\mu)$ and the result of $f_{\mu\nu}$, expanded into a series, will remain the same. The d orbitals are even functions, therefore all of the C^n coefficients with odd n will be zero. The maximum number of terms in a series over n is equal to three ($n=0, 2, 4$) for d orbitals.

Denoting the radial integral in (I.3) $\int_0^\infty R_\mu^*(r)R_\nu(r)j_n(2\pi Hr)\times r^2\,dr$ as $\langle j_n\rangle_{\mu\nu}$, the expression for $f_{\mu\nu}$ can be rewritten as

$$f_{\mu\nu}=\sum_{n=0,2,4}^{\infty} i^n(2(2n+1))^{1/2}C^n(l_\mu m_\mu, l_\nu m_\nu)\langle j_n\rangle_{\mu\nu}$$
$$\times \mathscr{I}_n^{m_\nu-m_\mu}(\beta)\exp[i(m_\nu-m_\mu)\gamma]. \qquad \text{(I.6)}$$

The $\langle j_n\rangle_{\mu\nu}$ values for HF atomic orbitals have been tabulated by Ibers and Hamilton (1974). The expressions for $\mathscr{I}_n^{m_\nu-m_\mu}(\beta)$ for $n=0$, 2, and 4 are listed in table I.2.

All d orbital scattering amplitudes can be derived from expression (I.6) in complex form. In order to obtain real scattering amplitudes the linear combinations should be composed according to the rules

$$d_{yz}=i(d_1+d_{-1})/2^{1/2} \qquad d_{zx}=-(d_1-d_{-1})/2^{1/2}$$
$$d_{xy}=-i(d_2-d_{-2})/2^{1/2} \qquad d_{x^2-y^2}=(d_2+d_{-2})/2^{1/2} \qquad d_{z^2}=d_0.$$

The general expression for all real orbital scattering amplitudes can be presented as follows (Tanaka 1988a):

$$f_{\mu\nu}=\langle j_0\rangle+\{A_0(2\cos^2\beta-\sin^2\beta)+A_1\sin\beta\cos\beta+A_2\sin^2\beta\}\langle j_2\rangle$$
$$+\{B_0(35\cos^4\beta-30\cos^2\beta+3)+B_1\sin\beta\cos\beta(7\cos^2\beta-3)$$
$$+B_2\sin^2\beta(7\cos^2\beta-1)+B_3\sin^3\beta\cos\beta$$
$$+B_4\sin^4\beta\}\langle j_4\rangle. \qquad \text{(I.7)}$$

The expressions for the basic scattering factors A_i and B_i are given in table I.3.

The expression (I.7) can easily be adapted to the calculation of x-ray scattering by d electrons in a crystal field with any symmetry. Some of the applications are presented in subsection 3.5.1.

In order to calculate the orbital scattering amplitudes, the expansions of the radial parts of the atomic wave-functions over Slater-type orbitals (STOs)

Table I.2 The expressions for the $\mathscr{I}_n^{m_\nu-m_\mu}(\beta)$ function.

$m_\nu-m_\mu$	$n=0$	$n=2$	$n=4$
0	$1/\sqrt{2}$	$(\sqrt{5}/\sqrt{8})(3\cos^2\beta-1)$	$(3/\sqrt{128})(35\cos^4\beta-30\cos^2\beta+3)$
∓ 1		$(\sqrt{15}/2)\sin\beta\cos\beta$	$(3\sqrt{5}/\sqrt{32})\sin\beta\cos\beta(7\cos^2\beta-3)$
∓ 2		$(\sqrt{15}/4)(\sin^2\beta)$	$(3\sqrt{5}/8)\sin^2\beta(7\cos^2\beta-1)$
∓ 3			$(3\sqrt{35}/\sqrt{32})(\sin^3\beta\cos\beta)$
∓ 4			$(3\sqrt{35}/16)\sin^4\beta$

Table I.3 Expressions for d electron basis scattering factors (Tanaka 1988a). c and s represent $\cos(n\gamma)$ and $\sin(n\gamma)$, respectively.

f_{ij}	A_0	A_1	A_2	B_0	B_1	B_2	B_3	B_4
$f_{yz,yz}$	$-\frac{5}{14}$		$15c_2/14$	$-\frac{3}{14}$		$-15c_2/14$		
$f_{yz,zx}$			$15s_2/14$			$-15s_2/14$		
$f_{yz,xy}$		$-15c_1/7$			$-15c_1$		$-15c_3/4$	
f_{yz,x^2-y^2}		$-15s_1/7$			$-15s_1/28$		$-15s_3/4$	
f_{yz,z^2}		$5\sqrt{3}s_1/7$			$-15\sqrt{3}s/14$			
$f_{zx,zx}$	$-\frac{5}{14}$		$-15c_2/14$	$-\frac{3}{14}$		$15c_2/14$		
$f_{zx,xy}$		$15s_1/7$			$15s_1/28$		$15s_3/4$	
f_{zx,x^2-y^2}		$-15c_1/7$			$-15c_1/28$		$15c_3/4$	
f_{zx,z^2}		$-5\sqrt{3}c_1/7$			$15\sqrt{3}c_1/14$			
$f_{xy,xy}$	$\frac{5}{7}$			$\frac{3}{56}$				$-15c_4/8$
f_{xy,x^2-y^2}								$-15s_4/8$
f_{xy,z^2}			$5\sqrt{3}s_2/7$		$15\sqrt{3}s_1/28$			
$f_{x^2-y^2,x^2-y^2}$	$\frac{5}{7}$			$\frac{3}{56}$				$15c_4/8$
$f_{x^2-y^2,z^2}$			$5\sqrt{3}c_2/7$		$15\sqrt{3}c_1/28$			
f_{z,z^2}	$-\frac{5}{7}$			$\frac{9}{28}$				

or Gaussian-type orbitals (GTO)s can also be used (see appendix C). This method, proposed by McWeeny (1953) and by Stewart (1969), is used in some quantum chemical models considered in subsection 3.5.2. In the case of AO expansion over STOs or GTOs at the same centre, the radial integral is the linear combination of terms of the form

$$\langle j_n \rangle_{\mu\nu} = N^S N^S \int_0^\infty r^{(n_\mu + n_\nu)} \exp[-(\xi_\mu + \xi_\nu) r] j_n(2\pi H r)\, \mathrm{d}r \qquad \text{(I.8)}$$

or

$$\langle j_n \rangle_{\mu\nu} = N^G N^G \int_0^\infty r^{(n_\mu + n_\nu)} \exp[-(\alpha_\mu + \alpha_\nu) r^2] j_n(2\pi H r)\, \mathrm{d}r \qquad \text{(I.9)}$$

where N^S and N^G are the corresponding normalization factors for STOs and GTOs. Thus, the problem is reduced to the calculation of the integrals

$$S_n(H) = \int_0^\infty r^N \exp(-Zr) j_n(2\pi H r)\, \mathrm{d}r \qquad \text{(I.10)}$$

and

$$G_n(H) = \int_0^\infty r^N \exp(-Zr^2) j_n(2\pi H r)\, \mathrm{d}r \qquad \text{(I.11)}$$

in which $N = n_\mu + n_\nu$, $N \geqslant 2$, and $Z = \xi_\mu + \xi_\nu$ (in I.10) or $Z = \alpha_\mu + \alpha_\nu$ (in I.11), $n \geqslant 0$. The first of these integrals has been calculated by Stewart *et al* (1965):

$$S_n(H) = \frac{2^{n+(1/2)}(N+n)!}{(2\pi H)^{N+1}(2n+1)!!\beta^{n+(1/2)}(1+\alpha^2)^{(N+(1/2))/2}} \times {}_2F_1[N+\tfrac{1}{2}, -N+\tfrac{1}{2}, n+\tfrac{3}{2}, (1+\beta^2)^{-1}] \qquad \text{(I.12)}$$

$\beta = \alpha + (1+\alpha^2)^{1/2}$, $\alpha = Z/2\pi H$. The hypergeometric function in (I.12) is (Arfken 1985)

$${}_2F_1(a, b, c, x) = 1 + \frac{ab}{c}\frac{x}{1!} + \frac{a(a+1)b(b+1)}{c(c+1)}\frac{x^2}{2!} + \ldots = \sum_{k=0}^{\infty} \frac{\{(a)_k (b)_k\}}{(c_k)}\frac{(x^k)}{k!}.$$

This is a rapidly converging series. The integral (I.11) can be found according to Watson (1966)

$$G_n(H) = \{\Gamma[(N+n+1)/2](2\pi H)^n / 2Z^{(N+n+1)/2}(2n+1)!!\} \exp(-\pi^2 H^2/Z) \; {}_1F_1[1+(n-N)/2, n+\tfrac{3}{2}, \pi^2 H^2/Z]. \qquad \text{(I.13)}$$

Here

$${}_1F_1(a, c, x) = 1 + \frac{a}{c}\frac{x}{1!} + \frac{a(a+1)}{c(c+1)}\frac{x^2}{2!} + \ldots = \sum_{k=0}^{\infty} \frac{(a)_k}{(c)_k}\frac{(x^k)}{k!}$$

is a degenerate hypergeometric function, $\Gamma[(N+n+1)/2]$ is a factorial gamma function (Arfken 1985).

If the atomic orbitals are centred at different spatial points, the calculation of $f_{\mu\nu}$ can be performed by the approximate method of McWeeny (1953) and Stewart (1969), which uses the AO expansion over GTOs. The GTO property (see appendix C) makes it possible to replace the product of the two Gaussians by a single Gaussian. This simplies the integration in (I.1). If one chooses the origin at the point $\boldsymbol{R}/2$, the orbital scattering amplitudes for s and p AOs are given as follows (Stewart 1969):

$$X_{ss}(\boldsymbol{H}, \boldsymbol{R}) = 2^{3/2} \sum_{\mu} \sum_{\nu} d_\mu d_\nu \frac{\alpha_\alpha^{3/4} \alpha_\nu^{3/4}}{(\alpha_\mu + \alpha_\nu)^{3/2}} \exp\left[-\frac{(\boldsymbol{A}_\mu \boldsymbol{A}_\nu)}{(\alpha_\mu + \alpha_\nu)} \right] \tag{I.14}$$

$$X_{sp}(\boldsymbol{H}, \boldsymbol{R}) = -2^{5/2} \sum_{\mu} \sum_{\nu} d_\mu d_\nu \frac{\alpha^{3/4} \alpha^{5/4}}{(\alpha_\mu + \alpha_\nu)^{5/2}} \exp\left[-\frac{(\boldsymbol{A}_\mu \boldsymbol{A}_\nu)}{(\alpha_\mu + \alpha_\nu)} \right] (\boldsymbol{\delta}_\nu \boldsymbol{A}_\nu) \tag{I.15}$$

$$X_{pp}(\boldsymbol{H}, \boldsymbol{R}) = 2^{5/2} \sum_{\mu} \sum_{\nu} d_\mu d_\nu \frac{\alpha_\mu^{5/4} \alpha_\nu^{5/4}}{(\alpha_\mu + \alpha_\nu)^{5/2}} \exp\left[-\frac{(\boldsymbol{A}_\mu \boldsymbol{A}_\nu)}{(\alpha_\mu + \alpha_\nu)} \right] \times \left[(\boldsymbol{\delta}_\mu \boldsymbol{\delta}_\nu) - \frac{2(\boldsymbol{\delta}_\mu \boldsymbol{A}_\nu)(\boldsymbol{\delta}_\nu \boldsymbol{A}_\mu)}{(\alpha_\mu + \alpha_\nu)} \right]. \tag{I.16}$$

Here $\boldsymbol{A}_\mu = \alpha_\mu \boldsymbol{R} - \mathrm{i}\pi \boldsymbol{H}$, $\boldsymbol{A}_\nu = \alpha_\nu \boldsymbol{R} + \mathrm{i}\pi \boldsymbol{H}$, $\boldsymbol{\delta}_\mu$ and $\boldsymbol{\delta}_\nu$ are dimensionless unit vectors directed along p orbitals of μ and ν atoms, and vector $\boldsymbol{R}$ is directed from the centre of μ towards the ν atoms.

The integral

$$f(H) = 4\pi \mathrm{i}^l \int_0^\infty R_l(r) j_l(2\pi H r) r^2 \, \mathrm{d}r \tag{I.17}$$

is found in the multipole models considered in subsection 3.4.5. This integral is the Fourier–Bessel transform of ED, approximated by a single function, but not by the wave-function product as is the case in quantum chemical models. If the radial function $R_{nl}(r)$ is approximated by the STO-type function $R_{nl}(r) = (1/4\pi)[\alpha^{n+3}/(n+2)!]r^n \exp(-\alpha r)$, the expression (I.17) can be presented as

$$f_{nl}(H) = \mathrm{i}^l \frac{(2\pi H/\alpha)^l (n+l+2)!}{(n+2)!(2l+1)!![1+(2\pi H/\alpha)^2]^{n+2}} \times {}_2F_1[(l-n-1)/2, (l-n)/2, l+\tfrac{3}{2}, -(2\pi H/\alpha)^2]. \tag{I.18}$$

Since $n \geqslant l$, in order to satisfy the Coulomb potential r^{-1} (Stewart 1977b), the hypergeometric function (I.18) is a small finite polynomial.

APPENDIX J

LEAST SQUARES IN CRYSTALLOGRAPHIC STRUCTURAL MODEL REFINEMENT

The methods of crystal structure determination (see e.g. Dunitz 1979) give an approximate set of crystal structural model parameters: positions of atoms and atomic thermal vibration tensor components. They are further completed by electron occupancies and exponents in the radial functions of multipoles, scale factors, etc. The parameter estimates are varied in order to obtain a best fit between experimental, F, and model, F_{model}, structure amplitudes. The latter is the estimator of the structural model under consideration and the adjusting of the parameters of a model to the experimental data is called refinement. In x-ray and neutron crystallography, the refinement procedure is mainly based on least squares (LS) when the weighted deviance $\hat{D}_w = \hat{d}^{\mathrm{T}} \hat{W} d$ (H.13) is minimized. The LS is the best method in the case when the error distribution of deviates, $d_j = F_j - F_{model,j}$, is unknown.

The crystallographic structural models nonlinearly depend on structural and electronic parameters. Therefore, the weighted deviance is minimized by iteratively linearizing F_{model} with a Taylor expansion at approximate parameter values. Thus, parameter shifts are determined rather than the parameters themselves.

In the linearized case minimization of weighted deviance $\hat{D}_w$ results in the well known matrix normal equation (Hamilton 1964)

$$\hat{N}\hat{X} = \hat{D}. \tag{J.1}$$

Here $\hat{N} = \hat{A}^{\mathrm{T}}\hat{A}$ is the symmetric square normal equation matrix of order $m \times m$, m being the number of parameters refined; $\hat{A}$, $\hat{X}$, and $\hat{D}$ are the design matrix and matrices of solutions and initial (experimental) data (as deviates), respectively. All matrices are normalized in order to eliminate the problem of different dimensions of the parameters.

The solution (estimates of the model parameters) of (J.1) is

$$\hat{X} = \hat{N}^{-1}\hat{A}^{\mathrm{T}}\hat{W}\hat{D}. \tag{J.2}$$

Minimal variances of the estimates are obtained if the weight matrix $\hat{W}$ is chosen as the inverse of the variance–covariance matrix $\hat{V}$ of the multivariate PDF of the observations (H.5). The inverse of the normal equation matrix is an unbiased estimate of the variance–covariance of the model parameters if, and only if, $\hat{W}=\hat{V}^{-1}$. In practive, $\hat{V}$ is usually chosen as a diagonal matrix and the variance of parameter estimates has the form

$$\sigma^2(x_i)=N_{ii}^{-1}D_w/(n-m) \tag{J.3}$$

where n is the number of symmetry-independent x-ray reflections. $\hat{W}$ is also a diagonal matrix (D_w includes the reflection variances). Off-diagonal terms in $\hat{W}$ may be important if non-random errors (such as TDS error) are present or if the measurements have been systematically altered. In this case the variances of model parameters may be seriously underestimated.

As can be seen from (J.2), the solution of the normal LS equation includes inversion of the matrix $\hat{N}$. If its determinant is zero or nearly zero this matrix is ill conditioned and it cannot be inverted. As a result, either the solution of equation (J.2) cannot be obtained or the parameter estimates have enormous variances. Usually such a situation is realized if linear or near-linear correlation dependences among model parameters exist. Estimation of such correlations is possible by means of the covariance

$$\mathrm{cov}(x_i, x_j)=N_{ij}^{-1}D_w/(n-m) \tag{J.4}$$

which includes the correlation coefficient ρ_{ij}, as can be seen from (H.6). In complicated multiparametrical crystal structural models correlation dependences nearly always exist. Even if such dependences do not result in divergence of the refinement process, it proceeds in an unstable fashion. Moreover the physical meaning of the model parameters can be distorted.

There are some other difficulties of LS application. First, there is the dependence of the $\hat{N}$ matrix conditional number not only on the model adequacy but also on the experimental data errors. Further, an ideal functional of (H.13) type only approximates an unknown true functional. It can have a number of local minima which are far from the correct minimum. Besides, the set of model parameters is usually determined supposing the linearity of functional (H.13), so there is no guarantee that the stationary point found corresponds to the minimum. Hence, the refinement of the crystallographic structural models by LS is a non-evident and non-trivial problem.

Tichonov and Arsenin (1986) worked out the general method of the solution of linear algebraic equation systems in the presence of errors. Their approach is called regularization; it results from the theory of ill posed problems. According to this theory, the stable solution of equation

$$(\hat{N}+\lambda\hat{I})\hat{X}=\hat{D} \tag{J.5}$$

which depends on parameter $\lambda>0$, can be taken as an approximate solution of the initial system. The parameter $\lambda>0$ is a regularization parameter which depends on the level of errors in the initial data. If λ tends to a limit, the succession of solutions of (J.5) should approach the correct solution of the initially ill posed problem in question. There are different ways of choosing λ (Arsenin and Zyabrev 1977).

An optimal, in the sense of minimal LS error, solution of equation (J.5) can be constructed in the form of an expansion over $\hat{N}$ matrix eigenvectors. This permits one to avoid the necessity of inverting it in the LS procedure. The algorithm of singular matrix decomposition, a reliable method for finding matrix eigenvalues (see e.g. Forsythe *et al* 1977), is suitable for this purpose. The singular decomposition of the symmetric matrix $\hat{N}$ will be its factorization, of the form

$$N = \hat{U}'\hat{S}\hat{U}. \tag{J.6}$$

$\hat{S}$ is a diagonal matrix and its elements s_j are called singular numbers. The numbers s_j are eigenvalues of the non-negative determined matrix $\hat{N}$ and the columns of the orthogonal matrix $\hat{U}$ will be the eigenvectors of this matrix. Using (J.6) the kth solution of the (J.5) system can be presented by a series

$$x_k = \sum_{j=1}^{m} \frac{c_j}{s_j + \lambda} u_{kj} \qquad k = 1, \ldots, m \tag{J.7}$$

where c_j are components of the vector $\hat{C} = \hat{U}\hat{D}$. In order to find an optimal approximation to (J.7), one replaces the set of one-parameter solutions of the form (J.7) by the m-parameter family of $\hat{X}^{\Lambda}$ solutions ($\Lambda = \lambda_1, \ldots, \lambda_m$) and obtains a set of m parameters, λ_j, from the condition of the minimum mean square error in the solution vector:

$$\sigma^2(\hat{X}) = \left\langle \sum_{k=1}^{m} (x_k^{\Lambda} - x_k^0)^2 \right\rangle. \tag{J.8}$$

Here $\hat{X}^0$ is the precise solution of the initial task. It has been shown by Arsenin and Zyabrev (1977) that X^{Λ} determine the stable optimum solution, the kth component of which has the form

$$x_k^{opt} = \sum_{j=1}^{m} \frac{(c_j^0)^2}{(c_j^0)^2 + \sigma^2(c_j)} \frac{c_j}{s_j} u_{kj}. \tag{J.9}$$

This expression contains unknown coefficients c_j^0 which correspond to the 'true' initial data. The values $\sigma^2(c_j)$ are the variances of c_j. In order to obtain the solution one should replace c_j^0 by known coefficients c_j. This is permissible only when c_j have small errors, e.g. when $\sigma(c_j) \ll |c_j|$; in other cases the algorithm would be unstable. In order to provide stability of solution one usually ignores those terms in the series (J.9) for which c_j are comparable with $\sigma(c_j)$. Then the optimal approximate solution of the normal equation can be composed in the regularized form

$$x_k = \sum_{j=1}^{m} M_j \frac{c_j}{s_j} u_{kj}. \tag{J.10}$$

The terms of series (J.10) proved to be linearly weighted, the weight factors dependent on the noise level of the initial data. This procedure can be called filtration. M_j, the filtering function, has the form (Streltsov *et al* 1988a)

$$M_j = \begin{cases} c_j^2/(c_j^2+\sigma^2(c_j)) & |c_j| \geqslant (\beta+1)\sigma(c_j) \\ [c_j^2/(c_j+\sigma^2(c_j))][|c_j|/\sigma(c_j)-\beta] & \beta\sigma(c_j) < |c_j| < (\beta+1)\sigma(c_j) \\ 0 & |c_j| \leqslant \beta\sigma(c_j). \end{cases} \qquad \text{(J.11)}$$

β is the filter parameter which is dependent on the error distribution law in $\hat{C}$ and on the smoothness of the solution searched for. The filtering function of (J.11) form provides continuous solution dependence on the error level in experimental data.

Variances $\sigma^2(c_j)$ of independent random values c_j in optimization of crystallographic structural models with x-ray diffraction data can be evaluated taking into account that $\hat{C}=\hat{U}\hat{D}$ and $\hat{D}$ vector components in (J.5) are dependent on structure amplitudes. Then

$$\sigma^2(c_j) \cong S^2 s_j \qquad \text{(J.12)}$$

where the goodness of fit, S^2 (H.17), gives an estimate of the adjustment of model parameters to experimental data.

The variances of the parameters refined, accounting for equation (J.10), are given by

$$\sigma^2(x_k) = \sum_{j=1}^{m} M_j \frac{\sigma^2(c_j)}{s_j} u_{kj}^2. \qquad \text{(J.13)}$$

As can be seen from (J.10) and (J.13) the filtration influences both the parameter values ('signal') and their variances ('noise'). The useful information can be more clearly seen on the noise background.

The structural models of a crystal are usually constructed on some physical basis. Therefore the correlation coefficients between model parameters usually have values of 0.4–0.8. However, even if near-linear dependences in the initial normal equation matrix exist and some of these coefficients were initially close to unity, regularized LS allows one to obtain the estimates for each parameter value. (J.8) is used for this purpose as the additional condition. The region of possible solutions is decreased and parameter estimates have smaller uncertainties with the regularized LS. The degree of approximation of the solution is determined by the accuracy of the x-ray diffraction data used and by the level of correlation between parameters.

The value

$$\text{cond}(\hat{N}) = s_{max}/s_{min} \qquad (s_{min} \neq 0) \qquad \text{(J.14)}$$

called the spectral conditional number for the $\hat{N}$ matrix, characterizes the degree of solution distortion. $\text{cond}(\hat{N})=1$ is for orthogonal matrices. The linear dependences in the refined model lead to an increase of $\text{cond}(\hat{N})$. The

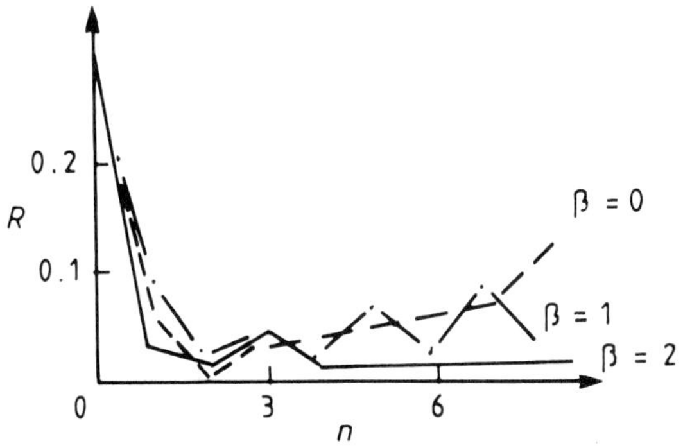

Figure J.1 The dependence of the discrepancy factor on number of iterations of regularized LS refinement of the fluorite, CaF_2, structural $\varkappa$-model at different values of filter parameter, β.

regularization decreases the conditional number of the normal equation matrix and, therefore, makes the refinement of the structural models more reliable.

Some points about the β coefficient in (J.11). Streltsov *et al* (1988a), using structure amplitudes from different precision measurements for the fluorite crystal, CaF_2, investigated the convergence of refinement of the $\varkappa$-model, taking into account the anharmonicity of F atom thermal vibrations (see subsection 3.4.4). Typical results for one of the structure amplitude sets used are presented in figure J.1. The numerical experiments have shown that for a more precise data set convergence is reached with a smaller β value. A β value ranging from one to two was recommended by the authors cited: stable convergence is provided and, at the same time, the information loss, because of the rejection of terms with $|c_j| \leqslant \beta\sigma(c_j)$ in (J.11), is minimal.

APPENDIX K

NEUTRON SCATTERING IN ELECTRON DENSITY STUDIES

The peculiarities of neutron diffraction which make them useful in ED studies are given in this appendix.

A neutron interacts with an atom by means of forces of various natures. The most important among them are nuclear and magnetic forces.

The nuclear scattering amplitudes due to the specificity of the interaction of a neutron with a nucleus via the nuclear forces possess some special properties. They have no regular dependence on the atomic number and are different for different isotopes of the same element. They can be positive, negative, and complex. The nuclear scattering amplitude does not depend on scattering angle, because the wavelength of thermal neutrons used in structure analysis is 10^5 times larger than the dimension of the scatterer. Thus, the nuclear scattering is spherically symmetric.

Each of these properties is used to obtain unique results in structure analysis. Values of nuclear scattering amplitudes for practically all known isotopes can be found in tables (Wilson 1992).

The wave–particle duality permits the realization of two methods in diffraction experiments. Indeed, there are two variables in the Bragg equation $n\lambda = 2d \sin \theta$. Bragg reflection detection is possible either keeping λ constant and varying scattering angle or making θ constant and measuring λ for every reflection. The majority of diffraction studies employ the first procedure with stationary-power reactors. This requires a monochromatic incident beam and a sophisticated system for determining diffraction angle. The second method requires short neutron impulses of white radiation and complicated electronics for determining the wavelength of neutrons taking part in a diffraction. This can be realized with impulse neutron sources, i.e. impulse reactors or, especially, with charged particle accelerators, which are becoming increasingly common worldwide. Because the wavelength measured here is determined by the time a neutron takes to pass a certain path length the method is called time-of-flight.

The same diffractometers as in x-ray structure analysis are mostly used with neutrons for single-crystal measurements. Kinematic theory is also used in data treatment. There is some specificity in TDS correction for neutron diffraction. This is described in appendix G. The absorption of neutrons is quite small, as a rule: for radiation wavelength $\lambda \approx 1$ Å linear absorption coefficient $\mu \approx 10^{-2}$ cm^{-1}. The exceptions are Li, B, Cd, and some rare-earth elements and isotopes ($\mu \approx 10^3$–10^5 m^{-1}), as well as proton containing compounds. Extinction with neutrons is more severe than in x-ray structure analysis because of the large sample size used. Nevertheless, the same methods (see subsection 3.3.3) are used in neutron studies; some specific pecularities have been discussed by Becker (1980).

The neutron diffraction data treatment results in the scattering density Fourier transform which is the nuclear density smeared by thermal vibrations. The nuclei positional and thermal parameters can be determined as usual by LS refinement of some structural model.

The use of neutron diffraction in combination with the x-ray diffraction in the studies of ED results from the fact that the coherent nuclear scattering is completely independent of any model of the atomic electron shell. That is why both the nuclei positional and thermal vibration parameters can be determined with neutrons without any concerns about the ED model. Assuming that the nucleus position coincides with the atomic electron cloud centre and that the vibrations of nucleus and electron shells are the same (Born–Oppenheimer and convolution approximations) the structural parameters obtained with neutrons may be introduced into the x-ray structural model as non-refined data. As a result, the number of model parameters to be refined is decreased drastically and the statistical reliability of their determination is increased. Besides, neutron structural parameters can be used in the deformation ED calculation (see subsection 3.6.2).

As well as nuclear, there is also magnetic interaction between a neutron and the atomic magnetic moment. The latter arises because of the electron shell properties. The scatterers' dimensions coincide with the wavelength of the scattering radiation in this case. Therefore, intra-atomic interference takes place and the magnetic scattering amplitudes are angle dependent.

The magnetic moment of an atom is mainly caused by electron spin and orbital states. One-electron, atomic magnetic moment projections on a fixed z axis due to spin, m_z^s, are given by

$$m_z^s = g(e\hbar/2m)m_s = 2(e\hbar/2m)m_s = 2\mu_b m_s \tag{K.1}$$

where $\mu_b = (e\hbar/2m) = 9 \times 10^{-27}$ J T^{-1} is the Bohr magneton, g is the Landé factor, and m_s is the spin magnetic quantum number, $m_s = \pm\frac{1}{2}$. The other notations are generally accepted. For electron spin $g = 2$ so $m_z^s = \pm\mu_b$.

The orbital (angular) magnetic moment of the one-electron atom is caused by the orbital electron motion. Its projection on the z axis, m_z^l, is $m_z^l = g\mu_b m_l = \mu_b m_l$. Here $g = 1$ and m_l is the magnetic angular quantum number ($m_l = 0, \pm 1, \pm 2, \ldots, \pm l$).

The magnetic moment of a many-electron atom with Russell–Saunders coupling is given by the vector model of an atom. The magnetic moment projection on the z axis is $m_z^J = g\mu_b m_J$, where the total quantum number J is the sum $\boldsymbol{J} = \boldsymbol{L} + \boldsymbol{S}$, and $\boldsymbol{L} = \Sigma \boldsymbol{l}_i$, $\boldsymbol{S} = \Sigma \boldsymbol{s}_i$ are the total angular and spin quantum numbers respectively. In the general case the Landé factor depends on J, S, and L. Together with the principal quantum number, n, these quantum numbers define the energy state of an atom because of the spin–orbit coupling depending on the degree of degeneracy of a level. These quantum numbers determine the ground state wave-function too.

Generally, both the orbit and spin contribute to the atomic magnetic moment, but in the simplest case only electron spin contributes to the atomic magnetic moment as in many 3d elements and their compounds. In this case the neutron magnetic scattering amplitudes are the Fourier transforms of the so-called atomic spin density (chapter 5), which describes the distribution of those electrons in an atom whose spins are not coupled. When both spin and orbit contribute to atomic magnetic moment, as in the majority of f elements and their compounds, the neutron magnetic scattering becomes more complicated. Calculation of the scattering amplitudes or, in other words, the atomic form factors, $f(H)$, demands knowledge of atomic ground state wave-functions. These are not always known because of either the energy degeneracy or insignificant differences in the energy levels of different quantum states. Calculation methods for magnetic form factors are described in detail by Wilson (1992). Experimental measurement of form factors and their comparison with calculations make it possible to define the atomic electron ground state and, then, the electron configuration of the scattering atom.

Neutron magnetic scattering by magnetics is of vector nature because both the magnetic crystal structure itself and the magnetic neutron moment are vectors.

The magnetic neutron scattering process is much more complicated, but opens the way to study the spin and/or magnetization density distribution.

For our purposes we will restrict ourselves to the experimental geometry when the magnetic field magnetizes the sample perpendicular to the scattering vector. This is usually possible in two important cases: first, when a relatively weak external field aligns the ferromagnetic domains along the field direction; second, and this is important to us, when a strong magnetic field magnetizes weak magnetics, such as paramagnetic and diamagnetic substances. The ED which is responsible for substance magnetization can be studied in this way.

The scalar atomic amplitude of the neutron magnetic scattering, $p(H)$, is given by the equation

$$p(H) = 2e^2/(mc^2)\gamma S f(H) = 0.54 S f(H) \times 10^{-14}\ \mathrm{m}. \qquad \text{(K.2)}$$

The neutron magnetic moment magnitude is $\gamma = -1.913\,07 \pm 0.000\,06\ \mu_N$ (nuclear magnetons). The negative sign means that the neutron magnetic moment is antiparallel to its angular momentum vector. In this equation, besides some universal constants, S is the effective spin quantum number and $f(H)$ is the atomic magnetic neutron scattering form factor. This factor accounts for the intra-atomic interference in magnetic scattering. If both orbit and spin take part in atomic magnetism $p(H)$ has the form

$$p(H)=e^2/(mc^2)\gamma gJf(H)=0.54\,gJf(H)\times 10^{-14}\ \mathrm{m} \tag{K.3}$$

where J is the total atomic quantum number.

The scalar form of $p(H)$ is quite appropriate for the description of neutron scattering by a free atom, the electron charge and spin density distribution of which have no fixed orientation, but this orientation becomes essential in the case of scattering by a crystal. Instead of scalars S (or J) vectors $\boldsymbol{S}$ (or $\boldsymbol{J}$) should be considered now. The form factor of magnetic scattering will also be a vector, $\boldsymbol{p}(H)$. Using $\boldsymbol{p}(H)$, the vector of magnetic (static) structure amplitude, $\boldsymbol{F}_M(H)$, can be written as

$$\boldsymbol{F}(\boldsymbol{H})=\sum_{\mu}\boldsymbol{p}_{\mu}(\boldsymbol{H})\mathrm{e}^{2\pi \boldsymbol{H}\cdot\boldsymbol{r}_{\mu}}. \tag{K.4}$$

The intensity of magnetic reflection depends also on the mutual orientation of vectors $\boldsymbol{F}(\boldsymbol{H})$ and unit scattering vector $\boldsymbol{H}/H$ (figure K.1)

$$\mathscr{F}(\boldsymbol{H})=\boldsymbol{F}(\boldsymbol{H})-((\boldsymbol{H}/H)\boldsymbol{F}(\boldsymbol{H}))\boldsymbol{H}/H \tag{K.5}$$

and

$$|\mathscr{F}(\boldsymbol{H})|^2=|\boldsymbol{F}(\boldsymbol{H})|^2-|(\boldsymbol{H}/H)\boldsymbol{F}(\boldsymbol{H})|^2. \tag{K.6}$$

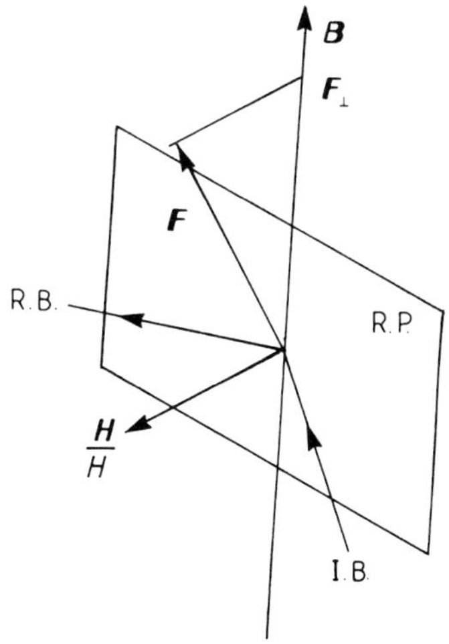

Figure K.1 Relations between vectors $\boldsymbol{H}/H$ (unit scattering vector), magnetization direction $\boldsymbol{B}$, $F_{\perp}$, and $\boldsymbol{F}$. IB and RB are incident and reflected beams, RP is the reflecting plane.

The latter quantity is the main one in the magnetic scattering intensity equation. If the magnetization is perpendicular to the scattering vector (or lies in the scattering plane), then

$$|\mathscr{F}(H)|^2 = |F(H)|^2. \tag{K.7}$$

If the direction of magnetization does not coincide with the external field the general formula (K.5) should be used.

Magnetic structure investigation is mainly carried out with polarized neutron beams. In these beams neutron spins lie in one plane which can be characterized by the unit vector $\boldsymbol{\lambda}$. The total structure amplitude $F(\boldsymbol{H})$ contains both nuclear and magnetic contributions and, in addition, the interference term

$$F^2 = F_N^2 + 2F_M F_N(\boldsymbol{\lambda}\cdot\boldsymbol{m}) + F_M^2 \tag{K.8}$$

where $\boldsymbol{m}$ is the unit vector of the sample magnetization direction. In an experiment with $\boldsymbol{\lambda}$ and $\boldsymbol{m}$ either parallel or antiparallel, this means that (with the polarization changing to the opposite direction) the interference term changes sign and

$$F^2 = F_N^2 \pm F_M^2. \tag{K.9}$$

In a modern magnetic scattering experiment the so-called flipping ratio, R, is measured, R being the $(F^2_{\uparrow\uparrow}/F^2_{\uparrow\downarrow})$ ratio with $\boldsymbol{\lambda}$ up and down. Then

$$R = (F_N^2 + 2F_N F_M + F_M^2)/(F_N^2 - 2F_N F_M + F_M^2) = ((1+\mu)/(1-\mu))^2 \tag{K.10}$$

where μ is the (F_M/F_N) ratio. From (K.9) μ could be found as

$$\mu = (R + 1 \pm 2\sqrt{R})/(R-1). \tag{K.11}$$

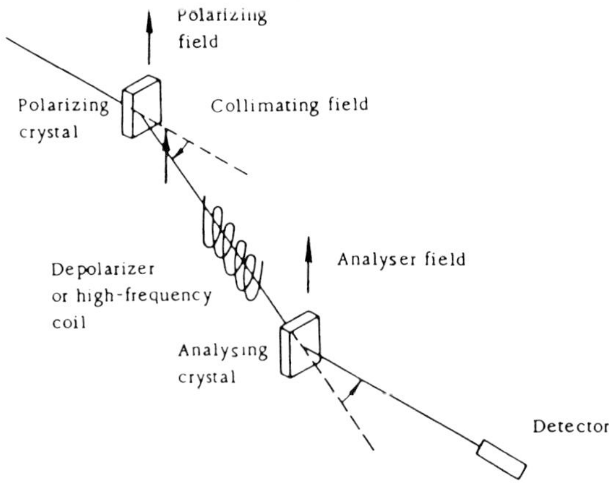

Figure K.2 A schematic drawing of the set-up of polarized neutron scattering.

The equation (K.10) makes possible the measurement of F_M in F_N units by determining the flipping ratio for every neutron reflection.

The layout of the apparatus for investigations of magnetization density distribution by means of polarized neutrons is presented in figure K.2. The neutron beam from a reactor falls onto the crystal which is a monochromator and polarizer simultaneously. The crystal under investigation, mounted on the goniometer head, is placed in the beam. The magnetic field, usually produced by two superconductor coils, is applied (in the majority of experiments) perpendicular to the scattering vector. Neutrons reflected at a certain angle are registered by a detector. The flipper, installed between the monochromator and the crystal sample, is used to change the neutron polarization direction. As has been mentioned, the flipping ratio is measured when the position of a crystal sample is constant and the flipper is turned on and off. The experiments in spin density studies (chapter 5) were carried out according to this scheme. Magnetic structure amplitudes measured in such experiments were then used for magnetization density reconstruction in the same way as x-ray amplitudes are used in calculations of the ED.

REFERENCES

Abrahams S C 1973 *Acta Crystallogr.* A **29** 111–6

Abrahams S C and Keve E 1971 *Acta Crystallogr.* A **27** 157–65

Abramov Yu A, Avilov A S, Belokoneva E L, Kitaneh R, Feil D, Tsirelson V G and Okamura F P 1995 *Asian Crystallogr. Congress* (*Bangkok, 1995*) p 1A23

Achmanov S A, Meisner L B, Parinov S T, Saltiel S M and Tunkin V G 1977 *Zh. Eksp. Teor. Fiz.* **73** 1710–28 (in Russian)

Aldred J P L and Hart M 1973a *Proc. R. Soc.* A **332** 223–38

—— 1973b *Proc. R. Soc.* A **332** 239–54

Aleksandrov Yu V, Tsirelson V G, Reznik I M and Ozerov R P 1989 *Phys. Status Solidi* B **155** 201–7

Alkire R W and Yelon W B 1981 *J. Appl. Crystallogr.* **14** 362

Alkire R W, Yelon W B and Schneider J R 1982 *Phys. Rev.* B **26** 3097–104

Allan N L, Cooper D L and Mackrodt W C 1990 *Mol. Simul.* **4** 269–84

Allen-Williams A J, Delaney W T, Furina R, Maslen E N, O'Connor B H, Varghese J N and Yang F H 1975 *Acta Crystallogr.* A **31** 101–15

Ambrosch-Draxl C, Blaha P and Schwarz K 1994 *J. Phys. Condens. Matter* **6** 2347–56

Amstutz R, Laube T, Schweizer W B, Seebach D and Dunitz J D 1984 *Helv. Chim. Acta* **67** 224–36

Angonoa G, Dovesi R, Pisani C and Roetti C 1981 *Phil. Mag.* B **44** 413–8

Anisimov V I, Gubanov V A, Ellis D E and Kurmaev E Z 1981 *J. Phys. F: Met. Phys.* **11** 405–18

Antipin M Yu, Ellern A M, Sukhoverkhov V F and Struchkov Yu T 1988 *Zh. Neorg. Khim.* **33** 307–11 (in Russian)

Antipin M Yu, Polyakov A V, Kapphahn M, Tsirelson V G, Grushin V V and Struchkov Yu T 1990 *Metalloorg. Chem.* **3** 927–32 (in Russian)

Antipin M Yu, Tsirelson V G, Flügge M, Struchkov Yu T and Ozerov R P 1987 *Koord. Chem.* **13** 121–9 (in Russian)

Arfken G 1985 *Mathematical Methods for Physicists* (Orlando, FL: Academic)

Arii T, Uyeda R, Terasaki O and Watanabe D 1973 *Acta Crystallogr.* A **29** 295

Arsenin V Ya and Zyabrev N B 1977 *Preprint Inst. Appl. Math.* 127 (Moscow: Institute of Applied Mathematics) (in Russian)

Ashcroft N W and Mermin N D 1976 *Solid State Physics* (New York: Holt, Rinehart and Winston)

Aslanov L A, Fetisov G V, Laktionov A V, Markov V T, Chernyshov V V, Zhukov S G, Nesterenko A P, Chulichkov A I and Chulichkova N M 1989 *Precision X-ray Diffraction Experiment* (Moscow: Moscow University) (in Russian)

Assing G and Monkhorst H J 1993 *Int. J. Quantum Chem. Symp.* **27** 81–9

Avery J 1978 *Acta Crystallogr.* A **34** 582–4

—— 1979 *Int. J. Quantum Chem.* **16** 1265–77

Avery J, Larsen P S and Grodzicki M 1984 *Local Density Approximations in Quantum*

Chemistry and Solid State Physics ed A P Dahl and J Avery (New York: Plenum) pp 733–50
Avery J and Ormen P-J 1979 *Acta Crystallogr.* A **35** 849–52
Avery J, Ormen P-J and Mullen D 1981 *Int. J. Quantum Chem. Symp.* **15** 477–86
Avilov A S 1976 *Sov. Phys. Crystallogr.* **21** 646–8
Avilov A S, Kitaneh R and Tsirelson V G 1996 *Krystallographiya* submitted
Avilov A S and Parmon V S 1990 *Sov. Phys.–Crystallogr.* **35** 733–7
Avilov A S, Parmon V S, Semiletov S A and Sirota M I 1984 *Krystallografiya* **29** 11–5
Avilov A S, Semiletov S A and Storozhenko V V 1989 *Sov. Phys.–Crystallogr.* **34** 110–5
Azavant P, Lichanot A, Rerat M and Chaillet M 1994 *Theor. Chim. Acta* **89** 213–26
Babenko V V, But'ko V G and Reznik I M 1988 *Preprint* 88-19 (Donetsk: Physico-Technical Institute of the Academy of Science, Ukraine (in Russian)
Bachelet G B, Ceperley D M and Chiocchetti M G B 1989 *Phys. Rev. Lett.* **62** 2088–91
Bachelet, G B, Hamann D. R and Schlüter M 1982 *Phys. Rev.* B **26** 4199–205
Bacon G E 1975 *Neutron Diffraction* 3rd edn (Oxford: Clarendon)
Bader R F W 1975a *Int. Rev. Sci.: Phys. Chem.* Ser. II, vol 1, ed A D Buckingham and C A Coulson (London: Butterworths) pp 43–78
—— 1975b *Account Chem. Res.* **8** 34
—— 1980 *J. Chem. Phys.* **73** 2871–83
—— 1981 *The Force Concept in Chemistry* ed B M Deb (New York: Van Nostrand Reinhold) pp 39–136
—— 1985 *Account Chem. Res.* **18** 9–15
—— 1990 *Atoms in Molecules: a Quantum Theory* (Oxford: Oxford University Press)
—— 1991 *Chem. Rev.* **91** 893–928
—— 1994 *Phys. Rev.* B **49** 13 348–56
Bader R F W and Becker P 1988 *Chem. Phys. Lett.* **148** 452–8
Bader R F W and Beddall P M 1972 *J. Chem. Phys.* **56** 3320–9
Bader R F W and Essen H 1984 *J. Chem. Phys.* **80** 1943–60
Bader R F W and MacDougall P J 1985 *J. Am. Chem. Soc.* **107** 6788
Bader R F W, MacDougall P J and Lau C D H 1984 *J. Am. Chem. Soc.* **106** 1594–605
Bader R F W and Nguyen-Dang T T 1981 *Adv. Quantum Chem.* **13** 63–124
Bader R F W, Nguyen-Dang T T and Tal Y 1981 *Rep. Prog. Phys.* **44** 893–48
Bader R F W and Preston H J T 1969 *Int. J. Quantum. Chem.* **3** 327–47
Bader R F W, Slee T S, Cremer D and Kraka E 1983 *J. Am. Chem. Soc.* **105** 5061–8
Bader R F W, Srebrenik S and Nguyen-Dang T T 1978 *J. Chem. Phys.* **68** 3680–91
Bader R F W, Tang T-H, Tal Y and König F W 1982 *J. Am. Chem. Soc.* **104** 946–52
Baerends E J, Ellis D E and Ros P 1973 *Chem. Phys.* **2** 41–51
Baerends E J, Vernooijs P, Rozendaal A, Boerrigter P M, Krijn M, Feil D and Sundholm D 1985 *J. Mol. Struct.* **26** 147–59
Baert F, Coppens P, Stevens E D and Devos L 1982 *Acta Crystallogr.* A **38** 143–51
Bagayoko D, Laurent D G, Singhal S P and Callawey J 1980 *J. Physique Lett.* A **76** 187
Bagus P S and Wahlgreen U J 1976 *Comput. Chem.* **1** 95–9
Balasubramanian K 1989 *J. Phys. Chem.* **93** 6585–96
Baldereschi A 1972 *Phys. Rev.* B **7** 2512–5
Baldereschi A, Maschke K, Milchev A, Pickenheim R and Unger K 1981 *Phys. Status Solidi* b **108** 511–20
Bamsai A S and Deb B M 1981 *Rev. Mod. Phys.* **53** 95–126
Barnes L A, Chandler G S and Figgis B N 1989 *Mol. Phys.* **68** 711

Bartell L S and Brockway L O 1953 *Phys. Rev.* **90** 833
Bartlett R J (ed) 1985 *Comparison of Ab Initio Quantum Chemistry with Experiment for Small Molecules* (Dordrecht: Reidel)
Basch H 1970 *Chem. Phys. Lett.* **5** 337–9
Batterman B 1990 *15th Congr. IUCrystallography* (*Bordeaux, 1990*) Abstracts, Main Lecture, 01.1
Becke A D 1989 *Int. J. Quantum Chem. Symp.* **23** 599–609
Becker P (ed) 1980 *Electron and Magnetization Densities in Molecules and Crystals* (New York: Plenum)
Becker P 1988 *Chemical Crystallography with Pulsed Neutrons and Synchrotron X-rays* (*NATO ASI Series*) ed M A Corrondo and G B Jeffrey (Dordrecht: Reidel) pp 217–43
Becker P and Al Haddad M 1989 *Acta Crystallogr.* A **45** 333–7
Becker P and Coppens P 1974a *Acta Crystallogr.* A **30** 129–47
—— 1974b *Acta Crystallogr.* A **30** 148–53
—— 1975 *Acta Crystallogr.* A **31** 417–25
—— 1985 *Acta Crystallogr.* A **41** 177–82
—— 1990 *Acta Crystallogr.* A **46** 254–8
Becker P, Coppens P and Ross F K 1973 *J. Am. Chem. Soc.* **95** 7604–9
Belokoneva E L, Yakubovich O V, Tsirelson V G and Urusov V S 1990 *Dokl. Akad. Nauk* **26** 595–601 (in Russian)
Benard M 1982 *Electron Distributions and the Chemical Bond* ed P Coppens and M B Hall (New York: Plenum) pp 221–53
Bernard M, Coppens P, DeLucia M L and Stevens E D 1980 *Inorg. Chem.* **19** 1924–30
Bendt P and Zunger A 1982 *Phys. Rev.* B **26** 3114–37
Benesh R and Smith V H 1970 *Acta Crystallogr.* A **26** 586–94
Bentley J 1979 *J. Chem. Phys.* **70** 159–64
Bentley J J and Stewart R F 1974 *Acta Crystallogr.* A **30** 60–7
—— 1975 *J. Chem. Phys.* **63** 3794–803
—— 1976 *Acta Crystallogr.* A **32** 910–4
Berglund B, Lindgren J and Tegenfeldt J 1978 *J. Mol. Struct.* **43** 179
Beri A C, Lee T, Das T P and Sternheimer R M 1983 *Phys. Rev.* B **28** 2335–51
Berkovitch-Yellin Z 1985 *J. Am. Chem. Soc.* **107** 8239–53
Berkovitch-Yellin Z, Ariel S and Leiserowitz L 1983 *J. Am. Chem. Soc.* **105** 765–7
Berkovitch-Yellin Z and Leiserowitz L 1980 *J. Am. Chem. Soc.* **102** 7677–90
—— 1982 *J. Am. Chem. Soc.* **104** 4052–64
—— 1984 *Acta Crystallogr.* B **40** 159–65
Berkovitch-Yellin Z, van Mil J, Addadi L, Idelson M, Lahav M and Leiserowitz L 1985 *J. Am. Chem. Soc.* **107** 3111–22
Berlin T 1951 *J. Chem. Phys.* **19** 208–13
Bersuker G I, Peng C and Boggs J E 1993 *J. Phys. Chem.* **97** 9323–9
Bersuker I B 1976 *Electronic Structure and Properties of Coordination Compounds* (Leningrad: Chimia) (in Russian)
—— 1984 *Jahn–Teller Effect and Vibronic Interactions in Modern Chemistry* (New York: Academic)
Bersuker I B, Budnikov S S and Leizerov B A 1977 *Int. J. Quantum Chem.* **11** 541–59
Bersuker I B and Ogurtsov I Ya 1979 *Methods in Quantum Chemistry* (Shernogolovka: Institute of Chemical Physics) pp 70–81 (in Russian)

Berthler G 1980 *Chim. Phys. Inf. J. Gif-sur-Yvette* (*Paris*) s.a. 8–13
Berzelius J J 1820 *Versuch über die Theorie der Chemischen Proportionen und über die Chemischen Wirkungen der Electrizität* (Dresden: Arnold)
Bethe H 1928 *Ann. Phys., Lpz.* **87** 55–69
Bethe H A and Salpeter E E 1957 *Quantum Mechanics of One- and Two-Electron Atoms* (Berlin: Springer)
Bevington P R 1969 *Data Reduction and Error Analysis for the Physical Sciences* (New York: McGraw-Hill)
Bianchi R, Pilati T and Simonetta M 1984 *Helv. Chim. Acta* **67** 1707–12
Biedenharn L C and Louk J D 1981 *Angular Momentum in Quantum Physics* (Reading, MA)
Bilderback D H and Colella R 1975 *Phys. Rev.* B **11** 793–7
Bingel W 1963 *Z. Naturf.* a **18** 1249–53
Bishop D 1964 *J. Chem. Phys.* **40** 1322
Blackman M 1939 *Proc. R. Soc.* A **173** 68
Blaha P, Schwarz K and Herzig P 1985 *Phys. Rev. Lett.* **54** 1192–5
Blessing R H 1987 *Crystallogr. Rev.* **1** 3–58
—— 1995 *Acta Crystallogr.* B **51** 816–23
Blessing R H, Coppens P and Becker P J 1974 *J. Appl. Crystallogr.* **7** 488–93
Bloembergen N 1965 *Non-linear Optics* (New York: Benjamin)
Blume M 1964 *Phys. Rev.* A **133** 1361
Boek E S, Feil D, Briels W J and Bennema P 1991 *J. Cryst. Growth* **114** 389–410
Boese R 1994 *IUCr-Symposia* vol 7, ed D Jones and I Katrusiak (Oxford: Oxford University Press) pp 20–37
Böhm H-J and Alrichs R 1982 *J. Chem. Phys.* **77** 2028
Bolotin H H, Stuchbery A E and Amos K 1978 *Nucl. Phys.* A **311** 75–92
Bolotovsky R, White M A, Darovsky A and Coppens P 1995 *J. Appl. Crystallogr.* **28** 86–95
Bordeaux D, Boucherle J X, Delley B, Gillon B, Ressouche E and Schweizer J 1993 *Z. Naturf.* a **48** 120–2
Born M 1942–1943 *Rep. Prog. Phys.* **9** 294–333
Born M and Huang K 1954 *Dynamical Theory of Crystal Lattices* (New York: Oxford Univeristy Press)
Borovik-Romanov A S, Kreines N M, Pankow A A and Talalaev M A 1973 *Zh. Eksp. Theor. Fiz.* **37** 890 (in Russian)
Borovskaya T N, Gerr R G, Tsirelson V G, Struchkov Yu T and Ozerov R P 1989 *Accurate Structure Studies of Crystals* ed A Mishnev (Riga: Institute of Organic Synthesis) pp 44–56 (in Russian)
Boscovish R 1758 *Theoria Phylosophica Naturalis Reducta ad Unicam Legem Virium in Natura Existentium* (Vienna)
Boucherle J X, Gillon B, Ressouche E, Rey P and Schweizer J 1992 *Physica* B **180 & 181** 135–8
—— 1993 *Z. Naturf.* a **48** 117–9
Bouteiler Y, Mijoule C, Nizam M, Barthelat J C, Daudey J P, Pelissier M and Silvi B 1988 *Mol. Phys.* **65** 295–312
Boyd R J 1977a *J. Chem. Phys.* **66** 356–8
—— 1977b *Can. J. Phys.* **55** 452–5
Breitenstein M, Dannöhl H, Meyer H, Schweig A, Seeger R, Seeger U and Zittlau W 1983 *Int. Rev. Phys. Chem.* **3** 335–91

Breitenstein M, Dannöhl H, Meyer H, Schweig A and Zittlau W 1982 *Electron Distribution and the Chemical Bond* ed P Coppens and M B Hall (New York: Plenum) pp 255–81
Brown A S and Spackman M A 1990 *Acta Crystallogr.* A **46** 381–7
—— 1991 *Acta Crystallogr.* A **47** 21–9
—— 1994 *Mol. Phys.* **83** 551–66
Brown D and Fatemi M 1971 *J. Appl. Phys.* **45** 1544–54
Brown P J 1972 *Phil. Mag.* **26** 1377–94
—— 1980 *Electron and Magnetization Densities in Molecules and Crystals* ed P Becker (New York: Plenum) pp 271, 723
Brown P J and Forsyth B N 1988 *Mater. Sci. Forum* **27/28** 19–22
Buckingham A D 1937 *Proc. R. Soc.* A **160** 94–112
—— 1959 *Q. Rev.* **13** 183–214
—— 1978 *Intermolecular Interactions: from Diatomic to Biopolymers* ed B Pullman (New York: Wiley)
Buckingham A D and Fowler P W 1983 *J. Chem. Phys.* **79** 6426–8
Bunge C F, Barrientos J and Bunge A V 1993 *At. Data Nucl. Data Tables* **53** 113–62
Burgi H B, Dunitz J D and Shefter E 1974 *Acta Crystallogr.* B **30** 1517–27
Burke V M and Grant I P 1967 *Proc. Phys. Soc.* **90** 297
Burkel E, Dorner B, Illini Th and Peisl J 1989 *Rev. Sci. Instrum.* **60** 1671–3
Buschmann J, Koritsansky T, Kuschel R, Luger P and Seppelt K 1991 *J. Am. Chem. Soc.* **113** 233–8
Cade P 1972 *Trans. Am. Crystallogr. Assoc.* **8** 1–20
Calais J-L 1991 *Sagamore X. Abstracts* ed M Springborn, A Saenz and W Weyrich (Konstanz: University of Konstanz) p 23
Cao W L, Gatti C, MacDougall P J and Bader R F W 1987 *Chem. Phys. Lett.* **141** 380–5
Carlile C J and Willis B T M 1989 *Acta Crystallogr.* A **45** 708–15
Carrol M T, Bader R F W and Vosko S H 1987 *J. Phys. B: At. Mol. Phys.* **30** 3599
Carus L C 1951 *The Nature of the Universe* transl. R Latham (Harmondsworth: Penguin)
Catlow C R 1990 *Application of Synchrotron Radiation* ed C R Catlow and G N Greaves (Glasgow and London: Blackie) ch 2
Causa M, Dovesi R and Roetti C 1991 *Phys. Rev.* B **43** 11 937–47
Causa H, Dovesi R, Roetti C, Kotomin E and Saunders V R 1987 *Chem. Phys. Lett.* **140** 120–3
Challacombe M and Cioslowsky J 1994 *J. Chem. Phys.* **100** 464–72
Chalvet O, Daudel R, Diner S and Malrien J P (ed) 1975 *Localization and Delocalization in Quantum Chemistry* vol I (Boston, MA: Reidel)
Chandler G S, Figgis B N, Phillips R A, Reynolds P A and Mason R 1982a *Proc. R. Soc.* A **384** 31
Chandler G S, Figgis B N, Philips R A, Reynolds P A, Mason R and Williams G A 1982b *Proc. R. Soc.* A **384** 39–56
Chandler G S and Phillips R A 1986 *J. Chem. Soc. Faraday Trans. II* **82** 573–92
Chandler G S and Spackman M A *Acta Crystallogr.* A **34** 341–3
—— 1982 *Acta Crystallogr.* A **38** 225–39
Chang S-L 1987 *Crystallogr. Rev.* **1** 59
Chelikowski J R 1990 *Solid State Commun.* **33** 1201–4

Chidambaram R 1981 *Computing in Crystallography* ed R Diamond, S Ramaseshan and K Venkatesan (Bangalore: Indian Academy of Science) ch 2

Chipman D R 1969 *Acta Crystallogr.* A **25** 209–14

Chulichkov A I, Chulichkova N M, Fetisov G V, Pyt'ev Yu P and Lyapun Yu V 1987 *Sov. Phys.-Crystallogr.* **32** 649–53

Cioslowski J 1988 *Phys. Rev. Lett.* **60** 2141–3

Clementi E and Raymondi D L 1963 *J. Chem. Phys.* **38** 2686–9

Clementi E and Roetti C 1974 *At. Data Nucl. Data Tables* **14** 177–478

Clinton W L, Frishberg C A, Goldberg M J, Massa L Y and Oldfield P A 1983 *Int. J. Quantum Chem. Symp.* **17** 517–25

Clinton W L, Frishberg C A, Massa L Y and Oldfield P A 1973 *Int. J. Quantum Chem. Symp.* **7** 505–14

Clinton W L, Galli A J and Massa L Y 1969 *Phys. Rev.* **177** 7–12

Clinton W L and Massa L Y 1972 *Phys. Rev. Lett.* **20** 1363–6

Cohen L 1979 *J. Chem. Phys.* **70** 788–9

Cohen M L 1981 *Structure and Bonding in Crystals* vol 1, ed M O'Keefe and A Navrotsky (New York: Academic) pp 25–48

Coleman A J 1963 *Rev. Mod. Phys.* **35** 668–89

—— 1972 *J. Math. Phys.* **13** 214

—— 1978 *Int. J. Quantum Chem.* **13** 67–82

Collins D M 1982 *Nature* **298** 49–51

—— 1993 *Z. Naturf.* a **48** 68–74

Collins D M, Makar M S and Whitehurst F M 1983 *Acta Crystallogr.* B **39** 303–6

Condon E U and Shortley G H 1959 *The Theory of Atomic Spectra* (Cambridge: Cambridge University Press)

Cook M and Karplus M 1987 *J. Phys. Chem.* **91** 31–7

Cooper D L, Gerratt J and Raimondi M 1984 *Faraday Symp. Chem. Soc.* **19** 149

—— 1987 *Adv. Chem. Phys.* **69** 319–97

—— 1989 *J. Chem. Soc. Perkin Trans.* II **8** 1187–97

—— 1991 *Chem. Rev.* **91** 929–64

Cooper M J 1979 *Acta Crystallogr.* A **35** 176–80

Cooper M J and Rouse K D 1971 *Acta Crystallogr.* A **27** 622–8

Coppens P 1974 *Acta Crystallogr.* B **30** 255–61

—— 1975 *Phys. Rev. Lett.* **35** 98–101

—— 1977 *Angew. Chem.* (English edn) **89** 33–42

—— 1978 *Neutron Diffraction* ed H Dachs (Berlin: Springer) pp 71–111

—— 1982 *Electron Distribution and the Chemical Bond* ed P Coppens and M B Hall (New York: Plenum)

—— 1984 *Acta Crystallogr.* A **40** 184

—— 1992 *Synchrotron Radiation Crystallography* (New York: Academic)

Coppens P and Becker P 1992 *International Tables for Crystallography* vol C, ed A J C Wilson (Dordrecht: Kluwer)

Coppens P, Boehme R, Price P F and Stevens E D 1981a *Acta Crystallogr.* A **37** 857–63

Coppens P and Guru Row T N 1978 *Ann. NY Acad. Sci.* **313** 214

Coppens P, Guru Row T N, Leung P, Stevens E D, Becker P and Yang Y W 1979 *Acta Crystallogr.* A **35** 63

Coppens P and Hall M B 1982 (ed) *Electron Distribution and the Chemical Bond* (New York: Plenum)

Coppens P and Hamilton W C 1968 *Acta Crystallogr.* B **24** 925–9
—— 1970 *Acta Crystallogr.* A **26** 71–83
Coppens P and Li L 1984 *J. Chem. Phys.* **81** 1983–93
Coppens P, Moss G and Hansen N K 1981b *Computing in Crystallography* (Bangalore: IUC) ch 16
Coppens P, Paulter D and Griffin J E 1971a *J. Am. Chem. Soc.* **93** 1051–8
Coppens P and Stevens E D 1977 *Isr. J. Chem.* **16** 175–9
Coppens P, Willoughby T V and Csonka L 1971b *Acta Crystallogr.* A **27** 248–56
Coulson C A, March N H and Altmann S 1952 *Proc. Natl Acad. Sci. USA* **38** 372–8
Coulson C A and Moffitt 1949 *Phil. Mag.* **40** 1–35
Cowley J M 1975 *Diffraction Physics* (Amsterdam: North-Holland)
Cowley J M (ed) 1992 *Electron Diffraction Techniques* (New York: Oxford University Press)
Cowley J M and Smith D J 1987 *Acta Crystallogr.* A **43** 737–51
Cowley R A 1963 *Adv. Phys.* **12** 421–80
Cox D E 1991 *Handbook on Synchrotron Radiation* vol 3, ed G S Brown and D E Moncton (Amsterdam: North-Holland) p 155
Cox D E, Hastings J B, Cardoso L P and Finger L W 1986 *Mater. Sci. Forum* **9** 1–20
Cox D E, Toby B H and Eddy M M 1988 *Aust. J. Phys.* **41** 117–31
Craven B M and McMullan R K 1979 *Acta Crystallogr.* B **35** 934–45
Craven B M and Weber H P 1983 *Acta Crystallogr.* B **39** 743–8
Creagh D C 1985 *Aust. J. Phys.* **38** 371–404
—— 1988 *Aust. J. Phys.* **41** 487–501
Creagh D C and Hubbell J H 1987 *Acta Crystallogr.* A **43** 102–12
Cremer D and Kraka E 1984a *Croat. Chem. Acta* **57** 1259–81
—— 1984b *Angew. Chem.* **96** 612–3
Criado A, Conde A and Marquez R 1985 *Acta Crystallogr.* A **41** 158–63
Cromer D T, Larsen A C and Stewart R F 1976 *J. Chem. Phys.* **65** 336–49
Cromer D T and Liberman D 1970 *J. Chem. Phys.* **53** 1891–8
—— 1981 *Acta Crystallogr.* A **37** 267
—— 1983 *J. Appl. Crystallogr.* **16** 437
Crow M L, Schupp G, Yellon W B, Mullen J G and Djedid A 1987 *Acta Crystallogr.* A **43** 638–45
Cruickshank D W J 1949 *Acta Crystallogr.* **2** 65–82
Cummings S and Hart M 1988 *Aust. J. Phys.* **41** 423–31
Dahl J P and Avery J (ed) 1984 *Local Density Approximations in Quantum Chemistry and Solid State Physics* (New York: Plenum)
Dam J, Harkema S and Feil D 1983 *Acta Crystallogr.* B **39** 760
D'Arco Ph, Sandrone G, Dovesi R, Orlando R and Saunders V R 1993 *Phys. Chem. Miner.* **20** 407–14
Darovsky A, Bolotovsky R and Coppens P 1996 *Acta Crystallogr.* at press
Darwin W 1992 *Phil. Mag.* **43** 800–29
Das G and Wahl A C 1966 *J. Chem. Phys.* **44** 87
Daul C A, Day P, Figgis B N, Gudel H U, Herren F, Ludi A and Reynolds P A 1988 *Proc. R. Soc.* A **419** 205–19
Davidson E R 1976 *Reduced Density Matrices in Quantum Chemistry* (New York: Academic)
Davidson E R and Feller D 1986 *Chem. Rev.* **86** 681–96

Davis C L and Maslen E N 1978 *Acta Crystallogr.* A **34** 743–6
Davis C L, Maslen E N and Varghese J N 1978 *Acta Crystallogr.* A **34** 371–7
Dawson B 1967a *Proc. R. Soc.* A **298** 255–63
—— 1967b *Proc. R. Soc.* A **298** 264–88
—— 1975 *Advances in Structure Research by Diffraction Methods* vol 6, ed R Mason and W Hoppe (Braunschweig: Vieweg)
Deeth R J, Figgis B N, Forsyth J B, Kucharski E S and Reynolds P A 1989 *Proc. R. Soc.* A **421** 153–68
Delaplane R G, Tellgreen R and Olovsson I 1985 *Sagamore VIII Conf. on Charge, Spin and Momentum Densities* Abstracts Sanga-Säby p F8
Delfs C D and Figgis B N 1990 *Mol. Phys.* **69** 401–9
Delley B 1991 *Physica* B **172** 185–93
Denne W A 1972 *Acta Crystallogr.* A **28** 192–201
—— 1977 *Acta Crystallogr.* A **33** 438–40
Destro R, Bianchi R and Morosi G 1989 *J. Phys. Chem.* **93** 4447–57
Destro R and March R E 1987 *Acta Crystallogr.* A **43** 711–8
Destro R, March R E and Bianchi R 1988 *J. Phys. Chem.* **92** 966–73
de Vries R Y, Briels W J and Feil D 1996 *Phys. Rev. Lett.* submitted
Dewar M J S, Suck S H and Weiner P K 1976 *Chem. Phys. Lett.* **38** 228–9
Dorfman J G 1961 *Diamagnetism and the Chemical Bond* (Moscow: Physmathgiz) (in Russian)
Dorner B, Burkel E, Illini Th and Peisl J 1987 *Z. Phys.* B **69** 179
Dorner B and Peisl J 1983 *Nucl. Instrum. Method* **208** 587
Dovesi R 1984 *Int. J. Quantum Chem.* **26** 197–212
Dovesi R, Cause M and Angonoa G 1981 *Phys. Rev.* B **24** 4177–83
Dovesi R, Cause M, Orlando R, Roetti C and Saunders V R 1990 *J. Chem. Phys.* **92** 7402–11
Dovesi R, Pisani C, Ricca F and Roetti C 1980 *Phys. Rev.* B **22** 5936–44
—— 1982 *Phys. Rev.* B **25** 3731–9
Dovesi R, Pisani C, Roetti C and Silvi B 1987 *J. Chem. Phys.* **86** 6967–71
Dovesi R, Roetti C and Saunders V R 1992 *CRYSTAL 92: an Ab Initio Hartree–Fock LCAO Program for Periodic Systems* User Documentation (Torino: University of Torino)
Downs J W and Gibbs G V 1987 *Amer. Mineral.* **72** 769–77
Dreizler R M and da Providencia J (ed) 1985 *Density Functional Methods in Physics* (New York: Plenum)
Dreizler R M and Gross E K 1990 *Density Functional Theory* (Berlin: Springer)
Duckworth J A K, Willis B T M and Pauley G S 1970 *Acta Crystallogr.* A **22** 263–71
Dudka A P, Rabadanov M H and Loshmanov A A 1989 *Kristallographiya* **34** 818–23 (in Russian)
Dunitz J D 1979 *X-ray Analysis and the Structure of Organic Molecules* (Ithaca, NY: Cornell University Press)
Dunitz J D, Schomaker V and Trueblood K N 1988 *J. Chem. Phys.* **92** 856–67
Dunitz J D, Schweizer W B and Seiler P 1983 *Helv. Chim. Acta* **66** 123–33
Dunitz J D and Seiler P J 1983 *J. Am. Chem. Soc.* **105** 7056–8
Durbin J and Watson G S 1950a *Biometrika* **37** 409–28
—— 1950b *Biometrika* **38** 159–78
Dvorjankina G G and Pinsker Z G 1958 *Sov. Phys.–Crystallogr.* **3** 439–44

Eckardt H, Fritshe L and Noffke J 1984 *J. Phys. F: Met. Phys.* **14** 97
Edgecombe K E and Boyd R J 1986 *Int. J. Quantum Chem.* **29** 959–73
Edgeworth F Y 1905 *Proc. Camb. Phil. Soc.* **20** 36–141
Eisenberger P and Platzman P M 1970 *Phys. Rev.* A **2** 415–23
Eisenstein M 1979 *Acta Crystallogr.* B **35** 2614–25
—— 1988 *Int. J. Quantum Chem.* **33** 127–58
Eisenstein M and Hirshfeld F L 1983 *J. Comput. Chem.* **4** 15–20
El-Brahimi M and Durand J 1986 *Rev. Chim. Minerale* **23** 146
Elcombe M M and Pryor A W 1970 *J. Phys. C: Solid State Phys.* **3** 492–9
Ellis D E 1990 *Portugal Phys.* **19** 1–20
—— (ed) 1995 *Density Functional Theory of Molecules, Clusters and Solids* (Dordrecht: Kluwer)
Ellis D E, Guenzburger D and Jansen H B 1983 *Phys. Rev.* B **28** 3697–705
English C A and Venables J A 1974 *Proc. R. Soc.* A **340** 57–80
Ensllin N, Bertozzi W, Kowalski S, Sargent C P, Turchinetz W, Williamson C F, Fivozinsky S P, Lightbody J W and Penner S 1974 *Phys. Rev.* C **9** 1705–17
Epstein J 1982 *Moment Wave Functions, 2nd Meeting (Adelaide, 1982)* (New York: American Institute of Physics) pp 101–14
Epstein J, Ruble J R and Craven B M 1982 *Acta Crystallogr.* B **38** 140–9
Epstein J and Stewart R F 1977 *J. Chem. Phys.* **66** 4057–64
—— 1979 *J. Chem. Phys.* **70** 5515–21
Epstein J and Swanton D 1982 *J. Chem. Phys.* **77** 1048–60
Epstein S T 1974 *The Variational Method in Quantum Chemistry* (New York: Academic)
Erdahl R and Smith V H 1987 *Density Matrices and Density Functionals* (Dordrecht: Reidel)
Evarestov R A 1982 *Quantum-Chemical Methods in Solid State Theory* (Leningrad: Izdatelstvo Leningradskogo Universiteta) (in Russian)
Ewald P P 1917 *Ann. Phys., Lpz.* **54** 519–97
Ewald P P and Hönl H 1936 *Ann. Phys., Lpz.* **25** 281
Fabelinskii I L 1968 *Molecular Scattering of Light* (New York: Plenum)
Feil D 1977a *Isr. J. Chem.* **16** 103–10
—— 1977b *Isr. J. Chem.* **16** 149–53
—— 1990 *Portugal Phys.* **19** 21–42
Feil D and Moss G 1983 *Acta Crystallogr.* A **39** 14–21
Feinberg M J and Ruedenberg K 1971 *J. Chem. Phys.* **54** 1495–511
Feller D F and Ruedenberg 1979 *Theor. Chim. Acta* **52** 231–51
Fender B E F, Figgis B N and Forsyth J B 1986a *Proc. R. Soc.* A **404** 139–45
Fender B E F, Figgis B N, Forsyth J B, Reynolds P A and Stevens E 1986b *Proc. R. Soc.* A **404** 127–38
Figgis B N, Forsyth J B, Kucharsky E S, Reynolds P A and Tasset F 1990 *Proc. R. Soc.* A **428** 113–27
Figgis B N, Forsyth J B, Mason R and Reynolds P A 1985 *Chem. Phys. Lett.* **115** 454
Figgis B N, Forsyth J B and Reynolds P A 1987a *Inorg. Chem.* **26** 101–5
Figgis B N, Kucharsky E S and Reynolds P A 1989 *Acta Crystallogr.* B **45** 240–7
Figgis B N and Reynolds P A 1985 *Inorg. Chem.* **24** 1864–73
—— 1986 *Int. Rev. Phys. Chem.* **5** 265–72
—— 1990b *J. Phys. Chem.* **94** 2211–5

Figgis B N, Reynolds P A and Mason R 1983a *J. Am. Chem. Soc.* **105** 440–3
Figgis B N, Reynolds P A and White A H 1987b *J. Chem. Soc. Dalton Trans.* N **7** 1737–45
Figgis B N, Reynolds P A and Williams G A 1980a *J. Chem. Soc. Dalton Trans.* N **12** 2339–47
—— 1980b *J. Chem. Soc. Dalton Trans.* N **12** 2348–53
Figgis B N, Reynolds P A, Williams G A and Lehner N 1981 *Aust. J. Chem.* **34** 993
Figgis B N, Reynolds P A, Williams G A, Mason R, Smith A R P and Varghese J N 1980c *J. Chem. Soc. Dalton Trans.* N **12** 2333–8
Figgis B N, Reynolds P A and Wright S 1983b *J. Am. Chem. Soc.* **105** 434–40
Finger L W 1989 *Reviews in Mineralogy* vol 20, ed D L Bish and J E Post (Washington DC: Mineralogical Society of America) p 309
Fink M, Moore P G and Gregori D 1979a *J. Chem. Phys.* **71** 5227–37
Fink M, Schmiedekamp C W and Gregory D 1979b *J. Chem. Phys.* **71** 5238–42
Finklea S L, Cathey and Amma E L 1976 *Acta Crystallogr.* A **32** 529
Flack H D 1984 *Methods and Applications in Crystallographic Computing* ed S R Hall and T Ashida (Oxford: Oxford University Press) pp 41–55
Fliszar S 1983 *Charge Distributions and Chemical Effects* (New York: Springer)
Flygare W H 1978 *Molecular Structure and Dynamics* (Englewood Cliffs, NJ: Prentice-Hall)
Foresman J B and Frisch A 1993 *Exploring Chemistry with Electronic Structure Methods: a Guide to Using Gaussian* (Pittsburgh, PA: Gaussian)
Forsyth B N 1979 *At. Energy Rev.* **172** 345–412
Forsythe G E, Malcolm M A and Moler C B 1977 *Computer Methods for Mathematical Computations* (New York: Prentice-Hall)
Fouler P W and Madden P A 1983 *Mol. Phys.* **49** 913–23
Fox A G 1984 *Phil. Mag.* B **50** 477
Fox A G and Fisher R M 1986 *Phil. Mag.* A **53** 815
—— 1987 *Acta Crystallogr.* A **43** 260
—— 1988 *Aust. J. Phys.* **41** 461
Fox A G, O'Keefe M A and Tabbernor M A 1989 *Acta Crystallogr.* A **45** 786–93
Fox A G and Tabbernor M A 1991 *Acta Metall. Mater.* **39** 1991
Fox A G, Tabbernor M A and Fisher R M 1990 *J. Phys. Chem. Solids* **51** 1323
Fraga A and Malli G 1968 *Many-Electron Systems: Properties and Interactions* (Toronto: Saunders)
Freeman and Watson R E 1961a *Acta Crystallogr.* **14** 231–4
—— 1961b *Phys. Rev.* **124** 1439–54
Frishberg C A and Massa L J 1982 *Acta Crystallogr.* A **38** 93–8
Frost A A 1962 *J. Chem. Phys.* **37** 1147–8
Fujimoto I 1974 *Phys. Rev.* B **9** 591–9
Fukamachi T 1971 *Technical Report of the Institute of Solid State Physics* Ser. B, No 12 (Tokyo: University of Tokyo)
Fukui K 1971 *Accounts Chem. Res.* **4** 57
Gadre S R and Bendale R D 1986 *Chem. Phys. Lett.* **130** 515–21
Gadre S R and Srivastava R K 1991 *J. Chem. Phys.* **94** 4384–90
Gaspar R and Gaspar R Jr 1986 *Int. J. Quantum Chem. Symp.* **19** 279–84
Gatti C, Bianchini R, Destro R and Merati F 1992 *J. Mol. Struct.* **255** 409–34
Gatti C, Fanticci P and Pacchioni F 1987 *Theor. Chim. Acta* **72** 433–58

Gatti M, Valerio G and Dovesi R 1995 *Phys. Rev* B **51** 7741–50
Gatti M, Valerio G, Dovesi R and Causa M 1994 *Phys. Rev.* B **49** 14 179–89
Gazquez J L and Silverstone H J 1977 *J. Chem. Phys.* **67** 1887
Gell-Mann M and Bruckner K A 1957 *Phys. Rev.* **106** 364–9
Gerr R G 1985 *14th Conf. on Application of X-rays to the Study of Materials. Abstracts* ed T Malinovsky (Kishinev: Stiintsa) pp 36–7 (in Russian)
Gerratt J 1971 *Adv. At. Mol. Phys.* **7** 141–221
Ghermani N-E, Bouhmaida N and Lecomte C 1993 *Acta Crystallogr.* A **49** 781–9
Ghosh S K and Parr R G 1985 *J. Chem. Phys.* **82** 3307–15
Gilbert T L 1975 *Phys. Rev.* B **12** 2111–20
Gillespie R J 1992 *Chem. Soc. Rev.* **21** 59–70
Gjonnes J and Hoier R 1971 *Acta Crystallogr.* A **27** 313
Glebov A S 1990 *PhD Thesis* Mendeleev Institute of Chemical Technology, Moscow (in Russian)
Goddard W A 1967 *Phys. Rev.* **158** 73–80
Godby R W, Schluter M and Sham L J 1988 *Phys. Rev.* B **37** 10 159–75
Gong P P and Post B 1983 *Acta Crystallogr.* A **39** 719–24
Goodenough J B 1971 *Prog. Solid State Chem.* **5** 145–399
Goodman P and Lehmful G 1967 *Acta Crystallogr.* **22** 14
Gordon R G and Kim Y S 1972 *J. Chem. Phys.* **56** 3122–33
Goscinski O and Delhalle J 1989 *Int. J. Quantum Chem.* **35** 761–7
Göttlicher S 1968 *Acta Crystallogr.* B **24** 122–9
Göttichler S and Wölfel E 1959 *Z. Elektrochem.* B **63** 891–901
Graf H A and Schneider J R 1986 *Phys. Rev.* B **34** 8629
Grant D F and Gabe E J 1978 *J. Appl. Crystallogr.* **11** 114–21
Grant R W, Housley R M and Gonser U 1969 *Phys. Rev.* **178** 523–30
Green S 1974 *Adv. Chem. Phys.* **25** 179–209
Grelland H H 1985 *Acta Crystallogr.* A **41** 301–3
Gubanov V A, Ivanovsky A L and Ryzhkov M V 1987 *Quantum Chemistry in Material Science* (Moscow: Nauka) (in Russian)
Guidatti C, Arrighini G P and Marinelli F 1979 *Theor. Chim. Acta* **53** 165–73
Gusev A A, Reznik I M and Tsitrin V A 1995 *J. Phys.: Condens. Matter* **7** 4855–63
Hadring M M 1995 *Acta Crystallogr.* B **51** 432–46
Hafner S and Raymond M 1968 *J. Chem. Phys.* **49** 3570–9
Hagler A T 1977 *Isr. J. Chem.* **16** 202–12
Hahn H and Ludwig W 1961 *Z. Phys.* **161** 404–23
Hall M B 1986 *Chem. Scr.* **26** 389–94
Hamilton W C 1957 *Acta Crystallogr.* **10** 629–34
—— 1964 *Statistics in Physical Science* (New York: Roland)
Handy N C 1984 *Faraday Symp. Chem. Soc.* **19** 17–37
Handy N C, Marron M T and Silverstone H J 1969 *Phys. Rev.* **180** 45
Hansen N K 1988 *Acta Crystallogr.* A **44** 1097
Hansen N K and Coppens P 1978 *Acta Crystallogr.* A **34** 909–21
Hansen N K, Protas J and Marnier G 1991 *Acta Crystallogr.* B **47** 660–72
Hansen N K, Schneider J R and Larsen F K 1984 *Phys. Rev.* B **29** 917
Hansen N K, Schneider J R, Yelon W B and Pearson W H 1987 *Acta Crystallogr.* A **43** 763

Hanson J C, Waterpaugh K D, Sicker L and Jensen L H 1979 *Acta Crystallogr.* A **35** 616–21
Harada J 1988 *Aust. J. Phys.* **41** 351–7
Hargittai I and Hargittai M (ed) 1988 *Stereochemical Application of Gas-phase Electron Diffraction, Part A: The Electron Diffraction Technique* (*Methods of Stereochemical Analysis*) (Weinheim: VCH) p 563
Harkema S, Dam J, van Hummel G J and Reuvers A J 1980 *Acta Crystallogr.* A **36** 433–5
Harriman J E 1981 *Phys. Rev.* A **24** 680–2
—— 1986 *Phys. Rev.* A **34** 29–39
—— 1990 *Adv. Quantum Chem.* **21** 27
Harris F E 1976 *Theoretical Chemistry: Advances and Perspectives* vol 1, ed H Eyring and D Henderson (New York: Academic) pp 147–218
Harrison W A 1980 *Electronic Structure and the Properties of Solids* (San Francisco: Freeman)
Hasnain S S, Helliwell J R and Kamitsubo H (ed) 1994 *J. Synchrotron Radiat.* Inaugural issue
Hattori H, Kurijama H and Kato N 1965 *J. Phys. Soc. Japan* **20** 1047–50
Hauser U, Destreich V and Rohrweck H D 1977 *Z. Phys.* A **280** 17–25
Hayter 1978 *Neutron Diffraction* ed H Dachs (Berlin: Springer) pp 41–67
Hazell R G and Willis B T M 1978 *Acta Crystallogr.* A **34** 809–11
He X M, Swaminathan S, Craven B M and McMullan R K 1988 *Acta Crystallogr.* B **44** 271–81
Heaton R A, Harrison J G and Lin C C 1983 *Phys. Rev.* B **28** 5992–7
Hedin L and Lundqvist B I 1971 *J. Phys. C: Solid State Phys.* **4** 2064
Hedvig P 1975 *Experimental Quantum Chemistry* (Budapest: Akademiai Kiado)
Hehre W J, Ditchfield R, Stewart R F and Pople J A 1970 *J. Chem. Phys.* **52** 2769–73
Hehre W J, Stewart R F and Pople J A 1969 *J. Chem. Phys.* **51** 2657–64
Heijser W 1979 *PhD Thesis* Free University, Amsterdam
Heijser W, Baerends E J and Ros P 1980a *Faraday Discuss. Symp.* **14** 211–34
—— 1980b *J. Mol. Struct.* **63** 109–22
Heine V and Weaire D 1970 *Solid State Physics* vol 24, ed H Ehrenreich, F Seitz and D Turnbull (New York: Academic) p 170
Heitler W 1954 *The Quantum Theory of Radiation* 3rd edn (Oxford: Clarendon)
Heitler W and London F 1927 *Z. Phys.* **44** 455–72
Helgaker T 1992 *Lecture Notes in Chemistry. European Summer School in Quantum Chemistry* ed B O Roos (Berlin: Springer) pp 325–412
Helliwell J R 1992 *Macromolecular Crystallography with Synchrotron Radiation* (Cambridge: Cambridge University Press)
Hellner E 1977 *Acta Crystallogr.* B **33** 3813–6
Helmholdt R B, Braam A W M and Vos A 1983 *Acta Crystallogr.* A **39** 90–4
Helmholdt R B and Vos A 1977 *Acta Crystallogr.* A **33** 38–45
Henderson J A and Maslen E N 1987 *Acta Crystallogr.* A **43S** c-102
Hermansson K 1985 *Acta Crystallogr.* B **41** 161–9
—— 1987 *Acta Chim. Scand.* A **41** 513–26
Hermansson K and Olovsson I 1984 *Theor. Chim. Acta* **64** 265–76
Hester J R, Maslen E N, Spadaccini N, Ishizawa N and Satow Y 1993 *Acta Crystallogr.* B **49** 842–6
Hewat A W 1970 *Acta Crystallogr.* A **35** 248–52

Hino K, Saito Y and Benard M 1981 *Acta Crystallogr.* B **37** 2164–70
Hirshfeld F L 1971 *Acta Crystallogr.* B **27** 769–81
—— 1976 *Acta Crystallogr.* A **32** 239–44
—— 1977 *Theor. Chim. Acta* **44** 129–38
—— 1984a *Acta Crystallogr.* B **40** 484
—— 1984b *Acta Crystallogr.* B **40** 613–5
—— 1991 *Cryst. Rev.* **2** 169–204
Hirshfeld F L and Hope H 1990 *Acta Crystallogr.* B **36** 406–15
Hirshfeld F L and Mirsky K 1979 *Acta Crystallogr.* A **35** 366–70
Hirshfeld F L and Rzotkiewicz S 1974 *Mol. Phys.* **27** 1319–43
Hoch D E and Harriman J E 1995 *J. Chem. Phys.* **102** 9590–97
Hoche H R, Schulz H, Weber H-P, Belzner A, Wolf A and Wulf R 1987 *Acta Crystallogr.* A **43** 106–10
Hodges C H 1973 *Can. J. Phys.* **51** 1428–37
Hoffman-Ostenhof M, Hoffman-Ostenhof T and Thirring W 1978 *J. Phys. B: At. Mol. Phys.* **11** L571
Hohenberg and Kohn 1964 *Phys. Rev.* B **136** 864–71
Holladay A, Leung P and Coppens P 1983 *Acta Crystallogr.* A **39** 377–87
Hope H and Otterson R 1979 *Acta Crystallogr.* B **35** 370–2
Hsu L Y and Williams D E 1979 *Inorg. Chem.* **18** 79
Hund F 1929 *Ergeb. Exact. Naturwiss.* **8** 147–201
Hunt M J 1974 *J. Magn. Reson.* **15** 113–21
Hunter G 1986 *Int. J. Quantum Chem.* **29** 197–205
Huzinaga S (ed) 1984 *Gaussian Basis Sets for Molecular Calculations* (Amsterdam: Elsevier)
Ibers J A and Hamilton W A (ed) 1974 *International Tables for X-ray Crystallography* vol IV (Birmingham: Kynoch)
Ihm J and Joannopoulos J D 1981 *Phys. Rev.* B **24** 4191–7
Irngartinger H, Kallfas D, Prinzbach H and Klinger O 1989 *Chem. Ber.* **122** 175–9
Ivanov-Smolenskii G A, Tsirelson V G and Ozerov R P 1983 *Acta Crystallogr.* A **39** 411–5
—— 1984 *Acta Crystallogr.* A **30S** C-177
Iversen B B, Larsen F K, Souchassou M and Tanaka M 1995 *Acta Crystallogr.* B **51** 581–91
Iwata M 1977 *Acta Crystallogr.* B **33** 59–69
Iwata M and Saito 1973 *Acta Crystallogr.* B **29** 822–5
Iwata S 1982 *Quantum Chemistry Literature Data Base* ed K Ohno and K Morokuma (New York: Elsevier) pp 427–59
Izyumov Yu A, Naish V E and Ozerov R P 1991 *Neutron Diffraction of Magnetic Materials* (New York: Consultants Bureau)
Izyumov Yu A and Ozerov R P 1969 *Magnetic Neutron Diffraction* (London: Plenum)
Jaffe J and Zunger A 1983 *Phys. Rev.* B **28** 5822–47
James R W 1948 *Optical Principles of the Diffraction of X-Rays* (London: Bell)
—— 1967 *The Optical Principles of the Diffraction of X-Rays* (Ithaca, NY: Cornell University Press)
Jauch W, Schultz A J and Schneider J R 1989 *J. Appl. Crystallogr.* **21** 975–79
Jeans J 1947 *The Mathematical Theory of Electricity and Magnetism* (Cambridge: Cambridge University Press)

Jennings L D 1970 *Acta Crystallogr.* A **26** 613–22

Jensen, Larsen F K, Pressprich M R, Gao Y, Hansen N K, Su Z and Coppens P 1994 *Sagamore XI. Collected Abstracts* ed G Loupias and S Rabii (Brest: Laboratoire Mineralogie-Cristallographie, Université P et M Curie) pp 189–90

Johnson C K 1969 *Acta Crystallogr.* A **25** 187–94

—— 1980 *Computing in Crystallography* ed R Diamond, S Ramaseshan and K Venkatesan (Bangalore: Indian Academy of Science) ch 14

Johnson K H 1973 *Adv. Quantum Chem.* **7** 143–85

Jolly W L, Avanzino S C and Reitz R R 1977 *Inorg. Chem.* **16** 964

Jones D S, Pautler D and Coppens P 1972 *Acta Crystallogr.* A **28** 635–42

Jones W and Young W H 1971 *J. Phys. C: Solid State Phys.* **4** 1322–30

Jorgensen W J and Salem L 1973 *The Organic Chemist's Book of Orbitals* (New York: Academic)

Kaplan I G 1985 *Theory of Molecular Interactions* (Amsterdam: Elsevier)

Kapphahn M, Tsirelson V G and Ozerov R P 1988 *Portugal Phys.* **19** 213–6

—— 1989 *Dokl.–Phys. Chem.* **303** 1025–8

Kara M and Merisalo M 1982 *Phil. Mag.* B **45** 25–30

Kashiwase Y 1965 *J. Phys. Soc. Japan* **20** 320–5

—— 1973 *J. Phys. Soc. Japan* **34** 1303–13

Kato N 1976 *Acta Crystallogr.* A **32** 453, 458–66

—— 1979 *Acta Crystallogr.* A **35** 9–16

—— 1980 *Acta Crystallogr.* A **36** 763–9

—— 1988 *Aust. J. Phys.* **41** 337–49

—— 1991 *Acta Crystallogr.* A **47** 1–11

—— 1994 *Acta Crystallogr.* A **50** 17–20

Kato T 1957 *Commun. Pure Appl. Math.* **10** 151–77

Kawamura T and Kato N 1983 *Acta Crystallogr.* A **39** 305–10

Kelires P C and Das T P 1987 *Hyperfine Interact.* **34** 285–8

Kelires P C, Mishra K C and Das T P 1987 *Hyperfine Interact.* **34** 289–92

Kendall M G and Stuart A 1969 *The Advanced Theory of Statistics* vol 1 (London: Griffin)

Khramov D A and Polosin A B 1983 *Fiz. Tverd. Tela* **25** 2769–71 (in Russian)

Kikkawa T, Ohba S, Saito Y, Kamata S and Iwata S 1987 *Acta Crystallogr.* B **43** 83–5

Kikoin I K (ed) 1976 *Tables of Physical Values* (Moscow: Atom)

King R 1983 *Chemical Application of Topology and Graph Theory* ed R King (Amsterdam: Elsevier) pp 117–47

Kirfel A and Eichhorn K 1990 *Acta Crystallogr.* A **46** 271–84

Kirfel A, Gupta A and Will G 1979 *Acta Crystallogr.* B **35** 1052

Kirfel A and Will G 1980 *Acta Crystallogr.* B **36** 2881–90

Kirfel A, Will G and Stewart R F 1983 *Acta Crystallogr.* B **39** 175–85

Kirkwood J G 1932 *Phys. Z.* **33** 57–60

Kirzhnits D A 1957 *Sov. Phys.–JETP* **32** 115–23

Kirzhnits D A, Lozovik Yu E and Shpatakovskaya G V 1975 *Sov. Phys.–Usp.* **117** 3–47

Kissel L and Pratt R N 1990 *Acta Crystallogr.* A **46** 170–5

Kissel L, Pratt R H and Roy S C 1980 *Phys. Rev.* A **22** 1970–2004

Klahn B and Bingel W A 1977 *Theor. Chim. Acta* **44** 9

Klein D J and Trinojstic N (ed) 1990 *Valence Bond Theory and Chemical Structure* (Amsterdam: Elsevier)

Klooster W T 1992 *PhD Thesis* Twente University
Koelling D D 1981 *Rep. Prog. Phys.* **44** 140–212
Koelling D and Arbman G 1975 *J. Phys. F: Met. Phys.* **5** 2041–55
Kohn P 1983 *Phys. Rev. Lett.* **51** 1596
Kohn W and Sham L J 1965 *Phys. Rev.* A **140** 1133–8
—— 1966 *Phys. Rev.* **145** 561
Kohn W and Vashishta P 1983 *Theory of Inhomogeneous Electron Gas* ed S Lundqvist and N H March (New York: Plenum)
Kohout M, Savin A and Press H 1991 *J. Chem. Phys.* **95** 1928–42
Kok R A and Hall M B 1985 *J. Am. Chem. Soc.* **107** 2599–601
Kolmanovich V Yu and Reznik I M 1981 *Dokl. Akad. Nauk* **285** 1100–2
—— 1984 *Solid State Commun.* **50** 117–20
Konobeevskij S T 1948 *Dokl. Akad. Nauk* **59** 33–6 (in Russian)
—— 1951 *Usp. Fiz. Nauk* **44** 21–32 (in Russian)
Korolkova O V, Tsirelson V G, Ozerov R P and Rez I S 1985 *Preprint of Mendeleev Institute of Chemical Technology* N7684-85 (Moscow: Viniti) (in Russian)
Korolkova O V, Voloshina I V, Tsirelson V G and Ozerov R P 1986 *All-Union Conference on Crystal Chemistry of Inorganic and Coordination Compounds. Abstracts* (Buchara: Buchara Technological Institute) 9 188 (in Russian)
Kossel W 1916 *Ann. Phys., Lpz.* **49** 229–362
Krakauer H, Pickett W E and Cohen R E 1988 *J. Supercond.* **1** 111–41
Krijn M P C M and Feil D 1986 *J. Chem. Phys.* **85** 319–23
—— 1987a *J. Chem. Phys.* **91** 540
—— 1987b *J. Chem. Phys.* **91** 540
—— 1988a *J. Chem. Phys.* **89** 4199–208
—— 1988b *J. Chem. Phys.* **89** 5787–93
—— 1988c *Chem. Phys. Lett.* **150** 45
Krijn M P C M, Graafsma H and Feil D 1988 *Acta Crystallogr.* B **44** 609–16
Krivoglaz M A 1983 *Diffraction of X-rays and Neutrons on Non-ideal Crystals* (Kiev: Naukova Dumka) (in Russian)
Krivoglaz M A and Techonova E A 1961 *Sov. Phys.–Crystallogr.* **6** 399–403
Krueger H and Meyer-Berkhout K 1952 *Z. Phys.* **132** 171–8
Kryachko E S and Ludeña E V 1990 *Energy Density Functional Theory of Many-Electron Systems* (Dordrecht: Kluwer) p 850
Kudritskaya Z G and Danilov V I 1976 *Theor. Biol.* **59** 303–18
Kuhs W F 1984 *Acta Crystallogr.* A **40** 133–7
—— 1988 *Aust. J. Phys.* **41** 369–82
—— 1992 *Acta Crystallogr.* A **48** 80–8
Kulda J 1987 *Acta Crystallogr.* A **43** 167–73
—— 1988a *Acta Crystallogr.* A **44** 283–5
—— 1988b *Acta Crystallogr.* A **44** 286–90
Kunze K L and Hall M B 1987 *J. Am. Chem. Soc.* **109** 7617–23
Kupcik V, Wendeschuh-Josties W, Wolf A and Wulf R 1986 *Nucl. Instrum. Methods Phys. Res.* A **246** 624
Kurki-Suonio K 1968 *Acta Crystallogr.* A **24** 379–89
—— 1977a *Isr. J. Chem.* **16** 115–23
—— 1977b *Isr. J. Chem.* **16** 132–6
Kurki-Suonio K, Merisalo M and Peltonen M 1979 *Phys. Scr.* **19** 57–63

Kurki-Suonio K and Sälke R 1984 *Local Density Approximations in Quantum Chemistry and Solid State Physics* ed A P Dahl and J Avery (New York: Plenum) p 713

Kutoglu A, Scheringer C, Meyer H and Schweig A 1982 *Acta Crystallogr.* B **38** 2626–32

Kvardakov V V, Somenkov V A and Shilstein S Sh 1988 *Mater. Sci. Forum* **27/28** 221–2

Kvick A 1988 *Chemical Crystallography with Pulsed Neutrons and Synchrotron X-rays* (*NATO ASI Series*) ed M A Corrondo and G A Jeffrey (Dordrecht: Reidel) pp 187–201

Kyogoku Y, Lord R C and Rich A J 1967 *J. Am. Chem. Soc.* **89** 496–504

Ladd M F C and Palmer (ed) 1980 *Theory and Practice of Direct Methods in Crystallography* (New York: Plenum)

Laidig K and Frampton C S 1991 *Sagamore X Collected Abstracts* ed M Springborg, A Saenz and W Weyrich (Konstanz: University of Konstanz) p 71

Lally J S, Humphreys C J, Methereol A J and Fisher R M 1972 *Phil. Mag.* **25** 321

Lanczoc C 1957 *Applied Analysis* (New York: Prentice-Hall)

Landau L 1937 *Phys. Z. Sowjet* **12** 133–43

Langmuir I 1919 *J. Am. Chem. Soc.* **41** 868–934

Larkins F P 1971 *J. Phys. C: Solid State Phys.* **4** 3065–78

Larsen F K 1995 *Acta Crystallogr.* B **51** 468–82

Larsen F K and Hansen N K 1984 *Acta Crystallogr.* B **40** 169–79

Larsen F K, Lehmann M S and Merisalo M 1980 *Acta Crystallogr.* A **36** 159–63

Lecomte C, Blessing R H, Coppens P and Tabbard A 1986 *J. Am. Chem. Soc.* **108** 6942–50

Lecomte C, Chadwick P, Coppens P and Stevens E 1983 *Inorg. Chem.* **22** 2982–92

Lee Y S and McLean A D 1982 *J. Chem. Phys.* **76** 735

Lehmann M S 1980a *Electron and Magnetization Density in Molecules and Crystals* ed P Becker (New York: Plenum) pp 287–322

—— 1980b *Electron and Magnetization Density in Molecules and Crystals* ed P Becker (New York: Plenum) pp 355–72

Lehmann M S and Coppens P 1977 *Acta Chem. Scand.* A **31** 530–4

Lehmann M S and Larsen F K 1974 *Acta Crystallogr.* A **30** 580–4

Leib E H 1983 *Int. J. Quantum Chem.* **24** 243–77

Leibfried G and Ludwig W 1961 *Solid State Physics* vol 12, ed P Seitz and D Turnbull (New York: Academic) p 275

Lennard-Jones J E 1929 *Trans. Farad. Soc.* **25** 668

Levin A A, Syrkin Ya K and Dyatkina M E 1969 *Usp. Chim.* **38** 193–221 (in Russian)

Levine B F 1969 *Phys. Rev. Lett.* **22** 787–90

—— 1973a *Phys. Rev.* B **7** 2600–26

—— 1973b *J. Chem. Phys.* **59** 1463–86

—— 1974 *Phys. Rev.* B **10** 1655–64

Levy M 1979 *Proc. Natl Acad. Sci. USA* **76** 6062

—— 1982 *Phys. Rev.* A **26** 1200–8

—— 1990 *Adv. Quantum Chem.* **21** 69–96

Levy M and Goldstein J A 1987 *Phys. Rev.* **56** 7887–90

Levy M and Perdew J P 1985 *Phys. Rev.* A **32** 2010

Lewis G N 1916 *J. Am. Chem. Soc.* **38** 762–85

Lewis J and Schwarzenbach D 1981a *Acta Crystallogr.* A **31S** C-138

—— 1981b *Acta Crystallogr.* A **37** 507–10
—— 1982 *Acta Crystallogr.* A **38** 733–9
Lewis J, Schwarzenbach D and Flack H D 1982 *Acta Crystallogr.* A **38** 733–9
Li N, Coppens P and Landrum J 1988 *Inorg. Chem.* **27** 482–8
Li N, Su Z, Coppens P and Landrum J 1990 *J. Am. Chem. Soc.* **112** 7294–8
Lichanot A, Azavant P and Peitsch U 1996 *Acta Crystallogr.* B **52** (submitted)
Lieb E 1983 *Int. J. Quantum Chem,* **24** 243–77
Lobanov N N 1986 *PhD Thesis* Mendeleev Institute of Chemical Technology, Moscow (in Russian)
Lobanov N N, Butman L A and Tsirelson V G 1989 *Zh. Struct. Khim.* **30** 113–22 (in Russian)
Lobanov N N, Tsirelson V G and Belokoneva E L 1988 *Zh. Neorg. Khim.* **33** 3021–5 (in Russian)
Lobanov N N, Tsirelson V G and Ozerov R P 1984 *Solid State Commun.* **50** 129–31
Lobanov N N, Tsirelson V G and Shchedrin B M 1990 *Kristallographia* **35** 589–95 (in Russian)
Locus T L 1967 *Augmented Plane Wave Method* (Amsterdam: Benjamin)
Loudon R 1973 *The Quantum Theory of Light* (Oxford: Clarendon)
Low A A and Hall M B 1990 *J. Phys. Chem.* **94** 628–37
Löwdin P-O 1955 *Phys. Rev.* **97** 1474–94
—— 1970 *Adv. Quantum Chem.* **5** 185–99
Lu Z W and Zunger A 1992 *Acta Crystallogr.* A **48** 545–54
Luck W A P 1976 *The Hydrogen Bond II. Structure and Spectroscopy* ed P Schuster, E Zundel and C Sandorfy (Amsterdam: North-Holland) pp 527–62
Luger P 1993 *Cryst. Res. Technol.* **28** 767–94
Lundgren J-O 1979 *Acta Crystallogr.* B **35** 1027–33
Lundqvist S and March N H (ed) 1983 *Theory of Inhomogeneous Electron Gas* (New York: Plenum)
Luque F J, Illas F and Orozco M 1990 *J. Comput. Chem.* **11** 416–30
MacDonald A H, Daams J M, Vosko J H and Koelling D D 1982 *Phys. Rev.* B **25** 713; 1982 *Phys. Rev.* B **26** 3473
MacDougall P J and Bader R F W 1986 *Can. J. Chem.* **64** 1496–508
Mackay A 1990 *Symmetry* **1** 3–17
Mackenzie J K and Maslen V W 1968 *Acta Crystallogr.* A **24** 628
Mackrodt W C, Harrison N M, Saunders V M, Allaln N P, Towler M D, Apra E and Dovesi R 1992 *Preprint* DL/SCI/P851T Daresbury Laboratory
Mair S L 1980 *J. Phys. C: Solid State Phys.* **13** 2857–68
Mair S L and Wilkins S W 1976 *J. Phys. Condens. Matter* **9** 1145–57
Malli G L (ed) 1983 *Relativistic Effects in Atoms, Molecules and Solids* (New York: Plenum)
Mallinson P R, Koritsansky T, Elkaim E and Coppens P 1988 *Acta Crystallogr.* A **44** 336–42
Maradudin A A and Ambegaokar V 1964 *Phys. Rev.* **135** A1071–80
Maradudin A A and Flinn P A 1963 *Phys. Rev.* **129** 2529–47
March N H 1983 *Theory of Inhomogeneous Electron Gas* ed S Lindqvist and N H March (New York: Plenum) p 1
—— 1989 *J. Mol. Struct. Theochem.* **199** 75–83
—— 1992 *Electron Density Theory of Atoms and Molecules* (New York: Academic)

March N H and Deb B M (ed) 1987 *The Single-Particle Density in Physics and Chemistry* (London: Academic) chs 8–10
March N H, Yang W and Sampanthar S 1967 *Many-body Problem in Quantum Mechanics* (New York: Plenum)
Margaritondo G 1995 *J Synchrotron Radiat.* **2** 148–54
Margl P, Schwarz K and Blochl P 1993 *IBM Research Report* RZ 2397, No 80630
Marshall W and Lovesey S W 1971 *Theory of Thermal Neutron Scattering* (Oxford: Clarendon)
Martin M, Rees B and Mitschler A 1982 *Acta Crystallogr.* B **38** 6–15
Maslen E N 1968 *Acta Crystallogr.* B **24** 1172–6
—— 1988 *Acta Crystallogr.* A **44** 33–7
Maslen E N and Spackman M A 1985 *Aust. J. Phys.* **38** 273–87
Maslen E N and Spadaccini N 1989 *Acta Crystallogr.* B **45** 45–52
—— 1992 *Asian Crystallogr. Assoc. Inaugural Conf. AsCA '92 Programme and Abstracts* 15V-63
Maslen E N and Streltsov V A 1992 *Asian Crystallogr. Assoc. Inaugural Conf. AsCA '92, Programme and Abstracts* 15V-68
Maslen E N and Streltsova N R 1992 *Asian Crystallogr. Assoc. AsCA '92 Programme and Abstracts* 15V-69
Mason R 1986 *Int. Rev. Phys. Chem.* **5** 259
Mason W P and Thurston R N (ed) 1964 *Physical Acoustics* (New York: Academic)
Mathieson A McL 1984 *Aust. J. Phys.* **37** 55–61
—— 1989 *Acta Crystallogr.* A **45** 613–20
Mathieson A M and Stevenson A W 1986 *Acta Crystallogr.* A **42** 223–30
Matthews D A, Stucky G D and Coppens P 1972 *J. Am. Chem. Soc.* **94** 8001–8
Maverick E, Seiler P, Schweizer W B and Dunitz J B 1980 *Acta Crystallogr.* B **36** 615–20
McCandlish L E, Stout G H and Andrews L C 1975 *Acta Crystallogr.* A **31** 245–9
McHenry M E, O'Handley R C and Johnson K H 1987 *Phys. Rev.* B **35** 3555–9
McIntyre G J, Ptasiewicz-Bak H and Olovsson I 1990 *Acta Crystallogr.* B **46** 27–39
McLean A D and McLean R S 1981 *At. Data Nucl. Data* **26** 197–381
McMaster B N, Smith V H and Salahub D R 1982 *Mol. Phys.* **46** 449–63
McWeeny R 1951 *Acta Crystallogr.* **4** 513–19
—— 1953 *Acta Crystallogr.* **6** 631–7
—— 1960 *Rev. Mod. Phys.* **32** 335–69
—— 1990 *Int. J. Quantum Chem,* **24** 733–52
McWeeny and Pickup B T 1980 *Rep. Prog. Phys.* **43** 68–144
McWeeny R and Sutcliffe B T 1976 *Methods of Molecular Quantum Mechanics* 2nd edn (New York: Academic)
Meisner L B 1975 *Zh. Eksp. Teor. Fiz.* **69** 2101–8 (in Russian)
Menschung L, von Niessen W, Valtazanos P, Ruedenberg K and Schwarz W H E 1989 *J. Am. Chem. Soc.* **111** 6933–41
Merisalo M and Kurittu J 1978 *J. Appl. Crystallogr.* **11** 179–83
Mermin N D 1965 *Phys. Rev.* A **137** 1441–3
Messiah A 1966 *Quantum Mechanics* (New York: Wiley)
Mestechkin M M 1977 *Density Matrix Method in Theory of Molecules* (Kiev: Naukova Dumka) (in Russian)
—— 1990 *Self-Consistent Field. Theory and Application* ed R Carbo and

M Klobukowski (Amsterdam: Elsevier)
Metzger H, Behr H and Peisl J 1981 *Solid State Commun.* **40** 522–9
Meyer H, Schweig A and Zittlau W 1982 *Chem. Phys. Lett.* **92** 637–41
Mills D and Batterman B W 1980 *Phys. Rev.* B **46** 2887–93
Mirsky K V and Cohen M D 1978 *Chem. Phys.* **28** 193–204
Moeckli P, Schwarzenbach D, Bürgi H-B, Hauser J and Delley B 1988 *Acta Crystallogr.* B **44** 636–45
Møller C and Pesset M S 1934 *Phys. Rev.* **46** 618–22
Monkhorst H J 1979 *Phys. Rev.* B **20** 1504–13
Monkhorst H J and Pack J D 1979 *Solid State Commun.* **29** 675–7
Morokuma K 1971 *J. Chem. Phys.* **55** 1236–44
Morrell M, Parr R G and Levy M 1975 *J. Chem. Phys.* **62** 549–54
Morrison J C and Froese-Fischer C 1987 *Phys. Rev.* A **35** 2429–39
Morrison R C 1988 *Int. J. Quantum Chem. Symp.* **22** 43–9
Moss G 1982 *Electron Distributions and the Chemical Bond* ed P Coppens and M B Hall (New York: Plenum) pp 383–411
Moss G and Coppens P 1980 *Chem. Phys. Lett.* **75** 298–302
—— 1981 *Chemical Applications of Atomic and Molecular Electrostatic Potentials* ed P Politzer and D G Truhlar (New York: Plenum) pp 427–43
Moss G and Feil D 1981 *Acta Crystallogr.* A **37** 414–21
Moss G R, Souchassou M, Blessing R H, Espinosa E and Lecomte C 1995 *Acta Crystallogr.* B **51** 650–60
Mougenot P, Demuynck J and Benard M 1988 *J. Phys. Chem.* **92** 571–6
Moullet I and Martins J L 1990 *J. Chem. Phys.* **92** 527–35
Mulhausen G and Gordon R G 1981a *Phys. Rev.* B **24** 2147–60
—— 1981b *Phys. Rev.* B **24** 2161–7
Mullen D and Hellner E 1978 *Acta Crystallogr.* B **34** 2789–94
Mullen D and Scheringer C 1978 *Acta Crystallogr.* A **34** 476–7
Mulliken R S 1928 *Phys. Rev.* **32** 186–222
—— 1955 *J. Chem. Phys.* **23** 1833–40
—— 1962 *J. Chem. Phys.* **36** 3428–39
Murphy D R 1981 *Phys. Rev.* A **24** 1682–8
Murrell J N, Kettle S F A and Tedder J M 1978 *The Chemical Bond* (Chichester: Wiley)
Nada R, Catlow C R A, Dovesi R and Pisani C 1990 *Phys. Chem. Miner.* **17** 353–62
Nagata F and Fukuhara A 1967 *Japan. J. Appl. Phys.* **6** 1233
Nagel S 1985a *J. Phys. C: Solid State Phys.* **18** 3673–85
—— 1985b *J. Chem. Phys. Solids* **16** 905–19
Nelmes R J and Tun Z 1987 *Acta Crystallogr.* A **43** 635–8
—— 1988 *Acta Crystallogr.* A **44** 1098
Nemoshkalenko V V, Krasovskii A E, Antonov V N 1983 *Phys. Status Solidi* b **120** 283–96
Nielsen F S, Lee P and Coppens P 1986 *Acta Crystallogr.* B **42** 359–60
Niu J E 1994 *PhD Thesis* Siegen University
Noda Y, Ohba S, Sato S and Saito Y 1983 *Acta Crystallogr.* B **39** 312–17
Nozik Ju Z, Ozerov R P and Hennig K 1979 *Neutrons and Solids. 1. Structure Analysis* ed R P Ozerov (Moscow: Atomizdat) (in Russian)
Nye J F 1962 *Physical Properties of Crystals. Their Representation by Tensors and Matrices* (Oxford: Clarendon)

Oddershede J and Sabin J R 1982 *Chem. Phys.* **71** 161–71
Ohba S and Saito Y 1982 *Acta Crystallogr.* A **38** 725–9
Ohba S, Saito Y and Wakoh S 1982 *Acta Crystallogr.* A **38** 103–8
Ohgaki M, Marumo F and Tanaka K 1992 *Report of the Research Laboratory of Engineering Materials, Tokyo Institute of Technology* N 17, p 1
O'Keefe M and Hyde B G 1981 *Structure and Bonding in Crystals* vol I, eds M O'Keefe and A Navrotsky (New York: Academic) pp 227–54
Okuda M, Ohba S, Saito Y and Ito T 1990 *Acta Crystallogr.* B **40** 343–7
Olehnovich N M 1973 *Dokl. Akad. Nauk* **213** 560–2 (in Russian)
Olehnovich N M and Markovich V L 1980 *Acta Crystallogr.* A **36** 989–96
Oliver G L and Perdew J P 1979 *Phus. Rev.* A **20** 397
Olovsson I 1980 *Electron and Magnetization Densities in Molecules and Crystals* ed P Becker (New York: Plenum) p 831
Olovsson I and Jonsson P-G 1976 *Hydrogen Bond Recent Developments in Theory and Experiment* ed P Schuster *et al* (Amsterdam: North-Holland) p 396
Orechov S V and Avilov A S 1989 *12th Eur. Crystallogr. Meeting (Moscow) Coll. Abstr.* **3** 136
Orlando R, Dovesi R, Roetti C and Saunders V R 1990 *J. Phys. Condens. Matter* **2** 7769–89
Otterson T, Almlöf J and Carle J 1982 *Acta Chem. Scand.* A **36** 63–8
Otterson T and Hope H 1979 *Acta Crystallogr.* B **35** 373–8
Ozerov R P and Datt I D 1975 *At. Energy Rev.* **13** 651–94
Ozerov R P, Tsirelson V G, Korkin A A, Ionov S P, Zavodnik V E and Fomicheva E B 1981 *Sov. Phys.–Crystallogr.* **26** 20–4
Pack R T and Brown W B 1966 *J. Chem. Phys.* **45** 556
Palmer A 1993 *PhD Thesis* Hahn-Meitner-Institute
Palmer A and Jauch W 1993 *Phys. Rev.* B **48** 10 304–10
Parini E V 1988 *PhD Thesis* Mendeleev Institute of Chemical Technology Moscow (in Russian)
Parini E V, Tsirelson V G and Ozerov R P 1985 *Sov. Phys.–Crystallogr.* **30** 497–502
Parr R G and Yang W 1989 *Density-Functional Theory of Atoms and Molecules* (New York: Oxford University Press)
Pauling L 1931 *J. Am. Chem. Soc.* **53** 1367–400
—— 1960 *The Nature of the Chemical Bond* 3rd edn (Ithaca, NY: Cornell University Press)
Pearson E W, Jackson M D and Gordon R G 1984 *J. Phys. Chem.* **88** 119–28
Penn D R 1962 *Phys. Rev.* **128** 2093–7
Perdew J P 1995 *Density Functional Theory of molecules, Clusters and Solids* ed D E Ellis (Dordrecht: Kluwer) pp 47–66
Perdew J P and Levy M 1983 *Phys. Rev. Lett.* **51** 1888–3
Perdew J P and Zunger A 1981 *Phys. Rev.* B **23** 5048
Pertsin A I and Kitaigorodsky A I 1986 *The Atom–atom Potential Method: Application to Organic Molecular Solids* (Berlin: Springer)
Petkov I Zh, Stoitsov M V and Kryachko E S 1986 *Int. J. Quantum Chem.* **29** 149–61
Phillips J C 1970 *Covalent Bonding in Crystals, Molecules and Polymers* (Chicago: University of Chicago Press)
Pickenheim P and Milchev A 1970 *Phys. Status Solidi* **76** 571
Pickett W E 1989a *Comput. Phys. Rep.* **9** 115–98

—— 1989b *Rev. Mod. Phys.* **81** 433–512
Pietsch U 1980 *Phys. Status Solidi* b **102** 127–33
—— 1981 *Phys. Status Solidi* a **68** 321–7
—— 1982a *Phys. Status Solidi* b **111** K7–12
—— 1982b *Phys. Status Solidi* b **113** 203–7
—— 1993 *Z. Naturf.* a **48** 29–37
Pietsch U and Hansen N K 1996 *Acta Crystallogr.* B **52** submitted
Pietsch U, Mahlberg I and Unger K 1985 *Phys. Status Solidi* b **131** 621–7
Pietsch U Tsirelson V G and Ozerov R P 1986a *Phys. Status Solidi* b **137** 441
—— 1986b *Phys. Status Solidi* b **138** 47–52
Pinsker Z G 1953 *Electron Diffraction* (London: Butterworth)
Pisani C 1995 *Lecture on Network School 'Hartree–Fock Theory of the Electronic Structure of Solids' (Villa Gualino, 1995)*
Pisani C, Dovesi R and Roetti C 1988 *Hartree–Fock Ab Initio Treatment of Crystalline Systems* (*Lecture Notes in Chemistry 48*) (Berlin: Springer)
Pitzer R M, Kern C W and Lipscomb W N 1962 *J. Chem. Phys.* **37** 267–74
Pohler R F and Hanson H P 1965 *J. Chem. Phys.* **42** 2347–53
Politzer P 1976 *J. Chem. Phys.* **64** 4229–30
—— 1980 *Isr. J. Chem.* **19** 224–32
—— 1987 *Single-Particle Density in Physics and Chemistry* ed N H March and B M Deb (London: Academic) pp 59–72
Politzer P and Daiker K C 1981 *The Force Concept in Chemistry* ed B M Deb (New York: Van Nostrand Reinhold)
Politzer P and Parr R G 1974 *J. Chem. Phys.* **61** 4258–68
—— 1976 *J. Chem. Phys.* **64** 4634–7
Politzer P and Truhlar D G (ed) 1981 *Chemical Applications of Atomic and Molecular Electrostatic Potentials* (New York: Plenum)
Politzer P and Zillis B A 1984 *Croat. Chem. Acta* **57** 1055–64
Popa N C 1987 *Acta Crystallogr.* A **43** 304–16
Pople J A, Binkley J S and Sceger R 1976 *Int. J. Quantum Chem.* S **10** 1–19
Post B 1983 *Acta Crystallogr.* A **39** 711–18
Pound R V 1950 *Phys. Rev.* **79** 685
Prencipe M, Zupan A, Dovesi R, Apra E and Saunders V R 1995 *Phys. Rev.* B **51** 3391–6
Preston H J T 1969 *Thesis* (McMaster University, Hamilton
Price P F and Maslen E N 1978 *Acta Crystallogr.* A **34** 173–83
Primorac M, Kavačevič K and Maksiič Z B 1989 *Croat. Chem. Acta* **62** 561–77
Pryor A W 1966 *Acta Crystallogr.* 20 138–40
Ptasiewicz-Bak H, Olovsson I and McIntyre G J 1993 *Acta Crystallogr.* B **49** 192–201
Pullman B (ed) 1978 *Intermolecular Interactions from Diatomics to Biopolymers* (New York: Wiley)
Pyykkö P 1988 *Chem. Rev.* **88** 563–94
Raccah P M, Euewma R N, Stukel D J and Collins T C 1970 *Phys. Rev.* B **1** 756–63
Rae A D 1978 *Acta Crystallogr.* A **34** 719–24
Raffenetti R C 1973 *J. Chem. Phys.* **59** 5936–49
Rajagopal A K 1980 *Adv. Chem. Phys.* **41** 59–193
Ramsey N F 1956 *Molecular Beams* (London: Oxford University Press)
Rannev N V, Ozerov R P, Datt I D and Kshnjakina A N 1966 *Kristallographiya* **11** 175 (in Russian)

Ransil B J and Sinai J J 1972 *J. Am. Chem. Soc.* **94** 7268–76
Rauch H and Petraschek D 1978 *Dynamical Neutron Diffraction and its Application in Neutron Diffraction* ed H Dachs (Berlin: Springer)
Redondo A 1989 *Phys. Rev.* A **39** 4366–76
Reed W A and Eisenberger P 1972 *Phys. Rev.* B **6** 4596–604
Rees B 1976 *Acta Crystallogr.* A **32** 483–8
—— 1977a *Isr. J. Chem.* **16** 154–8
—— 1977b *Isr. J. Chem.* **16** 180–6
—— 1978 *Acta Crystallogr.* A **34** 254–6
Rees B and Mitschler A 1976 *J. Am. Chem. Soc.* **98** 7918–24
Reid J S 1987 *Acta Crystallogr.* A **43** 98–102
Reid J S and Pirie J D 1980 *Acta Crystallogr.* A **36** 957–65
Reisland J A 1973 *The Physics of Phonons* (London: Wiley)
Renninger M 1937a *Phys. Z.* **106** 141
—— 1937b *Naturwissenschaft* **25** 43
Ressouche E 1991 *Thesis* Université Grenoble (cited by Brown P J 1991 *X Sagamore Abstracts* (Konstanz: University of Konstanz)
Ressouche E, Boucherle J X, Gillon B, Ressouche E, Rey P and Schweizer J 1993a *J. Am. Chem. Soc.* **115** 3610–7
Ressouche E, Zheludev A, Boucherle J X, Gillon B, Ressouche E, Rey P and Schweizer J 1993b *Mol. Cryst. Liq. Cryst.* **232** 13–26
Restori R and Schwarzenbach D 1986 *Acta Crystallogr.* B **42** 201–8
Reynolds P A, Kjems J K and White J K 1974 *J. Chem. Phys.* **60** 824–34
Reznik I M 1982 *Preprint* 82-32 (Donetsk: Physico-Technical Institute of the Academy of Science of the Ukraine, SSR) (in Russian)
—— 1986 *Ukr. Fiz. Zh.* **31** 150–3 (in Russian)
—— 1988 *Fiz. Tverd. Tela* **30** 3496–8 (in Russian)
—— 1989 *J. Structurnoy Chem.* **30** 3–6 (in Russian)
—— 1993 *Electron Density in Theory of Ground State Crystal Properties* (Kiev: Naukova Dumka)
Reznik I M and Shatalov V M 1981 *Ukr. Fiz. Zh.* **7** 1093–100 (in Russian)
Reznik I M and Tolpygo K B 1983 *Preprint* 83-15 (Donetsk: Physico-Technical Institute of the Academy of Science of the Ukraine) (in Russian)
Ricart J M, Dovesi R, Saunders V R and Roetti C 1995 *Phys. Rev.* B **52** 2381–89
Rigoult P J 1979 *J. Appl. Cryst.* **12** 116–8
Ritchie Y P and Bachrach S M 1987 *J. Am. Chem. Soc.* **109** 5909–16
Roberto J B and Batterman B W 1970 *Phys. Rev.* B **2** 3220
Roberto J B, Batterman B W and Keating D T 1974 *Phys. Rev.* B **9** 2590
Rocher A M and Jouffrey B 1972 *Electron Microscopy and Analysis* (Bristol: Institute of Physics)
Roos B O and Siegbahn P E M 1977 *Methods of Electron Structure Theory* (*Modern Theoretical Chemistry 3*) ed H F Schaefer III (New York: Plenum) pp 277–318
Ros P, Snijders J G and Ziegler T 1980 *Chem. Phys. Lett.* **69** 297–300
Rosen A and Ellis D E 1975 *J. Chem. Phys.* **62** 3039–49
Roux M, Besnainou S and Daudel R 1956 *J. Chim. Phys.* **53** 218–23
Rozendaal A and Baerends E J 1986 *Acta Crystallogr.* B **42** 354–8
Rozendaal A and Ros P 1982 *Acta Crystallogr.* A **38** 372–7
Ruedenberg K 1951 *J. Chem. Phys.* **19** 1433–4
—— 1962 *Rev. Mod. Phys.* **34** 326–76

Ruedenberg K and Schwarz W H E 1990 *J. Chem. Phys.* **92** 4956–69
Sabine T M 1988 *Acta Crystallogr.* A **44** 368–73
Saenz A and Weyrich W (ed) *1994 Workshop in Density Matrix (Brest) Collected Abstracts*
Saethre L J, Siggel M R F and Thomas T D 1991 *J. Am. Chem. Soc.* **113** 5224–30
Sagar R P, Ku A C T, Smith V H and Simas A M 1988 *J. Chem. Phys.* **88** 4367–74
Saka T and Kato N 1986 *Acta Crystallogr.* A **42** 469–78
Sakabe N, Sakabe K, Higashi Y, Nakagawa A and Watanabe N 1992 *Asian Crystallogr. Assoc. AsCA '92 (Singapore) Programme and Abstracts* A4A-1
Sakata M and Harada J 1976 *Acta Crystallogr.* A 32 426–33
Sakata M and Sato M 1990 *Acta Crystallogr.* A **46** 263–70
Sambe H and Felton R A 1975 *Chem. Phys.* **62** 1122–6
Sasaki S, Fujino K and Takeuchi Y 1979 *Proc. Japan. Acad.* B **55** 43–8
Sasaki S, Fujino K, Takeuchi Y and Sadanaga R 1980 *Acta Crystallogr.* A **36** 904–15
Sasaki S and Takeuchi Y 1982 *Z. Kristallogr.* **158** 279–97
Satow Y and Iitaka Y 1989 *Rev. Sci Instrum.* **60** 2390–93
Savariault J M and Lehmann M S 1980 *J. Am. Chem. Soc.* **102** 1298–303
Sayre D (ed) 1985 *Computational Crystallography* (New York: Oxford University Press)
Schneider J R, Bouchard R, Graf H A and Nagasawa H 1992 *Acta Crystallogr.* A **48** 804–19
Scheringer C 1977 *Acta Crystallogr.* A **33** 879–84
—— 1980 *Acta Crystallogr.* A **36** 205–10
—— 1988 *Acta Crystallogr.* A **44** 343–9
Scheringer C and Fadini A 1979 *Acta Crystallogr.* A **35** 610–3
Scheringer C, Katoglu A and Mullen D 1978 *Acta Crystallogr.* A **34** 481–3
Scheringer C and Reitz H 1976 *Acta Crystallogr.* A **32** 271–3
Schevyrev A A and Simonov B I 1981 *Sov. Phys.–Crystallogr.* **26** 17–20
Schmider H, Edgecombe K E, Smith V H and Weyrich W 1992a *J. Chem. Phys.* **94** 8411–19
Schmider H, Sagar R P and Smith V H 1991 *J. Chem. Phys.* **94** 8627–9
Schmider H, Smith V and Weyrich W 1992b *J. Chem. Phys.* **96** 8986–94
Schmider H and Weyrich W 1988 *Sagamore IX Conf. on Charge, Spin and Momentum Densities Abstracts* (Luso-Bussaco: University of Coimbra) p 1.18
Schneider J R 1974a *J. Appl. Crystallogr.* **7** 541–6
—— 1974b *J. Appl. Crystallogr.* **7** 547–54
—— 1975a *J. Appl. Crystallogr.* **8** 195–201
—— 1975b *J. Appl. Crystallogr.* **8** 530–4
—— 1976 *J. Appl. Crystallogr.* **9** 394–402
—— 1977 *Acta Crystallogr.* A **33** 235–43
—— 1988a *Portugal Phys.* **19** 103–18
—— 1988b *Portugal Phys.* **19** 409–20
Schneider J R, Bouchard R, Graf H A and Nagasawa H 1992 *Acta Crystallogr.* A **48** 804–19
Schneider J R, Hansen N K and Kretschmer H 1981 *Acta Crystallogr.* A **37** 711–22
Schneider J R, Pattison P and Graf H A 1978 *Phil. Mag.* B **38** 141–54
Schofield P and Willis B T M 1987 *Acta Crystallogr.* A **43** 803–9
Schülke W 1988 *Portugal Phys.* **19** 421–8

Schülke W and Mourikis S 1986 *Acta Crystallogr.* A **42** 86–98
Schwarz K 1994 *Phase Transitions* **52** 109–22
Schwarz K and Blaha P 1988 *Portugal Phys.* **19** 159–71
—— 1991 *Sagamore X Abstracts* ed M Springborg, A Saenz and W Weyrich (Konstanz: University of Konstanz) p 114
Schwarz K, Blaha P and Ambrosch-Draxl C 1990 *Int. J. Quantum Chem. Symp.* **24** 339–47
Schwarz K, Blaha P and Haas H 1994 *Sagamore XI. Collected Abstracts* ed G Loupias and S Rabii (Brest: Laboratorie Mineralogie-Cristallographie, Université P et M Curie) pp 94–5
Schwarz M E 1970 *Chem. Phys. Lett.* **6** 631–6
Schwarz W H E 1990 *Theoretical Models of Chemical Bonding. Part 2. The Concept of the Chemical Bond* ed Z B Maksic (Berlin: Springer) pp 593–643
—— 1991 *Sagamore X Collected Abstracts* ed M Springborg, A Saenz and W Weyrich (Konstanz: University of Konstanz) p 115
Schwarz W H E, Lin H L, Irle S and Niu J E 1992 *J. Mol. Struct.* **255** 435–59
Schwarz W H E, Menschung L, Ruedenberg K, Jacobson R and Miller L L 1988 *Portugal Phys.* **19** 185–9
Schwarz W H E and Müller B 1990 *Chem. Phys. Lett* **166** 621–6
Schwarz W H E, Ruedenberg K and Mensching L 1989 *J. Am. Chem. Soc.* **111** 6926–33
Schwarz W H E, Valtazanos P and Ruedenberg K 1985 *Theor. Chim. Acta* **68** 471–506
Schwarzenbach D, Abrahams S C, Flack H D, Gonschorek W, Hahn Th, Huml K, March R E, Prince E, Robertson B E, Rollett J S and Wilson A J C 1989 *Acta Crystallogr.* A **45** 63–75
Schwarzenbach D and Lewis J 1982 *Electron Distributions and the Chemical Bond* ed P Coppens and M B Hall (New York: Plenum) pp 413–30
Schwarzenbach D and Thong N 1979 *Acta Crystallogr.* A **35** 652–8
Schweizer J 1980 *Electron and Magnetization Densities in Molecules and Crystals* ed P Becker (New York: Plenum) pp 479, 501
Schwinger J 1948 *Phys. Rev.* **73** 407
—— 1951 *Phys. Rev.* **82** 914
Scrocco E and Tomasi J 1978 *Adv. Quantum Chem.* **11** 115–202
Seiler P 1987 *Chimia* **41** 104–16
—— 1991 *Sagamore X Collected Abstracts* ed M Springborg, A Saenz and W Weyrich (Konstanz) pp 118–9
Seiler P and Dunitz J D 1978 *Acta Crystallogr.* A **34** 329–36
—— 1986 *Helv. Chim. Acta* **69** 1107–12
Seiler P, Schweizer W B and Dunitz J D 1984 *Acta Crystallogr.* B **40** 319–27
Sellar J R, Imeson D and Humphreys C J 1980 *Acta Crystallogr.* A **36** 686
Seminario J M and Politzer P (ed) 1995 *Modern Density Functional Theory* (Amsterdam: Elsevier)
Sen K D (ed) 1993 *Chemical Hardness* (*Structure and Bonding 80*) (Berlin: Springer)
Sen K D and Jorgensen C K 1987 *Electronegativity* (New York: Springer)
Sen K D and Narasimhan P T 1977 *Phys. Rev.* B **16** 107–14
Sham L J and Schluter M 1983 *Phys. Rev. Lett.* **51** 1884
Shannon R 1976 *Acta Crystallogr.* A **32** 451–67
Sharma R R 1971 *Phys. Rev. Lett.* **26** 563–5
Sharma R R and Das T P 1964 *J. Chem. Phys.* **41** 3581–91

Shavitt I 1962 *Methods in Computational Physics* vol 2, ed B Alder *et al* (New York: Plenum) pp 1–45
—— 1977 *Method of Electron Structure Theory* (*Modern Theoretical Chemistry 3*) ed H F Schaefer III (New York: Plenum) pp 198–276
Sherwood D E and Hall M B 1983 *Inorg. Chem.* **22** 93–100
Shi Z and Boyd R J 1988 *J. Chem. Phys.* **88** 4375–7
Shibata S and Hirota F 1988 *Stereochemical Application of Gas-phase Electron Diffraction* part A, ed I Hargittai and M Hargittai (New York: VCH) p 107
Shin C C 1976 *Phys. Rev.* A **14** 919–21
Shirley E L, Allan D C, Martin R M and Joannopoulos J D 1989 *Phys. Rev.* B **40** 3652–60
Shirley E L, Martion R M, Bachelet G B and Ceperley D M 1990 *Phys. Rev.* B **42** 5057–66
Shull C G 1958 *Phys. Rev. Lett.* **21** 1589
Shukla R C and Hübschle H 1989 *Phys. Rev.* B **40** 1555–9
Sidgwick N Y 1927 *The Electronic Theory of Valence* (Oxford: Oxford University Press)
Sihai S, Bagus P S and Ladik J 1982 *Chem. Phys.* **68** 467–71
Silberbach H 1991 *J. Chem. Phys.* **94** 2977–85
Silver D M, Wilson S and Nieuwpoort W C 1978 *Int. J. Quantum Chem.* **14** 635
Silverstone H J 1968 *J. Chem. Phys.* **48** 4098
Silvi B, Allavena M, Hannachi Y and D'Arco Ph 1992 *J. Am. Ceram. Soc.* 1239–46
Silvi B, Bouaziz A and D'Arco Ph 1993 *Phys. Chem. Miner.* **20** 333–40
Silvi B, Causa M, Dovesi R and Roetti C 1989 *Mol. Phys.* **67** 891–901
Singh S, Bonner W A, Potopowicz J R and Van Uitert L C 1970 *Appl. Phys. Lett.* **17** 292–4
Sirota N N 1962 *Dokl. Akad. Nauk* **142** 1278–9 (in Russian)
Sirota N N, Gololobov E M, Olechnovich N M and Sheleg A U 1972 *Chemical Bonds in Semiconductors and Semimetals* ed N N Sirota (Minsk: Nauka i Technika) pp 62–7 (in Russian)
Sirota N N and Sheleg A U 1963 *Dokl. Akad. Nauk* **152** 81–3 (in Russian)
Skriver H L 1983 *The LMTO Method* (Berlin: Springer)
Slater J C 1951 *Phys. Rev.* **81** 385–92
—— 1974 *The Self-Consistent Field for Molecules and Solids* (New York: McGraw-Hill)
Smart D J and Humphreys C J 1980 *Electron Microscopy and Analysis 1979* ed T Mulvey (Bristol: Institute of Physics)
Smith V H 1977 *Phys. Scr.* **15** 147–62
Smith V H 1980 *Electron and Magnetization Density in Molecules and Crystals* ed P Becker (New York: Plenum) pp 27–47
Smith V H 1982 *Electron Distributions and the Chemical Bond* ed P Coppens and M B Hall (New York: Plenum) pp 3–59
Smith V H and Absar I 1977 *Isr. J. Chem.* **16** 87–102
Snijders J C and Baerends E J 1977 *Mol. Phys.* **33** 1651–62
—— 1978 *Mol. Phys.* **36** 1789
—— 1982 *Electron Distributions and the Chemical Bond* ed P Coppens and M B Hall (New York: Plenum) pp 111–30
Snijders J G, Baerends E J and Ros P 1979 *Mol. Phys.* **38** 1909

Sommer-Larsen P and Avery J 1987 *Acta Crystallogr.* A **37** C109

Sommer-Larsen P, Kadziola A and Gajhede M 1990 *Acta Crystallogr.* A **46** 343–51

Souchassou M 1991 *Sagamore X, Collected Abstracts* ed M Springborg, A Saenz and W Weyrich (Konstanz: University of Konstanz) p 126

Souchassou M, Espinosa E, Lecomte C and Blessing R H 1995 *Acta Crystallogr.* B **51** 660–9

Spackman M A 1986a *J. Chem. Phys.* **85** 6579–86

—— 1986b *J. Chem. Phys.* **85** 6587–601

—— 1986c *Acta Crystallogr.* A **42** 271–81

—— 1987 *J. Chem. Phys.* **91** 3179–86

—— 1992 *Chem. Rev.* **92** 1769–97

Spackman M A, Hill R J and Gibbs G V 1987 *Phys. Chem. Miner.* **14** 139–50

Spackman M A and Maslen E N 1985 *Acta Crystallogr.* A **41** 347–53

—— 1986 *J. Phys. Chem.* **90** 2020–7

Spackman M A and Stewart R F 1981 *Chemical Applications of Atomic and Molecular Electrostatic Potentials* ed P Politzer and D G Truhlar (New York: Plenum) pp 407–25

—— 1984 *Methods and Application of Crystallogr. Comput (Kyoto); Int. Summer School on Crystallographic Computings (Oxford, 1984)* ed S R Hall and T Ashida (Oxford: Oxford University Press) p 27

Spackman M A, Weber H-P and Craven B M 1988 *J. Am. Chem. Soc.* **110** 775–82

Spence J C H 1993 *Acta Crystallogr.* A **49** 231–60

Srebrenik S 1975 *Int. J. Quantum Chem. Symp.* **9** 375

Srebrenik S and Bader R F W 1975 *J. Chem. Phys.* **63** 3945–61

Srebrenik S, Bader R F W and Nugyen-Dang T T 1978 *J. Chem. Phys.* **68** 3667–79

Srebrenik S, Weinstein H and Pauncz R 1974 *J. Chem. Phys.* **61** 5050–3

—— 1975 *Chem. Phys. Lett.* **32** 420–5

Srivastava G P 1982 *Phys. Rev.* B **25** 2815–20

Steiner E and Sykes S 1972 *Mol. Phys.* **23** 643–56

Stefanov B B and Cioslowsky J 1995 *J. Comput. Chem.* **16** 1394–404

Stephens M E and Becker P J 1983 *Mol. Phys.* **49** 65–89

Sternheimer R M 1950 *Phys. Rev.* **80** 102–3

—— 1972 *Phys. Rev. A* **6** 1702

Stevens E D 1974 *Acta Crystallogr.* A **30** 184–9

—— 1978 *Acta Crystallogr.* B **34** 544–51

—— 1979 *Mol. Phys.* **37** 27–45

—— 1980 *Acta Crystallogr.* B **36** 1876–86

Stevens E D and Coppens P 1975 *Acta Crystallogr.* A **31** 612–7

—— 1980 *Acta Crystallogr.* B **36** 1864–76

Stevens E D, DeLucia M L and Coppens P 1980 *Inorg. Chem.* **19** 813–20

Stevens E D and Hope H 1975 *Acta Crystallogr.* A **31** 494–8

—— 1977 *Acta Crystallogr.* A **33** 723–9

Stevens E D, Rys J and Coppens P 1977 *Acta Crystallogr.* A **33** 333–8

Stevenson A W and Harada J 1983 *Acta Crystallogr.* A **39** 202–7 (Erratum: *Acta Crystallogr.* A **40** 162)

Stewart R F 1968 *J. Chem. Phys.* **48** 4882–9

—— 1969 *J. Chem. Phys.* **51** 4569–73

—— 1970 *J. Chem. Phys.* **53** 205–13

—— 1972 *J. Chem. Phys.* **57** 1664–8
—— 1973a *J. Chem. Phys.* **58** 4430–8
—— 1973b *Acta Crystallogr.* A **29** 602–5
—— 1976 *Acta Crystallogr.* A **32** 565–74
—— 1977a *Isr. J. Chem.* **16** 137–43
—— 1977b *Isr. J. Chem.* **16** 124–31
—— 1977c *Chem. Phys. Lett.* **49** 281–4
—— 1979 *Chem. Phys. Lett.* **65** 335–42
—— 1982 *God. Jugosl. Cent. Kristallogr.* **17** 1–24
—— 1983 unpublished results
—— 1991 *The Application of Charge Density Research in Chemistry and Drug Design* (*NATO-ASI Series*) ed G A Jeffrey and J F Piniella (New York: Plenum) p 63
Stewart R F, Bentley J and Goodman B 1975 *J. Chem. Phys.* **63** 3786–93
Stewart R F, Davidson E R and Simpson W T 1965 *J. Chem. Phys.* **42** 3175–87
Stewart R F and Feil D 1980 *Acta Crystallogr.* A **36** 503–6
Stewart R F and Spackman M A 1981 *Structure and Bonding in Crystals* vol 1, ed M O'Keefe and A Navrotsky (New York: Academic) p 279
Storozhenko V V, Avilov A S and Semiletov S A 1984 *Izv. Akad. Nauk* **48** 1753–7 (in Russian)
Streltsov V A 1986 *PhD Thesis* Mendeleev Institute of Chemical Technology Moscow (in Russian)
Streltsov V A, Belokoneva F L, Tsirelson V G and Hansen N K 1993 *Acta Crystallogr.* B **49** 147–53
Streltsov V A and Maslen E N 1992 *Acta Crystallogr.* A **48** 651–3
Streltsov V A, Tsirelson V G, Krasheninnikov M V and Ozerov R P 1985 *Sov. Phys.–Crystallogr.* **30** 32–4
Streltsov V A, Tsirelson V G and Ozerov R P 1990 *Problems of Crystal Chemistry* ed M A Porai-Koshitz (Moscow: Nauka) pp 8–47 (in Russian)
Streltsov V A, Tsirelson V G, Ozerov R P and Golovanov O A 1988a *Sov. Phys.–Crystallogr.* **33** 49–52
Streltsov V A, Tsytsenko A K, Antipin M Yu, Fundamenskii Y S, Tsirelson V G, Frank-Kamenetskaya O V and Struchkov Yu T 1988b *Zh. Struct. Chim.* **29** 238–41 (in Russian)
Streltsov V A and Zavodnik V E 1989 *Sov. Phys.–Crystallogr.* **34** 1369
—— 1990 *Sov. Phys.–Crystallogr.* **35** 281
Stucky G D, Phillips M L F and Gier T E 1989 *Chem. Mater.* **1** 492–509
Su Z and Coppens P 1992 *Acta Crystallogr.* A **48** 188–97
—— 1994 *Sagamore XI. Collected Abstracts* ed G Loupias and S Rabii (Brest: Laboratorie Mineralogie–Cristallographie, Université P et M Curie) p 198
—— 1995 *Acta Crystallogr.* A **51** 27–32
Suortti P 1982 *Acta Crystallogr.* A **38** 648–56
Susuki K, Onishi S, Koide T and Seki B 1956 *Bull. Chem. Soc. Japan* **29** 127–31
Swaminathan S, Craven B M and McMullan R K 1985 *Acta Crystallogr.* B **41** 113–22
Sweet R M and Woodhead A D (ed) 1989 *Synchrotron Radiation in Structural Biology* (New York: Plenum)
Szacz L 1985 *Pseudopotential Theory of Atoms and Molecules* (New York: Wiley)
Tabbernor M A and Fox A G 1990 *Phil. Mag. Lett.* **62** 291
Tabbernor M A, Fox A G and Fisher R M 1990 *Acta Crystallogr.* A **62** 165

Takagi 1969 *J. Phys. Soc. Japan* **26** 1239–45
Takama T and Harima H 1994 *Acta Crystallogr.* A **50** 239–46
Takama T and Sato S 1981 *Japan. J. Appl. Phys.* **20** 1183–9
Takama T, Tsuchiya K, Kobayashi K and Sato S 1990 *Acta Crystallogr.* A **46** 514–7
Takazawa H, Ohba S and Saito Y 1988 *Acta Crystallogr.* B **44** 580–3
—— 1989 *Acta Crystallogr.* B **45** 432–7
Takeda T 1979 *J. Phys. F: Met. Phys.* **9** 815–29
Takemura S and Kato N 1972 *Acta Crystallogr.* A **28** 69–80
Tal Y, Bader R F W and Erkku J 1980 *Phys. Rev.* A **21** 1–11
Tanaka K 1979 *Acta Crystallogr.* B **34** 2487–94
—— 1988a *Acta Crystallogr.* A **44** 1002–8
—— 1988b *J. Cryst. Soc. Japan* **30** 119–28
—— 1993 *Acta Crystallogr.* B **49** 1001–10
Tanaka K, Elkaim E, Li L, Jue Z N, Coppens P and Landrum J 1986 *J. Chem. Phys.* **84** 6969–75
Tanaka K and Marumo F 1983 *Acta Crystallogr.* B **38** 1422–7
Tanaka K and Saito J 1975 *Acta Crystallogr.* A **31** 841–5
Tavard C, Ronald M, Roux M and Cornille M 1964 *J. Chim. Phys.* **61** 1324–9
Taylor J R 1982 *An Introduction to Error Analysis* (Mill Valley: University Science Books)
Tegenfeldt J and Hermansson K 1985 *Chem. Phys. Lett.* **118** 293–8
Teller E 1962 *Rev. Mod. Phys.* **34** 627–33
Templeton L K and Templeton D H 1988 *Acta Crystallogr.* A **44** 1045–51
Teophilou A K 1979 *J. Phys. C: Solid State Phys.* **12** 5419
Thomas J O, Tellgren R and Almlöf J 1975 *Acta Crystallogr.* B **31** 1946–55
Thomas L E, Shirley C G, Lally J S and Fisher R M 1974 *High Voltage Electron Microscopy* (New York: Academic)
Thong N and Schwarzenbach D 1979 *Acta Crystallogr.* A **35** 658–64
Thornley F R and Nelmes R J 1974 *Acta Crystallogr.* A **30** 748–57
Thosar B V, Iyengar P K, Srivastava J K and Bhargava S C (ed) 1983 *Advances in Mössbauer Spectroscopy* (Amsterdam: Elsevier)
Tichonov A N and Arsenin V Ya 1986 *Methods of Solutions of the Ill-Posed Problem* (Moscow: Nauka) (in Russian)
Tichonov A N and Tschedrin B M 1986 *Crystallography and Crystal Chemistry* ed B K Vainstein (Moscow: Nauka) pp 79–90 (in Russian)
Tischler J Z and Batterman B W 1984 *Phys. Rev.* B **30** 7060–6
Tokagi S 1969 *J. Phys. Soc. Japan* **26** 1239–53
Topiol S, Zunger A and Ratner M A 1977 *Chem. Phys. Lett.* **49** 367–77
Towler M D 1995 *Lecture on Network School 'Hartree–Fock Theory of the Electronic Structure of Solids' (Villa Gualino, 1995)*
Towler M D, Dovesi R and Saunders V R 1995 *Phys. Rev.* B **52** 10 150
Trefry M G, Maslen E N and Spackman M A 1987 *J. Phys. C: Solid State Phys.* **20** 19–28
Treworte R and Bonse U 1984 *Phys. Rev.* B **29** 2102–8
Trucano P and Batterman B W 1972 *Phys. Rev.* B **6** 3659–66
Tsarkov A G and Tsirelson V G 1990 *Preprint* Mendeleev Institute of Chemical Technology N 6205-V90 (Moscow: Viniti) (in Russian)
—— 1991 *Phys. Status Solidi* b **167** 417–28

Tschinke V and Ziegler T 1990 *J. Chem. Phys.* **93** 8051–60

Tsirelson V G 1978 *2nd All-Union Conf. on Organic Crystal Chemistry, Abstracts* (Chernogolovka: Institute of Chemical Physics) p 139 (in Russian)

—— 1985 *14th All-Union Conf. on Application of X-rays to the Study of Materials. Abstracts* (Kishinev: Stiintsa) p 27 (in Russian)

—— 1989 *Methods of Structure Analysis* vol 1, ed B K Vainstein (Moscow: Nauka) pp 37–52 (in Russian)

—— 1993 *Advances in Science and Technology* (*Ser. Crystal Chemistry*) vol 27 (Moscow: Viniti) (in Russian)

—— 1994 *Sagamore XI. Collected Abstracts* ed G Loupias and S Rabii (Brest: Laboratoire Mineralogie–Cristallographie, Université P et M Curie) p 252

Tsirelson V G and Antipin M Yu 1989 *Problems of Crystal Chemistry* ed M A Porai-Koshitz (Moscow: Nauka) pp 119–60 (in Russian)

Tsirelson V G, Antipin M Yu, Gerr R G, Ozerov R P and Struchkov Yu T 1985a *Phys. Status Solidi* a **87** 425–33

Tsirelson V G, Antipin M Yu, Streltsov V A, Ozerov R P and Struchkov Yu T 1988a *Sov. Phys.–Dokl.* **33** 89–91

Tsirelson V G, Baskakov A A, Zorky P M and Ozerov R P 1984a *4th All-Union Conf. on Organic Crystal Chemistry. Abstracts* (Chernogolovka: Institute of Chemical Physics) p 98

Tsirelson V G, Belokoneva E L, Evdokimova O A and Urusov V S 1990 *Phys. Chem. Miner.* **17** 275–92

Tsirelson V G and Belyaev V 1992 *6th Conf. on Crystal Chemistry of Inorganic and Coordination Compounds* (Lvov: Uzhgorod University) p 259

Tsirelson V G, Korolkova O V, Ozerov R P and Rez I S 1984b *Phys. Status Solidi* b **122** 599–612

Tsirelson V G and Lobanov N N 1986 unpublished

Tsirelson V G, Lobanov N N, Antipin M Yu and Ozerov R P 1988b *Sov. Phys.–Crystallogr.* **33** 171–4

Tsirelson V G, Lobanov N N and Ozerov R P 1987a *Phys. Met. Metallogr.* **63** 18–24 (in Russian)

Tsirelson V G, Mesteshkin M M and Ozerov R P 1977 *Dokl. Akad. Nauk* **233** 108–10 (in Russian)

Tsirelson V G and Nozik Ju Z 1982 *Kristallographia* **27** 661–3 (in Russian)

Tsirelson V G and Ozerov R P 1979 *Kristallographia* **24** 1156–63 (in Russian)

—— 1992 *J. Mol. Struct.* **255** 335–92

Tsirelson V G, Parini E V and Ozerov R P 1980a *Preprint* Mendeleev Institute of Chemical Technology 3600–80

Tsirelson V G, Reznik I M and Ozerov R P 1992 *Crystall. Rev.* **3** 31–110

Tsirelson V G, Sokolova E V and Urusov V S 1986 *Geochimija* N **8** 1170–80 (in Russian)

Tsirelson V G, Streltsov V A, Makarov E F and Ozerov R P 1987b *Sov. Phys.–JETP* **65** 1065–9

Tsirelson V G, Streltsov V A and Ozerov R P 1985b *Sov. Phys.-Dokl.* **30** 111–12

Tsirelson V G, Streltsov V A, Ozerov R P and Yvon K 1989 *Phys. Status. Solidi* a **115** 515–21

Tsirelson V G, Zavodnik V E, Fomicheva E B, Ozerov R P, Kuznetsova L I and Rez I S 1980b *Sov. Phys.–Crystallogr.* **25** 326–9

Tsirelson V G, Zhurova E A and Tsarkov A G 1991 *Problems of Crystal Chemistry* ed M A Porai-Koshitz (Moscow: Nauka) pp 111–86 (in Russian)
Tsirelson V G, Zou P F, Tang T H and Bader R F W 1995 *Acta Crystallogr.* A **51** 143–53
Tuinstra F and Fraase Storm G M 1978 *J. Appl. Crystallogr.* **11** 257–9
Unsöld A 1927 *Z. Phys.* **43** 388
Utemisov K, Shilstein S Sh and Somenkov V A 1981 *Sov. Phys.–Crystallogr.* **26** 101
Utemisov K, Somenkova V P, Somenkov V A and Shilstein S Sh 1980 *Sov. Phys.–Crystallogr.* **25** 484–7
Vainstein B K 1960 *Q. Rep.* **14** 105–32
—— 1964 *Structure Analysis by Electron Diffraction* (Oxford: Pergamon)
Vainstein B K and Dvorjankin V F 1956 *Kristallographiya* **1** 626–30 (in Russian)
van der Wal H R, Boer de J L and Vos A 1979 *Acta Crystallogr.* A **35** 685–8
van der Wal H R and Vos A 1979 *Acta Crystallogr.* B **35** 1804–9
van der Wal H R, Vos A and Kirfel A 1987 *Acta Crystallogr.* B **43** 132–43
van Nes G J H 1978 *Thesis* University of Groningen
van Nes G J H and Vos A 1978 *Acta Crystallogr.* B **34** 1947–50
—— 1979a *Acta Crystallogr.* B **38** 2580–93
—— 1979b *Acta Crystallogr.* B **38** 2593–601
Van Vechten J A 1969 *Phys. Rev.* **182** 891–905
Varghese J N and Mason R 1980 *Proc. R. Soc.* A **372** 1
Varnek A A 1985 *PhD Thesis* Mendeleev Institute of Chemical Technology Moscow (in Russian)
Varnek A A, Krasheninnikov M V, Tsirelson V G and Ozerov R P 1982 *Russ. Chem. Phys.* No 6, 716–20
Varnek A A, Tsirelson V G and Ozerov R P 1981 *Dokl. Akad. Nauk* **257** 382–5 (in Russian)
Velders G and Feil D 1989 *Acta Crystallogr.* B **45** 359–64
Vidal-Valat G, Vidal J P and Kurki-Suonio K 1978 *Acta Crystallogr.* A **34** 594–602
Voloshina I V, Gerr R G, Antipin M Yu, Tsirelson V G, Pavlova N I, Rez I S and Ozerov R P 1985 *Kristallographiya* **30** 668–76 (in Russian)
von Barth U 1986 *Chem. Scr.* **26** 449–61
von Barth U and Pedroza A C 1985 *Phys. Scr.* **32** 353
Wakoh S and Kubo Y 1980 *J. Phys. F: Met. Phys.* **10** 2707–15
Wang C S and Klein B M 1981 *Phys. Rev.* B **24** 3393–416
Wang L-C and Boyd R J 1989 *J. Chem. Phys.* **90** 1083–90
Wang Y, Angermund K, Goddard R and Kruger K 1987 *J. Am. Chem. Soc.* **109** 587–9
Wang Y, Blessing R H and Coppens P 1976 *Acta Crystallogr.* B **32** 572–8
Watson G N 1966 *A Treatise on the Theory of Bessel Functions* (Cambridge: Cambridge University Press)
Weber H-P and Craven B M 1987 *Acta Crystallogr.* B **43** 202–9
—— 1990 *Acta Crystallogr.* B **46** 532–8
Weber H-P, Ruble J R, Craven B M and McMullan R K 1980 *Acta Crystallogr.* B **36** 645–9
Weiss R J and Freeman A J 1959 *J. Chem. Phys. Solids* **10** 147–61
Weiss K J 1966 *X-ray Determination of Electron Distribution* (Amsterdam: North-Holland)
Weizsaker von C 1935 *Z. Phys.* **96** 431

Wells A F 1985 *Structural Inorganic Chemistry* 5th edn (Oxford: Clarendon)
Weyrich K H 1988 *Phys. Rev.* B **37** 10 269–82
Wigner E 1934 *Phys. Rev.* **46** 1002–10
Willis B T M 1969 *Acta Crystallogr.* A **25** 277–300
—— 1970 *Acta Crystallogr.* A **26** 396–401
—— 1984 *Modern Aspects of Diffraction Physics* (Lausanne: Lausanne University)
Willis B T M and Howard J A K 1975 *Acta Crystallogr.* A **31** 514–20
Willis B T M and Pryor A W 1975 *Thermal Vibrations in Crystallography* (Cambridge: Cambridge University Press)
Wilkins SW, Varghese J N and Lehmann M S 1983 *Acta Crystallogr.* A **39** 47–60
Wilson A J C 1979 *Acta Crystallogr.* A **35** 122–30
—— (ed) 1992 *International Tables for Crystallography* vol C (Dordrecht: Kluwer)
Wilson C W and Goddard W A 1972a *Theor. Chim. Acta* **26** 210–30
—— 1972b *Theor. Chim. Acta* **26** 195–210
Wilson E B 1962 *J. Chem. Phys.* **36** 2232–3
Wilson J N and Curtis R M 1970 *J. Chem. Phys.* **74** 187–96
Wilson S 1987 *Ab Initio Methods in Quantum Chemistry* ed K P Lawley (New York: Wiley) pp 439–500
Wolfsberg M and Helmholtz L 1952 *J. Chem. Phys.* **20** 837
Wooster W A 1962 *Diffuse X-ray Reflections from Crystals* (Oxford: Clarendon)
Zachariasen W H 1967 *Acta Crystallogr.* 23 558–64
—— 1968 *Acta Crystallogr.* A **24** 425–7
Zahradnik R (ed) 1988 *The Intermolecular Interactions; Chem. Rev.* **88** special issue
Zalepuchin M V, Kvadrakov V V, Somenkov V A and Shilstein S Sh 1989 *Sov. Phys.–JETP* **68** 883–6
Zavodnik V E and Stash A I 1996 *Kristallographiya* **41**, 1–10
Zavodnik V E, Stash A I, Tsirelson V G and Feil D 1994 *Sagamore XI. Collected Abstracts* ed G Loupias and S Rabii (Brest: Laboratoire Mineralogie–Cristallographie, Université P et M Curie) pp 53–4
Zernike F and Midwinter J E 1973 *Applied Non-linear Optics* (New York: Wiley–Interscience)
Zhao Q and Parr R G 1993 *J. Chem. Phys.* **98** 543–8
Zheludev A, Bonnet M, Ressouche E, Schweizer M, Wan M and Wang H 1994a *J. Magn. Mater.* **135** 147–60
Zheludev A, Ressouche E, Schweizer J, Turek J, Wan M and Wang H 1994b *Solid State Commun.* **90** 233–5
Zhurova E A, Zavodnik V E, Ivanov S A, Syrnikov P P, Tsirelson V G and Ozerov R P 1991 *Sagamore X, Collected Abstracts* ed M Springborg, A Saenz and W Weyrich (Konstanz: University of Konstanz) p 150
Ziegler T, Snijders J G and Baerends E J 1981 *J. Chem. Phys.* **74** 1271
Zobel D, Luger P and Dreissig W 1991 *Sagamore X Abstracts* ed M Springborg, A Saenz and W Weyrich (Konstanz: University of Konstanz) pp 151–2
—— 1993 *Z. Naturf.* a **48** 53–4
Zou P F and Bader R F W 1994 *Acta Crystallogr.* A **50** 714–25
Zucker U H and Schulz H 1982 *Acta Crystallogr.* A **38** 568–76
Zumbach G and Maschke K 1983 *Phys. Rev.* A **28** 544
Zumsteg F C, Bierlein J D and Gier T K 1976 *J. Appl. Phys.* **47** 4980–3

Zunger A 1979 *Phys. Rev.* B **21** 4785–96
—— 1980 *Phys. Rev.* B **21** 4785
—— 1986 *Int. J. Quantum Chem. Symp.* **19** 629–53
Zunger A and Cohen M L 1978 *Phys. Rev.* B **18** 5449–72
—— 1979a *Phys. Rev.* B **20** 4082–108
—— 1979b *Phys. Rev.* B **20** 1189–93
Zunger A and Freeman A J 1977a *Phys. Rev.* B **15** 4716–37
—— 1977b *Phys. Rev.* B **16** 2901–26
—— 1977c *Phys. Rev.* B **15** 5049
Zuo J M, Hoier R and Spence J C H 1989a *Acta Crystallogr.* A **45** 839
Zuo J M, Spence J C H, Fox A G and Tabbernor M A 1990 *Acta Crystallogr.* A **46** Supplement C-13
Zuo J M, Spence J C H and Hoier R 1989b *Phys. Rev. Lett.* **62** 547
Zuo J M, Spence J C H and O'Keefe M 1988 *Phys. Rev. Lett.* **61** 353

INDEX